Guide to Wireless Network Security

Guide to Wireless Network Security

Guide to Wireless Network Security

by

John R. Vacca
USA

Springer

John R. Vacca
Author and IT Consultant
34679 TR 382
Pomeroy, Ohio 45769
e-mail:jvacca@hti.net
http://www.johnvacca.com

Library of Congress Control Number: 2005939009

Guide to Wireless Network Security
By John R.Vacca

ISBN-13: 978-0-387-95425-7
ISBN-10: 0-387-95425-2
e-ISBN-13: 978-0-387-29845-0
e-ISBN-10: 0-387-29845-2

Vacca: Guide to Wireless Network Security

First Indian Reprint 2009

ISBN 978-81-8489-125-6

This edition is manufactured in India for sale in India, Pakistan, Bangladesh, Nepal and Sri Lanka and any other country as authorized by the publisher.

This edition is published by Springer (India) Private Limited,
A part of Springer Science+Business Media, Registered Office: 906-907, Akash Deep Building, Barakhamba Road, New Delhi – 110 001, India.

Printed in India by Rakmo Press (P) ltd., New Delhi

Dedication

This book is dedicated to
Hunter.

Contents

Preface

1. INTRODUCTION

With the increasing deployment of wireless networks (802.11 architecture) in enterprise environments, IT enterprises are working to implement security mechanisms that are equivalent to those existing today for wire-based networks. An important aspect of this is the need to provide secure access to the network for valid users. Existing wired network jacks are located inside buildings already secured from unauthorized access through the use of keys, badge access, and so forth. A user must gain physical access to the building in order to plug a client computer into a network jack. In contrast, a wireless access point (AP) may be accessed from off the premises if the signal is detectable (for instance, from a parking lot adjacent to the building). Thus, wireless networks require secure access to the AP and the ability to isolate the AP from the internal private network prior to user authentication into the network domain.

Furthermore, as enterprises strive to provide better availability of mission-critical wireless data, they also face the challenge of maintaining that data's security and integrity. While each connection with a client, a supplier or a enterprise partner can improve responsiveness and efficiency, it also increases the vulnerability of enterprise wireless data to attack. In such an environment, wireless network security is becoming more important every day.

Also, with the growing reliance on e-commerce, wireless network-based services and the Internet, enterprises are faced with an ever-increasing responsibility to protect their systems from attack. Intrusion detection systems (IDSs) and intrusion prevention systems (IPSs)--used in conjunction with information warfare countermeasures--are the latest and most powerful tools for identifying and responding to network- and host-based intrusions.

With that in mind, this book begins by discussing the basic access control methods that form the basis of the 802.11 architecture. These methods are best suited to small wireless networks with low-to-medium security requirements. The book then presents the more-robust virtual private network (VPN)-based security solution that provides better security and scales well to large networks. The book concludes with possible future solutions based on the 802.1X security standard, which enables port-level access control.

2. PURPOSE

The purpose of this book is to show experienced (intermediate to advanced) wireless network security professionals how to install and maintain the security of mission-critical wireless data and systems. It also shows through extensive hands-on examples, how you can install and configure firewalls, evaluate, implement and manage wireless secure remote access technologies, and deploy a variety of intrusion detection systems and intrusion prevention systems in conjunction with information warfare countermeasures.

3. SCOPE

Throughout the book, extensive hands-on examples will provide you with practical experience in installing, configuring and troubleshooting wireless network security applications and Internet and intranet Firewalls; Virtual Private Networks; intrusion prevention systems and intrusion detection systems. In addition to advanced wireless network security application technology considerations in commercial enterprises and governments, the book addresses, but is not limited to completing the following line items as part of installing wireless network security-based systems:

- Analyze network traffic and detect attacks using the latest tools and techniques.
- Authenticate remote users with passwords, security servers and digital certificates.
- Automate responses to detected intrusions.
- Be able to describe methods of advanced data modulation.

- Be able to describe methods of detection, disruption (denial of service or jamming), and interception and understand appropriate countermeasures.
- Be able to describe the use of wireless security technologies such as frequency hopping, time hopping, direct-sequence spread spectrum, etc.
- Build a firewall to protect your wireless network.
- Create an effective response strategy (via information warfare) based on your organizational needs
- Deploy and manage an IDS and IPS.
- Deploy Internet and intranet firewalls: hands-on.
- Deploying intrusion detection systems and intrusion prevention systems: hands-on.
- Design, configure and deploy an IDS and IPS; and, analyze your current wireless network security risks.
- Design, install and configure virtual private networks (VPNs).
- Detect and respond to wireless network- and host-based intruder attacks.
- Detect attacker scans and probes.
- Evaluate, install, configure and manage secure virtual private networks (VPNs) for remote users, sites and business partners.
- Gain extensive hands-on experience installing and configuring a firewall.
- Gain extensive hands-on experience using an IDS and IPS to identify and respond to intruder attacks.
- Gain hands-on experience with a range of security tools and techniques for maintaining the integrity of your wireless network security operations.
- Gain the skills to respond to potential attacks before they become problematic by recognizing the scans and probes used by a potential intruder.
- Identify buffer overruns, fragmentation and other attacks.
- Identify methods hackers use to break into wireless network systems.
- Implement information privacy using standardized encryption techniques.
- Implement information warfare countermeasures.
- Implement publicly accessible servers without compromising wireless network security, provide access to HTTP and FTP services on the Internet, and implement a firewall-to-firewall virtual private network (VPN).
- Integrate intrusion detection systems (IDSs) and intrusion prevention systems (IPSs) into your current network topology
- Install and configure proxy-based and stateful-filtering firewalls.
- Knowledge of how attackers break into wireless networks and how an IDS and IPS (used in conjunction with information warfare

countermeasures) can play a key role in detecting and responding to these events.

- Learn about a variety of technologies available, including software, hardware and firewall add-on products.
- Learn how to allow access to key services while maintaining your enterprise's security, as well as how to implement firewall-to-firewall virtual private networks (VPNs).
- Protect internal IP addresses with Network Address Translation (NAT) and deploy a secure DNS architecture.
- Select the best secure remote wireless access technologies for your organization.
- Understand basic electronic countermeasures and electronic counter-countermeasures for wireless communications.
- Understand issues of network communications such as service, confidentiality, authentication, reliability, access control, and availability.
- Understand the functions of the layers in a wireless communication system.
- Understand the relationship between network layers, network services and functions.
- Understand the security problems with wireless transmissions.
- Using router logging to detect a DoS attack.

This book will leave little doubt that a new architecture in the area of advanced wireless network security installation is about to be constructed. No question, it will benefit enterprises and governments, as well as their wireless network security professionals.

4. TARGET AUDIENCE

This book is primarily targeted toward domestic and international network and systems administrators; IT administrators; IT managers; wireless network security specialists; computer and network security personnel; security professionals; and, consultants and IT/IS directors who plan to select, implement and maintain secure wireless access solutions for an enterprise. Basically, the book is targeted for all types of people and enterprises around the world that are involved in planning and implementing wireless network security and other wireless Internet systems.

5. ORGANIZATION OF THIS BOOK

The book is organized into eight parts as well as an extensive glossary of security, wireless network and Internet networking terms and acronyms at the back. It provides a step-by-step approach to everything you need to know about wireless network security; as well as, information about many topics relevant to the planning, design, and implementation of intrusion detection systems and intrusion prevention systems; and, how to conduct information warfare. The following detailed organization speaks for itself.

5.1 Part I: Overview Of Wireless Network Security Technology

Part One discusses wireless network security fundamentals; types of wireless network security technology; standards; enhanced wireless network security; and, handling wireless private information.

Chapter 1, "Wireless Network Security Fundamentals," presents a classification of denial-of-service attacks according to the type of the target (firewall, Web server, router), a resource that the attack consumes (wireless network bandwidth, TCP/IP stack) and the exploited vulnerability (bug or overload).

Chapter 2, "Types of Wireless Network Security Technology," presents some common types of wireless network security technologies to help guide your path.

Chapter 3, "Standards," discusses the following wireless network security standards: WEP, IEEE 802.11b, IEEE 802.11i, IEEE 802.1X, Bluetooth, SSL, WTLS, WPA and WPA2.

Chapter 4, "Enhanced Security For Wireless Lans And Wans In The Enterprise: Hands On," helps managers get a grasp of basic WLAN and WWANs security issues.

Chapter 5, "Handling Wireless Private Information," covers the pitfalls of wireless- LANs and WANs, with regards to the security risks to private information.

5.2 Part II: Designing Wireless Network Security

The second part of this book discusses wireless network security- design issues; cost justification and consideration; standards design issues; and, authenticating architectural design issues.

Chapter 6, "Wireless Network Security- Design Issues," covers enterprise critical systems security, by first illustrating the importance of and potential difficulties in protecting information that traverses networks; and, then examining wireless network security as a holistic concept before focusing specifically on the IEEE 802.1X enterprise edge security standard.

Chapter 7, "Cost Justification And Consideration," assesses the costs associated with the security risks and vulnerabilities to the wireless network.

Chapter 8, "Standards Design Issues," discusses particular aspects of wireless security architecture standards design issues in detail, that can be used for enterprise wireless networks standards design.

Chapter 9, "Authenticating Architectural Design Issues," presents the architecture and the underlying mechanism of the WMTIP.

5.3 Part III: Planning For Wireless Network Security

Part Three covers the implementation plan development and wireless network security planning techniques.

Chapter 10, "Implementation Plan Development," describes the overall security implementation plan development for wireless networks.

Chapter 11, "Wireless Network Security Planning Techniques" briefly discusses wireless network security planning techniques by providing an overview of the security risks and technical challenges in this area, as well as summarizing key recommendations for secure wireless LANs and WWANs..

5.4 Part IV: Installing And Deploying Wireless Network Security

Part Four covers testing techniques; internetworking wireless security; installation and deployment; securing your wireless e-commerce storefront; and, certification of wireless network security performance.

Chapter 12, "Testing Techniques," focuses more on a cracker attempting to penetrate your wireless network and hacking one of the servers held therein.

Chapter 13, "Internetworking Wireless Security," focuses on the concept of performance enhancing proxies (PEPs), which were introduced in a working group of the Internet Engineering Task Force (IETF).

Chapter 14, "Installation And Deployment," eases your concerns about the security of wireless networks installations and deployments, by increasing your knowledge on the subject.

Chapter 15, "Securing Your Wireless E-Commerce Storefront," focuses on two "flavors" of wireless service, as provided by the WAP Forum and by NTT DoCoMo of Japan.

Chapter 16, "Certification Of Wireless Network Security Performance," focuses on IEEE 802.11 networks, and how the Service Set Identifier (SSID) is viewed by some security professionals as an unneeded advertisement of the wireless network to attackers. It also discusses how these professionals assert that all measures should be taken to hide the SSID.

5.5 Part V: Maintaining Wireless Network Security

Part Five covers configuring secure access; management of wireless network security; ongoing maintenance, standards development and ensuring site security.

Chapter 17, "Configuring Secure Access," focuses on how to configure your wireless network security access.

Chapter 18, "Management Of Wireless Network Security," focuses on the management of wireless network security.

Chapter 19, "Ongoing Maintenance," discusses the ongoing maintenance of integrated wireless network analyzers; and, how they have several advantages over laptop computers and handheld, personal digital assistant (PDA)-style devices, as well as centralized systems.

Chapter 20, "Standards Development," focuses on the development of wireless network security standards.

Chapter 21, "Ensuring Site Security," focuses wireless network site security and integrity.

5.6 Part VI: Information Warfare Countermeasures: The Wireless Network Security Solution

Part Six discusses wireless network security with regards to how to fight against macro threats–defensive strategies for governments and industry groups; the information warfare arsenal and tactics of the military; the information warfare arsenal and tactics of terrorists and rogues; the information warfare arsenal and tactics of private enterprises; the information warfare arsenal of the future; surveillance tools for information warfare of the future; and civilian causalities– the victims and refugees of information warfare.

Chapter 22, "Defensive Wireless Network Security Strategies For Governments And Industry Groups," is an in-depth examination of the implications of IW for the U.S. and allied infrastructures that depend on the unimpeded management of information that is also required in the fight against macro threats–defensive strategies for governments and industry groups.

Chapter 23, "The Information Warfare Wireless Network Security Arsenal And Tactics Of The Military," focuses on two goals. First, you need to find a way to protect yourself against catastrophic events. Second, you need to build a firm foundation on which you can make steady progress by continually raising the cost of mounting an attack and mitigating the expected damage of the information warfare arsenal and tactics of the military.

Chapter 24, "The Information Warfare Wireless Network Security Arsenal And Tactics Of Terrorists And Rogues," recommends a number of specific steps that could better prepare the U.S. military and private enterprises to confront "the new terrorism" and its information warfare arsenal and tactics.

Chapter 25, "The Information Warfare Wireless Network Security Arsenal And Tactics Of Private Enterprises," deals with the IW tools and strategies of private enterprises and how they're used against the aggressors. It will also help to realistically guide the process of moving forward in dealing with the information warfare arsenal and tactics of private enterprises.

Chapter 26, "The Information Warfare Wireless Network Security Arsenal Of The Future," discusses how the increasing dependence on sophisticated information systems brings with it an increased vulnerability to hostile elements, terrorists among them, in dealing with the information warfare arsenal of the future.

Chapter 27, "Wireless Network Security Surveillance Tools For Information Warfare Of The Future," discusses the basic concepts and principles that must be understood and that can help guide the process of moving forward in dealing with the surveillance tools for the information warfare of the future.

Chapter 28, "Civilian Casualties: The Victims And Refugees Of Information Warfare Wireless Network Security," considers the application of civilian information operations (CIOs) to the conventional warfare environment. Although the array of CIO tools and techniques has been presented as discrete elements in a schematic diagram, the CIO environment is complex, multidimensional, interactive, and still developing.

5.7 Part VII: Results and Future Directions

Finally, Part Seven discusses how to provide wireless ISP intranet, Internet and e-commerce solutions; enhance wireless web server security; wireless network security solutions for consideration and finally the summary, conclusions and recommendations.

Chapter 29, "Providing Wireless Network Security Solutions For ISP Intranet, Internet And E-Commerce," outlines the new security concerns for an enterprise to deploy Intranets and extranets.

Chapter 30, "Enhancing Wireless Web Server Security," discusses what you can do to protect your wireless Web server from wireless network security risks.

Chapter 31, "Wireless Network Security Solutions For Consideration," This chapter describes the various security challenges and solutions for consideration of Wi-Fi wireless LANs; attempts the industry has made to address those challenges; shortcomings of those initial attempts; and, the best possible practices, for enterprises and residential users who want to take advantage of the real benefits of WLANs..

Chapter 32, "Summary, Conclusions, and Recommendations," addresses at a summary level, the most significant security risks in the wireless computing environment. The chapter also introduces in a centralized fashion, the scope of the problem and the most significant talking points on the issue of wireless security and to summarize, conclude and recommend where the industry is in addressing these problems and where it is going. Finally, this chapter presents WPA, then end to end encryption, and finally the services appropriate for larger enterprises.

5.8 Part VIII: Appendices

Eight appendices provide additional resources that are available for computer forensics. Appendix A shows how to ensure built-in frequency hopping spread spectrum wireless network security. Appendix B shows how to configure wireless Internet security remote access. Appendix C covers wireless network security management, resiliency and security. Appendix D contains a list of top wireless network security implementation and deployment enterprises. Appendix E contains a list of wireless network security products. Appendix F contains a list of wireless network security standards. Appendix G contains a list of miscellaneous wireless network security resources. The book ends with Appendix H--a glossary of wireless network security and information-warfare-related terms and acronyms.

6. CONVENTIONS

This book uses several conventions to help you find your way around, and to help you find important sidebars, facts, tips, notes, cautions, and warnings. They alert you to critical information and warn you about problems.

John R. Vacca
Author and IT Consultant, e-mail: jvacca@hti.net
visit us at http://www.johnvacca.com/

Foreword

The use of wireless networks is increasingly popular among personal, academic, business, and government users. Everyone wants to be connected and everyone wants to be connected with out the need for a physical cable plugged into his or her technology. Mobility is a real requirement for business in the 21st century. Wireless networks offer a wide range of benefits to government federal agencies, private sector business and individual citizens. These include increased flexibility and ease of network installation.

However, without security electronic communications hold little value of in competitive arena of business management and operations. Wireless networks present significant security challenges, including protecting against attacks to wireless networks, establishing physical control over wireless-enabled devices, and preventing unauthorized deployments of wireless networks. Security professionals, application developers, along with IT and network staff in all types of organizations will eventually need to address wireless network security issues.

To secure wireless devices and networks and protect information and systems, it is crucial for user organizations to implement controls—such as developing wireless security policies, configuring their security tools to meet policy requirements, monitoring their wireless networks, and training their staffs in wireless security.

Ease of installation is often cited as a key attribute of wireless networks. Generally, deployments of wireless networks do not require the complicated undertakings that are associated with wired networks. The ability to connect the network without having to add or pull wires through walls or ceilings or modify the physical network infrastructure can greatly expedite the installation process. As a result, a wireless network can offer a cost-effective alternative to a wired network. In addition to their increased ease of installation, wireless networks can be easily scaled from small peer-to-peer networks to very large enterprise networks that enable roaming over a broad area.

This book provides an extensive analysis of wireless network security practices, procedures, and technologies. Design issues and architectures are also expertly covered. But this book goes beyond theory and analysis to explain numerous implementation issues.

This book is written for people that need to cut through the confusion about wireless network security and get down to adoption and deployment. The book starts with the basic concepts and takes readers through all of the necessary learning steps to enable them to effectively secure wireless networks.

Michael Erbschloe
Security Consultant and Author
St. Louis, Missouri

Acknowledgements

There are many people whose efforts on this book have contributed to its successful completion. I owe each a debt of gratitude and want to take this opportunity to offer my sincere thanks.

A very special thanks to my Springer Publishing Editor/CS, Susan Lagerstrom-Fife, without whose initial interest and support would not have made this book possible; and, for her guidance and encouragement over and above the business of being a Publishing Editor. And, thanks to Editorial Assistant, Sharon Palleschi of Springer, whose many talents and skills are essential to a finished book. Many thanks also to Deborah Doherty of Springer Author Support, whose efforts on this book have been greatly appreciated. Finally, a special thanks to Michael Erbschloe who wrote the foreword for this book.

Thanks to my wife, Bee Vacca, for her love, her help, and her understanding of my long work hours.

Finally, I wish to thank all the organizations and individuals who granted me permission to use the research material and information necessary for the completion of this book.

PART I

OVERVIEW OF WIRELESS NETWORK SECURITY TECHNOLOGY

Chapter 1

WIRELESS NETWORK SECURITY FUNDAMENTALS

1. INTRODUCTION

It's an epidemic waiting to happen to many security experts. While most IT managers today have their hands full securing wired networks, their companies have been spending billions of dollars on wireless. Businesses will have spent $60 billion on wireless communications services and wireless network security by the end of 2006; and, 80 percent of the U.S. work force will use some sort of wireless device, including cell phones, pagers and mobile computing devices. That's good news for employee productivity, but bad news for companies ill-prepared to head off wireless network security breaches and debilitating viruses [10].

There's been a lack of functionality and a lack of mature infrastructure globally. And, that's the only reason the wireless viruses of today have not been more damaging. But, that's about to change. Industry analysts predict dramatic increases in wireless handheld use and the proliferation of new mobile capabilities. They expect to see 5.9 billion handsets, personal digital assistants (PDAs) and Internet appliances equipped with wireless capabilities by the end of 2008—with that, you get a full scale epidemic in the works. Simply put: It's coming [10].

For many IT managers, the wireless world, with its often-incompatible alphabet soup of standards, may be new territory. Many enterprises have felt that protecting their wireless processes against viruses is one piece of the complicated puzzle they can afford to omit. They'll soon need to face threats that could wreak havoc on a very large scale.

Therefore, in order to fight the viruses and security breaches of the future, wireless network security vendors (even giants like IBM) are busy developing products. In addition, within applications and on devices, they are also heading off problems on a wireless network level.

2. OVERVIEW OF WIRELESS NETWORK SECURITY

In markets where wireless devices are more widely used, to date, most wireless attacks have happened outside the U.S. Nevertheless, one virus that did hit U.S. handhelds was known as the Liberty virus.

For example, some personal digital assistant users received what they thought was a program that would allow them to play a certain game for free. However, it launched a virus that erased all the data on the devices when they double-clicked on the link.

Nonetheless, the virus wasn't devastating for people who regularly back up their PDA information on their PCs. Nevertheless, more serious problems have occurred overseas in the form of viruses and/or malicious code that forced phones to dial particular numbers, intercepted transmissions and perpetuated the theft of data.

Disguised as a short message, one virus was distributed in Scandinavia. The virus rendered the buttons useless, when a user received the message. In order to get their phones fixed, users had to take them in to their service providers.

New types of malicious code have been written that force wireless devices to make phone calls, because many of them also have telephony capabilities. One incident in Japan caught the attention of wireless operators and software companies around the globe. Users of NTT DoCoMo's (http://www.nttdocomo.com/top.html) popular I-mode service [1] received an e-mail with what looked like an embedded Web site link. When customers clicked on the link, their phones automatically dialed Japan's emergency response number. Luckily, they could stop it before it got too bad; but, that could shut down a 911 system that could have life-and-death consequences.

For example, similar viruses could be unleashed that might flood a company's call center, or cause phones to dial a 900 number. If a virus that spread to all its mobile workers racked up significant charges, a corporation could be seriously affected.

The threat of data theft, perhaps, is more alarming to businesses. In order to prevent the interception of information as it's being transmitted, all wireless transmission standards have security built in, but they're known to be fallible. Encryption technology designed to head off the threat of sniffing, has been included by developers of standards such as Wireless Application Protocol (WAP) and the wireless LAN 802.11b standard [10].

Because the wireless network is essentially everywhere, sniffing is an inherent problem in wireless. Sniffers must have access to physical parts of the network in order to break in the wired world. The problem is, with wireless, they don't even have to be in the network. They can be in a van outside with a transmitter.

When researchers at the University of California at Berkeley figured out how to crack its built-in encryption, the widely used wireless LAN standard, 802.11, came under fire recently [10]. Because developers addressed wireless network security from the start and are working to beef it up before wireless LANs become more pervasive, there is still some hope,

Companies will also have to secure wireless transactions. There will be attacks on the devices themselves, but they quickly will be focused on transactions.

As devices develop more capabilities, these threats are expected to grow more serious and frequent. Typically, you should look to the past to predict the future. Also, new possible threats come into play every time there is a technology advancement.

Each time software companies release popular technologies in the PC environment, people use them to write malicious code. The same is expected with regard to wireless. For example, a Windows program can currently run on a Windows CE device, but CE doesn't yet support macros. So, the ability for viruses to spread is nil, because the device doesn't support macros.

Nevertheless, wireless devices are rapidly developing other capabilities. In the beginning the PDA was just something used to store contacts. But, today they are little computing devices. There's more of a chance of things being used improperly, as you create more functionality.

Most viruses have been regional so far. But, the threat of viruses spreading around the globe grows, as regions of the world begin to standardize wireless technologies. NTT DoCoMo, for example, plans to open its wireless network globally by 2006 [10]. This would then allow the worldwide spread of NTT DoCoMo threats.

In addition, there is a greater potential for viruses to spread between PCs and mobile devices (which could enable viruses to spread very quickly), because more of the capabilities are supported by devices. So that the same applications can run on PCs and handheld devices, Windows CE will support Java script in the future. Then viruses can spread easily via e-mail or programs that synchronize PCs and handheld devices. A version of Java is already supported by most wireless phones (including versions Nextel Communications sells primarily to businesses).

2.1 Firewall Building

What's really becoming readily available, are more wireless network security measures and products. Still, uncertainty about how to address potential threats is preventing some enterprises from deploying wireless.

Still though, many companies have to contend with wired security issues. And, it is difficult for IT managers to stay on top of new developments, because both the wired and wireless worlds change quickly. It's a tremendous challenge for IT managers to understand the space and issues, and what the solutions are to address it.

Many enterprises aren't yet concerned about protecting against viruses, because wireless viruses haven't been widespread. Also, many enterprises haven't heard very much about wireless viruses, and so, it's not a real issue right now. Although the data from wireless devices passes through the corporate firewall [8], additional wireless network security isn't necessary, because the information wouldn't be valuable to anyone else.

For example, at Final Mile Communications (http://www.finalmilecom.com/), a professional services company, field service workers use Nextel phones to receive trouble tickets and report status back to the dispatch center [10]. They haven't seen any viruses yet, but when and if a virus does present itself, it will be a serious issue to be dealt with.

Warning: URLs may change without notice!

There seems to be more concern about the possibility of data being stolen by other companies. By determining what's acceptable and what isn't, the first decision an enterprise must address when implementing a secure wireless system is to define its security model.

It's ponderous; that's part of the problem. For example, in the wired world, encryption based on public key infrastructure [7] hasn't taken off because it is difficult to use. An enterprise that wants to give a field service worker access to important data is aiming to make that worker more efficient. However, you've negated the advantage of going after wireless network security, if it introduces more error and takes longer to use. You have to take a holistic approach.

Coming to market, are more products purporting to provide end-to-end wireless network security that starts with the device and includes transmission and the software that runs applications. Thus, giving companies more options that fit their specific needs. One of the simplest problems, though, has not been widely addressed: Should the device be lost or stolen, few mobile devices have mechanisms for protecting information stored on them. So that only the owner can access stored data [6], there are some early products that companies can add to user devices to encrypt data.

For example, F-Secure (http://www.f-secure.se/foretaget/) has encryption and antivirus software for Pocket PCs, and Palm and Symbian devices [10]. F-Secure also offers antivirus engines for WAP gateways at the operator level. A further example is Trend Micro (http://www.trendmicro.com/), which has antivirus software for devices and guards against all entry points, including beaming, synching, e-mail and Internet downloading.

Programs similar to those available on laptops could be developed to allow a user whose device is lost or stolen to remotely destroy information or make it useless to anyone else. So, to protect important information stored on devices, such capabilities can be crucial.

Although serious threats aren't expected for some time, activity has begun to create virus protection software that lives on devices. The consumer electronics (CE) environment will probably be the first to see viruses written on a constant basis. But, it probably won't happen on a regular basis until sometime in 2006 [10].

It isn't easy to create antivirus software for devices. On phones, the real estate is owned by the operating company and there's a turf battle over what software can do. Virus protection could be required for each piece of software. The patchwork on the phones makes it difficult for a virus scanner to cover all parts of the device.

> **Note:** For example, F-Secure's PDA solution includes antivirus software for PCs that are constantly updated via the Internet and uploaded to PDAs when users synch with their PCs.

Because short-range communications techniques such as Bluetooth or infrared connections bypass wireless networks, software on devices become an important component of an antivirus campaign. Without sending the data through a server, users can beam information (and viruses) directly to one another.

> **Note:** Phones sold by NTT DoCoMo since December incorporate software to defeat viruses like the one that commanded the handset to automatically dial the emergency phone number.

By buying handsets that have authentication technology built in, IT managers can also protect the devices of mobile workers. For example, in order to make phones capable of accepting digital signatures, RSA Security (http://www.rsasecurity.com/) is one company working with handset vendors [10]. Unless they have the IT manager's digital signature, IT managers that send out updates or regular messages to mobile workers, could program workers' devices not to accept messages.

Presently being built into handsets, consists of such authentication technology. And, phone makers are interested in adding such capabilities, because they represent the potential for additional revenue. For the complete authentication platform, a device manufacturer could charge an enterprise per user.

Another line of defense concerns the servers accessed by mobile workers—that being in addition to virus protection on user devices. Currently, most platforms that support mobile e-mail carry antivirus and antispam software. For example, as threats pass through firewalls using the same technology that's been used to detect viruses in the PC environment, Trend Micro helps enterprises and wireless operators detect threats. Trend Micro will give its customers updated tools that can block viruses (usually within an hour), as new threats arise. According to Trend Micro, the stopping of viruses in the infrastructure is the best way to block their spread. The last line of defense is the device.

But, that type of protection gets more difficult in the future environment of always-on communications that will come with next-generation of packet wireless networks. Unlike a mail client, applications like instant messaging (IM) and location- or presence-based applications are chatty. So, the number of messages and the frequency go up by a factor of 10 [10]. Antivirus efforts become more complicated, as the sheer number of applications and messages are being sent.

2.1.1 Paranoia

Some space-age solutions are already available for the seriously paranoid. For example, for phones that can encrypt conversations, Siemens (http://www.siemens.com/index.jsp) has introduced an add-on capability [10]. Such technologies are used primarily by government workers, but certain industry segments might be interested in such rigorous precautions. Oil companies, for example, spend billions on exploration and want that information kept confidential. Some of their wireless network security is more stringent than that of governments.

> **Note:** There have also been some products designed for Palm devices with fingerprint-identification technology.

In order to fit the wireless security needs of some corporations, even more far-out ideas could be morphed from their original intent. Some law enforcement agencies have been working on a technology that could render a handgun (such as one carried by a police officer) useless if someone else tries to use it. Such technology could be ported to a handheld device.

Still, no one actually believes that they'll stop the troublemakers, even after all of these impressive wireless network security solutions are created and deployed. This is not a sense of resignation. It's a battle of attrition. It requires constant understanding and vigilance [10].

In continuing with the preceding concept of sci-fi parania, let's look at how a wireless network attack intercepts wireless Internet messages.

2.1.2 Interception

It's the stuff of sci-fi thrillers! By using gear that can be picked up at any electronics store (and is an easily downloadable piece of freeware), a group of wireless network security researchers have discovered a simple attack that enables them to intercept Internet traffic moving over a wireless network.

The attack, accomplished by @Stake Inc. (http://www.atstake.com/), a wireless network security consulting company in Cambridge, Mass., affects a popular consumer version of Research In Motion Ltd.'s (http://www.rim.com/) BlackBerry devices as well as a variety of handhelds that send unencrypted transmissions over wireless networks such as Mobitex [11].

Like other wireless standards such as Global System for Mobile Communications and General Packet Radio Service, the Mobitex specification sends packets in unencrypted form by design. With customers using the wireless network for everything from point-of-sale verification to e-mail, the network, which handles data transmissions only, has been in operation since 1986 and has a large base of installed devices [11].

In other words, the attack is fairly simple. The problem is, this isn't a bug. It's part of the spec that data is transmitted in the clear, just like it's part of the spec that Internet data is transmitted in the clear. Thus, the risk depends on when and what data they're sending, and who is using the wireless network.

The researchers were able to intercept traffic destined for BlackBerry Internet Edition devices by using a scanner with a digital output, an antenna and freely downloadable software. And, the attackers can read the messages they intercept without further work, because the packets aren't encrypted.

> **Note:** The Internet Edition handhelds are sold mainly through co-branding relationships with ISPs such as AOL Time Warner Inc.'s America Online service, EarthLink Inc. and Yahoo Inc.

According to executives at Research In Motion Ltd (RIM), they don't see the attack as a problem because they have never touted the Internet Edition devices as being secure [11]. Internet traffic isn't supposed to be secure. It's kind of like a company making beer and cola; and, someone saying that there's alcohol in the company's drinks when the children are drinking cola.

Nevertheless, the attack serves as a reminder to users that e-mail and other Internet traffic is open to snooping and is inherently insecure. You can always figure that anything that's sent via e-mail can be read by at least hundreds of people which have either legitimate or compromised access to systems sitting between you and your recipient. All of this just adds another potential access point. The disappointment here is that they didn't make at least a modest attempt to obscure the content.

> **Note:** The messages are only as secure as the wireless networks of the ISPs that relay them, none of which provide encrypted e-mail.

The attack also applies to other devices on the Mobitex wireless network, many of which are proprietary solutions developed for in-house corporate uses [11]. This attack does not work on the BlackBerry Enterprise Edition, which uses Triple Data Encryption Standard encryption in addition to other wireless network security features.

Customers are typically advised by Mobitex operators, that they should choose the wireless network security scheme that fits their particular needs. It was a conscious decision not to put wireless network-level security in because customers have said that they don't want the overhead associated with security if they're just doing things like instant messages. Customers can absolutely add on their own encryption to whatever application they're using the wireless network for [11].

Now, let's look at a classification of wireless network denial-of-service- and distributed denial-of-service attacks; and, a classification of the defense mechanisms that strive to counter these attacks. Using both known and potential attack mechanisms, the attack classification is illustrated next. Along with this classification, a discussion ensues about important features of each attack category that will in turn define the challenges involved in combating these threats. Using only the currently known approaches, the defense system classification is illustrated in this part of the chapter. Thus, in order to gain a better understanding of challenges in thc distributed denial-of-service field, the goal of this part of the chapter is to impose some order to the multitude of existing attack and defense mechanisms.

2.1.3 Denial Of Service

Today, denial of service (DOS) and distributed denial of service (DDoS) attacks are still on the rise and getting worse. After the year 2000 first wave of DDoS attacks against several of the largest commercial Internet sites including Yahoo!, eBay, E'Trade, Amazon.com and Buy.com, major e-commerce and wireless information sites worldwide still remain very very vulnerable [12].

In a recent survey of 649 wireless network security administrators from industry, government and academia, conducted by the Computer Security Institute (CSI) of San Francisco and the FBI, 49 percent detected denial of service attacks. That number is up from 36 percent in 1999 and 38 percent in 2001 [12].

Denial of service attacks are becoming incessant. Recently, researchers at the University of California, San Diego, reported 23,916 observed DoS attacks over a recent three-week period. In more than 3,000 distinct domain name servers, there were more than 6,000 distinct victims [12].

DOS and DDoS attacks are also growing in ferocity. Most attacks usually last only a few minutes. However, for example, EFNet.org, a provider of IRC chat services, experienced an attack lasting almost 20 days in 2001, at times subjecting EFNet to almost 700 Mbps of traffic. During the attack, several ISPs null-routed EFNet's wireless networks, essentially preventing customers from reaching them [12].

In 2001, a worm-virus, dubbed Code Red, infected more than 400,000 systems worldwide in one day. This is an astounding propagation rate [12]. The amount of traffic that could be generated by such a veritable army of computers would certainly overwhelm any modern wireless network infrastructure, if not threaten the stability of the Internet itself.

Recently, a variant of Code Red has appeared, called Code Blue. Some observers consider this virus even more threatening than its predecessor. While computers affected by Code Red were largely able to continue functioning normally with no obvious degradation in system performance, Code Blue gradually increases its usage of system resources. If Code Blue is not stopped, it could bring computers running Windows NT or Windows 2000 or XP to a halt [12].

There are technologies available now that can detect a simple DoS- or DDoS attack, and protect a wireless network system from the increasing damage these malicious acts can bring. In other words, DDoS attacks themselves pose an immense threat to the Internet and wireless networks. Consequently, many defense mechanisms have been proposed to combat them. Attackers constantly modify their tools to bypass these wireless network security systems. Researchers in turn, modify their approaches to handle new attacks. The DDoS field is evolving quickly. It is becoming increasingly hard to grasp a global view of the problem. This part of the chapter strives to introduce some structure to the DDoS field by developing a classification of DDoS attacks and DDoS defense systems. The goal of this part of the chapter is to stimulate discussions that might lead to a better understanding of the DDoS problem and highlight the important features of both attack and wireless network security mechanisms [12].

The proposed classifications are complete in the following sense: the attack classification covers known attacks and also those that have not currently appeared, but are potential threats that would affect current defense mechanisms; and, the defense systems classification covers not only published approaches, but also some commercial approaches that are sufficiently documented to be analyzed. Along with classification, important features of each attack are emphasized; as well as, the defense system category. Representative examples of existing mechanisms are also provided. Even though some parts of the chapter might point out vulnerabilities of certain classes of defense systems, this part of the chapter does not propose or advocate any specific DDoS defense mechanism. The purpose here is to draw attention to these problems so that they might be solved, not to criticize.

2.1.4 Overview Of A DDoS Attack

A DoS attack is characterized by all explicit attempts by attackers to prevent legitimate users of a service from using that service. A DDoS attack deploys multiple machines to attain this goal. The service is denied by sending a stream of packets to a victim that either provides the attacker with unlimited access to the victim machine so he can inflict arbitrary damage or consumes some key resource, thus rendering it unavailable to legitimate clients.

In other words, a computer bombards another system with floods of packets in DoS attacks. The goal is to prevent legitimate users from accessing the target host or wireless network. In order to provide a cover for other cracking activities, crackers sometimes use DoS attacks.

DoS attacks are not a new phenomenon. One of the earliest DoS attacks was a late 1996 strike against Panix.com, an ISP in New York [2]. That attack used a method considered primitive by today's standards: A single computer sent Panix' systems thousands of copies of a simple message that computers use to start a two-way dialog. The Panix machines receiving the messages had to allocate computer capacity to handle the dialogs. As the requests continued to pour in, the Panix machines soon used up their resources, and were effectively crippled.

DoS techniques had grown more sophisticated by the fall of 1999: Crackers began infiltrating multiple computers on the Internet--using them to launch coordinated assaults against other computers.

In these DDoS attacks, a cracker installs a program on a machine that, later on, in conjunction with other wireless networking systems, will be called on to participate in an attack. Usually distributed geographically, these pawn-like computers that forward DDoS attacks are called zombies. After the zombie computers (as few as three but sometimes more than 200) have been infiltrated, a "wakeup" command causes them all to begin sending bogus messages to the ultimate target of the attack.

Because DDoS attacks deliver a much heavier volume of messages (basically they are many single DoS attacks occurring simultaneously), they have the potential to be much more disruptive than relatively simple DoS attacks originating from a single computer. Contributing to the prevalence of DDoS attacks, is the raw number of computer operating systems in use on the Internet.

Most of the systems connected over the public Internet were running Unix or other type of proprietary mini-computer OS in the early 1990s. Mostly based on the Linux and BSD kernels, since 1992, over 30 companies have released their own proprietary operating systems. This caused a proliferation of software modules (such as Telnetd, sshd, bind and httpd) that function in the same way but operate different software algorithms.

A daunting, if not impossible task, is assessing the wireless network security of hundreds of independent operating systems. Furthermore, the subsequent closure of many of those independent OS manufacturers has made the task of identifying and closing wireless network security holes all the more challenging. The result, is the availability of thousands upon thousands of computers connected to the Internet running insecure operating systems.

An attacker can build a veritable army of zombie systems to launch attacks at will, armed with little more than a port-scanner, a root-kit and a little hubris (extreme arrogance). Thus, with the preceding in mind, this part of the chapter will answer the following questions:

1. How do these attacks occur'?
2. Why do they occur?
3. What makes DDoS attacks possible'? [2]

2.1.4.1 Strategy of a DDoS Attack

The attacker needs to *recruit* the multiple agent (slave) machines, in order to perform a distributed denial-of-service attack. By looking for wireless network security holes that would enable subversion, this process is usually performed automatically through scanning of remote machines. In addition, by using the discovered vulnerability to gain access to the machine, they are also *infected* with the attack code by vulnerable machines that are *exploited*. The exploit/infection phase is also automated. And, the infected machines can be used for further recruitment of new agents (see discussion of propagation techniques later in this chapter). Agent machines perform the attack against the victim. Through spoofing of the source address field in packets, attackers usually hide the identity of the agent machines during the attack. Thus, the agent machines can be reused for future attacks [2].

2.1.4.1.1 The Workings of a DDoS

For resource-intensive computing tasks, including a massive DoS attack, DDoS leverages one of the inherent benefits of distributed computing. To distribute the work across many systems in a wireless network, is one of the best approaches to completing resource-intensive tasks.

Enlisting numerous computers in a DDoS assault makes it both more devastating and harder to stop, in the same way, due to DDoS's distributed nature. It also makes tracing the original source of the attack virtually impossible.

By spoofing, or faking, the source address or destination address fields of an IP packet, DDoS attacks exploit the inherent "trust" that wireless networked computers have for each other. Wireless Internet routers (trusting every packet that appears to be correctly formed) will route the packet to its marked destination. Thus, the receiving system (again trusting the data in the packet) will reply to the forged source address.

It's easy to launch a DDoS attack. To create a force of zombie computers, there are a growing number of attack tools that are downloadable from Internet sites. The attacker needs only to select a Web site to attack, once a zombie force is established. From a central "command console" which can activate zombies located anywhere in the world, the attack itself can be initiated from a single computer,.

One common variable is shared by many types of DDoS attacks: large wireless Internet traffic flows. The IP protocol itself is very trusting. Attackers are free to forge and change IP data with impunity. By fooling the victim's firewall or target host into accepting this seemingly legitimate traffic, the cracker uses the IP address of a trusted source [12].

2.1.4.1.2 Today's Practices

Being of little value in defending against DDoS attacks, wireless network firewalls are indispensable for countering many kinds of malicious incursions. Firewalls are designed to manage an environment where everyone outside the enterprise is untrusted and everyone inside the enterprise is trusted.

However, the model of trusted and untrusted wireless networks is no longer viable in today's wireless Internet, with service providers delivering the means for user access to the Web. Colocation and Web-hosting providers allow many different companies (each with its own wireless network security requirements) to reside within a single Internet data center. Because an ISP cannot inherently trust every host on their wireless network, this has diminished the effectiveness of firewalls for use in colocation and Web-hosting data centers.

ISPs and colocation providers created an environment where their systems could easily become the source of DoS and DDoS attacks, with little overall wireless network security. By implementing bi-directional firewall filtering (managing outbound traffic as vigilantly as inbound traffic), wireless network administrators in these organizations can lower the likelihood of these attacks. While bi-directional filtering is a good method to protect the wireless Net at large, unless it is implemented everywhere, there is still the threat of attacks from those wireless networks that are not well protected. Extensive filtering of this nature is rarely implemented, because filtering is cost heavy on a CPU (creating traffic bottlenecks) [12].

2.1.4.1.3 Fighting DDoS Attacks: The Technologies

Advanced implementation of intrusion detection system (IDS) technology is required as an effective defense against DDoS attacks. Intrusion detection systems fall into two broad categories: wireless network-based and host-based (see sidebar, "Pros And Cons Of Intrusion Detection Systems").

Normally, wireless network-based IDS systems are placed near the wireless system or systems being monitored. They examine the wireless network traffic and determine whether it falls within acceptable boundaries (looking deep into the payload portion of the packet for signs of malicious activity). In order to match suspicious traffic, signature files are compared to those of known attack patterns.

Signature-based systems must update their libraries of known attacks continuously, like software that protects against known viruses. They also cannot learn new traffic patterns. This allows a large number of attacks to go undetected and leaves the wireless network vulnerable [12].

Pros And Cons Of Intrusion Detection Systems

Intrusion detection systems (IDS) are more complicated than firewalls and can be pricey but they have definite benefits. Let's briefly look at the pros and cons of such systems.

Wireless network security is a key consideration for network design. IDS must strike a balance between wireless network security and network usability, for it to be useful in real world conditions.

IDS Defined And How It Differs From Traditional Firewall Technology

IDS is different from traditional firewalls because it involves the detecting of a security breach. In most cases, it translates into defending against a wireless network security breach. It's sort of part system administration. If you look at the way IDS evolved originally, the idea was that if you set up a particular probe or a particular application, you'd be able to detect unauthorized access. Most of those things prior to the creation of the wireless network security industry, were things that network engineers already implemented. So, a lot of the earlier tools such as Transmission Control Protocol (TCP) wrappers and the original ISS applications, and several others were originally developed and released as open-source tools. Ten or eleven years ago, IDS was very different than it today.

The Workings of IDS Technology

There are three approaches to how IDS technology works. The first is a blind a barricade, which is similar to a firewall. You implement a wrapper or some sort of a scanner that looks for specific events. When it sees the event, it correlates it and decides whether or not it's an acceptable event or an unacceptable event. So, if you try and log into a system and it tells you that you don't have the right credentials, when that failure occurs, an application will then alarm some other console to say that there is an unauthorized access.

Second Approach To How IDS Technology Works

The second approach is signature based. Signature based works on a host or a network basis where a particular type of packet or data stream is looked for. When the scanner sees it, it generates an alarm. The issue with signature based applications is that they're really only as good as the signatures are. So, in terms of total effectiveness, they leave a large area to be desired.

Last Approach To How IDS Technology Works

Profile or anomaly detection is the last approach to how IDS technology works, and the most recent. As an administrator, you create profiles of your system; you determine what their normal operational parameters are; and, you look at' events that are out of profile. So, for instance, if you have a Web server (and as a Web server), it only runs SSH. Then, one day all of a sudden, it starts running HTTP clients. That in itself should set up a major flag. Now, if you were running a Microsoft server for example, that you've got a Code Red like virus or something like that, it would be an indication.

Deep Packet Inspection And How It Relates To IDS

In order to be able to determine the potential attack, you have to eliminate the problem with most signature-based events; and, that basically is that they are looking at just the packet header. The idea of deep packet inspection now is that you can actually look inside the data-stream to determine whether or not an attack is really being perpetrated. This is a very relevant form of IDS that is just starting to be developed. Now, developers can write around firewalls and keep their applications in use, because they've become a lot smarter.

For example, if you are running in a firewall environment, and the firewall administrator doesn't want you to run AOL Instant Messenger (AIM), he or she doesn't permit the service to happen. If you turn on Secure Socket Layer (SSL) now, all of a sudden you can run AIM and the firewall administrator can't do anything about it. Because they will be interpreted as normal traffic, the idea is that this sort of work around will be incorporated into the tools that hackers make, in order to be able to completely evade IDS.

Factors That Should Be Considered When Tuning An IDS To Your Wireless Network

Factors that should be considered when tuning an IDS to your wireless network, is all based on success and cost. Success means that IDSes need to be tuned, and that they need an enormous amount of tuning. That is probably the biggest problem with them. Secondly, they cost an excessive amount of money. So, typically what happens, is that someone will buy an IDS and implement it, and it will be the alarm for everything, because it needs to be tuned. Of course, when it goes through the tuning process, they then have to make assumptions about what to permit and what not to permit. Then, as part of your wireless network security policy, you have to make these decisions about what you're going to permit.

The Result

What typically happens is that a company will decide that they need to implement IDS. So, they will go out and buy a product; and, they will put the product in and it will sit dormant, because nobody wants to take the time to actually fix it. Or, they'll have just spent a lot of money, and they'll implement it without dealing with all the underlying issues (which is not running a wireless secure environment to begin with).

IDS Architecture

IDS architecture is driven by cost, but the ideal architecture would be multi-faceted. It would start with a standard border filter where you're just filtering out all the noise, stuff that just happens randomly, and people looking for holes in your wireless network that shouldn't be there if you're running any type of security at all. The second component would typically be some sort of wireless network probe for looking at network-based events. The third level might be some sort of host or signature-based system. An anomaly-based system is also a big value, depending on what you are looking at and what you are defending.

IDS Market

The IDS market has had a lot of fluctuation. For example, Enterasys has been severely impacted in the market overall, so they are making a desperate attempt to try and grow their product out. It has a very big following; so, the product will survive and continue to grow. ISS has been very aggressively developing a globalized system, integrating technologies from other companies that they've acquired to build out their original product base. Cisco is radically redesigning its IDS package. The Cisco Security Policy Manager is growing out and they are beginning to integrate IDS functionality in both their switches and routers. So here, there is a potential growth market.

Emerging Trends In IDS Technology?

Centralized collection, analysis and reporting, are what people are going to be looking for. Also, people will be looking at things like profiler trending. That is looking at how the wireless network is behaving in the context of a multi-faceted system, and using that profiling to detect events that you wouldn't see with a signature based system. Finally, you can detect anomalies ahead of time before they have a significant impact on your wireless network, with profiling [4].

Similar to critical system files, host-based IDSes (HIDSes) run directly on the system being monitored and monitor log files. For instance, an HIDS would send alerts to wireless network security administrators if an attacker managed to compromise a system and tried to modify the *netstat* command. Often, in a DDoS attack, the unauthorized user or cracker has installed code that may enable the computer to be used later as an agent or zombie.

Anomaly-based detection engines are used by HIDSes. Anomaly detection identifies any unacceptable deviation from expected behavior. Expected behavior must be defined in advance by an automatically developed profile or by a manually developed profile. An anomaly can be an increase in traffic from a known good source or simply a synchronized attack from zombies sending malicious traffic.

False alarm problems with anomaly-based systems are usually very high, as is with signature-detection systems. A frequently-updated IDS can alert wireless network managers of known attacks. That's the good news!

Finally, the goal of a DDoS attack is to inflict damage on the victim. This could be either for personal reasons (a significant number of DDoS attacks are against home computers, presumably for purposes of revenge); for material gain (damaging competitor's resources); or, for popularity (successful attacks on popular Web servers gain the respect of the hacker community) [2].

2.1.5 DDoS Attacks Classification

In order to devise a classification of DDoS attacks, let's look at the types of DDoS attacks; *means* used to prepare and perform the attack; the *characteristics* of the attack itself; and, the *effect* it has on the victim. Fig.1-1 summarizes the classification [2].

2.1.5.1 Types Of DDoS Attacks

There are several common types of dedicated denial of service attacks (DDoS) attacks, including the following:

- Smurf or Reflector Attack.
- Amplified Reflector Attack.
- TCP/UDP Floods.
- Spoofed-Source SYN Attack [12].

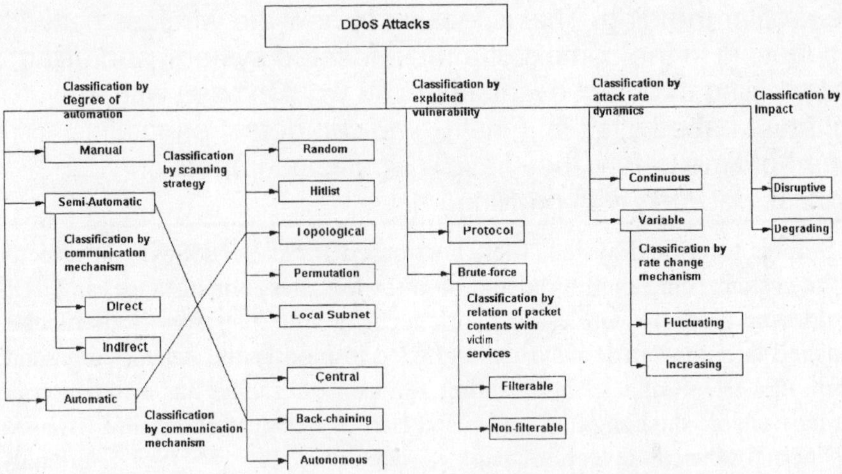

Figure 1-1. Classification of DDoS attacks.

2.1.5.1.1 Smurf or Reflector Attack

Here, an attacker spoofs Internet Control Message Protocol (ICMP) Echo Request (PING) packets with the IP address of the target host in the source address field of the IP packet. The destination field is chosen with a list of known reflectors (any device on the Internet that will respond to PING) or randomized. Each reflector responds to the intended target (the host that legitimately own the spoofed address) with an ICMP Echo Reply packet, thus overflowing the target wireless network with useless traffic, as these packets are flooded onto the Internet.

2.1.5.1.2 Amplified Reflector Attack

The amplified reflector attack is a variation on the Smurf or Reflector attack. The destination field is the broadcast address of an amplifier network. When the packet hits the reflector address, every device on the reflector-network will reply to the target. Ten (10) Echo Request packets become 1,000 Echo Reply packets, if 100 devices on the wireless network answer to the target [12].

2.1.5.1.3 TCP/UDP Floods

TCP/UDP floods are mainly used to disrupt wireless network communications or otherwise hinder performance. TCP and UDP traffic can be directed at specific ports on a host. But, to simply overwhelm a target wireless network with traffic, TCP and UDP traffic are more often than not used.

2.1.5.1.4 Spoofed-Source SYN Attack

An attacker randomizes the source address field and sets the destination address field of the IP packet as the target host. The target host will process each SYN packet, open a port and respond with a SYN_ACK to the address in the source field. So, in order to disable Web servers, this is the most common assault used.

2.1.5.2 Degree Of Automation Classification

The attacker needs to locate prospective agent machines and infect them with the attack code during the attack preparation. Based on the degree of automation of the attack, this part of the chapter differentiates between *manual, semi-automatic* and *automatic* DDoS attacks [2].

2.1.5.2.1 Manual Attacks

Only the early DDoS attacks belonged to the manual category. The attacker scanned remote machines for vulnerabilities, broke into them and installed the attack code, and then commanded the onset of the attack. During the lead to the development of semiautomatic DDoS attacks (the category where most contemporary attacks belong), all of these actions were soon automated.

2.1.5.2.2 Semi-Automatic Attacks

The DDoS wireless network consists of handler (master) and agent (slave, daemon) machines in semi-automatic attacks. The attacker deploys automated scripts for scanning and compromise of those machines and installation of the attack code. He or she then uses handler machines to specify the attack type and the victim's address and to command the onset of the attack to agents, who send packets to the victim. Based on the communication mechanism deployed between agent and handler machines, let's divide semi-automatic attacks into *attacks with direct communication* and *attacks with indirect communication.*

2.1.5.2.2.1 Direct Communication Attacks

The agent and handler machines need to know each other's identity in order to communicate during attacks with direct communication. This is achieved by hard-coding the IP address of the handler machines in the attack code that is later installed on the agent. Each agent then reports its readiness to the handlers, who store its IP address in a file for later communication. The obvious drawback of this approach is that discovery of one compromised machine can expose the whole DDoS wireless network. Also, agents and handlers are identifiable by wireless network scanners, since they listen to wireless network connections.

2.1.5.2.2.2 Indirect Communication Attacks

To increase the survivability of a DDoS wireless network, attacks with indirect communication deploy a level of indirection. Recent attacks provide the example of using Internet Relay Chat (IRC) channels for agent/handler communication. Since the IRC channel offers sufficient anonymity to the attacker, the use of IRC services replaces the function of a handler [2]. Agent communications to the control point may not be easily differentiated from legitimate wireless network traffic, since DDoS agents establish outbound connections to a standard service port used by a legitimate wireless network service. The agents do not incorporate a listening port that is easily detectable with wireless network scanners. An attacker controls the agents using IRC communications channels. Thus, discovery of a single agent may lead no further than the identification of one or more IRC servers and channel names used by the DDoS wireless network. From there, identification of the DDoS wireless network depends on the ability to track agents currently connected to the IRC server. There is nothing to prevent attackers from subverting other legitimate services for similar purposes, even though the IRC service is the only current example of indirect communication.

2.1.5.2.3 Automatic Attacks

By avoiding the need for communication between attacker and agent machines, automatic DDoS attacks additionally automate the attack phase. Preprogrammed in the attack code is the time of the onset of the attack, attack type, duration and victim's address. Since the attacker (he or she) is only involved in issuing a single command (the start of the attack script), it is obvious that such deployment mechanisms offer minimal exposure to the attacker. The hardcoded attack specification suggests a single-purpose use of the DDoS wireless network. However, by enabling easy future access and modification of the attack code, the propagation mechanisms usually leave the backdoor to the compromised machine open.

By deploying automatic scanning and propagation techniques, both semi-automatic and automatic attacks recruit the agent machines. Based on the scanning strategy, you should be able to differentiate between attacks that deploy *random scanning, hitlist scanning, topological scanning, permutation scanning* and *local subnet scanning.* Thus, by gaining a larger agent population, attackers usually combine the scanning and exploitation phases. Here, the description of scanning techniques relates to this model [2].

2.1.5.2.3.1 Random Scanning Attacks

By using a different seed during random scanning, each compromised host probes random addresses in the IP address space. Since many machines probe the same addresses, this potentially creates a high traffic volume. Here, random scanning was performed by Code Red (CRv2 or higher) [2].

2.1.5.2.3.2 Hitlist Scanning Attacks

From an externally supplied list, a machine performing hitlist scanning probes all addresses. It sends one half of the initial hitlist to the recipient and keeps the other half, when it detects the vulnerable machine. This technique allows for no collisions during the scanning phase and great propagation speed (due to exponential spread). An attack deploying hitlist scanning could obtain the list from netscan.org of domains that still support directed IP broadcast and can thus be used for a Smurf attack.

2.1.5.2.3.3 Topological Scanning Attacks

In order to select new targets, topological scanning uses the information on the compromised host. By exploiting the information from address books for their spread, all E-mail worms use topological scanning.

2.1.5.2.3.4 Permutation Scanning Attacks

All compromised machines share a common pseudo-random permutation of the IP address space during permutation scanning. In this permutation, each IP address is mapped to an index. By using the index computed from its IP address as a starting point, a machine begins scanning. It chooses a new random start point whenever it sees an already infected machine. While maintaining the benefits of random probing, this has the effect of providing a senti-coordinated, comprehensive scan.

2.1.5.2.3.5 Local Subnet Scanning Attacks

To preferentially scan for targets that reside on the same subnet as the compromised host, local subnet scanning can be added to any of the previously described techniques. A single copy of the scanning program can compromise many vulnerable machines behind a firewall by using this technique. Code Red II and Nimda Worm used local subnet scanning [2]. A differentiation is made here between attacks that deploy *central source propagation, back-chaining propagation* and *autonomous propagation* that is based on the attack code propagation mechanism [2].

2.1.5.2.3.6 Central Source Propagation Attacks

The attack code resides on a central server or set of servers during central source propagation. The code is downloaded from the central source through a file transfer mechanism after compromise of the agent machine. The li0n [2] worm operated in this manner.

2.1.5.2.3.7 Back-Chaining Propagation Attacks

The attack code is downloaded from the machine that was used to exploit the system during back-chaining propagation. During the next propagation step, the infected machine then becomes the source. Since back-chaining propagation avoids a single point of failure, it is more survivable than central-source propagation. Back-chaining propagation was used by the Ramen worm [2] and Morris Worm [2].

2.1.5.2.3.8 Autonomous Propagation Attacks

By injecting attack instructions directly into the target host during the exploitation phase, autonomous propagation avoids the file retrieval step. Autonomous propagation is used by Code Red [2], Warhol Worm [2] and numerous E-mail worms [2].

2.1.5.3 Exploited Vulnerability Classification

In order to deny the service of the victim to its clients, DDoS attacks exploit different strategies. There is a differentiation between *protocol attacks and brute-force attacks* that are based on the vulnerability that is targeted during an attack.

2.1.5.3.1 Attacks with Protocol

In order to consume excess amounts of its resources, protocol attacks exploit a specific feature or implementation bug of some protocol installed at the victim. Examples include the authentication server attack, the Transmission Control Protocol synchronous idle character. (TCP SYN) attack, and the Common Gateway Interface (CGI) request attack.

The exploited feature in the TCP SYN attack is the allocation of substantial space in a connection queue immediately upon receipt of a TCP SYN request. The attacker initiates multiple connections that are never completed, thus filling up the connection queue indefinitely. The attacker consumes the CPU time of the victim by issuing multiple CGI requests in the CGI request attack. Also, the attacker exploits the fact that the signature verification process consumes significantly more resources than bogus signature generation in the authentication server attack. He or she sends numerous bogus authentication requests to the server, thus tying up its resources.

2.1.5.3.2 Attacks with Brute-Force

By initiating a vast amount of seemingly legitimate transactions, brute-force attacks are performed. This exhausts the victim's resources, since an upstream wireless network can usually deliver higher traffic volume than the victim wireless network can handle. Here, based on the relation of packet contents with victim services *into filterable* and *nonfilterable* attacks, brute-force attacks are further divided [2].

2.1.5.3.2.1 Attacks Which are Filterable

For non-critical services of the victim's operation, filterable attacks use bogus packets or packets, and thus can be filtered by a firewall. Examples of such attacks are an Internet Control Message Protocol (ICMP) request flood attack on a Web server or a User Datagram Protocol (UDP) flood attack.

2.1.5.3.2.2 Attacks Which Are Non-Filterable

Packets that request legitimate services from the victim are used by non-filterable attacks. Thus, filtering all packets that match the attack signature would lead to an immediate denial of the specified service to both attackers and the legitimate clients. Examples are a domain name server (DNS) request flood targeting a name server or a HTTP request flood targeting a Web server.

The line between protocol and brute force attacks is thin. Protocol attacks also overwhelm a victim's resources with excess traffic. Badly designed protocol features at remote hosts are frequently used to perform "reflector" brute-force attacks, such as the DNS request attack [2] or the Smurf attack [2]. The difference is that a victim can mitigate the effect of protocol attacks by modifying the deployed protocols at its site. This takes place all the while it is helpless against brute-force attacks due to its own limited resources (a victim can do nothing about an attack that swamps its wireless network bandwidth) or due to their misuse of legitimate services (non-filterable attacks).

If you counter protocol attacks by modifying the deployed protocol, it pushes the corresponding attack mechanism into the brute-force category. For example, if the victim deploys TCP SYN cookies [2] to combat TCP SYN attacks, it will still be vulnerable to TCP SYN attacks that generate more requests than its wireless network can accommodate. However, to inflict damage at the victim, the brute-force attacks need to generate a much higher volume of attack packets than protocol attacks. So, by modifying the deployed protocols, the victim pushes the vulnerability limit higher. Evidently, classification of the specific attack needs to take into account both the victim's configuration and the attack mechanisms used.

> **Note:** It is interesting that the variability of attack packet contents is determined by the exploited vulnerability.

Packets comprising protocol and non-filterable brute force attacks must specify some valid header fields and possibly some valid contents. For example, unless they hijack connections from legitimate clients, TCP SYN attack packets cannot vary the protocol or flag field, and HTTP flood packets must belong to an established TCP connection and therefore cannot spoof source addresses.

2.1.5.4 Attack Rate Dynamics Classification

Now, let's differentiate here between *continuous rate* and *variable rate* attacks [2]. Of course, your ability to differentiate between the two all depends on attack rate dynamics.

2.1.5.4.1 Attacks With A Continuous Rate

A continuous rate mechanism is deployed by the majority of known attacks. Agent machines generate the attack packets with full force after the onset is commanded. This sudden packet flood disrupts the victim's services quickly, and thus leads to attack detection.

2.1.5.4.2 Attacks Which Are Of A Variable Rate

Variable rate attacks are more cautious in their engagement. They also vary the attack rate to avoid detection and response. You can also differentiate between attacks with *increasing rate and fluctuating rate* soley *based* on the rate change mechanism [2].

2.1.5.4.2.1 Attacks Which Are Of An Increasing Rate

A slow exhaustion of victim's resources are attacks that have a gradually increasing rate lead. A state change of the victim could be so gradual that its services degrade slowly over a long time period. This could result in delaying detection of the attack.

2.1.5.4.2.2 Attacks With A Fluctuating Rate

By occasionally relieving the effect to avoid detection, attacks that have a fluctuating rate adjust the attack rate based on the victim's behavior. There is also the example of *pulsing attacks* at the extreme end [2]. Agent hosts periodically abort the attack and resume it at a later time during pulsing attacks. The victim experiences periodic service disruptions if this behavior is simultaneous for all agents. The victim experiences continuous denial of service if, however, agents are divided into groups who coordinate so that one group is always active.

2.1.5.5 Impact Classification

Now, let's differentiate between *disruptive* and *degrading* attacks [2]. Of course, this again depends on the impact of a DDoS attack on the victim.

2.1.5.5.1 Attacks Which Are Disruptive

To completely deny the victim's service to its clients is the goal of disruptive attacks. All currently known attacks belong to this category.

2.1.5.5.2 Attacks Which Are Degrading

To consume some (presumably constant) portion of a victim's resources is the goal of degrading attacks. These attacks could remain undetected for a significant time period, since they do not lead to total service disruption. Damage inflicted on the victim could be immense on the other hand. For example, an attack that effectively ties up 40% of the victim's resources would lead to denial of service to some percentage of customers during high load periods, and possibly slower average service. Some customers dissatisfied with the quality would consequently change their service provider and the victim would thus lose income. To upgrade its servers and wireless networks, alternately, the false load could result in a victim spending money.

2.1.6 Classification Of DDoS Defense Mechanisms

The advent of numerous DDoS defense mechanisms, were the result of the seriousness of the DDoS problem and the increased frequency of DDoS attacks. Some of these mechanisms address a specific kind of DDoS attack such as attacks on Web servers or authentication servers. Other approaches attempt to solve the entire generic DDoS problem. If deployed in an environment where these requirements are not met, most of the proposed approaches require certain features to achieve their peak performance, and will perform quite differently.

There is no silver bullet against DDoS attacks, as is frequently pointed out,. Therefore you need to understand not only each existing DDoS defense approach, but also how those approaches might be combined together to effectively and completely solve the problem. The proposed classification, shown in Fig. 1-2, should help you reach this goal [2].

2.1.6.1 Activity Level Classification

Now, let's differentiate between *preventive* and *reactive* mechanisms [2]. This is based on the activity level of DDoS defense mechanisms.

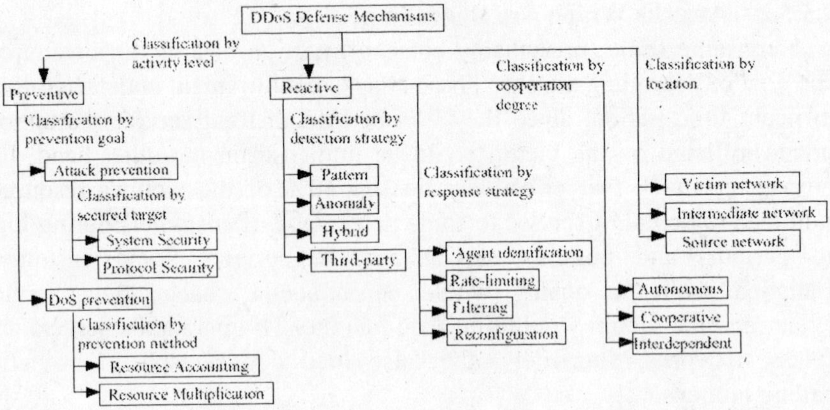

Figure 1-2. Classification of DDoS defense mechanisms.

2.1.6.1.1 Preventive Mechanisms Goal

To enable potential victims to endure the attack without denying services to legitimate clients or to eliminate the possibility of DDoS attacks altogether, is the goal of preventive mechanisms. According to these goals, let's further divide preventive mechanisms into *attack prevention, denial-qf-serv'ice prevention, intrusion prevention systems* mechanisms [2].

2.1.6.1.1.1 Modification Of Attack Prevention Mechanisms

In order to eliminate the possibility of a DDoS attack, attack prevention mechanisms modify the system configuration. Based on the target they secure, let's further divide them into wireless network security *system* and *protocol security* mechanisms [2].

2.1.6.1.1.1.1 Wireless Network Security System Mechanisms

Wireless network security system mechanisms increase the overall security of the system, guarding against illegitimate accesses to the machine, removing application bugs and updating protocol installations to prevent intrusions and misuse of the system. DDoS attacks owe their power to large numbers of subverted machines that cooperatively generate the attack streams. If these machines were secured, the attackers would lose their army and the DDoS threat would then disappear. On the other hand, systems vulnerable to intrusions can themselves become victims of DDoS attacks in which the attacker, having gained unlimited access to the machine, deletes or alters its contents. Potential victims of DDoS attacks can be easily overwhelmed if they deploy vulnerable protocols. Examples of wireless network security system mechanisms include monitored access to the machine; applications that download and install security patches; firewall systems; virus scanners; intrusion detection systems; access lists for critical resources; capability-based systems; and, client-legitimacy-based systems [2]. The history of computer wireless network security suggests that this approach can never be 100% effective. But, doing a good job here will certainly decrease the frequency and strength of DDoS attacks.

2.1.6.1.1.1.2 Protocol Wireless Network Security Mechanisms

The problem of bad protocol design is addressed by protocol wireless network security mechanisms. Many protocols contain operations that are cheap for the client but expensive for tile server. Such protocols can be misused to exhaust the resources of a server by initiating large numbers of simultaneous transactions. Classic misuse examples are the TCP SYN attack, the authentication server attack, and the fragmented packet attack, in which the attacker bombards the victim with malformed packet fragments forcing it to waste its resources on reassembling attempts. Examples of protocol wireless network security mechanisms include guidelines for a safe protocol design in which resources are committed to the client only after sufficient authentication is done, deployment of powerful proxy server that completes TCP connections or the client has paid a sufficient price, etc [2].

The victim is completely resilient to protocol attacks, by deploying comprehensive protocol and wireless network security system mechanisms. Also, these approaches are inherently compatible with and complementary to all other approaches.

2.1.6.1.1.2 Enabling Denial-Of-Service Prevention Mechanisms

Without denying service to legitimate clients, denial-of-service prevention mechanisms enable the victim to enable attack attempts. This is done either by enforcing policies for resource consumption or by ensuring that abundant resources exist so that legitimate clients will not be affected by the attack. Consequently, based on the prevention method, let's differentiate here between *resource accounting and resource multiplication* mechanisms [2].

2.1.6.1.1.2.1 Using Resource Accounting Mechanisms To Police

Based on the privileges of the user and his or her behavior, resource accounting mechanisms police the access of each user to resources. Such mechanisms guarantee fair service to legitimate well-behaving users. In order to avoid user identity theft [9], they are usually coupled with legitimacy-based access mechanisms that verify the user's identity [2].

2.1.6.1.1.2.2 Resource Multiplication Mechanisms Counter DDoS Threats

In order to counter DDoS threats, resource multiplication mechanisms provide an abundance of resources. The straightforward example is a system that deploys a pool of servers with a load balancer and installs high bandwidth links between itself and upstream routers. This approach essentially raises the bar on how many machines must participate in all attack to be effective. While not providing perfect protection, for those who can afford the costs, this approach has often proven sufficient. For example, Microsoft has used it to weather large DDoS attacks.

2.1.6.1.1.3 Intrusion Prevention Systems Mechanisms

Intrusion Detection Systems form a small but critical piece of the computer security jigsaw, alerting to intrusions and attacks aimed at computers or networks. They're not the computer security panacea. But, they are your eyes and ears, essential in knowing whether you are under attack. Intrusion Prevention Systems (IPS) mechanisms take this concept to the next level and sit inline blocking the packets you tell them to based on signatures as per the IDS. They can be highly effective as a defensive tool but need to be configured with great care and attention in stages.

IPS mechanisms sit inline on the network, statefully analyzing packet content and block certain packets that match a signature and alert on others. It is sometimes easier to explain what isn't an IPS mechanism, for instance, products that just block by port such as routers and many firewalls. Furthermore, the IPS mechanism must block the packet and not just use TCP resets, spoof reject packets from border devices or update border devices to shun addresses.

2.1.6.1.2 Reactive Mechanisms Alleviate The Impact Of An Attack

Reactive mechanisms strive to alleviate the impact of an attack on the victim. In addition, they need to *detect* the attack and *respond* to it in order to attain this goal [2].

To detect every attempted DDoS attack as early as possible and to have a low degree of false positives, is the goal of attack detection. Upon attack detection, steps can be taken to characterize the packets belonging to the attack stream and provide this characterization to the response mechanism. Based on the attack detection strategy into mechanisms that deploy *pattern detection, anomaly detection, hybrid detection,* and *third-party detection,* is the classification of reactive mechanisms [2].

2.1.6.1.2.1 Pattern Attack Detection Mechanisms

The signatures of known attacks in a database are stored by mechanisms that deploy pattern detection. Each communication is monitored and compared with database entries to discover occurrences of DDoS attacks. Occasionally, the database is updated with new attack signatures. The obvious drawback of this detection mechanism is that it can only detect known attacks, and it is usually helpless against new attacks or even slight variations of old attacks that cannot be matched to the stored signature. Nonetheless, no false positives are encountered and known attacks are easily and reliably detected.

2.1.6.1.2.2 Anomaly Attack Detection Mechanisms

Deployment of anomaly detection mechanisms have a model of normal system behavior, such as a model of normal traffic dynamics or expected system performance. The current state of the system is periodically compared with the models to detect anomalies. The advantage of anomaly detection over pattern detection is that unknown attacks can be discovered. However, anomaly-based detection has to address two issues: threshold setting and model update [2].

2.1.6.1.2.2.1 Setting The Threshold

When the current system state differs from the model by a certain threshold, anomalies are detected. While a high threshold reduces the sensitivity of the detection mechanism, the setting of a low threshold leads to many false positives.

2.1.6.1.2.2.2 Update The Model

With time, systems and communication patterns evolve. And, models need to be updated to reflect this change. Anomaly-based systems usually perform automatic model update using statistics gathered at a time when no attack was detected. The detection mechanism are made vulnerable by this approach to increasing rate attacks that can delay or even avoid attack detection and mistrain models.

2.1.6.1.2.3 Hybrid Attack Detection Mechanisms

Using data about attacks discovered through an anomaly detection mechanism to devise new attack signatures and update the database, have allowed many intrusion detection systems to use hybrid detection.

Mechanisms that deploy hybrid detection combine the pattern-based and anomaly-based detection.

Properly extracting a signature from a detected attack can be challenging, if these systems are fully automated. Also, the system itself can become a denial-of-service tool. Therefore, the system must be careful not to permit attackers to fool it into detecting normal behavior as all attack signature.

2.1.6.1.2.4 Third-Party Attack Detection Mechanisms

By relying on an external message that signals the occurrence of the attack and provides attack characterization, mechanisms that deploy third-party detection do not handle the detection process themselves. Easily found among traceback mechanisms, are examples of mechanisms that use third-party detection [2].

While imposing minimal collateral damage to legitimate clients of the victim, the goal of the attack response is to relieve the impact of the attack on the victim. The reactive mechanisms are classified here based on the response strategy into mechanisms that deploy *agent identification, rate-limiting, filtering* and *reconfiguration* approaches [2].

2.1.6.1.2.5 Mechanisms For Agent Identification

The identity of the machines that are performing the attack, are reveled by agent identification mechanisms that provide the victim with information. This information can then be combined with other response approaches to alleviate the impact of the attack. To eliminate spoofing, are agent identification examples that include numerous traceback techniques and approaches. So, for agent identification, this enables the use of the source address field [2].

2.1.6.1.2.6 Mechanisms For Rate-Limiting

On a stream that has been characterized as malicious by the detection mechanism, rate-limiting mechanisms impose a rate limit. Rate limiting is a lenient response technique that is usually deployed when the detection mechanism has a high level of false positives or cannot precisely characterize the attack stream. Even if all traffic streams are rate-limited, the disadvantage is that they allow some attack traffic through, so extremely high scale attacks might still be effective.

2.1.6.1.2.7 Mechanisms Filtering

In order to filter out the attack stream completely, filtering mechanisms use the characterization provided by a detection mechanism. Examples include dynamically deployed firewalls, and also a commercial system called TrafficMaster (http://www.mazunetworks.com/) [2]. Unless detection strategy is very reliable, filtering mechanisms run the risk of accidentally denying service to legitimate traffic. Thus, clever attackers might leverage them as denial-of-service tools. And, that is even a worse situation.

2.1.6.1.2.8 Mechanisms for Reconfiguration

The topology of the victim or the intermediate wireless network are changed by the reconfiguration mechanisms to isolate the attack machines or to either add more resources to the victim. Reconfigurable overlay wireless networks, resource replication services, attack isolation strategies, etc., are examples [2].

Detection and response can be performed by reactive DDoS defense mechanisms that are either in cooperation with other entities in the Internet or alone. So, based on the cooperation degree, let's differentiate between *autonomous, cooperative* and *interdependent* mechanisms [2].

2.1.6.1.2.9 Mechanisms Which Are Autonomous

Independent attack detection and response are performed by autonomous mechanisms. They are usually deployed at a single point in the Internet and act locally. Firewalls and intrusion detection systems provide an easy example of autonomous mechanisms.

2.1.6.1.2.10 Mechanisms Which Are Cooperative

Capable of autonomous detection and response, are what are known as cooperative mechanisms. But, they can also achieve significantly better performance through cooperation with other entities. Mechanisms deploying pushback [2] provide examples of cooperative mechanisms. They detect the occurrence of a DDoS attack by observing congestion in a router's buffer, characterize the traffic that creates the congestion, and act locally to impose a rate limit on that traffic. Nevertheless, if the rate limit requests can be propagated to upstream routers who otherwise may be unaware of the attack, they achieve significantly better performance.

2.1.6.1.2.11 Mechanisms That Are Interdependent

Interdependent mechanisms cannot operate autonomously. In other words, iInterdependent mechanisms rely on other entities either for attack detection or for efficient response. Traceback mechanisms [2] provide examples of interdependent mechanisms. Almost no benefit would be provided by a traceback mechanism deployed on a single router.

2.1.6.2 Deployment Location Classification

Now, let's look at deployment location. In other words, let's differentiate between DDoS mechanisms deployed at the victim, intermediate, or source wireless network.

2.1.6.2.1 Victim- Wireless Network Mechanisms

This network is protected from DDoS attacks by DDoS defense mechanisms deployed at the victim wireless network; as well as, a response to detected attacks by alleviating the impact on the victim. Historically, most defense systems were located at the victim since it suffered the greatest impact of the attack and was therefore the most motivated to sacrifice some resources for increased wireless network security. Examples of these systems are provided by resource accounting and protocol wireless network security mechanisms [2].

2.1.6.2.2 Intermediate- Wireless Network Mechanisms

Infrastructural service is provided to a large number of Internet hosts by DDoS defense mechanisms deployed at the intermediate wireless network. Victims of DDoS attacks can contact the infrastructure and request the service, possibly providing adequate compensation. Examples of intermediate-wireless network mechanisms, are pushback and traceback techniques [2].

2.1.6.2.3 Source- Wireless Network Mechanisms

To prevent customers that are using this network from generating DDoS attacks, is the goal of DDoS defense mechanisms deployed at the source wireless network. Such mechanisms are necessary and desirable, but motivation for their deployment is low since it is unclear who would pay the expenses associated with this service [2].

2.1.7 Traffic Limiting And Profiling IDS Technology: The Next Step

Finally, a particularly important development in anomaly-based intrusion detection system (IDS) technology for countering DDoS attacks is the capability to taking action against traffic flows that exceed admin-configured thresholds. This entails the use of traffic restriction and profiling measures that leverage a key feature of TCP/IP to validate whether or not the core wireless network is receiving real traffic from non-spoofed sources ' a requirement in order to take appropriate action.

The relevant TCP/IP feature is the way the protocol reframes to manage communication between two hosts. Legitimate TCP/IP hosts will react if the destination system slows down the rate of acknowledgments used in the communication protocol. Bad traffic (such as traffic originating from flooding hosts) cannot react and, therefore, is an anomaly.

With this implementation of IDS technology, network managers can configure custom policies that expose wireless network traffic anomalies. These policies may include data traffic thresholds, wireless network protocol, source and destination port, source and destination IP addresses, and timeout intervals. The IDS could then alert, throttle (traffic limit), redirect or block traffic based on the configured policy. All the while, legitimate traffic is allowed to pass with no added latency.

User-defined timeout intervals are used to progressively enforce the policy for a specified duration and then execute the next action. These policies may also be used to deny traffic from a set of source IP addresses for a user-specified amount of time after the attack subsides. Once the malicious traffic is no longer present, the IDS automatically removes the policies and filters that were applied during the attack and logs all actions taken during the attack.

A traffic-limiting and profiling IDS works optimally when implemented "inline" within the wireless network, not on the edge like many IDS products. Accordingly, its functioning does not involve logging into the network router. This is an important architectural consideration since an edge IDS product that sends a user name and password across a wire to a router, even if it is a secure login, presents a wireless network security risk.

An advanced implementation of traffic-limiting and profiling IDS technology can execute policies within one second of detecting an anomaly. As a result, wireless network managers have a powerful tool for countering:

- Reflector attacks.
- Flood attacks.
- Spoofed-Source attacks [12].

2.1.7.1 Reflector Attacks

Because all hosts answering an Echo Request will reply with their native IP address, the IDS will immediately deny traffic from any responding host. The IDS can be configured to dynamically adapt itself and deny traffic from all reflecting hosts.

2.1.7.2 Flood Attacks

Any IP host that refuses to lower its outbound traffic-rate to meet administrator-set thresholds is most likely running an attack tool of some type. Traffic from this host will be denied until the attack subsides.

2.1.7.3 Spoofed-Source Attacks

The IDS can complement existing edge devices to provide egress, or outbound, filtering of packets. This is essential to any ISP's Good-Neighbor policy, as it would prevent a wireless network from being the source of such an attack.

The increasing sophistication of DoS and DDoS attacks serves as a serious wakeup call to the importance of proactive information protection in today's Internet economy. As more types of attacks continue to be developed, it is more critical than ever that organizations take the appropriate actions to prevent their systems and wireless networks from being overwhelmed, while protecting the availability and integrity of their networks and Web sites. To this end, IDS technologies that utilize traffic restriction and profiling techniques can play a key role [12]. Finally, IDS technology will be discussed later in this chapter and in greater detail in Chapters 8, "Standards Design Issues;" 11, "Wireless Network Security Planning Techniques;" and 18, "Management Of Wireless Network Security."

Now, let's look at wireless security in business networks. First of all, let's begin by discussing the basic access control methods that form the basis of the 802.11 architecture. These methods are best suited to small networks with low-to-medium security requirements. Second, a more robust virtual private network (VPN)-based security solution is presented that provides better security and scales well to large networks. Finally, a possible future solution based on the upcoming 802.1X security standard is discussed, which enables port-level access control.

2.2 Business Wireless Network Security

IT organizations are working to implement security mechanisms that are equivalent to those existing today for wire-based networks, with the increasing deployment of 802.11 wireless networks in business environments. An important aspect of this is the need to provide secure access to the network for valid users. Existing wired network jacks are located inside buildings already secured from unauthorized access through the use of keys, badge access, and so forth. A user must gain physical access to the building in order to plug a client computer into a network jack. In contrast, a wireless access point (AP) may be accessed from off the premises if the signal is detectable (for instance, from a parking lot adjacent to the building). Thus, prior to user authentication into the network domain, wireless networks require secure access to the AP and the ability to isolate the AP from the internal private network.

2.2.1 802.11 Security Basics

To secure access to an AP that are built into 802.11 networks, there are currently three basic methods:

- Service set identifier (SSID).
- Media Access Control (MAC) address filtering.
- Wired Equivalent Privacy (WEP) [3][5].

One or all of preceding methods may be implemented, but all three together provide a more robust solution.

> **Note:** Network administrators may also choose not to implement any of these methods.

2.2.2 SSID

By using an SSID associated with an AP or group of APs, network access control can be implemented. The SSID provides a mechanism to "segment" a wireless network into multiple networks serviced by one or more APs. Each AP is programmed with an SSID corresponding to a specific wireless network. To access this network, client computers must be configured with the correct SSID. A building might be segmented into multiple networks by floor or department. So, for users who require access to the network from a variety of different locations, typically, a client computer can be configured with multiple SSIDs.

The SSID acts as a simple password and, thus, provides a measure of security, because a client computer must present the correct SSID to access the AP. However, this minimal security is compromised if the AP is configured to "broadcast" its SSID. When this broadcast feature is enabled, any client computer that is not configured with a specific SSID is allowed to receive the SSID and access the AP. In addition, because users typically configure their own client systems with the appropriate SSIDs, they are widely known and easily shared.

> **Note:** It is strongly recommended that APs be configured with broadcast mode disabled.

2.2.3 Filtering Of The MAC Address

A client computer can be identified by the unique MAC address of its 802.11 network card, while an AP or group of APs can be identified by an SSID. To increase the security of an 802.11 network, each AP can be programmed with a list of MAC addresses associated with the client computers allowed to access the AP. Therefore, the client is not allowed to associate with the AP, if a client's MAC address is not included in this list.

Improved security is provided by MAC address filtering (along with SSIDs). But, it is best suited to small networks where the MAC address list can be efficiently managed. Each AP must be manually programmed with a list of MAC addresses, and the list must be kept up-to-date. So, the scalability of this approach is limited by this administrative overhead.

2.2.4 Security Of The WEP-Based Protocol

Transmissions over wired networks are harder to intercept than wireless transmissions. The 802.11 standard currently specifies the WEP security protocol to provide encrypted communication between the client and an AP. Ron's Code 4 Pseudo Random Number Generator (RC4 PRNG) is employed by the WEP symmetric key encryption algorithm [3]:

All clients and APs on a wireless network use the same key to encrypt and decrypt data under WEP. The key resides in the client computer and in each AP on the network. The 802.11 standard does not specify a key management protocol, so all WEP keys on a network must be managed manually. Support for WEP is standard on most current 802.11 cards and APs. Ad hoc (or peer-to-peer) 802.11 networks that do not use APs, do not have WEP security available to them.

A 40-bit encryption key is specified by WEP. And, there are also implementations of 104-bit keys. The encryption key is concatenated with a 24-bit "initialization vector," resulting in a 64-or 128-bit key. This key is input into a pseudorandom number generator [3]. The data to be transmitted is encrypted by the resulting sequence.

WEP encryption has been proven to be vulnerable to attack. Scripting tools exist that can be used to take advantage of weaknesses in the WEP key algorithm to successfully attack a network and discover the WEP key. The industry and IEEE are working on solutions to this problem. And, encryption technology for WEP will be replaced by the Advanced Encryption Standard (AES).

In the meantime, WEP-based security can still be a component of the security solution used in small, tightly managed networks with low-to-medium security requirements, despite the weaknesses. In these cases, 128-bit WEP should be implemented in conjunction with MAC address filtering and SSID (with the broadcast feature disabled). To minimize risk, customers should change WEP keys on a regular schedule.

The VPN solution discussed next, is preferable for networks with high security requirements. The VPN solution is also preferable for large networks, in which the administrative burden of maintaining WEP encryption keys on each client system and AP; as well as, MAC addresses on each AP, makes these solutions impractical. In addition, a lost or stolen client system requires that all keys be changed, because all clients and APs use the same WEP encryption key.

Depending on the organization's ability to administer the network; the choice of security methods (SSID; WEP; MAC address filtering); and, its tolerance for risk, the number of wireless client systems becomes unmanageable and varies. If MAC address filtering is used on a wireless network, the fixed upper limit is established by the maximum number of MAC addresses that can be programmed into each AP used in an installation. In some cases, this upper limit is 255.

However, the manageable number of clients when using MAC address filtering is likely to be considerably less than 255 clients for many organizations. Fig. 1-3 depicts WEP-based security with MAC address filtering [3].

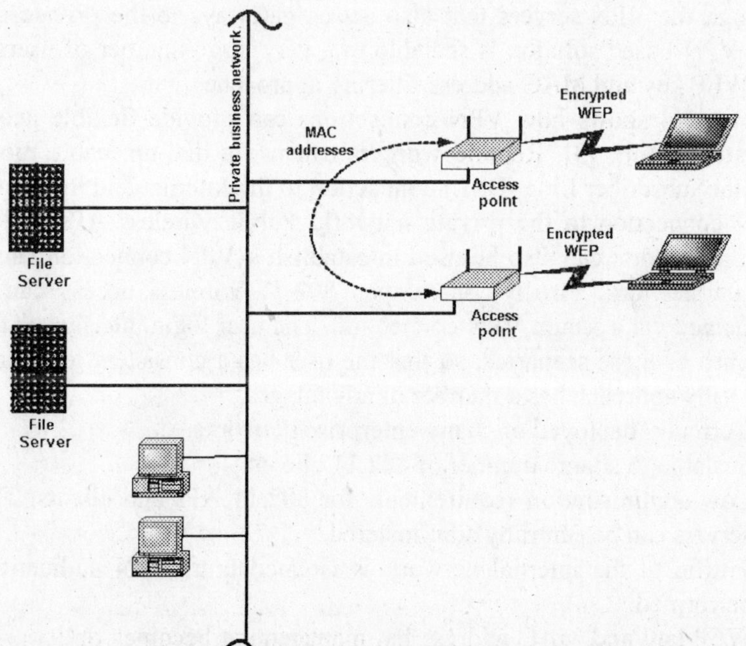

Figure 1-3. 802.11 Security Using SSID, MAC Address Filtering and WEP.

2.2.5 Wireless Security For VPN Solutions

A VPN solution for wireless access is currently the most suitable alternative to WEP and MAC address filtering for business networks. VPN solutions are already widely deployed to provide remote workers with secure access to the network via the Internet. In this remote user application, the VPN provides a secure, dedicated path (or "tunnel") over an "untrusted" network. The Internet Various tunneling protocols, including the Point-to-Point Tunneling Protocol (PPTP) and Layer 2 Tunneling Protocol (L2TP) are used in conjunction with standard, centralized authentication solutions, such as Remote Authentication Dial-In User Service (RADIUS) servers, in this particular case [3].

For secure wireless access, the same VPN technology can also be used. In this application, the "untrusted" network is the wireless network. The APs are configured for open access with no WEP encryption, but wireless access is isolated from the enterprise network by the VPN server and a VLAN between the APs and the VPN servers.

> **Note:** The APs should still be configured with SSIDs for network segmentation.

Authentication and full encryption over the wireless network is provided through the VPN servers that also act as gateways to the private network. The VPN-based solution is scalable to a very large number of users, unlike the WEP key and MAC address filtering approaches.

Fig. 1-4 shows how VPN connections can provide flexible access to a private network [3]. Remote workers can use a dial-up, cable modem, or Digital Subscriber Line (DSL) connection to the Internet and then establish a VPN connection to the private network. Public wireless APs in locations such as airports can also be used to establish a VPN connection back to the private network. Finally, on-campus 802.11 wireless access can be implemented via a secure VPN connection. The user login interface is the same for each of these scenarios, so that the user has a consistent login interface. The VPN approach has a number of advantages:

- Currently deployed on many enterprise networks.
- Scalable to a large number of 802.11 clients.
- Low administration requirements for 802.11 APs and clients. The VPN servers can be centrally administered.
- Traffic to the internal network is isolated until VPN authentication is performed.
- WEP key and MAC address list management becomes optional because of security measures created by the VPN channel itself.
- Consistent user interface in different locations such as at home, at work, and in an airport [3].

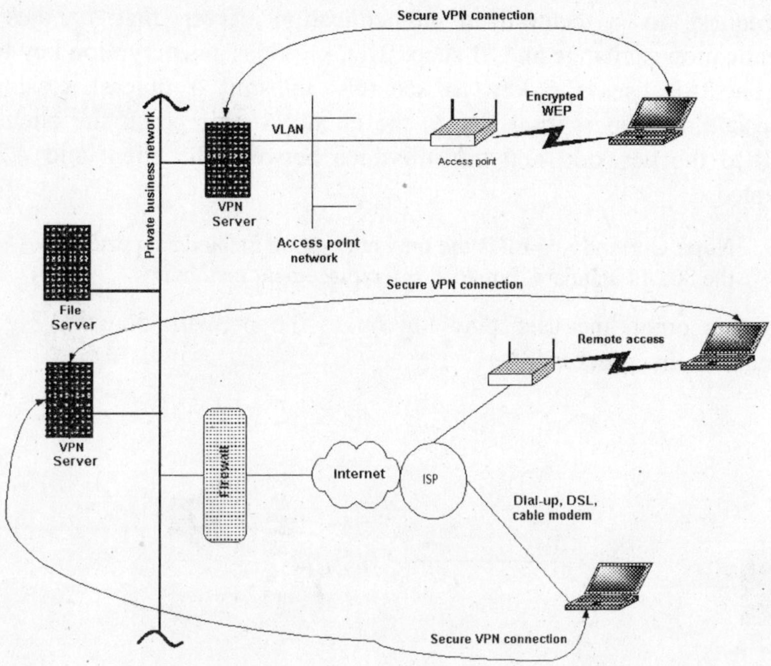

Figure 1-4. 802.11 VPN Wireless Security.

The lack of support for multicasting is a drawback to current VPN solutions. See "Wireless Clients Multicasting" later in this part of the chapter.

Roaming between wireless networks is not completely transparent. However, it is another minor drawback. When roaming between VPN servers on a network or when the client system resumes from standby mode, users receive a logon dialog.

All VPN client computers should be equipped with personal firewall protection. This is highly recommended. See "Protecting Firewalls" later in this part of the chapter.

2.2.6 Port-Based Network Access With 802.1X

The IEEE standard, 802.1X, for generic, extensible port-based authentication has now been approved. The specification is general: it applies to both wireless and wired Ethernet networks. In the context of an 802.11 wireless network, 802.1X is used to securely establish an authenticated association between the client and the AP. Generally, the scenario would be as follows. The user of an 802.11 wireless client system (rather than the client system itself) requests access to an AP. The AP passes the request to a centralized authentication server that handles the authentication exchange and, if successful, provides an encryption key to the AP. The AP uses the key to securely transmit a unicast session or multicast/global encryption key to the client. At this point, the client has access to the network, and transmissions between the client and AP are encrypted.

> **Note:** Currently, WEP is the only encryption method supported by the 802.11 standard, but AES is a replacement method.

At this point, the user may log on to the network domain. Fig. 1-5 summarizes this process [3].

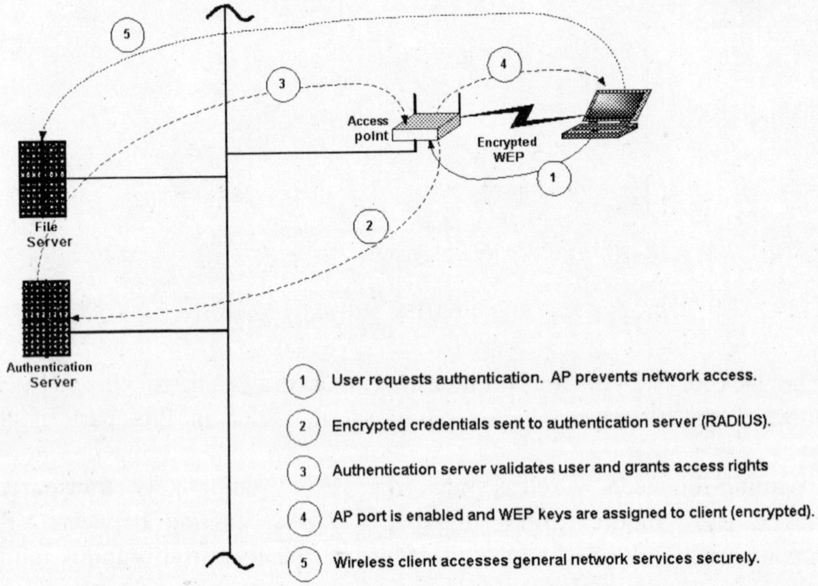

Figure 1-5. Possible Future 802.1X/RADIUS Architecture.

802.1X does not require a specific protocol for authentication. Instead, it specifies that the Extensible Authentication Protocol (EAP) will be used. EAP is an encapsulation protocol that allows different authentication protocols to be selected and used. Effectively, EAP serves as a conduit for other authentication protocols such as RADIUS, Kerberos, and Secure IDs.

The 802.1X standard also includes a management specification for complete end-to-end management of the 802.1X protocol. The 802.1X approach has the following advantages:

- Standards-based.
- Flexible authentication: administrators may choose the type of authentication method used.

- Scalable to large enterprise networks.
- Centrally managed.
- Encryption keys are dynamically generated and propagated.
- Because authentication is central, rather than at each AP, roaming can be made as transparent as possible. At most, the user may be asked for alternate credentials if an AP requires alternate identification [3].

> **Note:** The 802.1X standard was finalized in June 2001 and future Microsoft® Windows® operating systems continue to include native 802.1X support.

2.2.7 VPNs And 802.1X

With the preceding in mind, 802.1X is designed to authenticate and distribute encryption keys (currently WEP) between the wireless client and an AP. It is not designed to be a generalized VPN solution suitable for secure remote access. Thus, from remote or home offices, VPNs are still required for remote access using public APs (in airports or hotels).

2.2.8 Wireless Clients Multicasting

From one source to many users over a network, multicasting is a technique used to deliver data efficiently in real time. In some large networks, a portion of the network bandwidth is used for multicasting applications such as streaming audio and video applications (press conferences, training classes, and so forth) to groups of users. These data streams, if unicast (that is, transmitted individually) to each end user, would be a much greater burden on the network. Multicasting uses network bandwidth more efficiently by transmitting these types of data as one common data stream over the backbone. Network switches then deliver individual data streams to each client connected to them, in most cases.

When designing an 802.11 wireless network, multicasting requirements must be carefully considered. While multicasting is supported in 802.11 networks, current bandwidth limitations at the AP, as well as WEP key management issues, restrict the ability to deliver these data streams efficiently and securely to wireless clients. Current 802.11 b (also referred to as Wi-Fi) APs have a maximum theoretical bandwidth of 11 Mbps [3], which must be shared by all clients accessing the AP. Depending on its volume, multicasting traffic can reduce the AP's available bandwidth significantly. In addition, if WEP encryption is used to secure the wireless network, a separate key must be configured on each client to support multicasting (in addition to the key required for unicast traffic). Furthermore, current VPN solutions only support multicast by unicasting the data stream individually to each wireless VPN client. Multicasting to wireless clients can be problematic in real-world, secure deployments, because of these limitations.

The 802.1X standard addresses some of these multicasting issues by enabling dynamic encryption key management. So that dynamic multicast key distribution becomes possible in large organizations, keys can be issued and managed in real time (provided authentication succeeds).

2.2.9 Protecting Firewalls

All 802.11 client systems should be equipped with personal firewall software protection (see sidebar, "Firewall Bypass Exploitation"), in addition to the security solutions presented in this chapter so far. Like PCs with always *on [3]* cable modem and DSL Internet access connections, 802.11 clients can be vulnerable to hacker attacks from other users connected to the AP. Thus, the goal of firewall software is to protect the roaming user's confidential local data against as many possible attacks as are currently known and understood. Although, the nature of these attacks may vary.

Firewall Bypass Exploitation

A new and some say troubling trend is emerging among wireless network carriers who are enabling users to lift data remotely from corporate networks without IT oversight, according to industry observers. Dissatisfied with the slow pace of corporate adoption of wireless networks, carriers are taking a new route, going directly to employees and bypassing the IT departments.

Cingular Wireless has followed Sprint PCS' lead to become the second major wireless network carrier to give users access not only to e-mail, calendars, and contacts residing on the network, but to just about any file on any directory; as long as, a user's desktop or a delegated co-worker's desktop is active. It is sneaking into the firewall, but sometimes you get to the IT department by showing them how many different individual users are already using a technology.

Behind Cingular's service and Sprint's Business Connection Personal Edition is Redwood City, California.-based Seven, which offers its System Seven architecture in two flavors: one for IT departments and another for individuals. The Seven solution also supports Lightweight Directory Access Protocol (LDAP) access. It makes the cell phone a wonderful extension to the PC.

Seven establishes an outbound connection and gains access by using Port 443, the same Web link used to surf the Web and send email. The System Seven server registers itself as an available resource, allowing queries back to the desktop. Company officials insist that System Seven conforms to the highest levels of transport wireless network security.

But, unsupervised port access can be harmful. Network intrusions and lost or stolen devices could lead to information loss or theft. At the very least, companies need to acknowledge the issue. They need to find out what people are doing and put a policy or device restrictions in place [13].

As previously discussed earlier in the chapter, the threat of denial of service attacks has caused a backlash in the ISP market. As a result, ISPs are demanding higher levels of wireless network security and information warfare countermeasures from their customers—levels that seem unreasonable to some IT departments.

2.2.10 Information Warfare

Finally, the prevalence of denial-of-service (DoS) attacks shows that cyberterrorism is alive and well—and that e-businesses and their service providers aren't doing enough to stop it. Unfortunately, most of corporate America and ISPs seem to only focus on is who to blame. After an attack on Microsoft, access was shut off to everything from Expedia to Hotmail. The company attributed the problem to one employee's misconfiguration of a router. Yet, experts noted it was a failure to distribute DNS servers that made the company vulnerable to begin with.

If a private company is going to minimize the number and effect of DoS attacks, what's required is a spirit of cooperation between companies and their ISPs, as well as among the ISPs themselves. ISPs are starting to tackle the subject of wireless networkwide security, but they're doing it by laying out requirements for their corporate customers, rather than working *with* them. In many cases, customers either follow the ISP's wireless security guidelines or find a new ISP—there's no room for negotiation. It's high time ISPs and their clients start sharing information about what works (and doesn't work) in terms of network architecture, data access, wireless network security systems and information warfare (see Chapters 22 to 28 for a detailed discussion of information warfare).

2.2.11 The Danger Of DoS

As previously discussed earlier in the chapter, everyone along the e-business food chain has something to lose when a DoS attack succeeds. The site that's been hit loses traffic, revenue, and customer loyalty. The ISP loses customer confidence and spends significant resources combating the attack. Ultimately, every site that relies on the ISP must spend time and resources reassessing its wireless network security levels.

ISPs must ask themselves whether they're doing everything possible from a wireless network monitoring and warning perspective. They should be giving serious thought to the latest wireless network security tools that can stop DoS attacks at their routers. After all, once an attack gets through the ISP, it's a lot tougher for an individual site to fend it off.

ISPs must communicate the types of attacks they're experiencing. They also must be prepared to notify one another of attacks, and even coordinate their responses when they do get hit. With so much at risk, it's hard to imagine why these conversations haven't been taking place all along.

2.2.12 Reasonable Expectations?

Many ISPs have some kind of policy that dictates what companies can and can't do as customers, and the kinds of wireless network security systems that must be in place before they can purchase services. Service providers that have such a policy include Ameritech, AT&T, and CTC Communications, and national ISPs such as EarthLink, Exodus Communications, and PSINet.

These service providers want to see IT managers install information warfare (IW) countermeasures like encryption and authentication products (see sidebar, "Countermeasures"), firewalls that interact with intrusion detection software, dedicated servers, and VPN links to secure data. They also want IT shops to use tools such as antivirus software, specified intrusion detection systems and anti-spam content filtering. With the recent increase in cyberterror attacks, it would seem that the preceding information warfare countermeasures request from the ISPs is very reasonable.

Countermeasures

The various legal roadblocks and technical wizardry contrived by governments and law enforcement to block encryption's spread have, of course, curbed neither the need for the technology nor the ingenuity of privacy-loving programmers. As a result, a number of countermeasures have been engineered to augment or replace encryption. Among them are anonymizers, which conceal the identity of the person sending or receiving information, and steganography, which hides the information itself.

The need for anonymity in a democratic society has long been recognized, to shield whistleblowers and political dissenters from retaliation, to protect the records of medical patients, and so on. Less dramatic situations also justify anonymity, such as placing a personal ad or seeking employment through the Internet without jeopardizing one's current job. To be sure, anonymity can be exploited by sociopaths seeking to avoid accountability for their actions. But, in general, it serves a useful social function.

Anonymous and pseudonymous remailers are computers accessible through the Internet that launder the true identity of an e-mail sender. Most are operated at no cost to the user. A pseudonymous remailer replaces the sender's e-mail address with a false one and forwards the message to the intended recipient. The recipient can reply to the sender's pseudonymous address, which, in turn, forwards the response to the sender's true address.

Anonymous remailers come in three flavors: cypherpunk, mixmaster, and Web-based. Cypherpunk remailers strip away the message header, which describes where the message came from and how it got there, before forwarding the message to the recipient. Conceivably, someone with physical access to such a remailer's phone lines could correlate the incoming and outgoing traffic and make inferences.

Mixmaster remailers avoid that problem by using stronger encryption and tricks for frustrating traffic analysis, such as padding messages to disguise their true length. But even mixmasters can be compromised. For example, through a concerted effort, it would be possible to detect a correlation between Mr. A sending an encrypted message through a remailer, and Ms. B receiving a message at some variable time afterwards.

Web-based anonymizers range from sites offering conventional anonymizer services, to others where the connection between the user's computer and the anonymizer is itself encrypted with up to 128-bit encryption. The job is done using the standard Secure Socket Layer (SSL) encryption, built into all Web browsers of recent vintage.

For extra privacy, a message may be routed through a series of remailers. Two popular remailer software packages, Private Idaho and Jack B. Nymble, enable the sender to do this automatically.

The Onion Router project (see the site at http://www.onion-router.net) of the Naval Research Laboratory in Washington, D.C., offers another way to string together remailers. What's more, it allows anonymized and multiply encrypted Web browsing in real time.

Onion routing is a two-stage process. The initiator instructs router W (in this case, a proxy server at the firewall of a secure site) to create an onion, which consists of public-key-encrypted layers of instructions. Router X peels off the first layer of the onion, which indicates the next step in the path and supplies a symmetric decrypting key for use when the actual message comes through later.

The onion then goes to Routers Y and Z, depositing keys at each stop. Once the connection is established, the encrypted message is sent through and successively decrypted, arriving at the recipient as plaintext. To respond, the recipient sends the message to Router Z, which encrypts the text, onion-style, and sends it back through the already established path.

A good example of this emerging trend is CTC Communications in Waltham, Mass., one the largest Competitive Local Exchange Carriers (CLECs) in New England. CTC bans the use of spamming and pornography; and, doesn't let its customers use a router for more than one core application.

Note: Of course, there are free speech issues here; but, some CLECs and/or ISPs in their Terms Of Use contract with

> customers, reserve the right in some states to filter out spamming and pornography. Customers reserve the right not to do business with these CLECs and/or ISPs—and, take their business elsewhere.

The company also demands its customers protect all servers with firewalls. Companies either comply or they don't become a CTC customer.

Precisely when ISPs started getting tougher is hard to determine. But it's clear that ISPs weren't making these types of wireless network security demands before the denial-of-service (DoS) attacks occurred. The attacks made service providers aware that it's not just corporate customers getting hacked. The ISPs' systems were commandeered and used to launch virus attacks and DoS attacks; as well as, for vandalism and theft. That's why some of the most common requirements of the ISPs have to do with customers' outgoing traffic, which can directly affect the ISPs. From the ISPs' point of view, their own customers or prospective customers are now secu wireless network security threats.

This new approach by ISPs is being felt by IT shops. When an IT manager at CLD Consulting Engineers Inc. in Manchester, N.H., wanted to lease a T1 line from CTC in May 2001, CTC wouldn't agree to deliver service until the civil engineering firm beefed up its wireless network security. CLD spent nearly $14,000 to build a hardware firewall, change the IP address scheme to be much more secure, and install separate routers--one dedicated to Internet access, the other for branch office access.

CLD's experience is not unique. Customers of Boston ISP Breakaway aren't allowed to have dial-up Internet access to company databases from offsite locations, and they must use a VPN or dedicated T1/partial T1 link. This makes it tough on smaller companies, many of whose employees access company data remotely via a dial-up Internet connection. Analysts also say many companies use basic wireless network security mechanisms such as IDs and passwords to secure this access. But, standard wireless network security methods don't cut it for companies that want to sign on with Breakaway.

Now, all architecture has to be approved by the wireless network security desk before services are offered, and a customer with single-tier access won't be approved, even though it may want to be. They have to lease a site-to-site VPN or a dedicated T1 link. The VPN (the cheaper of the two options) costs an additional $570 per month.

The ISP demands don't always result in higher costs for corporate customers. Sometimes providers want to look under the hood, which can be just as objectionable to some companies.

But, ISPs' fears may be justified. According to a survey in 2004 by the Computer Securities Institute and the San Francisco FBI Computer Intrusion

Squad, 83 percent of CSI's 1,000 member companies detected unauthorized use of their systems over the previous year, up from 87 percent in 2003. The rate of hacking is growing faster than e-commerce itself. CSI has 1,080 security professionals onboard (more than one IT person per company in some cases), and 96 percent worry about disgruntled employees as the biggest security threat. In the past year, ISPs have set up entire departments devoted to fielding phone calls and handling subpoenas from individuals and companies, claiming that ISP customers are spamming, sending viruses, vandalizing Web sites and launching DoS attacks.

One example is the Policy Enforcement Group at Exodus Communications. The group has an official abuse policy that covers unintentional as well as intentional abuses. It specifically supplies evidence of wrongdoing so Exodus can shut down services if a customer is abusing the service, even if it's doing so unwittingly. For instance, if Exodus is monitoring a company's wireless network and sees traffic patterns that look like a DoS attack, it can shut down the company's network to clean machines and do whatever else is necessary.

> **Note:** Cleaning machines means that all applications have to be reloaded.

Service providers, in effect, are establishing rules that make it clear that they no longer want to bear the burden of the risks corporate customers are willing to take. They say it's no longer up to clients to determine how risk-free they want to be when it comes to e-commerce. Ultimately, companies that want ISPs to deliver any service at all (even a simple pipe to the Internet) will pay more in hard costs, internal policy changes, infrastructure and business processes.

How much more depends on how secure the ISP thinks its customer's wireless network should be. But, the ISP in many cases is dictating the terms. And whether the customer buys the needed wireless network security technology from the ISP or elsewhere, it will have to be bought before network services begin. This could mean a huge cash outlay before service even starts.

> **Note:** Service providers, in effect, are establishing rules that make it clear that they no longer want to bear the burden of the risks corporate customers are willing to take.

2.2.13 Fair Is fair

IT managers may be told to spend more on wireless network security prerequisites, but there's still room for negotiation. This is an emerging trend, not a government regulation. So it's entirely fair for IT managers to bark back, particularly if the ISP doesn't deliver the level of wireless network security it requires from customers.

If customers are required to protect the wireless network, it begs the question: What level of wireless network security does the ISP itself offer corporate customers? And, if ISPs are demanding customers walk into the relationship with higher levels of wireless network security, corporate customers should demand the same of the ISPs.

The ISPs must ensure that its wireless network security mechanisms will work, or be responsible for damages, so ask about the ISP's wirelessnetwork monitoring tools and alert mechanisms. Once companies open up the conversation to include both sides, it becomes more of a negotiation, less of an ultimatum.

Before you give an ISP access to your entire wireless network for inspection, you need to determine if the ISP will actually manage every aspect of the network. The ISP does not need access to the portions of your wireless network it doesn't manage. It may prove to be more trouble for ISPs to deliver wireless network security if parts of a potential customer's network are unknown to them. But, that's the ISP's problem. In the end, it's the ISP's responsibility to monitor your outgoing traffic, so the ISP already has access to what it needs to know to protect itself. The ISP should have monitoring tools robust enough to give it intelligent reporting, traffic analysis, and alerts to red flags.

The best way to protect the company is to handle these issues in the service level agreement (SLA). The ISPs have the leverage to force customers to implement wireless network security and information warfare countermeasures, but customers have a certain leverage. The ISP market is more competitive than ever.

According to Gartner Group, there are now tens of thousands of full-service ISPs, up from less than 1,000 since 1999. The competition means it's in the ISP's interest to offer corporate customers as much value-add as possible. But, there's one caveat: The idea is to get the ISP to concede some points—for example, help with making the company's wireless network compatible with the ISP's and/or onsite tech support.

The reality is that even if ISPs can dictate wireless network security policies, it will be eager to offer value-added services. If you do the negotiating in the context of drafting an SLA agreement, it shows the ISP you're a serious customer and gives them an opportunity to offer you fee-based services over the long term.

2.2.14 Gateways And Gatekeepers

A good SLA won't get your out of paying more for ASP and ISP services. In fact, it can end up costing more. But, it will at least get you the most bang for the buck. The truth is that ISPs will dictate how much wireless network security customers will have, because they can. They are the conduits to the wireless networks.

IT managers should understand that pushing back at the ISPs will only do so much. This is a trend that's here to stay. The ISPs started the trend, but it won't end with them. Business partners and regulators will step in and give the wireless network security push even more teeth, including standards such as best practices and default security requirements.

Ultimately, the ISPs will protect themselves from outgoing traffic by shutting Web sites down that have been commandeered for DoS and other attacks. The ISPs that survive will start offering wireless network security services. It's only during this interim period that the onus will be on companies to pick up the slack. Whether the time period is six months or six years is hardly important. Companies that want to do business with top-tier providers had better get serious about wireless network security and information warfare countermeasures.

There's a lot more to come on information warfare (IW) in Part VI: "Information Warfare Countermeasures: The Wireless Network Security Solution" (Chapters 22 to 28). Stay tuned!

3. SUMMARY AND CONCLUSIONS

Robust wireless network security solutions are required, as 802.11 networks proliferate and mature. The basic 802.11 wireless network security solutions that are available out of the box (SSID, MAC address filtering, and WEP), are suitable for small, tightly managed wireless networks with low-to-medium security requirements. For wireless networks with high security requirements, the weaknesses in WEP encryption require a more robust solution. In addition, the manual task of maintaining MAC addresses and WEP keys becomes overwhelming as the number of wireless network clients increase. For larger wireless networks, or for networks with high security requirements, a VPN solution based on currently available technology provides a very scalable solution for 802.11 wireless networks. VPN for wireless is also a logical extension of the remote access VPN capability found in most large businesses today. Here now, is 802.1X, a standards-based solution for port-level authentication for any wired or wireless Ethernet client system [3].

In addition, although distributed denial-of-service attacks have been recognized as a serious problem, there isn't any other attempt to introduce formal classification into the DDoS attack mechanisms. The reason might lay in the use of fairly simple attack tools that have dominated most DDoS incidents. Those tools performed full-force flooding attacks, using several types of packets. As defense mechanisms are deployed to counter these simple attacks, it is expected to be faced with more complex strategies.

This chapter presented a classification of denial-of-service attacks according to the type of the target (firewall, Web server, router), a resource that the attack consumes (wireless network bandwidth, TCP/IP stack) and the exploited vulnerability (bug or overload). This classification focuses more on the actual attack phase, but is not limited to the complete attack mechanism in order to highlight features that are specific to distributed attacks.

The Computer Emergency Response Team (CERT) Coordination Center is currently undertaking the initiative to devise a comprehensive classification of computer incidents as part of the design of common incident data format and exchange procedures. Unfortunately, their results are not yet available. Thus, as of this writing, there are no attempts to formally classify DDoS defense systems, although similar works exist in the field of intrusion detection systems.

Nevertheless, distributed denial of service attacks are a complex and serious problem, and consequently, numerous approaches have been proposed to counter them. The multitude of current attack and defense mechanisms obscures the global view of the DDoS problem. This chapter is a first attempt to cut through the obscurity and achieve a clear view of the problem and its solutions. The classifications described here are intended to help the wireless network community think about the threats they face and the IW measures they can use to counter those threats.

One positive benefit from the development of DDoS classifications is to foster easier cooperation among researchers on DDoS defense mechanisms. Attackers cooperate to exchange attack code and information about vulnerable machines, and to organize their agents into coordinated wireless networks to achieve immense power and survivability. The Internet community must be equally cooperative among itself to counter this threat. Good classifications for DDoS attack and defense mechanisms will facilitate communications and offer the community a common language to discuss their solutions. They will also clarify how different mechanisms are likely to work in concert, and identify areas of remaining weakness that require additional mechanisms. Similarly, the research community needs to develop common metrics and benchmarks to evaluate the efficacy of DDoS defense mechanisms, and these classifications can be helpful in shaping these tasks, as well.

The preceding classifications are not necessarily complete and all-encompassing. You should not be deceived by the simplicity of the current attacks; for the attackers this simplicity arises more from convenience than necessity. As defense mechanisms are deployed to counter simple attacks, you are likely to see more complex attack scenarios. Many more attack possibilities exist and must be addressed before you can completely handle the DDoS threat, and some of them are likely to be outside the current boundaries of the classifications presented here. Thus, these classifications are likely to require expansion and refinement as new threats and defense mechanisms are discovered [2].

The DDoS attack classification and DDoS defense classification outlined in this chapter are useful to the extent that they clarify your thinking and guide you to more effective solutions to the problem of distributed denial-of-service. The ultimate value of the wireless network security technology described in this chapter will thus be in the degree of discussion for the rest of this book.

4. REFERENCES

[1] John R. Vacca, *i-mode Crash Course,* McGraw-Hill Professional Book Group, 2001.

[2] Jelena Mirkovic, Janice Martin and Peter Reiher, "Ataxonomy of DDoS Attacks and DDoS Defense Mechanisms," Technical Report #020018, Computer Science Department, University of California, Los Angeles, 2002.

[3] "802.11 Wireless Security in Business," Dell Computer Corporation, One Dell Way, Round Rock, Texas 78682, United States, (©Copyright 2002, Dell Computer Corporation. All rights reserved), September, 2001.

[4] Kurt Ringleben, "Intrusion Detection Systems Grow Up," SearchNetworking.com, TechTarget, 117 Kendrick Street, Needham, MA 02494, (©Copyright 2002, TechTarget. All rights reserved).

[5] John R. Vacca, *Net Privacy: A Guide to Developing and Implementing an Ironclad ebusiness Privacy Plan,* McGraw-Hill, 2001.

[6] John R. Vacca, *The Essential Guide To Storage Area Networks,* Prentice Hall, 2002.

[7] John R. Vacca, *Public Key Infrastructure: Building Trusted Applications and Web Servicess,* Auerbach Publications, 2004.

[8] John R. Vacca, *Firewalls : Jumpstart for Network and Systems Administrators,* Digital Press, 2004.

[9] John R. Vacca, *Identity Theft,* Prentice Hall, 2002.

[10] Nancy Gohring, "Motion Sickness," eWeek, Copyright © 2002-2005 Ziff Davis Publishing Holdings Inc. All Rights Reserved. Ziff Davis Media Inc., Corporate Headquarters, 28 East 28th Street, New York, NY 10016, October29, 2001.

[11] Dennis Fisher and Carmen Nobel, "New Attack Intercepts Wireless Net Messages," eWeek, Copyright © 2002-2005 Ziff Davis Publishing Holdings Inc. All Rights Reserved.

Ziff Davis Media Inc., Corporate Headquarters, 28 East 28th Street, New York, NY 10016, March 11, 2002.

[12] Kieth Waldorf, "Flood Protection--Guard Your Networks From Denial of Service Attacks," Captus Networks Corp., Copyright © 2005 Captus Networks Corp. All Rights Reserved. Captus Networks Corp., 1680 Tide Court Suite B, Woodland, California 95776, 2005.

[13] Ephraim Schwartz and Brian Fonseca, "Wireless Carriers Exploit Firewall Bypass," InfoWorld Media Group., Copyright © 2005 InfoWorld Media Group. All Rights Reserved. InfoWorld Media Group 501 Second Street San Francisco, CA 94107 U.S.A, January 25, 2002.

Chapter 2

TYPES OF WIRELESS NETWORK SECURITY TECHNOLOGY

1. INTRODUCTION

Wireless networks are too inexpensive to ignore. But, security has stymied many network managers looking to bring wireless into the corporate fold. There's a lot of information and misinformation out there about types of wireless network security technologies. This chapter will help to clear up some of that confusion, and present some common types of wireless network security technologies to help guide your path.

The first thing you have to do is educate yourself. The Internet has a lot of data and opinions on wireless security technologies, but it's difficult to get a perspective on things without a good background primer (like this book). You need to put this into the context of corporate security. What threats are you worried about? How sensitive are the data on the wireless- local area network (LAN) or wide area network (WAN)? What vulnerabilities do you need to guard against? Sniffing? Denial of service? Freeloading? Impersonation? You'll never establish an appropriate 802.11 security policy for your corporate network if you don't think about these technologies now.

Second, you must do something to get started. Wired Equivalent Privacy (WEP) is still an awful technique to use. It's like giving everyone in the company the same password and never changing it. But, that doesn't mean you shouldn't use it. The theoretical attacks on WEP exploited by various tools are blocked by modern firmware. In some recent testing, using current releases of 10 different enterprise-class access points and eight different client cards, Initialization Vector-based attacks on WEP were no longer effective.

Third, you should arm yourself with wireless security tools. Most wireless security tools are fabulous for enterprise network managers. If you only have a few access points to worry about, a laptop with some public domain tools is a fine start. But, without at least some tools, you'll be left completely in the dark about the wireless data speeds that are beginning to surround your network [10].

Fourth, you should prepare your wireless security strategy. Today, the 802.1X-based authentication is up-and-running technology to help resolve basic wireless security problems. Or, you can go down the virtual private network (VPN) path and treat wireless users the same way you treat remote access VPN clients. Either works fine with off-the-shelf hardware [11].

Over the long run, the Institute of Electrical and Electronics Engineers (IEEE) 802.11i standard will lay out a path to higher security for wireless networks that combines 802.1X authentication with better key management than is available on WEP. But that standard is still being cooked, and it will be a year or more before things completely settle.

So, with the preceding in mind, many wireless networks are not properly secured or-even worse-are completely unsecured. Naturally, security is a top concern among those interested in deploying wireless networks. Fortunately, both user knowledge about security and the solutions offered by technology vendors are improving. Today's wireless networks feature comprehensive security capabilities and, when these networks are properly protected, enterprises can confidently take advantage of the benefits they offer. This first part of the chapter will help you gain a better understanding of wireless LAN security elements and best practices that can go a long way toward enabling you to reap the benefits of wireless networking. And, you get peace of mind, knowing your enterprise's data is secure.

2. WIRELESS NETWORK SECURITY TECHNOLOGIES

Vendors are doing a good job of improving security features, and users are getting an understanding of wireless security. "But, all threats are still considered important, and vendors continually need to address the lingering perception that wireless LANs are insecure.

Indeed, security is the biggest barrier to the adoption of wireless LANs. And, it's not just a big-enterprise worry. When it comes to wireless networking, security is still the number one concern for enterprises across all sizes.

Gaining a better understanding of wireless LAN security elements and employing some best practices can go a long way toward enabling you to reap the benefits of wireless networking. And, you get peace of mind, knowing your enterprise's data is secure.

2.1 Elements Of Wireless Security

Intentionally or not, enterprises and individuals may set up wireless networks with no security at all. That happens because most wireless access points come from the factory in open access mode by default, meaning that all security features are turned off. It's the buyer's responsibility to turn them on.

Three actions can help to secure a wireless network:

- Discouraging unauthorized users through authentication
- Preventing unofficial connections through the elimination of rogue access points
- Protecting data while it's being transmitted through encryption [3]

Not coincidentally, these are also important issues to companies.

The number one wireless LAN security concern is users from outside the company (illicitly or maliciously accessing the enterprise wireless LAN. Number two is internal rogue access points, and number three is encryption.

2.1.1 Using Authentication

When you want to make sure that the individuals who use a wireless network are authorized to do so, use authentication (sometimes called access control). Unique logins and passwords are the basis of authentication, but additional tools can make authentication more secure and reliable. The best authentication is per-user, per-session mutual authentication between the user and the authentication source.

2.1.2 Checking For Rogue Access Points

A well-meaning employee who enjoys a wireless network at home might want to enjoy the same freedom at work. He or she might purchase a cheap access point and plug it into a network jack without asking permission. These are known as rogue access points, and the majority of these are installed by employees—not malicious intruders. Even company-sanctioned access points, when configured improperly, can be security risks.

Checking for rogue access points isn't difficult. There are tools that can help, and checking can be done with a wireless laptop and software in a small building or by using a management appliance collecting data from your access points.

You can have technical personnel scan for new wireless access points. And, if they do a daily scan, they can pick these things up early.

2.1.3 Using Encryption

To make sure that data can't be read, and to protect data from being altered as it's transmitted between an access point and a wireless device, use encryption. In a basic sense, encryption is like secret code: It translates your data into gibberish that only the intended recipient understands. Encryption requires that both the sender and receiver have a key to decode the transmitted data. The most secured encryption uses very complicated keys, or algorithms, that change regularly to protect data.

2.2 Available Solutions For Wireless Security

Three solutions are available for secure wireless LAN encryption and authentication:

- Wi-FI Protected Access (WPA)
- Wi-Fi Protected Access 2 (WPA2)
- Virtual private networking (VPN) [3]

The solution you select is specific to the type of wireless LAN you're accessing and the level of data encryption required.

2.2.1 WPA and WPA2 Standards-Based Security Certifications

WPA and WPA2 are standards-based security certifications from the Wi-Fi Alliance for enterprise, SMB, and small office/home office wireless LANS that provide mutual authentication to verify individual users and advanced encryption. WPA provides enterprise-class encryption and WPA2. These are the next generation of WiFi security, which supports government-grade encryption.

It is recommended that WPA or WPA2 be used for enterprise and SMB wireless LAN deployments. WPA and WPA2 provide secure access control, strong data encryption; and, they protect the network from passive and active attacks.

2.2.2 Using VPN

VPN provides effective security for users wirelessly accessing the network while on the road or away from the office. With VPN, users create a secure tunnel between two or more points on a network using encryption, even if the encrypted data is transmitted over unsecured networks such as the public Internet. Home-based teleworkers with dial-up or broadband connections can also use VPN.

2.3 Policies For Wireless Security

In some cases, you may have different security settings for different users, or groups of users, on your network. These security settings can be established using a virtual LAN (VLAN) on the access point. For example, you can set up different security policies for distinct user groups within your enterprise such as finance, legal, manufacturing, or human resources. You can also set up separate security policies for customers, partners, or visitors accessing your wireless LAN. This allows you to cost effectively use a single access point to support multiple user groups with different security settings and security requirements—all while keeping your network secure and protected.

It is also important to consider wireless network security in the context of overall network security and network management. The majority of enterprises that deploy, or will deploy, wireless LANs want to do it in a way that complements the wired LAN. They want it integrated with common management.

A common management system increases efficiency for network administrators. Resource-strapped SMBs can use management tools to simplify and automate many repetitive and time-consuming administrative tasks.

Wireless LAN security (even when integrated with overall network management) only works if it's turned on and used consistently across the entire wireless LAN. That's why user policies are also an important part of good security practices. Resist the temptation to overact when setting a wireless LAN security policy. The first policy is often 'no wireless. The problem with that is, there are so many massive gains from having wireless in place.

The challenge is to devise a wireless LAN user policy that's simple enough that people will abide by it, but secure enough to protect the network. Today that's an easier balance to strike because WPA and WPA2 are built into WiFi certified access points and client devices.

Your wireless LAN security policy should also cover when and how employees can use public hot spots, the use of personal devices on the enterprise wireless network, the forbidding of rogue devices, and a strong password policy.

2.4 Taking Practical Steps

The first step in security a wireless network is to turn on the security features inherent in your access points and interface cards. This is usually done by running a software program that came with your wireless equipment.

The same program that turns on your wireless security features will probably also show what *firmware* version your access points use [3]. Check the device manufacturer's Web site for the most current firmware version, and update your access point if it's not current. Updated firmware will make your wireless network more secure and reliable. Also check to see what security resources your hardware vendor offers.

> **Note:** Firmware is software used by devices such as access points or routers.

Not everyone who wants the benefits wireless networking is capable of, or interested in, deploying and maintaining a secure wireless LAN. In such cases, value-added resellers, network implementers, or other suppliers of wireless networking gear can often help with these tasks.

Some SMBs choose to enlist the aid of an outsourced managed security service, many of which have wireless security offerings. According to Jupiter Research [4], a small but considerable segment of enterprises (13%), would outsource wireless security; compared to the 18% that would outsource their overall network management [3].

No matter how you proceed, do it in an organized fashion. Security is definitely something that has to be planned for, just like managing the network, providing coverage and access, and so forth. But, it shouldn't be a barrier to the deployment of a wireless LAN.

3. WIRELESS NETWORK SECURITY TECHNOLOGY PERSPECTIVES

Today, many enterprises are embracing wireless networking technologies to enhance productivity, provide better customer service, and even offer

Internet access to partners and on-site visitors. The emergence of new technologies, widespread cellular-data service, and an increasing number of wireless access points are making it easier for users to access information they need, when and where they need it.

With wireless hotspots available in coffee shops, airports, and restaurants, business travelers can work easily no matter where they are. In addition, many users are enjoying the convenience of wireless fidelity (Wi-Fi) connections in their hotel rooms and homes.

While providing users wireless access to file shares, applications, and other network resources offers many benefits, doing so can present security and manageability challenges. The multiplicity of connectivity options-cellular, local area networks (LANs), wireless local area networks (WLANs), and WiFi-can be difficult for IT departments to manage.

Because users need to access resources from both IT-managed devices such as corporate laptops and from unmanaged devices such as personal digital assistants (PDAs), many existing remote access solutions leave the network open to security threats from viruses, malware, and Trojan horses. In addition, the lack of interoperability among wireless vendors, an ever-evolving security framework for WLANs, and issues related to Internet and firewall traversal present further challenges [7].

3.1 Using SSL VPNs In Secure Wireless Networking

Enterprises are embracing wireless technologies to increase productivity, provide more flexible work arrangements for their employees, and work more closely with their business partners. Wireless technologies include both local area and wide area systems. However, the multiplicity of networking options as well as computing platforms creates significant security issues, including:

- An evolving security framework for WLANs and interoperability issues between vendors.
- Different native security options for wireless local area networks (WLANs) than cellular networks.
- Employees using both managed devices and unmanaged devices, such as home computers and public terminals.
- Internet traversal for many wireless remote-access solutions.
- Outdated WLAN equipment that is insecure.
- The danger of rogue access points [1].

The security architecture that addresses all these issues is an SSL virtual private network (VPN). SSL VPNs provide a means of protecting every node, whether internal or external to the enterprise, leading to the concept of an inverted security model that does not depend on a hardened perimeter. By taking advantage of installed browsers and the associated SSL security layer, enterprises can not only provide access through computers that have no VPN client software installed, but can also provide additional communications flexibility for systems with dynamically installed software.

3.1.1 Wireless Networks Prevalence

As previously explained, many enterprises are embracing wireless networking technologies to enhance the productivity of their workers, to improve customer service, and even to provide Internet access to visitors. Business travelers are taking advantage of wireless hotspots in public locations such as airports and restaurants, as well as enjoying the convenience of Wi-Fi in their hotel rooms and homes. They are also using cellular networks for communications from almost anywhere.

Although most wireless-data usage has been with Wi-Fi (based on the IEEE 802.11 family of standards), enterprises are increasingly using cellular-data services, which now offer a near-broadband experience over wide geographic areas. Cellular-data usage includes smartphones, PDAs and laptops with PC Card modems, and laptops using phones as modems by means of a cable or Bluetooth connection. Cellular-data networks encompass multiple technologies, the most prevalent of which today include Enhanced Data Rates for GSM Evolution (EDGE), Wideband CDMA (WCDMA), and the CDMA2000 group of technologies. Despite the alphabet soup of names, they all have a common capability—the ability to support IP-based packet communications from almost anywhere [1].

Emerging technologies such as WiMAX promise even higher performance over the wide area. Whereas cellular-data networks offer rates approaching 1 Mbps, WiMAX vendors are hoping to provide higher throughput rates [1].

Many professionals use a combination of Wi-Fi and cellular data networks. Wireless networking not only increases productivity, but it also enhances personal lifestyles—employees can telecommute not just from home but from practically anywhere.

The multiplicity of connectivity options, however, raises significant security challenges for your enterprise, which needs to secure these connections while accommodating a wide variety of mobile computing platforms, providing a simplified user experience, and limiting access to specific resources, all within a system that can be managed easily.

3.1.2 Wireless Connectivity Security Challenges

The number one concern expressed by IT managers regarding wireless networking is security. This is justifiable, because radio signals are inherently subject to eavesdropping due to their extended propagation.

Fortunately, there are many effective approaches for securing both Wi-Fi and cellular-data connections. To understand the benefits and limitations of the various approaches, you need to first consider the security issues in greater detail.

3.1.2.1 Issues For Wi-Fi Security

Initial implementations of Wi-Fi security, called Wired Equivalency Protocol (WEP), were completely inadequate, allowing any determined attacker to easily monitor connections or access the network. A new Wi-Fi security standard, IEEE 802.11i, addresses the security problems of WEP. This standard has come in two iterations: Wi-Fi Protected Access (WPA), which addresses all the deficiencies of WEP, and WPA2, which bolsters encryption by using the Advanced Encryption Standard (AES). IEEE 802.11i is based on IEEE 802.1X, a port-based security architecture where authentication is handled by using Extensible Authentication Protocol (EAP) methods in conjunction with authentication systems such as RADIUS. Most new equipment supports WPA or WPA2, both of which are considered reasonably secure [1].

A number of issues exist, however, for organizations using IEEE 802.11i for Wi-Fi security. First, IEEE 802.11i does not accommodate older deployed equipment; second, it applies only to access equipment in the organization's control; and third, the complexity of IEEE 802.11i–based security solutions is already raising interoperability concerns among different vendors' equipment.

To fill the Wi-Fi security gap, many Wi-Fi vendors have implemented security enhancements in their equipment. Many of these enhancements have required customers to buy cards and access points from the same vendor. This vendor dependence also applies to new WLAN architectures that employ centralized controllers to coordinate and manage access points. These controllers often include security functions, such as detecting rogue access points and providing VPN tunnel end points. However, any security benefits from these architectures apply only to the directly connected WLAN nodes and do not extend to other connections, such as Ethernet, or to WLAN connections in public places or employee homes.

3.1.2.2 Issues For Cellular-Data

With cellular-data connections, the security issues are somewhat different than for Wi-Fi. For example, whereas Wi-Fi attackers can use normal Wi-Fi hardware for their attacks, cellular-network attackers require specialized equipment to receive and decode the radio signal. The cost of such specialized equipment by itself, however, is not a sufficient deterrent. As a result, some (but not all) cellular networks encrypt the radio link. The general trend of cellular networks is for 3G technologies, with the most current generation of these technologies designed to offer strong encryption based on algorithms such as Kasumi and AES. Even after these next-generation networks are widely available, though, they are still likely to rely on previous-generation technology for coverage in less densely populated areas, where encryption is not always provided. And even if your home operator encrypts the link, you may roam onto a partner network that does not. The bottom line is that you cannot depend on the protection of the radio link.

Cellular-data connections share a common issue with Wi-Fi hotspots. Both primarily offer connectivity to the Internet, and even when they encrypt the radio signal, the IP traffic over the Internet portion remains unprotected. As an option, some cellular operators offer more secure back-end connectivity options to connect from the operator's core network to the customer network, including dedicated frame relay circuits or IPSec-based, network-to-network VPN connections. However, these arrangements incur additional costs, both through initial networking setup fees and then through recurring monthly fees.

3.1.2.3 Connections And Platforms Multiplicity

Although it is possible for you to implement specific security solutions for Wi-Fi and for cellular-data connections, each solution will be unique, and managing both is probably not practical. Another concern is that employees may use a variety of computing devices, including portable computers, smartphones, PDAs, home computers, and public systems. IT will have control over some, but not all, of these devices. Unmanaged devices, including home systems and public workstations may leave the network open to security risks.

3.1.3 Wireless Security Architecture Recommendations

There is a clear need for a security solution that embraces the world of mobile and wireless computing—with an approach that addresses all forms of connectivity, including Wi-Fi on premises, Wi-Fi off premises, cellular data, public kiosks, home access, and whatever else may become available. But before you can specify an effective security architecture, there are other important security features that you will probably need, including the ability to:

- Allow conformance with government regulations that protect items such as financial and medical information.
- Have control over the end-point node to check for proper software configuration such as virus protection, to scan the system for dangerous code, and to clear caches.
- Provide granular control to resources, rather than just providing access to a network.
- Support both managed and unmanaged nodes as well as accommodate a wide range of device types, including desktops, portable computers, PDAs, and smartphones [1].

The security architecture that meets all of these needs is an SSL-based VPN. SSL VPNs take advantage of the browsers and the SSL security layer that are available for nearly all computing platforms, including notebook platforms, PDAs, and smartphones. Fig. 2-1 shows an SSL appliance securing all forms of wireless access and shows how IP traffic is redirected into an SSL tunnel [1].

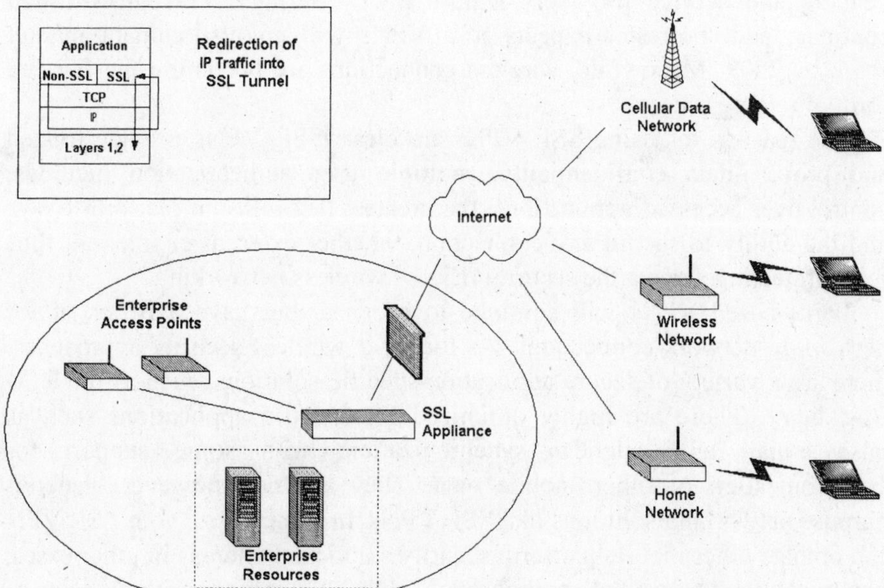

Figure 2-1. An SSL VPN can secure both external remote access and internal access.

3.1.3.1 The Inverted Security Model

Traditional security architectures are based on the concept of a perimeter, where firewalls mediate between nontrusted external networks and trusted internal networks. However, this model breaks down when a significant portion of your employees reside on public networks. The perimeter model also falls short when your network needs to allow select access by contractors, business partners, and customers.

The inverted security model assumes that you trust no node, whether internal or external. It shifts the emphasis to granular identity and access management. SSL VPNs enable this model by providing strong control of end points, strong user authentication, access to predefined resources, data integrity, nonrepudiation, and detailed auditing. By restricting access at the application layer, SSL VPNs shift security management from the network domain to the user domain.

3.1.3.2 Other Approaches Compared To VPN

According to Forrester Research [2], SSL VPNs are becoming the dominant remote access solution: By breathing new life into secure mobility and remote access, SSL VPNs will experience significant growth as both vendors and service providers mature their offerings. This growth will continue; and, Forrester expects SSL VPNs will enjoy dominant market share by 2009. Meanwhile, wireless connections are becoming the favored form of remote access.

The reasons for using SSL VPNs are clear. SSL VPNs provide trusted and proven data confidentiality, multiple user authentication methods, control over accessed applications, the greatest flexibility in platform types, and the ability to secure any connection, whether external or internal, thus fully protecting against the security risks of wireless networking.

IPSec-based VPNs will continue to be used, but have a better fit for network-to-network connections. As for other wireless security approaches, there are a variety of secure application-specific solutions, such as the RIM Blackberry. These are highly optimized for specific applications such as push e-mail and calendar synchronization, with some support for synchronization of other mobile data. They are not, however, general-purpose networking solutions like SSL VPNs. In some cases, your SSL VPN can obviate other mobile platform security solutions, whereas in other cases, you might want to use both approaches. Table 2-1 summarizes the aspects of different security approaches [1].

Table 2-1. Summary of Pros and Cons of Different Security Approaches

Type of Security Solution	**Pros**	**Cons**
IEEE 802.11i Security Standard	Comprehensive security framework for Wi-Fi security.	Available only in corporate settings. Does not address other wireless connections such as cellular. Does not support outdated hardware. Vendor interoperability issues.
Cellular Data Security Mechanisms	Primarily designed to protect operator from fraudulent use. Many networks, but not all, encrypt the radio link.	Only a partial security solution, and only if operator offers data encryption. Normal configuration has data passing the Internet in the clear. Does not address other wireless connections such as Wi-Fi.
IPSec VPN	Mature and available from multiple vendors.	Requires client code. Not necessarily available for all mobile platforms. Optimal for network access, not application level access.
Wireless-specific VPN	Efficient for wireless networking.	Specific to wireless networking and not necessarily applicable as a comprehensive enterprise security framework.
Application Specific Solutions (Wireless EMail)	Efficient for wireless networking. Limits access to specific resources.	Requires client code Limited to small application base such as email, calendar synchronization, and contact databases. Not a general-purpose, remote-access solution.
SSL VPN	Efficient for wireless networking. Enhanced features such as persistent sessions can address wireless challenges. Secures all forms of connections, whether WLAN, cellular, or other.	Not yet as widely adopted as IPSec VPNs. Market still evolving.

	Provides the greatest degree of control over access to resources. Greatest client flexibility between browser, browser with agent code, and full client version.	

Next, serious flaws have emerged in the basic Bluetooth specification and the Advanced Encryption Standard (AES). Now, let's look at the details on both of these vulnerabilities, as part of the wireless network security (WNS) technology perspectives; as well as, a look at the rest of the latest security threats out there. It's been a bad year for encryption all around—even when encryption technology wasn't in effect.

3.2 Wireless Network Security Encryption

Apparently, all of Bluetooth is divided at the core, according to Israeli security specialists who have recently reportedly found a serious core vulnerability in the basic Bluetooth specification. According to NewScientist.com (http://www.newscientist.com/home.ns), researchers have discovered a cryptographic flaw (the worst kind of flaw) in the Bluetooth standard that renders all Bluetooth implementations vulnerable to a fairly simple attack, making those implementations completely insecure.

Bluetooth is a short-range (about a 300-foot maximum) radio standard used by networks to feed data to printers, portable phones, laptops, and other electronic devices. The newly discovered decryption technique makes all Bluetooth communications insecure—even when the user has enabled all of the security features to the maximum levels.

This is not the same vulnerability that exists when Bluetooth devices initially negotiate their connection, which is a well-known threat that's rather difficult to exploit. Instead, the new threat lets attackers penetrate a Bluetooth network at any time and take over the connection, perhaps establishing a connection allowing unlimited long distance calls. Basically, the researchers have found a way to force Bluetooth devices into the initial pairing mode and thus decrypt the 128-bit key in well under a second, even using older PCs.

In addition, one basic design flaw in Advanced Encryption Standard (AES) allows a timing attack to recover AES keys from a remote server using OpenSSL AES. To make matters worse, this basic design flaw in AES is not limited to any particular implementation.

3.2.1 Serious Applicability Risk Level

The Bluetooth flaw affects any and all Bluetooth networks, and the AES vulnerability affects any and all AES encryption. Because these new vulnerabilities affect the overall technology rather than a specific implementation, the risk level is serious for both flaws.

With regards to the preceding, the FBI recently demonstrated how easily someone can break Wired Equivalent Privacy (WEP) encryption and gain access to a secured network. This part of the chapter examines the critical role that security plays in wireless technology and offers some suggestions for locking down your wireless network.

3.3 WEP Woes

As previously mentioned, the Federal Bureau of Investigation recently were able to conclusively demonstrate how insecure the majority of wireless networks really are. In addition, the agency announced that even 802.11b wireless access with Wired Equivalent Privacy (WEP) encryption (widely touted as the secure replacement for the 802.11a standard) is just as insecure.

Anyway, it took the FBI literally four minutes to demonstrate how to break WEP encryption and gain access to a secured network. The FBI's findings should serve as a warning to enterprises currently using wireless access, and it might prevent some enterprises from using wireless networks entirely. Regardless, enterprises need to be more aware of the critical role that security plays in wireless networks.

Whatever comes of the FBI demonstration, it's important that enterprises fully understand this concept: Unless you've deployed end-to-end data encryption, communication is never really secure, no matter how well-secured the wireless network. Despite advances in wireless technology, the security of a wireless network will never equal that of a wired network.

Unfortunately, most enterprises that have already deployed wireless access chose usability over security, just like most software enterprises. In addition, many enterprises don't consider the fact that wireless access doesn't really offer any advantages over wired access in many cases.

In fact, it can actually introduce new problems. Numerous 802.11b wireless network problems have been caused entirely by the use of 2.4-GHz wireless phones, often from wireless PBX systems.

Wireless networks are now in the enterprise environment, and enterprise deployments are increasing. However, it is strongly recommended that enterprises use this strategy when deciding whether to go wireless: Use wireless networking only in cases where wired access is impossible, not just as a simple or trendy alternative.

And, while security should be a primary factor in this decision, keep in mind that there are more than just security-related reasons for staying wired. For example, wired networks can handle significantly higher bandwidth, as well as offer better security, because they don't broadcast packets of information.

But, if bandwidth isn't a concern, and the powers-that-be are convinced that wireless is the way to go, rest assured that it is possible to make wireless access much more secure without depending on WEP. Two methods for accomplishing this include using protocols such as Point-to-Point Tunneling Protocol (PPTP) or Layer Two Tunneling Protocol (L2TP) and enforcing access controls with usernames and passwords or some other authentication method. Add IPSec to the mix, and you've got both access control and end-to-end encryption that's more secure than wired network access. But, keep in mind that this solution is still prone to interference.

Of course, some people will argue that 802.11i features all of this security provided by Wi-Fi Protected Access (WPA) (WEP's expected replacement); as well as, better interference control. While this is great news, 802.11i is no use to anyone until there are plans to replace all existing wireless networking equipment or upgrade the firmware, if that's even possible.

In addition, remember that no matter what security technologies or standards emerge, there will always be someone out there trying to break it—and that includes WPA. In any event, you can deploy Gigabit Ethernet access at a lower cost, so it can provide both superior security and bandwidth irrespective of data encryption.

If wireless access is your only alternative, explore the use of PPTP/L2TP and IPSec on your existing infrastructure before deciding to replace or upgrade existing 802.11a and 802.11b equipment. While it's not pretty from a technological point of view, it's quite functional, and it just might prove to be more secure than 802.11i [1].

Even with the heightened awareness of the security risks posed by wireless networks, some IT pros continue to overlook some of the essential safeguards the systems require. This next part of the chapter presents five practices you should avoid so that you can ensure the best protection for your wireless network.

4. WIRELESS NETWORKS AT RISK: ASSESSING VULNERABILITIES

Wireless networks require the same security measures as conventional networks, and then some. The issues that concern you in the wired realm should still concern you with wireless networks and devices: Keep the encryption strong, keep the certificates in place, and keep focused on security. Wireless security isn't a matter of *different* security, it's a matter of *more* security. Here are the most common security oversights and how you can avoid them.

- Don't breach your own
- Don’t spurn MAC
- Don’t spurn WEP or WPA
- Don’t allow unauthorized access points
- Don’t permit ad-hoc laptop communications [5]

4.1 Don't Breach Your Own Firewall

You've firewalled the network, wireless or not, and rightly so. However, you've done yourself no good if your configuration doesn't place your wireless system's access points outside the firewall. Make sure it does—otherwise, you're not only failing to create a necessary barrier, you're creating a convenient tunnel through one that was already there.

4.2 Don't Spurn MAC

Media Access Control (MAC) is often ignored because it's not spoof-proof. But it is another brick in the wall. It's essentially another address filter, and it clogs up the works for the potential hacker by limiting network access to registered devices you identify on address-based access control rosters. MAC also lets you turn the tables on the potential intruder.

Consider that the intruder must knock on the door before being denied. If you have MAC in place, the intruder will bump into it before realizing it's there and then must regroup to get past it. And, now your network knows what the intruder looks like. So think of your MAC list as creating three classes of visitors:

- Entities that aren't on the list but are known because they've tried to get in before, uninvited, and are now instantly identifiable if they approach again
- Friendly entities that are on the MAC list
- Unknown entities that aren't on the list and that knock by mistake [5]

In short, if you monitor your wireless network and watch for multiple attempts at access by entities not on the MAC list, you've spotted a potential intruder. And, he or she won't know you've seen him or her.

4.3 Don't Spurn WEP Or WPA

Wired Equivalent Privacy (WEP) is a protocol specific to wireless security, conforming to the 802.11b standard. It encrypts data as it goes wireless, over and above anything else you're using. Use it. But remember that it is key-based, so don't stay with the default key. You may even wish to create a unique WEP key for individual users when they first access the system. Don't rely on WEP alone. Even multiple layers of encryption don't make you hack-proof, so use WEP in combination with other wireless-specific security measures.

And, don't overlook WiFi Protected Access (WPA2), which addresses header weakness issues in WEP and is readily available to Windows XP users. WPA2 can be configured to rekey your encryption and is actually easier to use than WEP.

4.4 Don't Allow Unauthorized Access Points

Access points are incredibly easy to set up, and an over-burdened IT department might loosen the rules to allow them to be set up on an as-needed basis by anyone smart enough to run a VCR. But, don't succumb to this temptation. The access point is a primary target for an intruder. Implement a deployment strategy and procedure and stick to them. Carefully outline the correct guidelines for positioning an access point and be certain that anyone deploying an AP has those guidelines on hand. Then, put a procedure in place for noting the presence of the AP in your wireless network configuration for future reference and for appropriately distributing or making available the revised configuration. Regardless of who sets up the AP, have another person double-check the installation as soon as it's convenient. Is this a lot of trouble to go to? Yes. And, security penetrations due to rogue or leaky APs are even more trouble.

4.5 Don't Permit Ad-Hoc Laptop Communications

This is a tough one to enforce in any enterprise. Ad-hoc mode lets WiFi clients link directly to another nearby laptop, which is so darned convenient; you just can't imagine not using it. As part of the 802.11 standard, ad hoc mode permits your laptop's network interface card to operate in an independent basic service set configuration. This means that it can go peer-to-peer with another laptop via RF. When you're in ad hoc mode, you can spontaneously form a wireless LAN with other laptops.

At face value, this is such a cool trick that no one can resist trying it out. But, it permits access to the entire hard drive of the laptop. If you enable it and forget that it's enabled, your fly is open for all the world to see. And, the danger isn't only to your open machine. An intruder can also use the networked laptop as a doorway into the network itself. If you leave your machine in ad hoc mode and somebody sneaks in, you've exposed the entire network. Avoid this risky habit by never letting it develop in the first place. Just accept that it isn't worth the risk [5].

Finally, let's look at end point security control. This final part of the chapter explains the range of features that increase IT control and data protection by giving users access that's finely tuned to the risks of their environment.

5. END POINT SECURITY CONTROL

The widespread use of SSL VPNs for remote access enables more users to gain access to your network from far more places than they would if they were using a traditional IPSec VPN. Clearly, that enhances productivity; at the same time, security threats for both the end user and IT increase substantially as access extends to places that IT cannot possibly control. To effectively control these risks, managing access by user identity alone is no longer enough. You also need to focus on the safety of that user's environment.

For example, your enterprise needs to deliver secure anywhere access to network resources from the most dangerous places (New Orleans, from airport kiosks, employee-owned PCs, wireless hot spots, and unmanaged PDAs) without sacrificing the integrity of the enterprise network. In other words, your IT administrators should be able to differentiate a remote access policy based on end point security. Here's what you should be able to offer:

- Security anywhere access from multiple environments
- Policy Zones rather than an "access" policy
- Device Interrogation
- Control and ease of administration [6]

5.1 Security Anywhere Access From Multiple Environments

An executive might work remotely using a enterprise-issued laptop PC, then log into the network in the afternoon to check e-mail from a tradeshow kiosk. Later that day, she might update a presentation from her home PC. End point security control should be able to secure all these end points as appropriate based on the security policy you set. For example, this would mean limiting her access to certain applications or requiring advanced data protection on riskier end points.

5.2 Policy Zones Rather Than An Access Policy:

You want to deliver as much access as possible to your users, without compromising security. To do that, you need more options than simply allow access or don't allow access. If an end point is semi-trusted, for example, you want to create three more Policy Zones, for example trusted, semi-trusted, and non-trusted. This allows IT to offer users some degree of centrally managed access even if an untrusted environment doesn't warrant full access rights.

5.3 Device Interrogation:

You should be able to detect what is or isn't on an endpoint machine. For example, you should be able to automatically launch an agent prior to authentication, so login can be stopped if any malicious software (malware) is discovered. Based on what applications are found on the end point (for example, a predefined personal firewall or anti-virus application), you should be able to automatically classify the end point into one of the Policy Zones, so that the level of access granted is appropriate both to the user and the level of risk of the end point.

5.4 Control And Ease Of Administration

Your object-based policy model should be able to administer SSL VPN. This provides the ability to easily enforce policies from a single point to deliver access with maximum security.

5.5 User Authentication Focused Security

In the past, enterprises relied on VPN access from the relative safety of the enterprise laptop. The immobility of traditional client VPN technologies actually reduced IT fears of security risks such as malware damaging the network. IPSec clients could not be planted on a kiosk and were unlikely to be deployed on a home network. Because of these limitations, IT had little concern about non-enterprise machines and presumed that employees gained access solely from enterprise owned and-managed machines. Non-employees generally were not given any access, or enterprises risked giving them the same full access that their own employees had.

As a result of this model, IT security concerns focused solely on user authentication-identifying users as strongly as possible. Many large enterprises invested in two-factor authentication systems to identify users. These enterprises believed that if users always accessed the network from the same trusted location, the key issue centered on ensuring that the users were who they said they were.

5.6 Mobility Boosts Risk And Productivity

SSL VPNs from enterprises, provide anywhere access that increases employee productivity. In addition, SSL VPNs provide security benefits such as SSL encryption, protection from direct network access, full authentication support, and granular access control. However, some types of risks for end users and IT substantially increase as users demand access from places that IT cannot control. Today, the issue focuses not only on who the user is, but also how trustworthy the remote access environment is. An SSL VPN user can gain access from a range of end devices, all posing different levels of security threats. For instance, home PCs, PDAs, and tradeshow kiosks have a higher risk of hosting malware that could infect the network with viruses or Trojan horses, capture passwords for later reuse, or cache confidential information.

5.7 IT Must Address User Identity And Risk

IT managers dearly have concerns about their ability to protect enterprise assets when access can occur from any desktop or device. In addition to accessing the enterprise network via unauditable entry points, users may inadvertently leave behind information at a kiosk or hotel business center if downloaded files, viewed e-mail attachments, Web pages, and passwords are cached on the hard drive. For the user, the issue transcends security policy and the integrity of the corporate network; it becomes an issue of personal privacy [8].

As SSL VPN access becomes widespread, remote access policies will undergo scrutiny and must address the concerns of end users, auditors, business stakeholders, and IT. With recent press reports on incidents in which information was secretly captured via keystroke logging at public kiosks, this pressing security issue is just beginning to get public attention.

But, what can IT realistically do to overcome these threats? A few written IT policies ban kiosk access, but in practice it is a very hard policy to enforce. Plus, that type of restriction significantly reduces some of the productivity gains that SSL VPNs offer. So far, few enterprises have clearly thought out how to handle this concern.

End point security control enables IT to manage the risk of Web-based access. Now enterprises *will* no longer have to ban mobile access to enterprise assets or accept a high degree of risk in return for providing greater user productivity [6].

5.8 So What Is End Point Control?

End point security control is the ability to enforce policy based upon the level of trust that IT has for the user and his or her environment. IT organizations can establish and define Policy Zones, including untrusted machines such as kiosks, semi-trusted machines such as home PCs, and trusted enterprise assets like laptops. With, end point security control, they can then appropriately manage those zones with a simple set of parameters. In other words, end point security control provides a very high degree of granularity, so IT can reduce risk, provide access from more places at a lower cost to the enterprise, and control access by user location.

5.9 How To Deliver End Point Security Control

End point security control delivers unprecedented capability for enterprises to provide exactly the type of remote access they want. You can ensure secure access to resources by using three essential components:

- Device Interrogation
- Policy Zones
- Enhanced data protection and remediation [6]

5.9.1 Device Interrogation

End point security control automatically interrogates the end point when the user accesses the enterprise's SSL VPN in order to determine what is and what is not on the machine. You need to ensure that the access point is free of malware like keystroke loggers and Trojan horses. This happens prior to authentication so login can be stopped if any malware is discovered.

5.9.2 Policy Zones

Device interrogation looks for certain applications or watermarks on the end point. For example, if a specified antivirus product or a personal firewall is present, device interrogation may instantly classify the end point into one of the predetermined Policy Zones-such as trusted, semi-trusted, and non-trusted. Each zone enables a different level of access, appropriate to its level of risk.

5.9.3 Enhanced Data Protection And Remediation

The most flexible remote access options should be required in the semi-trusted zone. And, in a non-trusted zone such as kiosk access, ASD could be required for remediation.

5.10 Policy Zones Grant Access By Trust Level

Many enterprises want to support multiple access environments, but would like to differentiate that access for less trusted end points. This is where end point security control comes in. End point security control provides extensive access control and policy management flexibility, so administrators can plan ahead for any remote access scenario. Other SSL VPN solutions offer access/no access models, but that approach simply isn't sufficient to support today's broad range of access scenarios. By using the SSL VPN, you can easily define multiple Policy Zones that limit access to resources and ensure that sensitive information is not left behind. Unlike competitors' products, which associate data protection with group membership, **Policy** Zones give administrators highly granular access control over users and their remote access environments, such as:

- IT-managed laptops
- Home PCs
- Kiosks [6]

Essentially, having more zones makes your network more secure. For instance, IT-managed laptops have the highest level of trust and get complete access, because they are owned and provisioned by your enterprise and have predefined characteristics such as a personal firewall, virus checker, and a digital certificate. But, a category of devices with a lower trust level could be classified as semi-trusted. This Policy Zone includes devices like home PCs that are not owned and provisioned by IT. Users in this zone would be able to access only a subset of the privileges they get in the trusted zone. A third category of non-trusted devices includes kiosks found at hotels, airports, and convention centers. Users in this zone may have even more limited access, such as access only to e-mail. Defining different Policy Zones for each access scenario makes it easy to support this broad range of use cases.

5.10.1 Defining Policy Zones

Policy Zones are created by specifying one or more device profiles and indicating the level of data protection for the zone. Administrators can create up to 10 zones.

5.10.1.1 What Is A Device Profile?

A device profile is a set of characteristics that must be present on a device to assign zone classification. Attributes might include:

- Application /process
- Directory/file name
- Registry key (can be used to specify an antivirus signature update or an O/S Patch Level)
- Antivirus program
- Personal firewall
- Windows domain membership [6]

For example, an administrator could define a profile that would authenticate a device as trusted only after the access point meets designated conditions: such as having the appropriate antivirus solution, the appropriate personal firewall, and membership in the correct Windows domain (one device profile can include multiple domains to check against). So for greater flexibility, administrators can specify one or more device profiles for each Policy Zone, which can match more than one profile with an OR condition. Thus, a device with either the Business Partner device profile OR the Home PC device profile can be classified into a semi-trusted zone, which makes it extremely easy to organize a small number of zones to cover multiple access scenarios.

5.10.1.2 What Is Data Protection?

SSL VPN technology enables you to tie the level of data protection directly to the Policy Zone. For example, with an *IT* managed device, you might not require data protection; whereas, with a semi-trusted home PC, you might want to protect against the loss of sensitive information. Likewise, if a device or environment is considered risky, such as a kiosk, end point security control helps remediate the situation by increasing the level of data protection through encryption and deletion of confidential data on the endpoint machine.

5.11 Object-Based Model Promotes Granular Control

Based on the same security and management principles that underlie leading firewalls, you can gain a single view of all access control rules, which is far simpler and ultimately more secure than the typical flat policy management approach used by other vendors' products. Unlike those models, which become increasingly complex as you add groups and resources, access control rules should match users/groups to defined resources, which can then be made dependent upon the specified *Policy* Zones. In a single step, you can make object changes [6].

You now have the option of defining all users/groups, resources, and Policy Zones; and, combining them to create extremely granular access control rules. Resources, which include client/server and Web resources, as well as file shares are defined once, and are then related to multiple rules. A defined resource can relate to a single application or to all applications that exist within a domain, subnet, or IP range.

For greater security, Web resources can also be defined as aliases, so users cannot view private URLs. This is ideal for enterprises that grant access to specific resources to someone outside the enterprise, such as a business partner or customer [6].

Finally, for very granular access control, you can create advanced policy variables such as time of day. For example, you can limit a contract employee's network access to a specific application only from a trusted zone during standard business hours. Rules are checked sequentially for policy matches, similar to the way that policy models work in other perimeter security solutions, such as firewalls and content security products.

6. SUMMARY AND CONCLUSIONS

Although Wi-Fi technologies have significantly improved their security capabilities, many of the features and abilities are available only in newer equipment for IT-managed infrastructure. Meanwhile, cellular data networks rely on a completely separate security architecture that emphasizes protection of the radio link and does not provide end-to-end encryption.

By using an SSL VPN, you can secure all forms of wireless communication, both externally and internally. Moreover, this approach accommodates a wide range of user equipment [1].

Nevertheless, it's too soon to tell whether WNS encryption problems will turn out to be a tempest in a teapot or seriously exploited vulnerabilities. However, they do point out that you shouldn't rely too heavily on encryption or any other security technologies—you never know when a new discovery will compromise what was otherwise a relatively secure platform (see sidebar, "Tales FromThe Encrypt").

Tales From The Encrypt

Is merely possessing encryption software evidence of criminal intent? Apparently so: An appeals court has recently ruled that the judge in a criminal case was correct in permitting the prosecution to argue that the mere presence of PGP software on a computer implied criminal intent. Now, that's really scary! The guy on trial was in a really deep hole to begin with, and there was little he could offer in his own defense, and his lawyers created a side issue in his case. But that doesn't detract from the fact that a court has ruled the mere presence of PGP can be evidence of a guilty mind.

After all of the recent stories about various schools' carelessness with personal data, it's comforting to know that banks are more interested in protecting their customers' privacy—yeah, right! Citigroup, the world's largest bank, is blaming United Parcel Service for the loss of 4.1 million personal banking records stored on computer tapes. Even better, it took 29 days before anybody realized the tapes were missing.

While Citigroup appears to be placing all of the blame on UPS, the actual tapes were unencrypted, a fact that will be hard to pin on the shipping company. Can you say dumbest move ever by a big financial institution? And, this just months after Bank of America lost its own tapes, resulting in more than a 2 million missing records. It must have cost Citigroup far more to mail out notices to all those customers than it would have to encrypt everything a dozen times over.

Of course, the biggest problems occur when black hats find a way to crack encryption that a company has used to protect data in long-term storage [9]. They can then go back and dig out the data the organization thought it had securely protected [1].

Finally, how do you fix these new problems in Bluetooth or AES? Just don't use the technology.

7. REFERENCES

[1] Peter Rysavy, "Secure Wireless Networking Using SSL VPNs," Rysavy Research, [© 2005 Aventail Corp. All rights reserved. Aventail Corporation, 808 Howell St., Second Floor, Seattle, WA 98101], Rysavy Research, PO Box 680, Hood River, OR 97031 U.S.A., 2005

[2] Robert Whiteley, Stan Schatt and Benjamin Gray. "SSL Is The Future Of Remote Access VPNs," © 1997-2005, Forrester Research, Inc. All rights reserved. Forrester Research, Inc., 400 Technology Square, Cambridge, MA 02139, USA), June, 2004.

[3] Fred Sandsmark, "Securing Wireless Networks," Copyright © 2005 Cisco Systems, Inc. All rights reserved. Cisco Systems, Inc., 170 West Tasman Dr., San Jose, CA 95134, USA [iQ Magazine (vil. VI,No. 1)., 2005.

[4] Jupitermedia Headquarters, Jupiter Research, 23 Old Kings Highway South, Darien, CT 06820, 2005.

[5] Scott Robinson, "Strengthen Your Wireless Security By Avoiding These Missteps," Copyright ©2005 CNET Networks, Inc. All rights reserved. TechRepublic, 235 Second Street, San Francisco, CA 94105, 2005.

[6] "End Point Control: Secure Anywhere Access With Reduced Risk And Increased IT Control," © 2004 Aventail Corp. All rights reserved. Aventail Corporation, 808 Howell St., Second Floor, Seattle, WA 98101, 2004.

[7] John R. Vacca, *Firewalls : Jumpstart for Network and Systems Administrators,* Digital Press, 2004.

[8] John R. Vacca, *Net Privacy: A Guide to Developing and Implementing an Ironclad ebusiness Privacy Plan,* McGraw-Hill, 2001.

[9] John R. Vacca, *The Essential Guide To Storage Area Networks,* Prentice Hall, 2002.

[10] John R. Vacca, *Wireless Data Demystified (Mcgraw-Hill Demystified Series) (Paperback)s,* McGraw-Hill Professional, 2003.

[11] "Securing The Wireless LAN," Network World, Network World, Inc., 118 Turnpike Road, Southborough, MA 01772Copyright, 1994-2005 Network World, Inc. All rights reserved. August 12, 2002.

Chapter 3

STANDARDS

1. INTRODUCTION

True wireless network security means protecting every device with a wireless network card for every user--everywhere they go. But, first you need to know which security standards are implemented in your hardware and software.

Wireless technologies conform to a variety of standards and offer varying levels of security features. The principal advantages of standards are to encourage mass production and to allow products from multiple vendors to interoperate. For this chapter, the discussion of wireless standards is limited to:

- WEP
- IEEE 802.11b
- IEEE 802.11i
- IEEE 802.1X
- Bluetooth
- SSL
- WTLS
- WPA and WPA2 [1]

WLANs follow the IEEE 802.11 standards. Ad hoc networks follow proprietary techniques or are based on the Bluetooth standard, which was developed by a consortium of commercial companies making up the Bluetooth Special Interest Group (SIG). All of these standards are described next.

Furthermore, this part of the chapter discusses the built-in security features of 802.11. It provides an overview of the inherent security features to better illustrate its limitations and provide a motivation for some of the recommendations for enhanced security.

2. WEP

The IEEE 802.11 specification identified several services to provide a secure operating environment. The security services are provided largely by the Wired Equivalent Privacy (WEP) protocol to protect link-level data during wireless transmission between clients and access points. WEP does not provide end-to-end security, except for the wireless portion of the connection as shown in Fig. 3-1 [1].

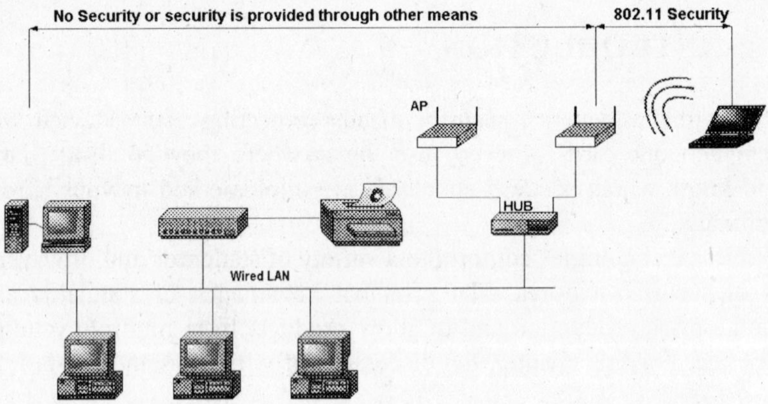

Figure 3-1. Wireless Security of 802.11 in Typical Network.

2.1 Security Features of 802.11 Wireless LANs Per the Standard

The three basic security services defined by IEEE for the WLAN environment are as follows:

1. Authentication: A primary goal of WEP was to provide a security service to verify the identity of communicating client stations. This provides access control to the network by denying access to client stations that cannot authenticate properly. This service addresses the question: Are only authorized persons allowed to gain access to my network?

2. Confidentiality: Confidentiality, or privacy, was a second goal of WEP. It was developed to provide privacy achieved by a wired network [5]. The intent was to prevent information compromise from casual eavesdropping (passive attack). This service, in general, addresses the question: Are only authorized persons allowed to view my data?
3. Integrity: Another goal of WEP was a security service developed to ensure that messages are not modified in transit between the wireless clients and the access point in an active attack. This service addresses the question: Is the data coming into or exiting the network trustworthy—has it been tampered with [1]?

> **Note:** The standard did not address other security services such as audit, authorization, and nonrepudiation. The security services offered by 802.11 are described next in greater detail.

2.1.1 Authentication

The IEEE 802.11 specification defines two means to validate wireless users attempting to gain access to a wired network: open-system authentication and shared-key authentication. One means, shared-key authentication, is based on cryptography, and the other is not. The open-system authentication technique is not truly authentication; the access point accepts the mobile station without verifying the identity of the station. It should be noted also that the authentication is only one-way: only the mobile station is authenticated. The mobile station must trust that it is communicating to a real access point (AP). A taxonomy of the techniques for 802.11 is depicted in Fig. 3-2 [1].

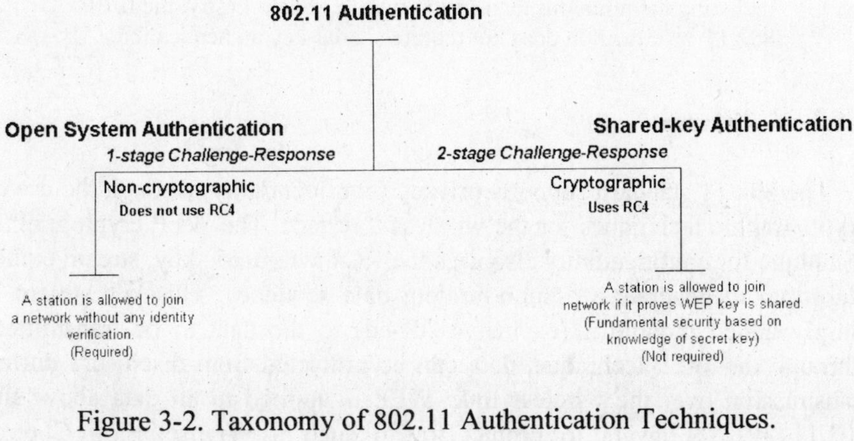

Figure 3-2. Taxonomy of 802.11 Authentication Techniques.

With Open System authentication, a client is authenticated if it simply responds with a Medium Access Control (MAC) address during the two-message exchange with an access point. During the exchange, the client is not truly validated, but simply responds with the correct fields in the message exchange. Obviously, without cryptographic validation, open-system authentication is highly vulnerable to attack and practically invites unauthorized access. Open-system authentication is the only required form of authentication by the 802.11 specification.

Shared key authentication is a cryptographic technique for authentication. It is a simple challenge response scheme based on whether a client has knowledge of a shared secret. In this scheme, as depicted conceptually in Fig. 3-3, a random challenge is generated by the access point and sent to the wireless client [1]. The client, using a cryptographic key that is shared with the AP, encrypts the challenge (or nonce, as it is called in security vernacular) and returns the result to the AP. The AP decrypts the result computed by the client and allows access only if the decrypted value is the same as the random challenge transmitted. The algorithm used in the cryptographic computation and for the generation of the 128-bit challenge text is the RC4 stream cipher developed by Ron Rivest of MIT.

> **Note:** The authentication method just described is a rudimentary cryptographic technique, and it does not provide mutual authentication. That is, the client does not authenticate the AP, and therefore there is no assurance that a client is communicating with a legitimate AP and wireless network.

> **Note:** Simple unilateral challenge-response schemes have long been known to be weak. They suffer from numerous attacks including the infamous man-in-the-middle attack. Lastly, the IEEE 802.11 specification does not require shared-key authentication.

2.1.2 Privacy

The 802.11 standard supports privacy (confidentiality) through the use of cryptographic techniques for the wireless interface. The WEP cryptographic technique for confidentiality also uses the RC4 symmetric key, stream cipher algorithm to generate a pseudo-random data sequence. This key stream is simply added modulo 2 (exclusive-OR-ed) to the data to be transmitted. Through the WEP technique, data can be protected from disclosure during transmission over the wireless link. WEP is applied to all data above the 802.11 WLAN layers to protect traffic such as Transmission Control

Protocol/Internet Protocol (TCP/IP), Internet Packet Exchange (IPX), and Hyper Text Transfer Protocol (HTTP).

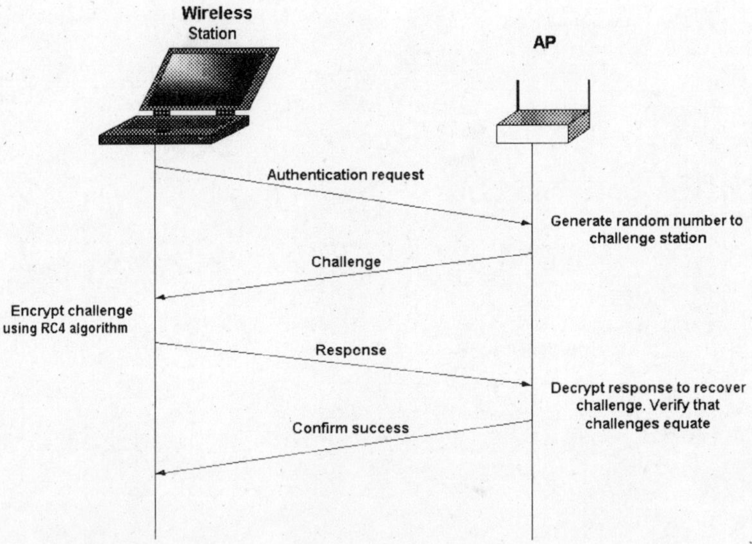

Figure 3-3. Shared-key Authentication Message Flow.

As defined in the 802.11 standard, WEP supports only a 40-bit cryptographic keys size for the shared key. However, numerous vendors offer nonstandard extensions of WEP that support key lengths from 40 bits to 104 bits. At least one vendor supports a keysize of 128 bits. The 104-bit WEP key, for instance, with a 24-bit Initialization Vector (IV) becomes a 128-bit RC4 key. In general, all other things being equal, increasing the key size increases the security of a cryptographic technique. However, it is always possible for flawed implementations or flawed designs to prevent long keys from increasing security. Research has shown that key sizes of greater than 80-bits, for robust designs and implementations, make brute-force cryptanalysis (code breaking) an impossible task. For 80-bit keys, the number of possible keys (a key space of more than 10^{26}) exceeds contemporary computing power. In practice, most WLAN deployments rely on 40-bit keys. Moreover, recent attacks have shown that the WEP approach for privacy is, unfortunately, vulnerable to certain attacks regardless of key size. However, the cryptographic, standards, and vendor WLAN communities have developed enhanced WEP, which is available as a prestandard vendor-specific implementations. The attacks previously mentioned are described later in this chapter and throughout the book. The WEP privacy is illustrated conceptually in Fig 3-4 [1].

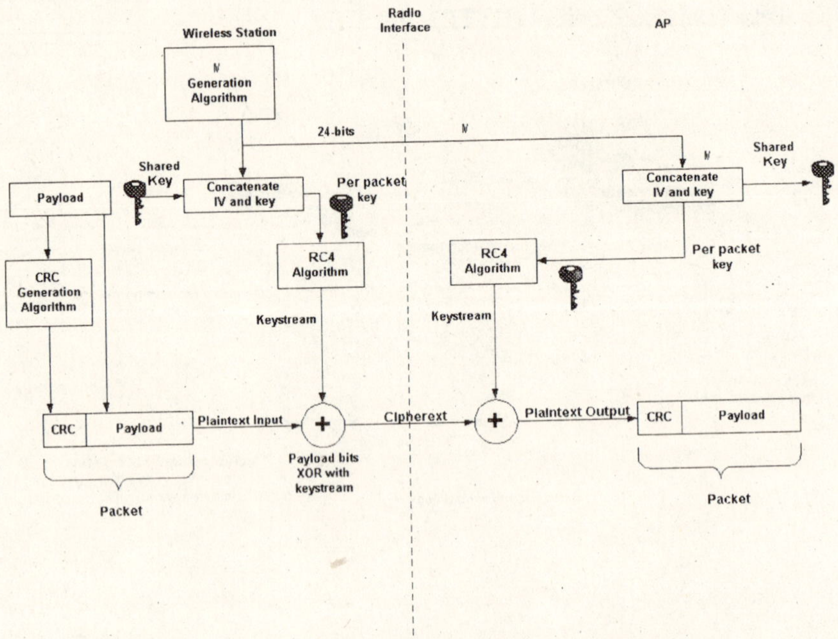

Figure 3-4. WEP Privacy Using RC4 Algorithm

2.1.3 Integrity

The IEEE 802.11 specification also outlines a means to provide data integrity for messages transmitted between wireless clients and access points. This security service was designed to reject any messages that had been changed by an active adversary in the middle. This technique uses a simple encrypted Cyclic Redundancy Check (CRC) approach. As depicted in Fig. 3-4, a CRC-32, or frame check sequence, is computed on each payload prior to transmission. The integrity-sealed packet is then encrypted using the RC4 key stream to provide the cipher-text message. On the receiving end, decryption is performed and the CRC is recomputed on the message that is received. The CRC computed at the receiving end is compared with the one computed with the original message. If the CRCs do not equal, that is, received in error, this would indicate an integrity violation (an active message spoofer), and the packet would be discarded. As with the privacy service, unfortunately, the 802.11 integrity is vulnerable to certain attacks regardless of key size. Basically, the fundamental flaw in the WEP integrity scheme is that the simple CRC is not a cryptographically secure mechanism such as a hash or message authentication code.

The IEEE 802.11 specification does not, unfortunately, identify any means for key management (life cycle handling of cryptographic keys and related material). Therefore, generating, distributing, storing, loading, escrowing, archiving, auditing, and destroying the material is left to those deploying WLANs. Key management (probably the most critical aspect of a cryptographic system) for 802.11 is left largely as an exercise for the users of the 802.11 network. As a result, many vulnerabilities could be introduced into the WLAN environment. These vulnerabilities include WEP keys that are non-unique, never changing, factory-defaults, or weak keys (all zeros, all ones, based on easily guessed passwords, or other similar trivial patterns). Additionally, because key management was not part of the original 802.11 specification, with the key distribution unresolved, WEP-secured WLANs do not scale well. If an enterprise recognizes the need to change keys often and to make them random, the task is formidable in a large WLAN environment. For example, a large campus may have as many as 26,000 APs. Generating, distributing, loading, and managing keys for an environment of this size is a significant challenge. It is has been suggested that the only practical way to distribute keys in a large dynamic environment is to publish it. However, a fundamental tenet of cryptography is that cryptographic keys remain secret. Hence, this is a major dichotomy. This dichotomy exists for any technology that neglects to elegantly address the key distribution problem.

This next part of the chapter discusses some known vulnerabilities in the standardized security of the 802.11 WLAN standard. As previously mentioned, the WEP protocol is used in 802.11-based WLANs. WEP in turn uses a RC4 cryptographic algorithm with a variable length key to protect traffic. Again, the 802.11 standard supports WEP cryptographic keys of 40-bits. However, some vendors have implemented products with keys 104-bit keys and even 128-bit keys. With the addition of the 24-bit IV, the actual key used in the RC4 algorithm is 152 bits for the 128 bits WEP key.

> **Note:** Some vendors generate keys after a keystroke from a user, which, if done properly, using the proper random processes, can result in a strong WEP key. Other vendors, however, have based WEP keys on passwords that are chosen by users; this typically reduces the effective key size.

2.2 Problems With the IEEE 802.11 Standard Security

Several groups of computer security specialists have discovered security problems that let malicious users compromise the security of WLANs. These include passive attacks to decrypt traffic based on statistical analysis, active attacks to inject new traffic from unauthorized mobile stations (based on known plain text), active attacks to decrypt traffic (based on tricking the access point), and dictionary-building attacks. The dictionary building attack is possible after analyzing enough traffic on a busy network.

Security problems with WEP include, first of all, the use of static WEP keys—many users in a wireless network potentially sharing the identical key for long periods of time, is a well-known security vulnerability. This is in part due to the lack of any key management provisions in the WEP protocol. If a computer such as a laptop were to be lost or stolen, the key could become compromised along with all the other computers sharing that key. Moreover, if every station uses the same key, a large amount of traffic may be rapidly available to an eavesdropper for analytic attacks, such as the second and third security problems with WEP, which are discussed next.

Second, the IV in WEP, as shown in Fig. 3-4, is a 24-bit field sent in the clear text portion of a message. This 24-bit string, used to initialize the key stream generated by the RC4 algorithm, is a relatively small field when used for cryptographic purposes. Reuse of the same IV produces identical key streams for the protection of data, and the short IV guarantees that they will repeat after a relatively short time in a busy network. Moreover, the 802.11 standard does not specify how the IVs are set or changed, and individual wireless NICs from the same vendor may all generate the same IV sequences, or some wireless NICs may possibly use a constant IV. As a result, hackers can record network traffic, determine the key stream, and use it to decrypt the cipher-text.

Third, the IV is a part of the RC4 encryption key. The fact that an eavesdropper knows 24-bits of every packet key, combined with a weakness in the RC4 key schedule, leads to a successful analytic attack, that recovers the key, after intercepting and analyzing only a relatively small amount of traffic. This attack is publicly available as an attack script and open source code.

Finally, WEP provides no cryptographic integrity protection. However, the 802.11 MAC protocol uses a noncryptographic Cyclic Redundancy Check (CRC) to check the integrity of packets, and acknowledges packets with the correct checksum. The combination of noncryptographic checksums with stream ciphers is dangerous and often introduces vulnerabilities, as is in the case for WEP. There is an active attack that permits the attacker to decrypt any packet by systematically modifying the packet and CRC sending it to the AP and noting whether the packet is acknowledged. These kinds of attacks are often subtle, and it is now considered risky to design encryption protocols that do not include cryptographic integrity protection, because of the possibility of interactions with other protocol levels that can give away information about cipher text. Some of the problems associated with WEP and 802.11 WLAN security are summarized in Table 3-1 [1].

> **Note:** Only one of the four problems previously listed depends on a weakness in the cryptographic algorithm. Therefore, these problems would not be improved by substituting a stronger stream cipher. For example, the third previously listed, is a consequence of a weakness in the implementation of the RC4 stream cipher that is exposed by a poorly designed protocol.

Table 3-1. Key Problems with Existing 802.11 Wireless LAN Security

Security Issue or Vulnerability	Remarks.
1. Security features in vendor products are frequently not enabled.	Security features, albeit poor in some cases, are not enabled when shipped, and users do not enable when installed. Bad security is generally better than no security.
2. IVs are short (or static).	24-bit IVs cause the generated key stream to repeat. Repetition allows easy decryption of data for a moderately sophisticated adversary.
3. Cryptographic keys are short.	40-bit keys are inadequate for any system. It is generally accepted that key sizes should be greater than 80 bits in length. The longer the key, the less likely a comprise is possible from a brute-force attack.
4. Cryptographic keys are shared.	Keys that are shared can compromise a system. As the number of people sharing the key grows, the security risks also grow. A fundamental tenant of cryptography is that the security of a system is largely dependent on the secrecy of the keys.
5. Cryptographic keys cannot be updated automatically and frequently.	Cryptographic keys should be changed often to prevent brute-force attacks.
6. RC4 has a weak key schedule and is inappropriately used in WEP.	The combination of revealing 24 key bits in the IV and a weakness in the initial few bytes of the RC4 key stream leads to an efficient attack that recovers the key. Most other applications of RC4 do not expose the weaknesses of RC4 because they do not reveal key bits and do not restart the key schedule for every packet. This attack is available to moderately sophisticated adversaries.
7. Packet integrity is poor.	CRC32 and other linear block codes are inadequate for providing cryptographic integrity. Message modification is possible. Linear codes are inadequate for the protection against advertent attacks on data integrity. Cryptographic protection is required to prevent deliberate attacks. Use of noncryptographic protocols often facilitates attacks against the cryptography.
8. No user authentication occurs.	Only the device is authenticated. A device that is stolen can access the network.

9. Authentication is not enabled; only simple SSID identification occurs.	Identity-based systems are highly vulnerable particularly in a wireless system because signals can be more easily intercepted.
10. Device authentication is simple shared-key challenge-response.	One-way challenge-response authentication is subject to "man-in the-middle" attacks. Mutual authentication is required to provide verification that users and the network are legitimate.
11. The client does not authenticate the AP.	The client needs to authenticate the AP to ensure that it is legitimate and prevent the introduction of rogue APs.

3. IEEE 802.11

WLANs are based on the IEEE 802.11 standard, which the IEEE first developed in 1997. The IEEE designed 802.11 to support medium-range, higher data rate applications, such as Ethernet networks, and to address mobile and portable stations.

802.11 is the original WLAN standard, designed for 1 Mbps to 2 Mbps wireless transmissions. It was followed in 1999 by 802.11a, which established a high-speed WLAN standard for the 5 GHz band and supported 54 Mbps. Also completed in 1999 was the 802.11b standard, which operates in the 2.4 - 2.48 GHz band and supports 11 Mbps. The 802.11b standard is currently the dominant standard for WLANs, providing sufficient speeds for most of today's applications. Because the 802.11b standard has been so widely adopted, the security weaknesses in the standard have been exposed. These weaknesses were previously discussed with regards to WEP. Another standard, 802.11g, operates in the 2.4 GHz waveband, where current WLAN products based on the 802.11b standard operate.

Two other important and related standards for WLANs are 802.1X and 802.11i. The 802.1X, a port-level access control protocol, provides a security framework for IEEE networks, including Ethernet and wireless networks. The 802.11i standard was created for wireless-specific security functions that operate with IEEE 802.1X. The 802.11i standard is discussed later in this chapter; but, first, let's discuss IEEE 802.11b.

3.1 IEEE 802.11b

In 1999 the IEEE completed and approved the standard known as 802.11b, and WLANs were born. Finally, computer networks could achieve connectivity with a useable amount of bandwidth without being networked via a wall socket. Suddenly connecting multiple computers in a house to share an Internet connection or play LAN games no longer required expensive or ugly cabling. Enterprise users could get up out of their chairs and sit in the sunshine while they worked. New generations of handheld devices allowed users access to stored data as they walked down the hall to a meeting. The dawn of networking elegance was upon us. Users could set their laptops down anywhere and instantly be granted access to all networking resources. This was, and is, the vision of wireless networks, and what they are capable of delivering [6].

Fast forward to today. While wireless networks have seen widespread adoption in the home user markets, widely reported and easily exploited holes in the standard security system have stunted wireless' deployment rate in enterprise environments. While many people don't know exactly what the weaknesses are, most have accepted the prevailing wisdom that wireless networks are inherently insecure and nothing can be done about it. Can wireless networks be deployed securely today? What exactly are the security holes in the current standard, and how do they work? Where is wireless security headed in the future? This chapter and the rest of the book attempts to shed light on these questions and others about wireless networking security in an enterprise environment.

3.1.1 WLAN Networks Themselves

WLAN networks exist in either infrastructure or ad hoc mode. Ad hoc networks have multiple wireless clients talking to each other as peers to share data among themselves without the aid of a central Access Point. An infrastructure WLAN consists of several clients talking to a central device called an Access Point (AP), which is usually connected to a wired network like the Internet or a corporate or home LAN. Because the most common implementation requiring security is infrastructure mode, most security measures center around this design, so securing an infrastructure mode wireless network will be the focus of this part of the chapter. 802.11b specifies that radios talk on the unlicensed 2.4GHz band on one of 15 specific channels (the US is limited to using only the first 11 of those 15 channels) [6]. Wireless network cards automatically search through these channels to find WLANs, so there is no need to configure client stations to specific channels. Once the Network Interface Card (NIC) finds the correct channel, it begins talking to the Access Point. As long as all of the security settings on the client and AP match, communications across the AP can begin and the user can participate as part of the network.

Bandwidth on an 802.11b network is limited to 11Megabits (Mb) per access point. This 11Mb is divided among all users on that access point. If ten people access the same AP, communication to the wired world will be limited to approximately the equivalent of a decent DSL line [6]. Because the 802.11b standard does not contain any specifications for load balancing across multiple access points, devices that strictly adhere to the standard have no answer if you find your network becoming over populated. The only way to manage this is to add another AP in the same area, but with a different network name and radio channel, effectively having more than one separate network (up to a maximum of three), in the exact same area. Some wireless vendors have proprietary solutions for load balancing, but discussing these initiatives falls outside the scope of this part of the chapter. Interested readers should look into individual enterprises' propaganda documentation before they deploy their wireless network if they feel they will need these services.

So, from its inception, the 802.11b standard was not meant to contain a comprehensive set of enterprise level security tools. Still, there are some basic security measures included in the standard which can be employed to help make a network more secure. With each security feature, the potential for making the network either more secure or more open to attack exists [6].

3.2 802.11i

After five years of rousing debate, the body responsible for the Wi-Fi standard is finally putting the finishing touches on its new security standard, IEEE 802.11i. Although this standard's Robust Security Network (RSN) feature will deliver the level of security the wireless world is clamoring for, don't be fooled: Your wireless network won't be secure until your transition to RSN is complete.

RSN defines two security methodologies--one for legacy hardware based on RC4 and one for new hardware based on Advanced Encryption Standard (AES). The standard also provides the flexibility to add new methodologies if the need arises. RSN uses the IEEE 802.1x port-authentication standard to authenticate wireless devices to the network and to provide the dynamic keys it requires. The task group does not specify any authentication method over 802.1x; it just defines the features such a method must provide. The idea here is to future-proof the RSN authentication process.

Throughout the process, the 802.11i Task Group remained painfully aware that the transition from old station hardware would be slow because of the installed base of wireless networking cards that conform to the Wi-Fi standard. RSN does provide for both RC4 and AES encryption, though it will take new WiFi adapters or fast stations to support the AES encryption. But the principal concession to the RSN migration effort is the inclusion of Transition Security Network (TSN), which is defined only to facilitate migration to an RSN, according to the standard. A TSN is insecure, since the pre-RSN equipment can compromise the larger network. The contradiction of an insecure security network comes from the ways broadcast and multicast traffic are (and aren't) protected and from the inclusion of preshared (manually configured) keys in RSN.

An access point sends broadcast and multicast frames encrypted with the weakest configured security method: WEP, Temporal Key Integrity Protocol TKIP (RSN with RC4) or CBC-MAC Protocol (CCMP (RSN with AES)). If the AP is configured for TSN, the WEP-encrypted broadcast frames will easily yield the WEP key, exposing all broadcast traffic even if no associated station is using WEP. And if RSN is being deployed with preshared keys because setting up Remote Authentication Dial-in User Service (RADIUS) and choosing a trustworthy authentication method are too difficult, chances are the same key is being used for WEP and preshared-key RSN! So much for robust security.

Sure, every security system has its weakest link. And, RSN does address all three aspects of a security system--authentication, key distribution and data confidentiality. But, it provides only legacy approaches to them all, and the legacy 802.11 Wi-Fi standard has shared keys for authentication; it has no key distribution and only weak data confidentiality. Using any of these legacy features in an RSN leaves the network compromised.

But, don't wait until RSN is a done deal to start your transition. Remove the risk of preshared keys in RSN by deploying RADIUS immediately. Many WEP-keyed products already support 802.1x and RADIUS, and several good authentication options are available. The Extensible Authentication Protocol-Wireless Transport Layer Security (EAP-WTLS) is getting most of the attention lately; but, the EAP-AKA) is also a viable alternative based on per-station shared secrets; and, the Wi-Fi Protected Access is an interim version of RSN based on a mid-2002 draft of 802.11i. Use any one of them to get your transition going. As long as legacy wireless stations are in your network, robust wireless security will remain out of reach.

So, unlike WiFi Protected Access (WPA) and 802.11i, 802.1x is available and is widely deployed on wireless networks today. There are three primary ways to authenticate using 802.1x: shared secrets (username/password), certificates, and SIM cards. While this part of the chapter focuses on the shared secrets method, each authentication method has advantages and disadvantages; and, the needs of individual deployments dictate which is used.

4. IEEE 802.1X STANDARD

Understanding 802.1x requires knowing the names of the different components that make up an 802.1x-secured wireless network. Fig. 3-5 shows the location role of each one of these terms in the authentication process [2]:

- Supplicant: End User System seeking access to the network
- Authenticator: Controls access to the network (access point)
- Authentication Server (RADIUS Server): Authenticates the end user, negotiates key material with the end user, and controls access to the network via the authenticator.
- Extensible Authentication Protocol (EAP): A secure protocol for negotiating other security protocols.
- EAP Over LAN (EAPOL): The version of EAP that is used over wireless networks.
- Port Access Entity (PAE): PAEs are similar to toggle switches. When the switch is open, no traffic is allowed to pass except for 802.1x traffic. After authentication is successful, the switch closes and user data is allowed to pass [2].

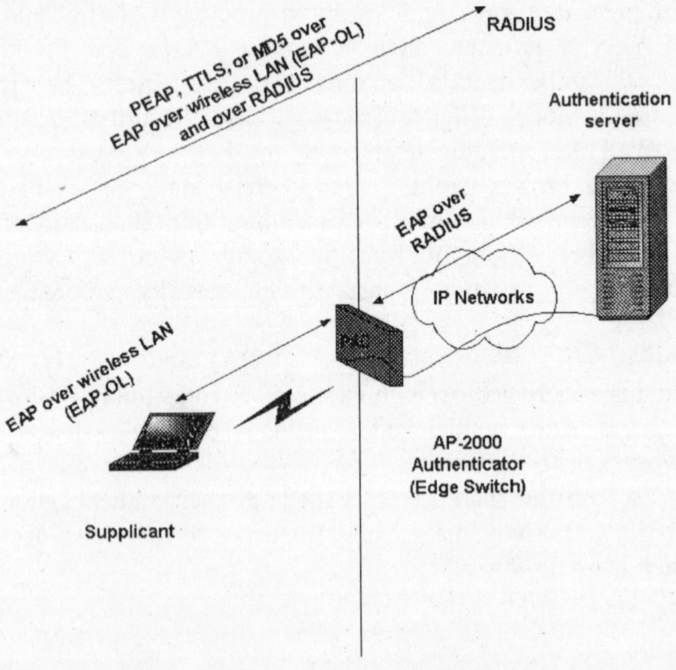

Figure 3-5. EAP and 802.1x

The supplicant negotiates the type of security protocol to be used with the authenticator using the EAP protocol. The properties of the different protocols that can be used across EAPOL and RADIUS are outlined in Table 3-2 [2]. The practical use of these protocols will be discussed later. Using the negotiated protocol, the supplicant provides credentials to the authentication server, and the authentication server provides credentials to the client. After each has been authenticated to the other, the security protocol is then used to negotiate session keys, which are used to encrypt user data.

Table 3-2. EAP Types.

			Authentication Credentials				
EAP Type	Open/ Propriety	Mutual/ Auth	Supplicant	Authenticator	Key Material	User Name	RFC
MD5	Open	No	Username/ Pwd	None	No	Yes	1321
TLS	Open	Yes	Certificate	Certificate	Yes	Yes	2716
TTLS	Open	Yes	Username/ Pwd	Certificate	Yes	No	IETF Draft
PEAP	Open	Yes	Username/ Pwd	Certificate	Yes	No	IETF Draft
SIM	Open/GSM	Yes	SIM		Yes	B	IETF Draft
AKA	Open/UMTS	Yes	USIM		Yes	B	IETF Draft
SKE	Open/CDMA	Yes			Yes	B	IETF Draft
LEAP	Proprietary	Yes	Username/ Pwd		Yes	Yes	N/A

The IEEE 802.1x, Port Based Network Authentication, uses the Extensible Authentication Protocol (EAP) as its authentication framework. EAP is a transport mechanism, and any defined EAP method can be used within EAP, enabling support for a wide variety of authentication credentials.

Common, standards based, non-proprietary EAP authentication methods on the market today include EAP-Transport Layer Security (TLS), EAP-Tunneled TLS (TTLS), and EAP-Protected EAP (PEAP). These methods support mutual authentication based on the two common ways to authenticate an end user or device: digital certificates and shared secrets (username/password).

EAP-PEAP is often the easiest to implement because of free client support from Microsoft, and can be just as secure as EAP-TLS if passwords are kept secure. EAP-PEAP does not require the use of client certificates.

EAP-TTLS is similar to EAP-PEAP because it does not require client certificates, but instead is based on client passwords. The disadvantage of

EAP-TTLS is that it is not free: server and individual client licenses must be purchased from vendors.

EAP-TLS requires certificates on both the RADIUS server and the wireless client. The distribution of certificates to each client can be challenging if the client-to-network administrator ratio is too high.

All three EAP types previously mentioned have been tested and deployed with access points and client cards. Other EAP types have also been tested and are in use, but they are not mentioned here because their use is not widespread yet.

Table 3-2 shows that some 802.1x-based systems pass the username in the clear. In these cases, end-user anonymity is not provided [2]. MD5 is particularly vulnerable because the username, machine name, and hashed password are sent in the clear. When a hash of the password is sent, data is vulnerable to an offline dictionary attack. Any EAP type that sends either username or password in the clear is neither secure nor recommended.

4.1 802.1x Encryption

Attacks are the most troublesome for 802.11 networks. However, they are also the easiest to prevent using two common mechanisms: weak key avoidance (WEPplus), together with the key rotation mechanism built into the 802.1x standard and access points, make it possible to create a secure wireless network.

In the existing pre-802.1x 802.11 specification, neither key distribution nor key rotation mechanisms are specified. With the exception of MD5, all EAP types listed in Table 3-2 provide a mechanism for the establishment of a session key at the station and the RADIUS server [2]. This session key provides a secure means to periodically transport new encryption keys to the station, so that the keys used to encrypt user data can continuously and securely change.

4.2 Attacks Against 802.1x

The session hijacking attack can only be performed on systems that are using 802.1x with encryption disabled. This is not a secure configuration. It is therefore recommended that encryption is always enabled with a key rotation period of less than 30 minutes. When the hijacked session attack is attempted on the EAP-TLS system, the attacker must

1. Wait until the client has successfully authenticated to the network.

2. Send a disassociate message to the client, on the legitimate access points behalf, using the MAC address of the access point.
3. Send frames to the valid access point, using the MAC address of the valid client [2].

The hijacked session attack assumes that no encryption is present, because if it were present, the radio perpetrating the attack would not be able to gain access to the network after the hijack, because the access point would reject all packets that did not match an encryption key corresponding to a known user. There is not an easy way to decrypt a WEP key generated using 802.1x, so the hijacker cannot create encrypted packets. When no encryption is present, this attack will succeed, allowing the attacker to use the session until the next re-authentication interval. At the next re-authentication time, the attacker would not be re-authenticated. He or she would then hijack another valid session. 802.1x wireless networks deployed with encryption enabled are not susceptible to this type of attack, and therefore it is not a concern.

4.2.1 Man-in-the-Middle Attack

The scenario used by the attacker to implement this attack is as follows:

1. Place a special rogue access point system to be within radio range of both a valid end user and a valid access point. This rogue system has the capability to simultaneously associate with a legitimate access point, while at the same time acting as an access point itself and allowing a legitimate user to associate to it.
2. Using the rogue system, associate to a valid access point as a client station.
3. Wait for a valid user to associate to the rogue system.
4. Transparently act as a repeater between the legitimate user and the legitimate access point, passing frame received from the user to the access point and vice-versa [2].

As noted earlier, all EAP types, except MD5, provide the ability to establish encrypted sessions. The man-in-the-middle attacker can observe this encrypted traffic, but cannot do anything malicious, because it is encrypted. Encrypted traffic is not compromised by this attack and the attacker does not gain access to the network. The attacker only gains the ability to target a particular user for the denial of service attack, which could be more easily perpetrated by a regular access point disconnected from any network.

When encryption is not used, the man in the middle will be able to see the user's traffic. This would have also been possible with a network sniffer. Network sniffers can see network traffic of other users, but if that traffic is encrypted, that traffic is useless to any hacker. Therefore, man-in-the-middle type attacks are not a concern.

5. BLUETOOTH

Bluetooth has emerged as a very popular ad hoc network standard today. The Bluetooth standard is a computing and telecommunications industry specification that describes how mobile phones, computers, and PDAs should interconnect with each other, with home and business phones, and with computers using short-range wireless connections. Bluetooth network applications include wireless synchronization, e-mail/Internet/intranet access using local personal computer connections, hidden computing through automated applications and networking, and applications that can be used for such devices as hands-free headsets and car kits. The Bluetooth standard specifies wireless operation in the 2.45 GHz radio band and supports data rates up to 720 kbps.*5* It further supports up to three simultaneous voice channels and employs frequency-hopping schemes and power reduction to reduce interference with other devices operating in the same frequency band. The IEEE 802.15 organization has derived a wireless personal area networking technology based on Bluetooth specifications.

6. SSL VPNS

The SSL VPN standard solution is designed to secure application streams between remote users and an SSL VPN gateway. In contrast with IPSec VPNs, which connect remote devices to trusted networks, SSL VPNs connect remote users (independent of device) to specific applications and network resources inside trusted networks. SSL VPNs are an elegant security solution for web-based traffic. The SSL client is pre-built into many of the web browsers common to today's operating systems, including Windows, Macintosh, Linux, Palm, Symbian, and Pocket PC. The SSL VPN standard solution is also well-suited for communicating to resources in a trusted network from non-corporate devices such as kiosks, Internet cafés, or an employee's own computer.

Though clearly convenient, these scenarios introduce a number of privacy- and security-related vulnerabilities. Connecting to the enterprise network from non-trusted devices leaves the user vulnerable to keyboard recording utilities. The user may also leave behind cookies and data that were cached during a browsing session. To address this vulnerability, most SSL VPN standard solutions use ActiveX or Java applet utilities to clean up after a user by deleting the local cache and cookies when a session ends. Enterprises also are susceptible to worms or Trojan horses that may have infected non-trusted equipment used during an SSL session. SSL VPN vendors are addressing these threats using cooperative enforcement with third-party client software such as anti-virus or personal firewall software [4]. Additional SSL VPN utilities (ActiveX or Java applets) are downloaded to ensure that the remote device is running the proper security software (checking for the latest anti-virus definition files, for example) before allowing access.

SSL VPN standard solution providers have been very responsive to addressing these vulnerabilities as their reach has expanded. At the same time, the complexity of SSL VPN deployments has increased to satisfy the requirements of a secure computing environment. The method for addressing these vulnerabilities is elegant in that it is performed without much user intervention and within a browser environment. The trade-off is that the browser must be enabled to support the download of ActiveX controls and Java applets, both of which have a number of documented vulnerabilities.

The allure of SSL VPN standard solutions has been their inherent simplicity. SSL is already part of the browser, no client installation is required and it's simple for mobile workers to use. Increasingly, though, such simplicity no longer is available. In fact, in an effort to address gaps in functionality and security, most SSL VPN vendors now provide a client as part of their clientless SSL VPN standard solution. Remote user access requirements continue to expand, and the need for access to legacy and other non-web based data sources is growing among mobile workers. As a result, new ActiveX, Java, and Win32 controls are available to download to the remote device. This client software typically requires some configuration to support the desired application and to port-forward the client-side traffic over the SSL tunnel. The solution is certainly no longer clientless—or simple.

7. WTLS

Wireless Transport Layer Security (WTLS) is designed by www.wapforum.org to provide security for WAP protocol. WTLS's design is based on TLS architecture and wireless computing environment requirements.

7.1 TLS and WTLS

WTLS originates from Transportation Layer Security (TLS), which is the Internet Engineering Task Force (IETF) standard for Internet transaction security. SSL is the implementation of TLS on Internet and is widely used in online business (E-business). WTLS follows the structure of TLS, but it must consider the specific features of wireless computing environment(PDA, Pocket PC, Handheld, etc…):

1. In a wireless computing environment, propagation delay is much more significant than transmission rate. So, WTLS must reduce the communication cost of security processing.
2. Wireless communication links have weaker reliability than wired links. So, WTLS must have mechanisms to allow certain extent of unreliability while still guaranteeing the security.
3. Mobile nodes have limited computation capability compared with desktop computers. So, the transmission rate on mobile node is much lower than desktop computers, either. SSL requires high computation capability to encrypt/decrypt data. So, WTLS must redefine the cypher suites to fit for this requirement. Also, mobile nodes have lower power and less memory [7].

From the aspect of encryption/decryption, WTLS is considered as a simplified SSL (so-called light-weight security protocol). But, WTLS supports secure connection resume, which allows mobile node to reuse the previous secure connection. This is a compromise between reliability and security.

The protocol stack of WAP is defined at www.wapforum.org. WTLS works between Wireless Transportation Protocol (WTP) and Wireless Datagram Protocol (WDP). WTP guarantees the reliability between peer nodes. WDP is a connectionless datagram protocol (like UDP in TCP/IP). So, you can see that WTLS works on an unreliable protocol, while SSL is based on TCP. This is a big issue for WTLS.

7.2 WTLS Architecture

The WAP gateway is the transition point between WAP client and Internet online servers. When a connection is established between WAP client and Internet online server, WTLS is used to establish the secure connection between the WAP client and WAP gateway, in which data is encrypted . The data is decrypted at the gateway. Then the gateway encrypts the data using SSL and relays the data to requested online servers. Here the WAP gateway is the point of security problems, because data is exposed at this point. At present, the WAP forum uses end-to-end security protocol to solve this problem. That is, there is a higher level of secure connection between the WAP client and online server. The WAP gateway is not involved in this secure connection. The idea of tunneling is used to implement this mechanism.

7.2.1 WTLS Protocol Stack

The protocol stack of WTLS has two layers. The lower layer is the Record Protocol, which is the part to encrypt/decrypt data using an agreed security policy(agreement on algorithms, keys and certificates). The upper layer has 4 different sub-protocols:

1. Handshaking Protocol: To establish/resume the secure connection between WAP client and WAP gateway.
2. Alert Protocol: To send urgent data or signals. The termination of secure connection uses alert message, which is a drawback of WTLS.
3. Change Cypher Protocol: To exchange keys on the fly to guarantee the security dynamically.
4. Application Protocol: To send data from application to Record Protocol and deliver the received data from Record Protocol to applications [7].

8. WPA AND WPA2 STANDARDS

If the confusion over the recent addition of four EAP authentication protocols to the WiFi Protected Access 2 (WPA2) standard wasn't already high enough, Gartner's recent attempt to bring clarity to the matter has exacerbated the problem. According to Gartner, Microsoft is somehow falling short on WPA2, because it's new WPA2 patch is missing the four new extended EAP types that were only recently added by the WiFi Alliance. Originally, the WPA and WPA2 standard only certified a single EAP protocol called EAP-TLS, which is universally supported by all manufacturers since it is the original EAP protocol used for wireless LAN authentication. Recently, the Wi-Fi Alliance added four additional EAP types to the WPA enterprise and WPA2 enterprise standards. An in-depth look at WPA/WPA2 and all five certified EAP protocols is discussed later in the chapter [3].

> **Note:** These changes to the WPA and WPA2 standard only pertain to the enterprise version and not the personal version. The WPA enterprise standard is meant for enterprises and mandates compliance with the five security-certified EAP authentication protocol, while the WPA personal standard is meant for the home or small office and only requires the use of a pre-shared key for authentication. The personal versions of WPA and WPA2 are also known as WPA-PSK and WPA2-PSK mode; and, you will often see this terminology in device configurations.

Since the addition of the four new EAP types to the WPA and WPA2 standard is not that old, one would hardly expect that the Microsoft WPA2 patch, which was probably in the works for quite some time, to magically be compliant with the new Extended EAP types. This is the exact reason the WiFi Alliance is giving vendors a grace period to comply with the modified WPA2 standard. Gartner is warning people to be very careful that the Microsoft WPA2 client is compatible with their existing infrastructure before committing to it, which sounds dire for the average IT manager. But, in proper context, the warning sounds pretty comical because the chances are about 99.99 percent that the Microsoft WPA2 client is compatible with their existing infrastructure [8].

Keep in mind that you get the WPA2 supplicant for the five-year-old Windows XP as a free download, so you're at liberty to use it or not. In contrast, Apple didn't even support WPA, until MAC OS 10.3 did, which occurred less than three years ago. Therefore, there is no single infrastructure in the world that can't support Microsoft's PEAP and EAP-TLS implementations, since they are the de facto standard given the enormous market share of Windows 2000 and XP [8]. Furthermore, the use of the word infrastructure in this context is highly dubious, since most of the wireless LAN infrastructure is composed of 802.11 access points and access point controllers that are EAP agnostic, since they act strictly as a pass-through for all EAP authentication protocols—including the ones certified for WPA and WPA2. In other words, no explicit compatibility is required. The only exceptions to this rule are the two proprietary Cisco EAP types LEAP and EAP-FAST, which only operate through Cisco Access Points, but neither EAP type is on the WPA or WPA2 standard. All standard EAP transactions will flow through a standard 802.1x-capable access point with ease. What dictates compatibility in EAP types is the supplicant (the client software on the user's computer) and authentication server (RADIUS server) to which the authentication attempt is passed after it passes through the access point.

The bottom line is that any of the WPA/WPA2 Extended EAP protocols are secure. Of the five certified EAP protocols, Microsoft already supports the two most common EAP types. Eventually, it would be nice to see Microsoft support the three remaining extended WPA/WPA2 EAP authentication protocols; and, once the grace period for the WPA2 Extended EAP types run out, Microsoft may have to add them or risk losing their WPA2 logo certification. For now, the biggest challenge for most enterprises is to move on to some form of enterprise WPA let alone WPA2. Most enterprises are still operating in dangerous waters with WEP encryption and some easy to crack EAP authentication protocol like LEAP. The last thing they need to be worried about is having to support for all five flavors of WPA/WPA2 EAP, since any one of them, including the two currently supported by Windows, will suffice in pretty much any situation [8].

9. EMERGING SECURITY STANDARDS AND TECHNOLOGIES

Finally, like the security industry, standards organizations have responded to the flurry over insecurities in 802.11 WLANs. Activity is occurring in the Internet Engineering Task Force (IETF) and the IEEE. The IEEE is currently working on three separate initiatives for improving WLAN security. The first involves the IEEE 802.11 Task Group i (TGi) which has proposed significant modifications to the existing IEEE 802.11 standard as a long-term solution for security. The TGi is defining additional ciphers based on the Advanced Encryption Standard (AES). The AES-based solution provides a highly robust solution for the future, but will require new hardware and protocol changes. TGi currently has design requirements to address many of the known problems with WEP, including the prevention of forgeries and detection of replay attacks.

The second initiative for improving WLAN security is the TGi's short-term solution (WiFi Protected Access (WPA)) to address the problems of WEP. The group is defining the Temporal Key Integrity Protocol (TKIP) to address the problems without requiring hardware changes—that is, requiring only changes to firmware and software drivers. The third initiative from IEEE is the introduction of IEEE 802.1X-2001, a generic framework for port-based network access control and key distribution, that was approved in June 2001. By defining the encapsulation of EAP (defined in RFC 2284) over IEEE 802 media, IEEE 802.1X enables an AP and station to mutually authenticate one another.

Since IEEE 802.1X was developed primarily for use with IEEE 802 LANs, and not for use with WLANS, the IEEE 802.11i draft standard defines additional capabilities required for secure implementation of IEEE 802.1X on 802.11 networks. These include a requirement for use of an EAP method supporting mutual authentication, key management, and dictionary attack resistance. In addition, 802.11i defines the hierarchy for use with the TKIP and AES ciphers and a four way key management handshake that is used to ensure that the station is authenticated to the AP and a back-end authentication server, if present. As a result, to provide adequate security, it is important that IEEE 802.1X implementations on 802.11 implement the IEEE 802.11i enhancements; as well as, the basic IEEE 802.1X standard.

IEEE 802.1X can be implemented entirely on the AP (by providing support for one or more EAP methods within the AP), or it can utilize a backend authentication server. The IEEE 802.1X standard supports authentication protocols such as RADIUS, Diameter, and Kerberos. RADIUS, described in RFC 2865-2869; and, RFC 3162, enables authentication, authorization, and accounting for Network Access Server (NAS) devices, including dial-up, xDSL, and 802.11.

The 802.1X standard can be implemented with different EAP types, including EAP-MD5 (defined in RFC 2284 and supporting only one-way authentication without key exchange) for Ethernet LANs and

EAP-TLS (defined in RFC 2716, supporting fast reconnect, mutual authentication and key management via certificate authentication). Currently a new generation of EAP methods are being developed within the IETF, focused on addressing wireless authentication and key management issues. These methods support additional security features such as cryptographic protection of the EAP conversation, identity protection, secure ciphersuite negotiation, tunneling of other EAP methods, etc. For the latest developments on the status of each specification, the reader is encouraged to refer to the IEEE 802.11 standards web site (http://standards.ieee.org/getieee802) for the latest developments on the IEEE 802.11 standards.

9.1 Updated WPA and WPA2 Standards

So, if the challenge of securing a wireless LAN wasn't already confusing enough, things have just gotten worse. The confusion started when the WiFi Alliance changed the WPA and WPA2 standards from supporting a single Extensible Authentication Protocol (EAP) standard to five EAP standards. Although this has broadened the WPA/WPA2 standards to be more inclusive, the decision of the WiFi Alliance to not rename the updated WPA and WPA2 standards, is causing mass confusion within the IT industry. This could have been avoided if the WiFi Alliance had named the updated standards WPA + Extended EAP and WPA2 + Extended EAP, because it would make it easy to differentiate between old WPA/WPA2 certified products from the newer WPA/WPA2 certified products, that will support five EAP types instead of one.

The WPA and WPA2 standards were created by the WiFi Alliance industry group that promotes interoperability and security for the wireless LAN industry. The Wi-Fi Alliance WPA and WPA2 standards closely mirrors the official IEEE 802.11i wireless LAN security standards group, but incorporates additional IETF EAP standards that the WiFi Alliance (http://www.weca.net/OpenSection/index.asp?noFlash=true) considers secure. The WPA and WPA2 standards have two components (encryption and authentication) that are crucial to a secure wireless LAN. The encryption piece of WPA and WPA2 mandates the use of TKIP or, because it's considered to be more secure than TKIP, preferably AES encryption. From an encryption standpoint, WPA leaves AES optional while WPA2 mandates both TKIP and AES capability. The authentication piece of WPA and WPA2 before the Extended EAP update, called for the use of a PSK (Pre-Shared Key) for personal mode and EAP-TLS for enterprise mode. After the Extended EAP update, there are now five EAP standards to choose from in WPA and WPA2 enterprise mode [3].

> **Note:** Besides the stricter encryption requirements, WPA2 also adds two enhancements to support fast roaming of wireless clients moving between wireless access points. Pair-wise Master Key ((PMK) used for each session between an access point and wireless client) caching support in WPA2 allows you to reconnect to an access point that you've recently connected to without the need to re-authenticate. Pre-authentication support in WPA2 allows a client to pre-authenticate with the access point toward which it is moving, while maintaining a connection to the access point it's moving away from. This new capability allows the roaming to occur in less than 1/10th of a second, while a traditional roam without PMK caching and pre-authentication would take more than one second. Timing-sensitive applications like Citrix, video, or VoIP will all break without fast roaming.

To give you some historical context on some significant EAP types, EAP-TTLS and PEAP were primarily created because the original EAP-TLS standard was deemed too difficult to deploy because of the need for a server-side x.509 digital certificate on the RADIUS authentication server and a client-side x.509 digital certificate on each and every client computer that needed to connect to the wireless LAN. While the server-side certificate requirement wasn't so bad because there are usually only a few RADIUS servers that need certificates, the client-side requirement was cause for major concern due to their sheer numbers. Because the client-side certificate required a PKI server infrastructure (rare for most enterprises) to be in place ahead of time or expensive third-party certificates, it automatically excluded EAP-TLS as a feasible option for most enterprises and forced them into using less secure forms of EAP such as Cisco's proprietary LEAP. EAP-TTLS and PEAP were created to eliminate the need for client-side certificates, but still leverage the server-side certificate to create a secure TLS tunnel to protect the inner authentication methods, such as EAP-MSCHAPv2 and EAP-GTC from eavesdropping and offline dictionary and brute force attacks. Conceptually, this works just like e-commerce security with SSL-enabled web sites where a web server's server-side certificate is leveraged to create a secure SSL tunnel even though the visitors to the secure web site don't have client-side digital certificates. The current WPA/WPA2 certified EAP standards are:

- EAP-TLS (originally certified protocol)
- EAP-TTLS/MSCHAPv2
- PEAPv0/EAP-MSCHAPv2
- PEAPv1/EAP-GTC
- EAP-SIM [3]

EAP-TLS is the original wireless LAN EAP authentication protocol. Although it's rarely implemented due to a steep deployment curve, it is still considered one of the most secure EAP standards available and is universally supported by all manufacturers of wireless LAN hardware and software including Microsoft. The requirement for a client-side certificate, however unpopular it may be, is what gives EAP-TLS its authentication strength. A compromised password is not enough to break into EAP-TLS enabled systems, because the hacker still needs to have the client-side certificate. When the client-side certificates are housed in smartcards, this offers the most secure authentication solution available because there is no way to steal a certificate from a smartcard without stealing the smartcard itself. Any physical theft of a smartcard would be immediately noticed and revoked and a new smartcard would be issued. Recently, this was the only EAP type vendors needed to certify for a WPA or WPA2 logo. There are client and server implementations of it in Microsoft, Cisco, Apple, Linux, and open source. EAP-TLS is natively supported in MAC OS 10.3 and above, Windows 2000 SP4, Windows XP, Windows Mobile 2003 and above, and Windows CE 4.2.

> **Note:** Although Windows 2000 supports EAP-TLS and PEAPv0/EAP-MSCHAPv2 authentication, it does not support WPA or WPA2 encryption, while all of the other newer operating systems mentioned support WPA. Windows XP with Service Pack 2 and the new WPA2 patch is currently the only operating system that natively supports WPA2.

EAP-TTLS was created by Funk software (http://www.funk.com/) and Certicom (http://www.certicom.com/index.php) and is primarily backed by Funk software and is supported by other third-party server and client software. Although it's a fine protocol and even better than PEAP in some ways, it isn't supported natively in Microsoft Windows clients such as Windows 2000, XP, Mobile 2003, or CE. Support on the server side is also lacking in Microsoft Windows 2003 server and Cisco ACS (Access Control Server). Where EAP-TTLS shines over PEAP authentication is that the username is not revealed in clear-text, which might avoid some DoS (Denial of Service) attacks where someone can maliciously log-in repeatedly with the right username and wrong password to lock out that user's account. PEAP authentication only protects the password portion with a strong TLS tunnel, but broadcasts the username in the clear [3].

PEAPv0/EAP-MSCHAPv2 is the technical term for what people most commonly refer to as PEAP. Whenever the word PEAP is used, it almost always refers to this form of PEAP, since most people have no idea there are so many flavors of PEAP. Behind EAP-TLS, PEAPv0/EAP-MSCHAPv2 is the second most widely supported EAP standard in the world. There are client and server implementations of it in Microsoft, Cisco, Apple, Linux, and open source. PEAPv0/EAP-MSCHAPv2 is natively supported in MAC OS 10.3 and above, Windows 2000 SP4, Windows XP, Windows Mobile 2003 and above, and Windows CE 4.2. The server side implementation of PEAPv0/EAP-MSCHAPv2, called Internet Authentication Service (IAS), is also included in Windows 2003 server. PEAPv0/EAP-MSCHAPv2 enjoys universal support and is known as the PEAP standard [3].

Finally, PEAPv1/EAP-GTC was created by Cisco as an alternative to PEAPv0/EAP-MSCHAPv2. It allows the use of an inner authentication protocol other than Microsoft's MSCHAPv2. Even though Microsoft (along with RSA and Cisco) co-invented the PEAP standard, Microsoft never added support for PEAPv1 in general, which means PEAPv1/EAP-GTC has no native Windows OS support. Since Cisco has always favored the use of its own less secure proprietary LEAP and EAP-FAST protocols over PEAP and markets them as simpler certificate-less solutions, standardized PEAP is rarely promoted by Cisco. Cisco stands to gain a monopoly in the access point market if LEAP or EAP-FAST is universally adopted. As a result, most Cisco customers run the less secure and proprietary LEAP or EAP-FAST authentication protocols, because they've swallowed the Cisco Kool-

Aid. With no interest from Microsoft to support PEAPv1 and little interest from Cisco to promote PEAP in general, PEAPv1 authentication is rarely used. There is no native OS support for this EAP protocol [3].

10. SUMMARY AND CONCLUSIONS

Although Wi-Fi technologies have significantly improved their security capabilities, many of the features and abilities are available only in newer equipment for IT-managed infrastructure. Meanwhile, cellular data networks rely on a completely separate security architecture that emphasizes protection of the radio link and does not provide end-to-end encryption.

Finally, attacks have proven WEP security provided by the 802.11 standard to be insecure. The WLAN industry has responded by creating WPA and 802.11i to address these issues in the long term, though these security solutions are not available today. Most of today's security requirements can be met with 802.1x, which provides a solution that is effective and has not yet been broken. Most wireless security vendors now offer 802.1x client and server solutions that are available today and provide security that is adequate for enterprise applications.

11. REFERENCES

[1] Tom Karygiannis and Les Owens, "Wireless Network Security 802.11, Bluetooth and Handheld Devices," Computer Security Division, Information Technology Laboratory, National Institute of Standards and Technology, U.S. Department of Commerce, Gaithersburg, MD 20899-8930, 2002.

[2] "Wireless Network Security," Proxim Corporation [© 2004 Terabeam Wireless. All rights reserved. Terabeam Wireless, 8000 Lee Highway, Falls Church, VA, 22042], 2003.

[3] George Ou, "Understanding the updated WPA and WPA2 standards," ZDNet, Copyright ©2005 CNET Networks, Inc. All Rights Reserved. CNET Networks, Inc., 235 Second Street, San Francisco 94105 [1808 Colonial Village Lane Lancaster, PA 17601], June 2, 2005.

[4] John R. Vacca, *Firewalls: Jumpstart for Network and Systems Administrators*, Digital Press, 2004.

[5] John R. Vacca, *Net Privacy: A Guide to Developing and Implementing an Ironclad ebusiness Privacy Plan*, McGraw-Hill, 2001.

[6] "Wireless Security Blackpaper," Copyright © 1998-2005 Ars Technica, LLC, Ars Technica, LLC, 45B West Wilmot Street, Suite 203, Richmond Hill, ON L4B 2P3, Canada, 2005.

[7] Yan Liu, "Wireless Transport Layer Security," Department of Computer Science, 14 MacLean Hall, Iowa City, IA 52242-1419, 2005.

[8] George Ou, "Gartner Issues False Alarm On Microsoft WPA2 Warning," ZDNet, Copyright ©2005 CNET Networks, Inc. All Rights Reserved. CNET Networks, Inc., 235 Second Street, San Francisco 94105 [1808 Colonial Village Lane Lancaster, PA 17601], May 27, 2005.

Chapter 4

ENHANCED SECURITY FOR WIRELESS LANS AND WANS IN THE ENTERPRISE: HANDS ON

1. INTRODUCTION

WLANs and WWANs are a common element in many enterprises' IT infrastructures, but the security picture leaves much to be desired. This chapter will help managers get a grasp of basic WLAN and WWANs security issues.

2. WIRELESS LOCAL AREA NETWORKS (WLANS)

Wireless LANs are a big part of the IT world. There are few IT managers in corporate America who haven't yet dipped their toes into the wireless networking waters. Take a look at the numbers: Sales of WiFi clients (mobile PCs, PDAs and phones) grew 77% in 2005, according to industry analysts. WiFi hardware (access points and switches) will surpass $7 billion in annual sales in 2006, and 91% of laptop PCs now are shipped with WLAN cards [1].

So, regardless of whether you are a newcomer to wireless LANs or are looking to upgrade, one of the biggest obstacles you will have to work through is security. New wireless standards, more awareness and new technologies are rapidly addressing that problem; but, at the end of the day, it will be your job to evaluate, implement and manage the wireless network.

Therefore, with the preceding in mind, let's briefly see whether you know your standards. Even though they were covered extensively in Chapter 3, an update on WiFi is in order here--the family of wireless protocols that forms the foundation of most WLANs.

2.1 Know Your Standards

For example, the newest WiFi flavor is 802.11g. It joins 802.11b and 802.11a. 802.11g operates in the same frequency band as 802.11b (2.4 GHz), so it's interoperable with b devices, but it allows for throughput of 54Mbit/sec. as opposed to b's 11Mbit/sec (See sidebar, "WiFi Types.") [1].

Note: With overhead, interference and attenuation factors, though, actual speeds are usually at least half the claimed peak.

WiFi Types

The following WiFi types are available now or will be soon:

- 802.11a: 5 GHz, maximum speed 54Mbit/sec. Doesn't work with 802.11b.
- 802.11b: 2.4 GHz, maximum speed 11Mbit/sec.
- 802.11e: Newer standard addresses voice and video QoS.
- 802.11g: 2.4 GHz, maximum speed 54Mbit/sec. Compatible with 802.11b.
- 802.11i: Newer standard that supports 128-bit+ encryption.
- 802.11n: Still in development; will top 100Mbit/sec.
- 802.11r: Still in development; will improve voice QoS and roaming abilities [1].

Meanwhile, 802.11a operates in a higher frequency band (5 GHz) and offers 54Mbit/sec. throughput. Being in a less crowded spectrum and having more channels available, it's less prone to interference than b and g [1]. On the downside, it has a shorter range (a drawback of higher frequencies) and a higher cost per device. The common rule of thumb is to use b/g products where greater coverage is needed and cost is a big factor; and, to consider

802.11a devices in small, congested areas where higher throughput is valued more than range.

The most popular devices sold today are b/g combinations; and, if you don't mind paying a price premium, you can get devices that support all three flavors. For the future, the IEEE is working on an 802.11n standard that promises speeds of 100Mbit/sec [1]. It's expected to be ratified near the end of 2007. In the meantime, several vendors are offering nonstandard, pre-N products with higher speeds, if you don't mind going with proprietary technology.

However, the recently ratified 802.11i standard may be the ticket for shops that are concerned about security. It provides a legitimized authentication and encryption approach and is starting to appear in some vendors' switches and other wireless products.

2.2 Wired Security and WLANs

The benefits of wireless LANs are undeniable, but the risks introduced by them are increasing exponentially as more enterprises adopt the technology (se sidebar, "Risky Business."). According to industry analysts, more than 86 million WiFi devices have been deployed worldwide, and another 5 million new WLAN devices are being shipped per month. Forty-eight percent (48%) of all U.S. enterprises with more than 200 employees have deployed a WLAN, with the following sectors leading the way:

- Education: 75% have WLANs; 34% plan to deploy by 2007.
- Retail: 68% have WLANs; 34% plan to deploy by 2007.
- Manufacturing: 68% have WLANs; 35% plan to deploy by 2007.
- Government: 65% have WLANs; 39% plan to deploy by 2007 [1].

Risky Business

WLANs provide an easy open door to the wired network. Through unintentional associations and ad hoc networks, unsecured wireless networks can be sniffed, acting as a launch pad to the wired network and an enterprise's corporate backbone. If accessed, an unsecured WLAN can compromise the following:

- Financial data, leading to financial loss.
- Reputation, damaging the efforts spent building the brand.

- Proprietary information, leaking trade secrets or patents.
- Regulatory information, forgoing customer privacy or ignoring government mandates.

All of these scenarios could have legal ramifications. As wireless networks become ubiquitous extensions of wired networks, the threat of intruders becomes more pervasive. Enterprises need to look beyond local APs and think globally to secure the air across the entire enterprise [1].

In addition, enterprises with 2,000 or more employees that have deployed WLANs, 77% have deployed to fewer than 40% of all employees, but about 21% have deployed to all of their employees [1]. Some enterprises think their investments in firewalls [6] and virtual private networks will protect them from the risks of WLANs. However, they don't realize that the WLAN signal bypasses all wired-side security and opens a back door for an intruder. Simply banning WLANs isn't an option either, because most laptops are shipped with built-in wireless cards. If enterprises were to ban wireless networks, they would need to ban the use of laptops, which is an impractical solution.

The fact is, any wireless device connected to a wired network essentially broadcasts an Ethernet connection and an on-ramp to the entire enterprise network. Unless they are properly secured and monitored across the global enterprise, these self-deploying, transient wireless devices and networks are dangerous to all enterprises. Intruders and hackers will use an unsecured WLAN to break into enterprise networks and compromise the integrity of financial data, customer information or even trade secrets. No longer should the security of wireless networks be a peripheral thought.

Another issue that concerns IT managers is legacy WLAN equipment. Larger enterprises that took an early plunge into the WLAN waters are now dealing with the fallout—thousands of wireless access points and devices that use older wireless standards that are not compatible with gear built on newer standards.

2.3 Dong Your Homework

According to industry analysts, enterprises really need to do their homework before merging the security and management of their wireless and wired networks. The first step should be a risk analysis of the security and management issues for the unified wireless and the wired networks (see sidebar, "Vulnerabilities of WLANs"). A key objective should be to determine a common set of authentication, access and authorization policies for all users.

Vulnerabilities Of WLANs

To understand the risk of WLANs, one must first understand the security vulnerabilities of all WLANs. WLANs face all of the security challenges of any wired network. In addition, risks are introduced by the nature of wireless technology.

First, the medium in which a WLAN operates is the air, an uncontrollable space. In addition, wireless devices self-deploy and have the capability to connect to strangers. Due to the growth of WLAN-enabled laptops and the increasingly wireless-friendly Windows XP operating system, laptops in the default setting automatically search for an access point (AP) to connect with. Lastly, wireless devices are transient in the way they connect. If a wireless device picks up a strong signal, it may connect with the new AP even if the AP is the laptop of an intruder in the parking lot. There are many ways in which WLANs can be compromised:

- Rogue Access Points
- Soft APs
- Accidental Associations
- Malicious Associations
- Ad Hoc Networks

Rogue Access Points

A rogue WLAN has traditionally been thought of as a physical AP unsanctioned by network administrators. Today, rogue WLANs are further defined as laptops, handhelds with wireless cards, bar code scanners, printers, copiers or any WLAN device. These devices have little to no security built in, making it easy for intruders to find an entry point. Rogue APs could be maliciously placed by intruders to hack into an enterprise, or they can be innocently deployed by employees for easy wireless access.

Soft APs

While hardware APs have been the focus of security concerns, wireless-enabled laptops are easily configured to function as APs with commonly available freeware such as HostAP or software from PCTel Inc. Known as soft APs, these laptops are harder to detect than rogue APs and are quite dangerous because they appear as user stations to all wire-side network scans.

Accidental Associations

Accidental associations are created when an AP across the street or on adjacent floors of a building bleeds over into another enterprise's airspace, triggering its wireless devices to connect. Once those devices connect with the neighboring network, the neighbor has access back into the enterprise. Accidental associations between a station and a neighboring WLAN are now being recognized as a security concern.

Malicious Associations

A malicious association is when an enterprise laptop is induced to connect with a malicious device such as a soft AP or laptop. The scenario also exists when a malicious laptop connects with a sanctioned AP. Once the association has been made, a hacker can use the wireless device to attack servers and other systems on the enterprise network.

Ad Hoc Networks

Ad hoc wireless networks, or peer-to-peer networking between two computers without connection to an access point, represent another major concern for WLAN security. These ad hoc networks can be self-deploying or intentional. In addition, such networks have little security in terms of authentication and encryption. Therefore, it's easy for an intruder to connect to innocent users' computers and copy private documents or sensitive information [1].

The impetus to merge wireless and wired networks into a common security and management infrastructure began in an effort to secure WLANs from the unique threats posed to mobile users accessing the flexible and wide-open wireless hot spots. Strong authentication has long been a requirement for wireless LANs because of the threats to wireless sessions. Now, this strong authentication; as well as, management for it, is being extended to the wired LAN side. Merging the two LAN architectures gives users a cost effective means to secure and manage two vastly different infrastructures.

2.4 Virtual LANs And VoIP

Common security and management architecture is still a work in progress for managing VoIP calls over WLANs. Many WLAN vendors are presenting virtual LANs (VLANs) as a solution of choice for VoIP traffic." VLANs enable network engineers to segregate traffic so users on a given VLAN see only the traffic on that VLAN.

VLANs are a good interim solution for creating subnets to segment certain types of LAN traffic, such as VoIP, because you don't have tons and tons of devices on the network, and it's all just getting started. But at some point, the VLAN runway runs out and an enterprise will need to look at other options.

Chief among the options for a unified management and security framework is providing better access control to sensitive applications and data. Every switch and access-point vendor has a security strategy that accounts for access and identity management.

For example, Cisco's Network Admission Control program, announced in 2004, as an effort to integrate security and configuration management information from WLAN vendors is one example [1]. Cisco created an umbrella security architecture for both wired and wireless networks that can provide the level of controls needed for compliance.

Furthermore, enterprises should be careful to not get too granular with VLAN deployments. Going to one VLAN per department gets counterproductive. It will put great management burdens on the network and have diminishing returns for security.

VLANs are a good technical solution once you determine your enterprise requirements. The cost of VLANs as a security solution needs to be weighted against the benefits it provides. You shouldn't be buying technology until you've done your homework and planned the relationship of the technology to the enterprise environment.

Once you determine who's going to be allowed on the network, how are you going to provision and control them? The ideal security solution for a unified wired/wireless architecture, will include some form of user policy management that controls access and authorizations for regulatory compliance and can be extended to give granular authorization for VoIP.

Some combination of role-based and rule-based security would be the best approach. VLANS alone are a role-based approach that should be augmented by rules for the different levels of permissions allowed between wired and wireless users. According to industry analysts, identity management is forecast to soar from $950 million worldwide in 2006 to $30.4 billion by 2010 [1].

VLANs do provide effective countermeasures against rogue access points and session spoofing--two WLAN security threats. VLANs centrally control 802.1x authentication and prevent a rogue access point from masquerading as an authorized WLAN onramp. VLANs also can thwart session spoofing with encrypted tunnels secured by the client and server both authenticating themselves with hashed values.

So, despite media hype, it is possible to secure your enterprise wireless network; however, it is important to understand that this is a process and not a simple step that you take once. You must be able to evaluate your WLAN's potential weak spots actively, at any given moment and minimize threats by deploying an effective intrusion-prevention system. This includes intelligent and active defenses that work to maintain and truly secure your enterprise wireless network. That being said, there is no silver bullet that will provide a 100% guarantee but there are smarter ways to approach security. Securing your wireless network requires a multi-tiered approach that includes:

- Encryption
- Intrusion prevention technology
- User education
- Securing all laptops and mobile devices
- Staying abreast of advances in technology that provide hackers with new opportunities [2]

This part of the chapter examines the intelligent and active defenses needed to maintain and truly secure your enterprise wireless network.

3. WIRELESS LAN SECURITY GOES FROM BAD TO WORSE?

There were 21 million unauthorized attempts in 2004 to get on the wireless network at the National Association of Securities Dealers [3]. But none was successful.

The biggest wireless worry came from a NASD employee who set up an MP3 server, thus potentially opening up the network to unauthorized access for people swapping non-approved files. In fact, a great deal of concern over wireless LAN security has led to smart technology and best practices that lead to some very secure networks.

To some extent, wireless gets a bad rap because it's implemented a lot of times before being completely understood. With all the scrutiny around wireless now, people have a tendency to make wireless networks even more secure than some of the wired networks.

Wireless LAN security polices for employees concerning rogue access points, are a necessary first step. NASD locks down all wireless-enabled laptops and publishes an appropriate use policy for employees.

Monitoring the wireless network, detecting intrusions and measuring the effectiveness of security efforts, is vital to ensuring a high level of security. It is recommended that your enterprise go with mature wireless vendors and service providers that have a track record of success. At this point, CIOs must think hard about wireless vendors' ability to meet an enterprise's unique needs [3].

3.1 What You Don't Know Will Hurt You

Outside of the IT department, enterprise employees seem to think that hacker is just the 21st Century word for the Boogeyman. Unfortunately, the threat is very real. Hackers have created a strong shadow-community that is so well organized they have their own conventions. DefCon, the largest of these, attracts thousands of hackers (and assorted types who like to keep an eye on hackers) every year. The official DefCon position is that the organizers do not condone criminal activity of any kind [3]. A brief tour around some of the hacker sites reveals that plenty of such discussions take place all the time in a large and very active community.

What do hackers want? As a group, they want freedom from the niggling little rules the rest of the world plays by. They often use the pirate's skull-and crossbones as a symbol of their swashbuckling quest for booty.

Note: DefCon's logo is a smiley-face-and-crossbones.

Booty, in the case of hackers, ranges from stealing free bandwidth from your WLAN, to theft of financial and other confidential information [8], to wholesale piracy of your network. At their most benign, hackers just want to prove they can penetrate your defenses. At their most malevolent, they can wreak millions of dollars worth of damage. Who wants to take the chance that the next war driver probing for an Access Point (AP) signal is just a harmless guy who gets his kicks out of filching a bit of your bandwidth?

Wireless connectivity opened up a whole new wonderful world to hackers. A wired network can only be hacked through a physical connection, usually through the Internet. It requires a fairly high level of skill to break through an enterprise's security gateway and firewall (unless the hacker has obtained a password and MAC address from an unwitting user–more on this later). A toolbox full of hacker freeware like Net-Stumbler quickly became available, enabling hackers to break encryption, detect wireless WLAN signals, and gain access to passwords, MAC addresses and Service Set Identifiers (SSIDs)—and full access to enterprise data. War drivers, cruising around with laptops and antennae, sniff out wireless networks and connect to them. War chalkers, like the hobos of the 1930's who chalked symbols on the houses where they could obtain a free lunch, chalk on the walls of buildings to indicate the presence of unsecured WLANs [3].

Perhaps the majority of these rogue users are merely stealing bandwidth. For obvious reasons, enterprise losses due to wireless network hacking are underreported. According to industry analysts, 13% of hackers attempted malicious activities (Denial of Service (DoS) attacks, destruction of data, espionage, theft of financial information or identity theft (see sidebar, "Military Records Ripped Off By Hacker")). The eight annual Computer Crime and Security Survey undertaken by the FBI and the Computer Security Institute of San Francisco (CSI), found that out of 614 U.S. computer security professionals polled, 91% reported network security breaches within the last 12 months of the survey, while 81% admitted to financial losses due to security breaches. The 55% of respondents who quantified financial losses reported a total of $556,959,000 lost [3].

Military Records Ripped Off By Hacker

Recently, a suspected hacker tapped into a military database containing Social Security numbers and other personal information for 44,000 Air Force officers and some enlisted personnel. That figure represents about 60% of the officers in the Air Force.

The case is currently under investigation. The Social Security numbers, birth dates and other information was accessed sometime in May or June of 2005, apparently by someone with the password to the Air Force computer system.

The people affected were notified of the steps they could take to protect their identity. But, as usual, the government/military decided to wait nearly 70 days (way too long to inform or take action) to inform those affected so as not to jeopardize the investigation and search for suspects (shades of hurricane Katrina that devastated New Orleans).

In the past, the military, while protecting classified information, has had trouble protecting data about its people. They have historically done much better at protecting operational systems, than at protecting administrative systems.

Hacking has been on the rise in commercial industry. Business leaders in July of 2005 announced an education campaign to better protect sensitive client information from hackers and other thieves, after a string of high-profile data thefts and losses. In June of 2005, CardSystems Solutions Inc. disclosed that a breach of its system that processes transactions between merchants and credit card issuers exposed 50 million accounts to possible fraud.

Happily, the nineth annual survey by CSI and the FBI shows that financial damages from attacks on networks have dropped for the fifth straight year in a row. The survey reported that one of the reasons was an increasing percentage of respondents (84%) have implemented intrusion detection/prevention technology. The previous year, only 71% reported implementing intrusion detection/prevention technology [3].

3.1.1 How Hackers Attack the Wireless Network

Being an innovative group, hackers are continually coming up with new ways to penetrate network defenses. But their attacks can generally be grouped into these categories:

- MAC spoofing
- Denial of Service
- Malicious association
- Man-in-the-middle attacks [3]

3.1.1.1 MAC Spoofing

Hackers use MAC spoofing to impersonate a legitimate network user. All Network Interface Cards (NICs), like PCMCIA or PCI cards, provide a mechanism for changing their MAC addresses, issued by the manufacturer, that identify a specific device. In many cases, the MAC address is used as an authentication factor in granting the device access to the network, or to a level of system privilege to a user.

The hacker will change his device's MAC address to that of a user and thus gain access to the network. If the hacker is using the MAC address of a top executive with access privileges to sensitive material, a great deal of damage can be done.

Attackers employ several different methods to obtain authorized MAC addresses from the network. A brute force attack uses software that will try a string of random numbers until one is recognized by the network. Another way is to monitor network traffic and fish out the authorized MAC addresses.

3.1.1.2 Denial of Service Attacks

DoS attacks may be launched merely as a form of vandalism, to prevent legitimate users from accessing the network, or they may be carried out to provide cover for another type of attack. In Layer 1 DoS attacks, the hacker uses a radio transmitter to jam the network by emitting a frequency in the 2.4GHz or 5GHz spectrum [3]. As 802.11 equipment operates at a certain signal-to-noise ratio, when the ratio drops below that threshold, the equipment will not be able to communicate.

In a more common type of DoS attack, the hacker uses a laptop or PDA with a wireless NIC to issue floods of associate frames to take up all available client slots in the AP, severing the AP's association with legitimate users. Alternatively, the hacker issues floods of de-association frames, forcing clients to drop their association with the AP. Either way, if the attack is successful, the hacker now controls access to the network.

3.1.1.3 Malicious Association

The hacker configures his or her device to behave as a functioning AP. When a user's laptop or station broadcasts a probe for an AP, it encounters the hacker's device, which responds with an association. At this point, the legitimate user's computer can be mined for any and all information, including MAC address, SSID, pass codes, etc.

3.1.1.4 Man-in-the-Middle Attacks

The hacker sends a de-authorization to a network device, which drops its association to its AP and begins searching for a new AP. It finds the hacker's station (configured to look like an AP), and associates with it. Using the information garnered from the legitimate device, the hacker's device now associates with the legitimate network AP and the network passes through the rogue user's device, allowing him or her to change or steal data at will.

3.1.2 Internal Vulnerabilities

It has been estimated that internal users create up to 45% of the vulnerabilities of the enterprise WLAN. There is the occasional bad apple that steals information or money from his or her employer, but the majority of internal users are not trying to harm the enterprise; they are just ignorant of the consequences of their actions or failure to take precautions.

For instance, there's the guy who takes pride in never following the rules and going his own way. This is the person who goes out and buys a cheap AP and sets it up so he can always run at peak performance and doesn't have to share bandwidth with others in his area. The problem with these rogue APs is that they are usually unsecured. They are shipped from the factory with default security settings or with the settings turned off, because he didn't reconfigure his AP when he plugged it in. This rogue AP creates a wide-open backdoor into your WLAN. Hackers have no difficulty at all dealing with default security settings, and they waltz right in.

Another potential weakness is the neighboring WLAN that may have one or more APs that associate with your network. Anyone on that neighboring network can peek into your WLAN at will. This includes any hacker that cracks the neighbor's WLAN. Even if you've taken all precautions, your neighbor's WLAN may not be secured.

Employees may telecommute from home using wireless technology, or they may take a laptop to a public hotspot like an Internet café to check email or catch up on work. Employees need the right MAC address and password to access the enterprise WLAN. However, if the employee's home computer or laptop is using the factory-set default security or no security, that device becomes a soft AP, broadcasting the MAC address and password freely to any war driver in the street, or the sniffer sipping a latte at the next table.

The SSID is the unique name of a WLAN. War drivers scan for SSIDs in an attempt to configure their device to look like a legitimate user of that network. If the network administrator has been careless about encryption, the rogue user can access the network. In newer wireless networks, some administrators disable the automatic SSID to prevent hackers from obtaining it.

Note: This is a deterrent only to inexperienced or casual hackers.

APs are shipped from the factory with default security settings, or security settings switched off. Every AP on the network must be individually configured to assure that the network is secure.

In an effort to combat rogue users, some network administrators have set up Virtual Private Networks (VPNs). Using advanced encryption techniques and the public telephone infrastructure, VPNs create an invisible data tunnel from one WLAN locale to another (say, between an enterprise's WLAN in New York and its London site), or between one Intranet and another. VPNs are highly secure. However, if one of the WLANs that it connects is unsecured, the entire system is vulnerable to any hacker that associates with the unsecured network.

3.1.3 Legal Implications

Damage to your enterprise data, theft and espionage are just the beginning of your potential problems once hackers get into your WLAN. Government-mandated legislation provides severe penalties for breach of privacy [7] of confidential records, or loss of mandated documentation.

3.1.3.1 The Gramm-Leach-Bliley Act (GLBA)

The Gramm-Leach-Bliley Act (GLBA) was enacted to assure that financial institutions protect the privacy of their customers' financial records. The act requires that such institutions take all reasonable measures to detect, prevent and respond to attacks, intrusions or other systems failures. Failure to do so can result in fines of up to $22,000 per day or $21,000 per violation [3].

3.1.3.2 The Health Insurance Portability and Accountability Act (HIPAA)

The Health Insurance Portability and Accountability Act (HIPAA) affects virtually all health care entities in the United States. It is (in part) intended to protect the privacy of individuals' health records. Any electronic transaction that contains confidential health information of an individual is covered by HIPAA. The act requires clear control of access, policies, procedures and technology to restrict who has access to the information, and requires establishment of security mechanisms to protect data that is electronically transmitted. Violations of HIPAA could cost you $36,000 per incident [3].

3.1.3.3 The Sarbanes-Oxley Act

The Sarbanes-Oxley Act (fondly known as SOX), affects every publicly held enterprise in the U.S. Section 404 of SOX requires enterprises to document their financial reporting controls and procedures such that it is comprehensively archived and readily retrievable [3]. Any document (even email) that relates to the auditing process must be archived, along with any data relating to material events. If this material is lost or damaged, SOX makes the CEO and CFO personally responsible. Enterprise officers potentially face stiff fines and jail sentences for lack of compliance.

Finally, there are liability issues. If a hacker uses your WLAN to launch a Denial of Service attack on another enterprise's network, or to distribute child pornography, your enterprise can be held liable for the crime.

3.1.4 Encryption

You can't prevent hackers from detecting your WLAN's RF signal. The use of directional antennas can help to control the shape of the signal, but they don't completely solve the problem. You can't prevent detection of your signal, but you can do something to make the contents unreadable by unauthorized listeners. This makes encryption the first, barebones line of defense. It comes built into your WLAN software. When network traffic is encrypted, it may deter the casual or less-adept intruder who is just looking for an easy target. But relying on encryption alone is risky. The initial standard encryption for wireless networking, Wired Equivalent Privacy (WEP) was hacked within weeks of its release.

The next attempt at standard encryption, IEEE's WiFi Protected Access (WPA) bridged the gap until IEEE's recent ratification of an improved 802.11i standard, WPA2. WPA2 employs a different encryption standard, Advanced Encryption Standard (AES), which reportedly meets Federal Information Processing Standard 140-S specifications for wireless security [3]. However, enterprises with existing LANs may have to purchase new hardware to support AES.

Internet Protocol Security (IPSec) and SSL (Secured Socket Layer) are both encryptions used to create VPNs. VPNs are often used to send enterprise data over the Internet in a secure tunnel that cannot be read or reproduced by unauthorized users. Web-based SSL uses the public-and-private key encryption system from RSA, which also includes the use of a digital certificate. As any Web-enabled machine can be used to access a SSL based VPN, two-way authentication is not available. Anyone with the correct username and password can access the SSL VPN from any PC connected to the Internet.

IPSec works at Layer 3 and secures everything in the network [3]. Unlike SSL, which is typically built into the Web browser, IPSec requires a client installation. IPSec facilitates two-way authentication using DES, a powerful block cipher that is highly resistant to attack because it requires an impractical expenditure of time and resources. Of course, if the VPN is unsecured at either the network where it begins or the network where it ends, a hacker can waltz right through.

It is the nature of hackers to work long and hard to break encryptions, so it is only a matter of time before someone successfully hacks AES and its successors. Encryption, while necessary, should be viewed as only beginning the task of securing the WLAN.

3.1.5 Intrusion Prevention

An effective intrusion prevention system monitors your WLAN to allow you to see what's happening in your airspace at all times; assures you that security policies are being followed; sounds the alarm when an intrusion or policy violation is detected; and, isolates and cuts off any unauthorized device or user. Wireless LAN infrastructure devices come with some built-in monitoring capabilities, but a specialized overlay monitoring product will provide a richer set of capabilities.

Being able to detect rogue and/or neighboring APs associated with your WLAN is essential. Once detected, it is critical to be able to physically locate these APs and take them out (or go have a polite discussion with the administrator of the neighboring WLAN). Locating a rogue AP requires triangulation, made possible by physically locating intelligent sensors around the area covered by the WLAN. Intelligent sensors detect the rogue AP or user, sound the alarm to the network administrator, and isolate the AP until it can be physically removed.

Intelligent remote sensors can also monitor users/devices and determine whether a user is inside or outside your network perimeter. If a user is outside the perimeter (whether the rogue user is a hacker or a neighbor who has unwittingly associated with your WLAN), the sensor will generate an alarm and isolate the rogue from the network.

One of the most useful aspects of a robust intrusion prevention system is the ability to set security protocols, and modulate the system's response to protocol violation. For instance, the system response to a low-level violation such as multipath detection, which could be merely a malfunctioning AP, might be just an email notification to the administrator, or a log notation. But, a high level violation such as a user detected outside the network perimeter might result in a phone call alert.

It's essential to be able to manage WLAN security from a centralized console, where all information is quickly available to the administrator. The system should provide detailed information, such as activity at any single sensor, specifics on any alarm, and a view of network traffic at any given time. Troubleshooting tools, detailed logs, and the ability to measure network performance against established standards are all necessary to good network security. It is especially important to get reports in highly granular detail, customized to the needs of the network administrator. These reports are key to studying network performance and usage over time, pinpointing and eliminating potential trouble spots *before* they become problems [3].

And, the network administrator needs flexibility from an intrusion prevention system. For instance, access privileges granted to a C-level executive need to travel with that executive when he or she works from home or leaves on a business trip. The administrator must be able to configure related alarms into a coordinated policy instead of being limited to a fixed hierarchy. When the network is expanded, the intrusion prevention system must be scaleable to grow with it.

Intrusion prevention is enhanced if the system has mobile detection ability. Wireless LANs are constantly in a state of flux by their very nature. Users move around from location to location. New devices can introduce RF signals that didn't exist the day before. Users can create dead spots in the network without realizing it. For all these reasons, it is sometimes necessary to be able to make an on-the spot investigation with a handheld extension of the intrusion prevention system that can perform real time readings and analysis.

Any wireless network, no matter how well protected, can be threatened by rogue devices. The key to preventing intrusion by rogues is the ability to detect, disable, and document every rogue before it can cause any damage.

3.1.6 User Education

There is no technology so sophisticated that human beings can't mess it up. This is particularly true of WLAN security, because users outside the IT department often don't realize that their actions may endanger the enterprise. A sales manager may know a great deal about negotiating deals, but may be unaware that his or her laptop is not securely configured, even though the manager has used the proper passwords and procedures to gain remote access to the WLAN.

So WLAN user education is an important leg on the security stool. Users need to know about security protocols and how to follow them. More critically, users need to know why these protocols exist. Some people believe that rules are made to be broken. Understanding that the use of a rogue AP may enable a hacker to access sensitive enterprise data (or steal a person's bank account number and identity) goes a long way toward gaining user compliance.

Though WLAN security protocol may be set by IT, education and enforcement is best implemented by Human Resources. HR is equipped to conduct training on the subject—and, to enforce the consequences for infringement. Every enterprise with WLAN technology must develop a serious user training program to assure that users understand that they each have a personal responsibility to keep the enterprise safe from intrusion. Part of that educational process is setting protocol for securing laptops and other wireless devices and banning unauthorized APs.

3.1.7 Keeping Up With The Hackers

Hackers may be at best mischievous individualists, and at worst criminals; but, they are one of the brightest and most creative groups of humans on the planet. As developers of legitimate technologies work hard to stay ahead of the hackers, the hackers are working harder to figure out how to break the new code.

If you are responsible for WLAN security in your enterprise, it behooves you to spend some time studying the enemy. The easiest way to do this is making routine visits to hacker web sites and to listen in on the chat. You might even attend DefCon or one of the other hacker confabs to get a lead on what hackers are doing.

Above all, the WLAN should be monitored 24x7. You must be able to view any AP's activities in detail, monitor individual devices associated with your network, and analyze network productivity. You should be able to get detailed information on any anomaly. A great deal rides on your ability to see problems and cut them off before damage can be done [2].

So, with the preceding in mind, Wireless Local Area Networks based on the IEEE 802.11 standard have proliferated in enterprises, homes, and public places. IEEE 802.11 a, b, and g are now considered as a de facto standard for WLANs. Embedding of wireless technology into laptops, personal digital assistants (PDAs), and phones has significantly increased the user base of WLAN devices. It is anticipated that majority of the laptops and notebook computers will have embedded WLAN capability. In addition, the mobile user base is anticipated to grow manifold in the coming years with many of them being WLAN users.'

While WLANs are known for convenience, flexibility, productivity, and low cost, enterprise networks are vulnerable due to security threats posed by the presence of these devices, irrespective of whether the enterprise has an officially deployed WLAN or not. Conventional firewalls, VPNs, and security mechanisms in the 802.11 standard are unable to alleviate these threats. This next part of the chapter describes these new security threats from WLANs and desirable features of a new type of security system-a WiFi Firewall-to prevent them.

3.2 Not Enough Security To Protect Your Wireless Network

Mis-configured APs can pose a variety of security threats. For example, intruders can eavesdrop on the wireless communication between a mis-configured authorized AP and an authorized client in the enterprise

WLAN. The intruder can read this communication if encryption is weak. If the mis-configured AP does not perform proper authentication of clients, the unauthorized user will be able to connect to the enterprise network through this AP. A recent hacking fad, called war-driving, involves using freely available tools on the Internet to discover and publicize APs whose signals spill in public places. Various reconnaissance tools such as Netstumbler, Wellenreiter, and others are freely available on the Internet.

3.2.1 Soft APs

With client cards and embedded WLAN radios in PDAs and laptops, a threat called soft AP is on the rise. A soft AP functions as an AP under software control and can be lunched inadvertently or through a virus program; unauthorized users can now connect to the enterprise network through soft APs using their radio spillage.

3.2.2 MAC Spoofing

APs in a WLAN, transmit beacons (or probe responses) to advertise their presence in the air. The beacons of an AP contain information about its MAC address, which is its identity, and SSID, which is the identity of the network it supports. Wireless clients listen to beacons from different APs in the vicinity. Clients typically connect to an AP that advertises the desired SSID and transmits a strong beacon signal. A number of WLAN AP models available in the market allow their MAC addresses and SSIDs to be user defined. APs as well as many software tools are also available that enable setting of MAC addresses and SSIDs of AP devices to virtually any user defined values.

In MAC Spoofing, the attacker programs the AP to advertise exactly the same identity information as that of the victim AP. A MAC spoofing AP can also launch disruptive attacks such as packet dropping and packet corruption and modification. A MAC Spoofing AP can even connect to the wired enterprise network as a Rogue AP and evade detection by conventional site survey tools.

A MAC Spoofing AP can lure authorized wireless clients in the enterprise WLAN into establishing a connection and providing confidential information to it. It can insert itself as a *man-in-the-middle* (described in more detail in the next paragraph) of an authorized communication [4].

3.2.3 Honeypot APs

Multiple wireless networks can coexist in the same space enabling users to connect to any available network, whether it is one's own network or some other network in the vicinity with overlapping radio coverage. This feature can be exploited by intruders who can set up an unauthorized wireless network with overlapping radio coverage with the enterprise wireless network. It requires powering on an AP in the vicinity (street or parking lot) of the enterprise wireless network. These APs can attract authorized enterprise clients into connecting to them by transmitting a stronger beacon signal and MAC spoofing. Such APs are called Honeypot APs or Evil Twins. An authorized user unwittingly connecting to a Honeypot AP creates security vulnerability by inadvertently providing sensitive information such as its identity to the Honeypot AP. The intruder can also act as a man-in-the-middle of a communication of an authorized client using Honeypot APs.

Authorized wireless clients in the enterprise WLAN can also accidentally connect to non-malicious neighboring APs (called client mis-associations). Nonetheless, this creates security vulnerability as the wireless clients may inadvertently provide confidential information to such APs. This can happen due to mis-configuration on clients and/or on neighboring APs.

3.2.4 Denial of Service

WLANs are being increasingly entrusted with carrying mission-critical applications such as database access, VoIP, e-mail and Internet access. These applications can be disrupted with a DOS attack causing network downtime, user frustration, and loss of productivity.

As 802.11 WLAN transmissions are a shared medium, they are easily susceptible to DOS attacks. Additionally, soft spots in the 802.11 MAC protocol can be easily exploited to launch DOS attacks. To name a few, DOS attacks such as authentication, association, de-authentication or disassociation flood, NAV attack, CTS flood, and EAP and EAPOL message floods are easy to launch and have the potential of bringing down the entire enterprise WLAN [4]. Unfortunately, a variety of DOS tools are freely available on the Internet including AirJack, FataJack, Void11 and Fake AP--just to name a few.

3.2.5 Ad Hoc Networks

The 802.11 WLAN standard has provisions for establishing peer-to-peer wireless connections between wireless clients. The wireless clients can therefore form an ad hoc network among themselves using this provision. However, the ad hoc networks can create security vulnerability. For example, an intruder on the street, parking lot, or neighboring premises can form a peer-to-peer ad hoc wireless connection with an authorized laptop in the enterprise premises. The intruder can then launch security attacks on the laptop using this wireless connection. For example, if the laptop has a setting to share certain resources (files, directories, etc.) with other authorized laptops in the enterprise, the intruder can also get access to these resources over the ad hoc wireless connection.

The seriousness of threats to enterprise network security from Rogue APs, Mis-configured APs, Soft APs and ad hoc networks cannot be underestimated. Unauthorized devices connecting to the enterprise network through such APs can engage in data theft, data rerouting, data corruption, impersonation, denial of service, virus injection, and many other types of attacks on the computer systems in the (wired) enterprise network. This vulnerability exists in enterprises that have official WLAN deployments, as well as those which have banned wireless on their premises [4].

3.2.6 Protecting Enterprise Networks from Wireless LAN Threats

The emergence of WLANs has created a new breed of security threats to enterprise networks, which cannot be mitigated by traditional firewall technologies and VPNs. The firewall is similar to a lock on the front door to block unauthorized wired traffic from reaching the internal trusted enterprise network. A VPN protects enterprise data traveling beyond the boundaries of the enterprise network into the public Internet. However, these technologies as well as the encryption and authentication mechanisms such as WEP, WPA, 802.ix, and 802.11i cannot plug the security holes created by Rogue APs, Mis-configured APs, and Soft APs. Conventional enterprise network security systems are not designed to detect and prevent threats from MAC spoofing, Honeypots, DOS, and ad hoc wireless networks. A new security solution called WiFi Firewall (as shown in Fig. 4-1 [4]) is therefore needed that:

- Monitors the wireless activity within and in the vicinity of the enterprise premises.
- Classifies WLAN transmissions into harmful and harmless.

- Prevents transmissions that pose a security threat to the enterprise network.
- Locates participating devices for physical remediation [4].

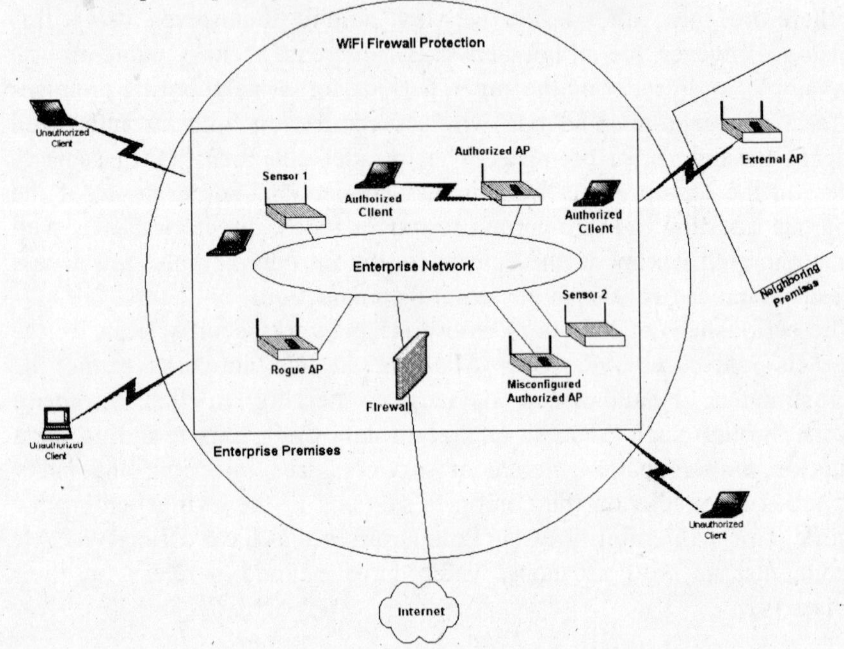

Figure 4-1. WiFi Firewall Protection.

The WiFi Firewall comprises of wireless sensor devices for wireless monitoring that are placed spatially to cover the enterprise premises. These sensors keep a constant vigil on the enterprise air and create an RF shield to alleviate security threats from WLANs.

3.2.7 A New Class of Security Threats to Enterprise Networks

The prevailing model of enterprise network security is rooted in the axiom that being physically inside is safe and outside is unsafe. Connecting to a network point within the enterprise premises is generally considered safe and is subject to weaker security controls. On the other hand, tight security controls are enforced at the network traffic entry and exit points using firewalls and Virtual Private Networks (VPNs).

A WLAN breaks the barrier provided by the building perimeter as the physical security envelope for a wired network. This is because invisible radio signals used by the WLAN cannot be confined within the physical perimeter of a building, and usually cut through walls and windows. This creates a backdoor for unauthorized devices to connect to the enterprise network. Some specific security threats from WLANs are described in Fig. 4-2 [4].

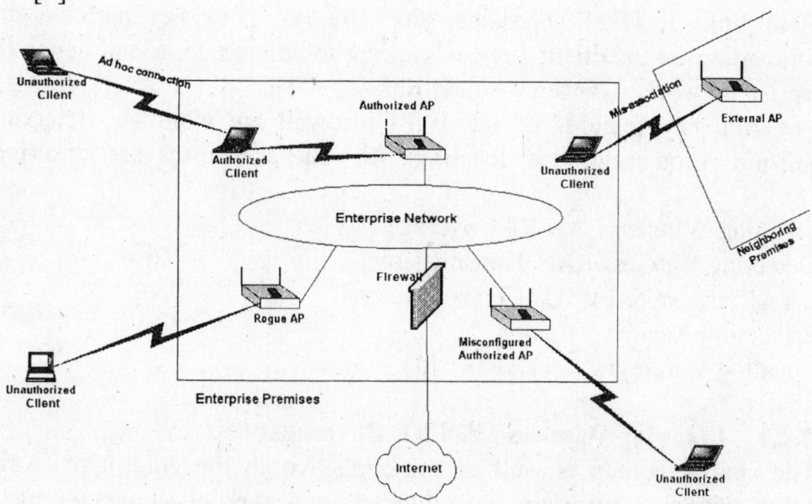

Figure 4-2. New security threats from WLANs.

3.2.7.1 Rogue APs

WLAN Access Points (APs) are inexpensive, easy to install, and small enough to be carried by a person. Unauthorized WLAN APs can be connected to an enterprise network unwittingly or with malicious intention without the knowledge of the IT administration. All it takes is to carry the device inside the enterprise premises, and connect it to an Ethernet port on the network.

Since Rogue APs are typically deployed by employees looking for quick wireless access, they are usually installed without any WLAN security controls (such as Access Control Lists, Wired Equivalent Protocol, 802.1x, 802.11i etc). As they can be connected to virtually any Ethernet port on the network, they can bypass existing WLAN security control points such as WiFi switches and firewalls. The radio coverage of Rogue APs cannot be confined within the building perimeter of the enterprise. Unauthorized users can now connect to the enterprise network through these Rogue APs using their radio spillage. The invisibility of wireless medium makes it difficult to prevent this undesirable activity.

3.2.7.2 Mis-configured APs

APs support a variety of security features and configuration settings. In many cases, the iT administration may have left the authorized APs to their factory default setting or may not set the configuration properly. This may result in no encryption or a weak form of encryption such as WEP on the wireless link. It is also possible that the AP does not perform any authentication on the client devices seeking to connect to it, and hence the enterprise network, over the wireless link.

The five key features of the WiFi Firewall are planning, detecting, classifying, protecting, and locating [4]. These features are described below:

1. Planning Wireless LAN RF Coverage
2. Detecting Wireless LAN Transmissions
3. Classifying Wireless LAN Transmissions
4. Protecting Against Intrusion
5. Locating Wireless LAN Devices [4]

3.2.7.2.1 Planning Wireless LAN RF Coverage

The spatial layout as well as materials within the enterprise (walls, columns, windows, furniture, etc.) interacts with the radio coverage of the sensor in a complex way creating a gap between rule-of-thumb for placing APs and reality. A systematic, scientific, and scalable RF planning process is therefore required for determining the right placement of access points and wireless sensors. This process must account for the spatial layout of the premises and indoor RF signal propagation characteristics. This ensures that there are no holes in the WiFi Firewall coverage through which undesirable wireless activity can go unabated [4].

3.2.7.2.2 Detecting Wireless LAN Transmissions

The WiFi Firewall needs to scan radio channels and capture any wireless activity detected on these channels using spatially distributed sensors. It is necessary to scan all the channels in the 2.4 GHz (b, b/g) and 5 GHz (a) band [4]. It needs to analyze, aggregate, and correlate information reported by different sensors.

3.2.7.2.3 Classifying Wireless LAN Transmissions

With increasing penetration of WLANs, there is a need to accurately and automatically sort harmful activity from the harmless activity in the shared wireless medium. For example, in enterprises with no official WLAN deployment, either any wireless activity detected in the air is due to a Rogue AP or it could be emanating from an external (neighbor's) wireless LAN. The WiFi Firewall must categorize it accordingly. In enterprises with official WLAN infrastructure, the WiFi Firewall must be able to differentiate between authorized, rogue, and external wireless activities.

This type of classification minimizes annoying false alarms and volumes of irrelevant alerts from the security standpoint, both of which make the security system unusable. The automatic classification also facilitates automatic intrusion prevention as described in the following paragraph [4].

3.2.7.2.4 Protecting Against Intrusion

The WiFi Firewall must automatically and instantaneously block harmful wireless activity detected by its wireless sensors until remediation. For example, the WiFi Firewall must block any client from connecting to a Rogue AP or a MAC spoofing AP, prohibit formation of ad-hoc networks, and mitigate any type of DOS attack. Further, it must block harmful wireless activity until physical remediation has taken place.

Prevention of harmful WLAN transmission (see Fig. 4-3 [4]) must be carried out with surgical precision without disturbing legitimate WLAN activities. It should not bring down the entire wireless network like some brute force methods such as radio jamming would do. The prevention should also be reliable to minimize false alarms and block every single unauthorized activity [4].

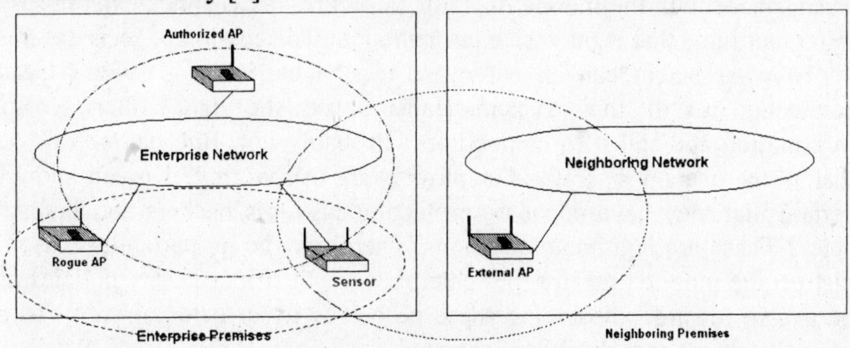

Figure 4-3. Classification of Wireless LAN Transmission.

3.2.7.2.5 Locating Wireless LAN Devices

Physical remediation (disconnecting and powering off the WLAN device(s) taking part in harmful activity) requires knowledge of the physical location of these devices. The WiFi Firewall must provide the location co-ordinates of such a device inside and around the perimeter of the enterprise premises. There should be no need for any specialized client side software or hardware [4].

Now, wide area wireless networks can be an enormous benefit to enterprises, because they have the potential to extend the reach of an enterprise application to a staggering proportion of the earth's surface.

However this expanded range also increases the vulnerability of the enterprise's devices, applications and data. In order to ensure their viability, you must validate the security of this new infrastructure.

Today's legacy and emerging Wireless Wide-Area Networks, such as GSM, GPRS,EDGE, UMTS and cdma2000, already include security provisions that are enforced by the mobile terminals and the base stations. However, there are still shortcomings in the security model that can only be addressed with an end-to-end approach. This final part of the chapter will explore the options an enterprise has at its disposal for securing remote connectivity over wireless WANs.

4. WIRELESS WIDE AREA NETWORKS (WWANS)

In addition to the challenge of securing data as it rides the airwaves, you also need to consider the vulnerabilities of the network between the base station and the application server. One approach could be to come to an arrangement with the mobile operator to secure this channel. This raises an important issue that is pervasive throughout any discussion of security: trust.

To what extent can an enterprise trust a carrier to provide a secure connection? Will they systematically abuse the data, filter sensitive information and sell it to competitors? Probably not. But, can you be sure that all the telecom operator's employees are trustworthy? Do you know for certain that they have adequate protection vis-à-vis hackers and industrial spies? These are legitimate questions. There may be no particular reason to distrust the mobile operator, but enterprises with strict security policies may be averse toward outsourcing these processes to an external entity whose operation is not completely transparent.

Consequently, you need to find connectivity options that do not rely on the security provisions of the mobile operator. A secure air interface is a nice benefit, but only a small piece of the puzzle of remote connectivity.

4.1 Overview of WWAN connectivity

Fundamentally, there are two different means a mobile network may offer to transfer data. It can provide a packet-data network or else it can use circuit-switched connections. A packet data network is simpler. CDPD, Mobitex and GPRS would all be examples of packet data networks [5]. In these cases, the mobile device has an IP address and it transfers data through the mobile network, which is connected to the Internet. No special configuration is typically required at the mobile end. Its data access is transparent. If the IP address given to the device is fixed, then a minimal amount of authentication is also implicit in any packets originating from it.

Data communication over primarily voice networks, such as GSM, IS-136 and IS-95, is not quite as straightforward [5]. Typically a Point-to-Point Protocol (PPP) connection must first be made from the device to a dial-in server. The dial-in server will assign an IP address and relay all the traffic between the device and any application servers. This implies some configuration at the mobile end. The user must specify a phone number and then authenticate to the dial-in server using an authentication protocol such as PAP, CHAP or MS-CHAP [5]. So, the dial-in server knows who the user is, but the application server does not. It cannot determine the phone number easily and the IP address is meaningless. If necessary, it would then re-authenticate the user, which means additional work for the user.

It would be possible to bypass the first authentication by storing the mobile phone number on the dial-in server and then comparing the caller-id of incoming calls. However, this would provide unlimited access to the enterprise network when a device was lost or stolen. Solutions to address this dilemma must combine security with ease of use; for example, by using biometric authentication.

They must also ensure (for example by encrypting the file system), that unauthenticated users cannot access any information on the device. It is then feasible to cache some of the network credentials on the device. Nonetheless, some authentication to the network should always be based on an action or token that is separate from the device.

4.2 Circuit-Switched Data

The practical steps required to set up a secure WWAN connection depend on which of the two categories of network that are being considered. Fig. 4-4 [5] illustrates a typical circuit switched connection, whereby the user connects via the mobile operator's network into the Public Switched Telephone Network. The path is relayed on to the private dial-up server, based on the phone number that was dialed.

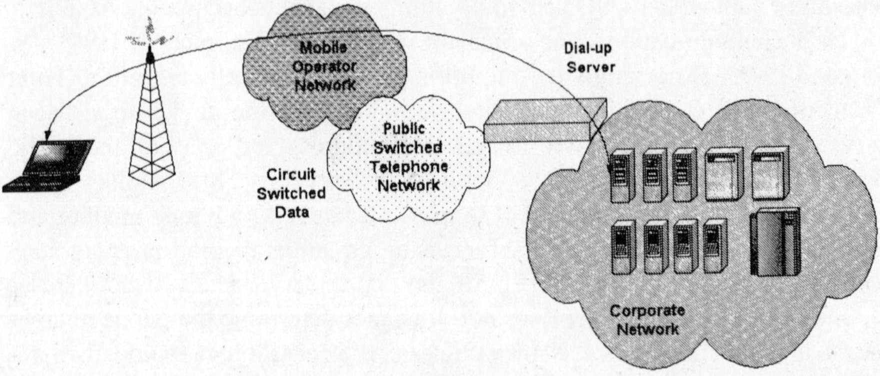

Figure 4-4. Dial-up to private network.

In the preceding case, you need to use a protocol that will encapsulate all the traffic between the mobile client and the enterprise dial-up server. In practice, this means you need to use Serial Line Internet Protocol (SLIP) or PPP. Both are communication protocols for serial data transmission between two devices. They allow a computer connected to a server via a serial line (with a modem) to gain access to the Internet.

SLIP is not an Internet standard, but is described in RFC1055 [5]. It is a simple framing scheme for putting IP packets on a serial line. SLIP's main advantage is its simplicity and consequently its implementation. Its drawback is ease of use. With SLIP, you have to know your own fixed IP address and that of the remote system you are dialing into. If IP addresses are dynamically assigned by your service provider, your SLIP software needs to be able to pick up the IP assignments automatically or else you have to setup them up manually. You may also need to configure such details as a maximum transmission unit (MTU), maximum receive unit (MRU), and the use of VJ compression headers, etc.

PPP is an Internet standard described in RFC1171 [5]. Unlike SLIP (which can only transport TCP/IP traffic), PPP is a multi-protocol transport mechanism that can accommodate various network protocols, like IP, IPX and Appletalk simultaneously. It does essentially the same thing SLIP does, but with a more complete set of features like error detection in every frame, IP address negotiation, automatic compression, login and connection configuration.

Most importantly, PPP is now supported and favored on almost all dial-up products through the industry. It can offer a relatively secure connection without any Internet exposure and supports multiple encryption algorithms such as RC4. PPP permits several authentication protocols including PAP,

CHAP, and MS -CHAP. These are not confined to PPP, or even to CSD connections [5].

4.3 Authentication Protocols

Authentication is at the foundation of any security scheme particularly when it involves remote access. The primary objective for an enterprise is to ensure that only legitimate users may access the resources and data on its network.

4.3.1 PAP: Password Authentication Protocol (RFC 1334)

PAP is the least sophisticated authentication protocol. It uses a simple, clear text authentication scheme. The authenticator requests the user's name and password, and PAP returns them in clear text (unencrypted). This authentication scheme is not secure, because a third party could capture the user's name and password and use it to get subsequent access to the authenticator and all of the resources provided by the authenticator. PAP provides no protection against replay attacks or remote client impersonation once the user's password is compromised.

Because PAP uses clear-text passwords, you would use PAP in only two circumstances: when you're dialing in to a Point-to-Point Protocol (PPP) server that does not support encrypted authentication and when you're dialing into a Serial Line IP (SLIP) server. Simply stated, you use PAP only when the client and server cannot negotiate a more secure form of authentication.

Tip: SLIP servers understand only clear-text passwords.

4.3.2 SPAP: Shiva Password Authentication Protocol

SPAP is Shiva's proprietary version of PAP. SPAP is more secure than PAP because SPAP uses a two-way (reversible) authentication method that encrypts passwords. Thus, SPAP offers a medium level of security for remote access. However, the proprietary nature of the protocol has hindered widespread adoption. It is mainly of historical interest for legacy implementations.

4.3.3 CHAP: Challenge-Handshake Authentication Protocol (RFC 1994)

CHAP provides a higher level of security for remote access than PAP. CHAP is an encrypted authentication mechanism that avoids transmission of the actual password on the connection. The authenticator sends a challenge to the remote client, consisting of a session ID and an arbitrary challenge string. The client then uses the MD5 one-way hashing algorithm to return the user's name and an encryption of the challenge, session ID, and the client's password.

CHAP is an improvement over PAP because the password is not sent over the link in the clear. Instead, the password is used to create an encrypted hash from the original challenge. The server knows the client's clear text password, and can replicate the operation and subsequently compare the result to the password sent in the client's response. CHAP protects against replay attacks by using an arbitrary challenge string for each authentication attempt. Furthermore, it protects against remote client impersonation by unpredictably sending repeated challenges to the remote client throughout the duration of the connection.

CHAP uses a three-way handshake to provide encrypted authentication. The authenticator first sends out a challenge to the client. The client responds with a one-way encrypted value. The authenticator checks to see whether the value matches. If it does, the authenticator acknowledges the authentication. CHAP then periodically verifies the client's identity. It changes the challenge value every time it sends out a message, which protects against playback attacks (a hacker records the exchange and plays back the message to obtain fraudulent access).

4.3.4 MS-CHAP: Microsoft Challenge-Handshake Authentication Protocol (RFC 2433)

MS-CHAP is the Microsoft version of CHAP, using Microsoft's version of RSA Data Security's MD10 standard. MS-CHAP uses a one-way hash function to produce a message-digest algorithm. A hash function takes a variable-size input and returns a fixed-size 128 -bit string [5]. This type of algorithm produces a secure checksum for each message, making it almost impossible to change the message if you don't know the checksum. MS-CHAP differs from the standard CHAP as follows:

- MS-CHAP does not require the authenticator to store a clear or reversibly encrypted password.

- MS-CHAP provides an authenticator-controlled authentication retry mechanism.
- MS-CHAP provides an authenticator-controlled change password mechanism [5].

Microsoft CHAP is an encrypted authentication mechanism very similar to CHAP. As in CHAP, the authenticator sends a challenge, which consists of a session ID and an arbitrary challenge string, to the remote client. The remote client must return the user name and an MD10 hash of the challenge string, the session ID, and the MD10-hashed password. This design, which manipulates a hash of the MD10 hash of the password, provides an additional level of security, because it allows the server to store hashed passwords instead of clear-text passwords. Microsoft CHAP also provides additional error codes, including a password expired code, and additional encrypted client-server messages that permit users to change their passwords. In Microsoft's implementation of Microsoft CHAP, both the Client and authenticator independently generate an initial key for subsequent data encryption by Microsoft's Point to Point Encryption (MPPE) [5].

4.3.5 MS-CHAP: Microsoft Challenge-Handshake Authentication Protocol Version (RFC 2759)

Microsoft CHAP offers improved security. These improvements include a server authentication scheme and a single change password packet. The most significant changes are:

- The weaker LAN Manager hash is no longer sent along with the stronger Windows NT hash. This thwarts automatic password crackers like L0phtcrack, which first breaks the weaker LAN Manager hash and then uses the information to break the stronger NT hash.
- An authentication scheme for the server has been introduced. This prevents malicious servers from impersonating legitimate servers.
- The change password packets from MS-CHAP have been replaced by a single change password packet. This addresses the active attack of spoofing MS-CHAP failure packets.
- MPPE uses unique keys in each direction. This is to prevent XORing the text stream in each direction to remove the effects of the encryption [5].

4.3.6 EAP: Extensible Authentication Protocol (RFC 2284)

The Extensible Authentication Protocol (EAP) is a PPP extension that provides support for additional authentication methods within PPP. Transport Level Security (TLS) provides for mutual authentication, integrity-protected negotiation, and key exchange between two endpoints.

EAP does not select a specific authentication mechanism at Link Control Phase, but rather postpones this until the Authentication Phase. This allows the authenticator to request more information before determining the specific authentication mechanism. This also permits the use of a back-end server which actually implements the various mechanisms while the PPP authenticator merely passes through the authentication exchange. The authenticator does not necessarily have to understand each request type and may be able to simply act as a pass-through agent for a back -end (RADIUS) server on another host. The device only needs to look for the success/failure code to terminate the authentication phase [5].

4.4 Packet Data Networks

Packet Data Networks are substantially different from Circuit Data Connections, since each packet is routed separately and therefore should be authenticated and encrypted individually. Since this would complicate the life of the network layer in an excruciating way, it makes sense to consolidate the authentication and encryption into a virtual connection using a tunnel, also called a virtual private network.

There are at least two ways in which you could approach VPNs in a mobile scenario. The first would be to create a dedicated tunnel between the mobile operator and the enterprise network as shown in Fig. 4-5 [5]. All traffic from the client is intercepted by the VPN on the mobile operator's network and tunneled over the public Internet into the VPN server on the enterprise network and likewise in reverse enterprise data passes through the tunnel on its way to the client.

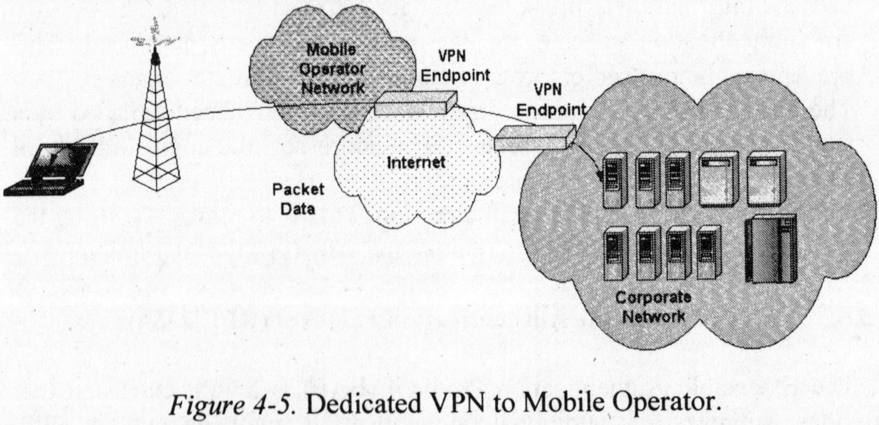

Figure 4-5. Dedicated VPN to Mobile Operator.

The advantage of this approach is that can be completely transparent to the user. No special software is necessary on the client or the enterprise servers. However, there are also some distinct disadvantages. This type of connection typically costs a lot to set up and requires major hardware investments. It is not a viable solution except for larger enterprises and even then it may not necessarily be the most cost-effective. It is also a restrictive approach since it binds the enterprise to a single mobile operator and implies that all users must have subscriptions with that carrier. In international scenarios, this can become complex to implement consistently. There are no worldwide mobile operators, so it would require a number of dedicated connections. And, roaming users would be forced to use inefficient routing topologies that looped through their home country.

Additionally, the enterprise must trust the mobile operator. If all the traffic is routed through the mobile operator in the clear, then this presents a security risk for the enterprise. They may choose to accept the risk, but at the very least they should carefully consider the implications and alternatives.

4.5 End-to-end Virtual Private Network

In addition to the dedicated VPN, there is also the possibility of a client VPN which extends to the perimeter of the enterprise network as shown in Fig. 4-6 [5]. It can encapsulate all the data over the mobile operator's network in addition to the public Internet and is therefore more secure than the dedicated VPN . It is also operator agnostic, since all the encryption and authentication take place on the device and the enterprise VPN server. Since this approach can use any available VPN technology, it can be implemented more simply and cost-effective, than the dedicated VPN described earlier.

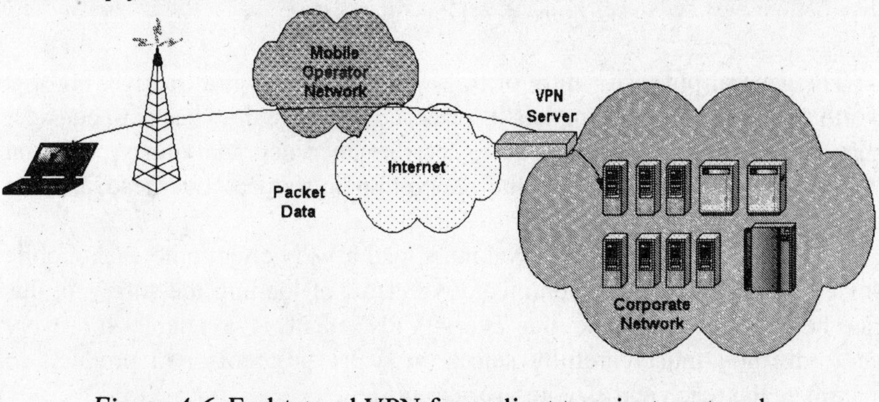

Figure 4-6. End-to-end VPN from client to private network.

A further advantage is that this can be used as a general-purpose remote access solution for all wireless and wired networks. Not only does it span all mobile operators running packet data networks, but it can also be used for circuit-switched data by allowing users to dial-into any Internet Service Provider and run a VPN over that connection.

Whether dialing up over GSM or a simple analog/ISDN line, users can connect to any ISP as shown in Fig. 4-7 [5]. From there, the user will have IP connectivity and can connect to any VPN server. To the VPN server, the users look the same whether they are connecting via a broadband cable-modem/DSL-modem, a GPRS packet data network, or any ISP through GSM or a phone line. It is just necessary to have Internet connectivity.

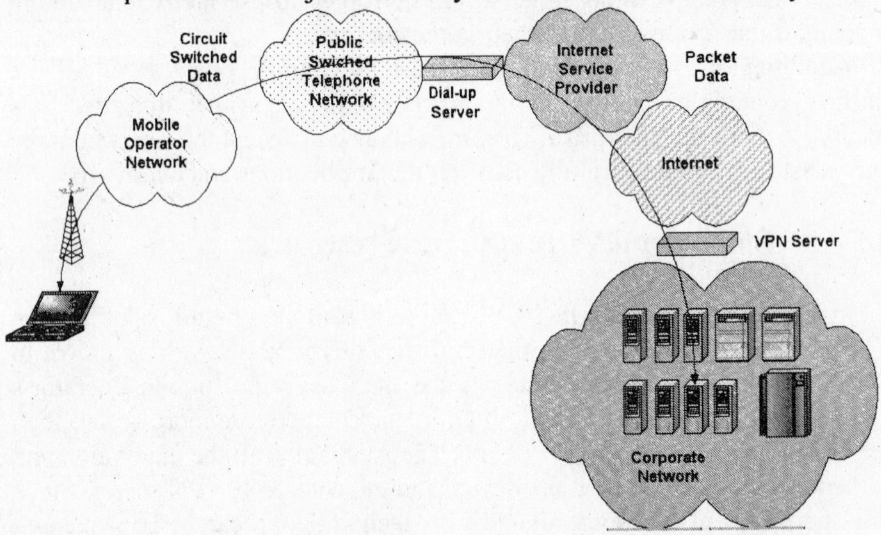

Figure 4-7. Dial-up via ISP.

The general-purpose nature of this approach makes it attractive, but it is worth noting that everything has its price, so there are drawbacks to consider here too. First of all, you will suffer performance degradation (typically on the order of 10%-30%) since the encryption introduces overhead into the transmission.

The second obstacle is that you must load a VPN client onto each mobile device. In addition to the administrative effort of loading the software, this can be a challenge since not every VPN client is available for every platform. You must carefully select the VPN protocols and products to maximize the reach across your user base [5].

4.6 VPN Protocols and Alternatives

Finally, there are several VPN protocols you can consider for mobile access. The industry protocols, such as PPTP, L2TP and IPsec, are well documented, so there is no need to repeat the discussion here. There are, however, some problems with standard VPNs that you should be aware of in a mobile environment, including their susceptibility to latency and incompatibility with network address translation (NAT).

5. SUMMARY AND CONCLUSIONS

Wireless LANs are being rapidly adopted due to the convenience and flexibility they provide. However, WLANs create a new set of security threats to enterprise networks such as Rogue APs, Mis-configured APs, Soft APs, MAC Spoofing, Honeypot APs, DOS, and Ad hoc Networks.

Neither traditional firewalls and VPNs nor IEEE 802.11 security standards such as WEP, WPA, 802.1x, and 802.11i can protect enterprise networks against over-the-air attacks from WLANs. So, a new and comprehensive security solution in the form of a WiFi Firewall is required to alleviate these new security threats. The WiFi Firewall must be able to provide RF planning, detection of RF activity, accurate classification of WLAN networks, automatic and reliable prevention from harmful WLAN transmissions, and precision location tracking of WLAN devices.

Now, in many ways, Wireless Wide Area Networks solutions can be integrated into a general purpose remote access solution for the enterprise. Whether they are circuit-switched dialup solutions or packet-data Internet connections, wired and wireless remote access can look the same to the enterprise perimeter.

Finally, they can use the same protocols and remote access products. In fact there is a compelling case for creating unified and simplified approach to remote access. However, there is need to ensure that wireless networks have unique requirements and therefore a traditional and standard solution may not necessarily be optimal.

6. REFERENCES

[1] "WLAN Security," Executive [Snapshot], Copyright © 2005 Computerworld Inc. All rights reserved. Computerworld Inc. One Speen Street, Framingham, MA 01701, 2005.

[2] Kathy Keenan, "What Hackers Don't Want You to Know About Your WLAN," Copyright © 2002-2005, AirMagnet, Inc. AirMagnet, Inc., 1325 Chesapeake Terrace, Sunnyvale, CA 94089, 2004.

[3] Tom Kaneshige, "Wireless Security: Why Such A Bad Rap?," SearchNetworking.com, TechTarget, 117 Kendrick Street, Needham, MA 02494, (©Copyright 2005, TechTarget. All rights reserved), July 28, 2005.

[4] "Wireless LAN Security--Why Your Firewall, VPN, and IEEE 802.11i Aren't Enough To Protect Your Network," Airtight Networks, Copyright © 2002-2005, Airtight Networks.

All rights reserved. 339 N. Bernardo Avenue, Suite 200, Mountain View, CA 94043, February, 2005.

[5] "Enterprise Wireless WAN Security," © 2005 Hewlett-Packard Development Company, L.P., Hewlett-Packard Company, 3000 Hanover Street, Palo Alto, CA 94304-1185 USA, 2005.

[6] John R. Vacca, *Firewalls: Jumpstart for Network and Systems Administrators*, Digital Press, 2004.

[7] John R. Vacca, *Net Privacy: A Guide to Developing and Implementing an Ironclad ebusiness Privacy Plan*, McGraw-Hill, 2001.

[8] John R. Vacca, *Identity Theft*, Prentice Hall, 2002.

Chapter 5

HANDLING WIRELESS PRIVATE INFORMATION

1. INTRODUCTION

While a wireless LAN can be installed by simply plugging an access point into an Ethernet port, an enterprise wireless LAN deployment requires a more thought-out plan that incorporates advanced security and management technologies; and, the handling of wireless private information. Recently, analysts and media have documented and publicized vulnerabilities of wireless- LANs and WANs, such as encryption that can be broken and rogue access points that allow intruders to connect to your network and steal private information.

> **Warning:** Through year-end 2006, the employees' ability to install unmanaged access points will result in more than 70 percent of enterprises exposing sensitive information through WLANs.

These reports focus on breaking encryption, the risk of unauthorized access points connected to the wired network, and the failure of enterprises to incorporate security into their wireless LANs. The attention on the pitfalls of wireless- LANs and WANs has inspired some enterprises to ban wireless-LANs and WANs altogether; but, any enterprise that utilizes laptop computers faces the risk of these easily becoming wireless stations that introduce security risks to private information.

However, security-conscious enterprises are fortifying their wireless LANs with a layered approach to security that resembles the accepted security practices of wired networks. This layered approach to security addresses all network components by locking down the wireless LAN's perimeter, securing communication across the wireless LAN, and monitoring network traffic. In fact, according to industry analysts, there are three must have requirements for enterprise wireless- LANs and WANs:

- Install a centrally managed firewall [2] on all laptops that are issued wireless network interface cards or are bought with built-in wireless capabilities. This protects against ad hoc WLAN and WWAN connections and Internet attacks when users connect to public hot spot Internet providers.
- Perform wireless intrusion detection to discover rogue access points, foreign devices connecting to enterprise access points and accidental associations to nearby access points in use by other enterprises.
- Turn on some form of encryption and authentication for supported WLAN and WWAN use [1].

2. SECURE WIRELESS LAN DEVICES

Like installing a door on a building to keep a passersby from wandering in, enterprises must control the perimeter of their enterprise networks. For the traditional wired LAN, this was accomplished by installing firewalls to control the entry point to the network. However, wireless- LANs and WANs present greater challenges from the hard-to-control nature of radio transmissions.

With data and network connections broadcasting private information [3] across the air and through windows, walls, floors, and ceilings, the perimeter of a wireless LAN can be as difficult to control as it to define. However, enterprises can control the perimeter of a wireless LAN by securing their WLAN devices that act as the endpoints of the network.

Perimeter control for the wireless LAN starts with deploying personal firewalls on every wireless-equipped laptop and also includes a deployment of enterprise-class access points that offer advanced security and management capabilities. The wireless LAN should be segregated from the enterprise wired network as part of a VLAN to allow for wireless-specific management and security policies that do not affect the wired network.

All access points should be completely locked down and reconfigured from their default settings. The SSIDs and passwords of the access points should be changed from their default names. Some enterprises choose to establish set channels of operation for each AP to identify all off-channel traffic as suspicious activity [1].

3. SECURE COMMUNICATION: AUTHENTICATION & ENCRYPTION

In deploying secure wireless- LANs and WANs, IT security and network managers face the most difficult decision in choosing how to secure WLAN and WWAN private communication with multiple forms of authentication and encryption. Like installing locks and keys on a door to control who can enter, the next layer of wireless LAN security is to control which users can access the wireless LAN. To provide basic authentication, most access points support simple MAC address filtering that maintains a list of approved stations' MAC addresses. While this is not foolproof, MAC address filtering provides basic control over which stations can connect to your network privately.

Enterprises that rely upon MAC address filtering for access control leave themselves vulnerable to simple identity thefts [4]. Larger enterprises with more complex wireless LANs with hundreds of stations and dozens of access points require more sophisticated access control through incorporating remote authentication dial-in service (RADIUS) servers.

As previously mentioned in earlier chapters, Cisco Systems, Microsoft, and Funk Software are recognized leaders in this area. In regards to industry standards, the IEEE introduced 802.1x to provide port-based access control, which incorporates a central authentication server. However, some versions of 802.1x have been shown to be vulnerable to hackers. Cisco introduced Lightweight Extensible Authentication Protocol (LEAP) as a proprietary authentication solution that is based on 802.1x, but adds proprietary elements of security (see Table 5-1 [1]). LEAP has its own security issues, and Cisco is moving away from LEAP toward Protected Extensible Authentication Protocol (PEAP).

Table 5-1. Description of private data protection technology.

Data Protection Technology	Description
WEP	Wired Equivalency Privacy: Original security standard for wireless LANs. Flaws were quickly discovered. Freeware, such as WEPCrack, can break the encryption after capturing traffic and recognizing patterns in the encryption. *(Industry standard)*
802.1X	As the IEEE standard for access control for wireless and wired LANs, 802.1x provides a means of authenticating and authorizing devices to attach to a LAN port. This standard defines the Extensible Authentication Protocol (EAP), which uses a central authentication server to authenticate each user on the network. *(Adopted industry*

	standard)
LEAP	Lightweight Extensible Authentication Protocol: Based on the 802.1x authentication framework, LEAP mitigates several of the weaknesses by utilizing dynamic WEP and sophisticated key management. LEAP also incorporates MAC address authentication as well. *(Developed by Cisco)*
PEAP	Protected Extensible Authentication Protocol: Securely transports authentication data, including passwords and encryption keys, by creating an encrypted SSL/TLS tunnel between PEAP clients and an authentication server. PEAP makes it possible to authenticate wireless LAN clients without requiring them to have certificates, simplifying the architecture of secure wireless LANs. *(Developed by Cisco, Microsoft, and RSA Security)*
WPA	Wi-Fi Protected Access: Subset of the future 802.11i security standard. Designed to replace the existing WEP standard. WPA combines Temporal Key Integrity Protocol (TKIP) and 802.1x for dynamic key encryption and mutual authentication. *(Industry standard adopted in 2003)*
TKIP	The Temporal Key Integrity Protocol, pronounced tee-kip, is part of the IEEE 802.11i encryption standard for wireless LANs. TKIP provides per-packet key mixing, a message integrity check and a re-keying mechanism, thus fixing the flaws of WEP. *(Industry standard)*

Encryption provides the core of security for wireless- LANs and WANs by protecting the data that crosses the airwaves. However, fail-proof encryption and authentication standards have yet to be implemented. Temporal Key Integrity Protocol (TKIP) has been introduced to address the flaws of WEP with per-packet key mixing, a message integrity check and a re-keying mechanism.

New industry standards and proprietary solutions are now being introduced to handle both encryption and authentication. Cisco, RSA Security, and Microsoft developed PEAP as one of these proprietary solutions [1]. However, Microsoft and Cisco have separated their PEAP development efforts and introduced their own versions of the protocol. Microsoft's version of PEAP does not work with Cisco's version of PEAP.

While Microsoft is bundling its version of PEAP on the desktop, Cisco's version of PEAP requires client software to be installed and managed on each WLAN user stations.

As previously mentioned in earlier chapters, in April 2003, the Wi-Fi alliance launched Wi-Fi Protected Access (WPA) as a subset of the 802.11i security standard based on TKIP [1]. Most vendors have announced that existing access points can be upgraded to support WPA with a firmware upgrade. However, new access points are needed for 802.11i.

Virtual Private Networks or WLAN gateways provide another alternative to standards-based encryption and authentication. Traditional firewall and VPN gateway vendors, such as Check Point and NetScreen Technologies, offer VPNs that funnel all traffic through their existing VPN gateway [1]. These VPN solutions are generally IPSec based and do not work well with wireless LANs where users roam between access points or signals may vary and drop off, which forces the user to re-authenticate and begin a new session.

Vendors, such as Bluesocket, ReefEdge, and Vernier Networks, offer wireless LAN gateways that include added features for network roaming and bandwidth management that are tailored to wireless LANs [1]. Another segment of wireless VPN vendors, including Fortress Technologies and Cranite Systems, offer more secure solutions with Layer 2 encryption [1]. While VPNs provide strong encryption and authentication, most require client-side software, which introduces management headaches.

4. WLAN MONITORING

Like a video camera that monitors all activity in a secure building 24 hours a day, a critical layer of wireless LAN security requires monitoring of the network to identify rogue WLANs, detect intruders and impending threats to private information, and enforce WLAN security policies. As an example of the need for monitoring, access points that are upgraded for WPA must be monitored to ensure the access point remains properly configured [1].

> **Tip:** WPA access points must be configured to disable legacy WEP security because the access points may still accept WEP client connections. Security is handled in the access point, reaffirming the need for validation of access-point implementation.

WLAN monitoring must scale to fit the specific needs of an enterprise. Some piece-meal solutions work for smaller enterprises, but do not scale for large enterprises with dozens or hundreds of locations around the world.

Large enterprises require a cost-effective solution that can be centrally managed and does not overtax personnel resources.

Manual site surveys are particularly unreasonable for enterprises operating dozens of offices around the country or retailers with hundreds of stores; even if these enterprises could feasibly devote a network administrator's full attention to survey each site on a daily, weekly, or monthly basis. Therefore, you must monitor the airwaves 24x7 to secure wireless LANs by identifying rogue WLANs, detecting intruders and impending threats to private information, and enforcing WLAN security policies.

> **Tip:** To truly secure wireless LANs, enterprises must monitor their airwaves to detect intruders and threats that can come from unscrupulous hackers and well-meaning employees. Monitoring the airwaves of a wireless LAN is an essential element of security that should also include advanced encryption and authentication.

You should also implement 24x7, stateful monitoring of wireless LANs with a distributed system of remote sensors that passively monitor all WLAN activity and report back to a central appliance that analyzes the traffic for threats, attacks, and policy violations. This approach scales to support wireless LANs in a single office or hundreds of access points in dozens of locations around the world [1].

5. WLAN MANAGEMENT REQUIREMENTS

Just as wireless LAN security mirrors security of the wired network, the same holds true for wireless LAN management. Network managers should already be familiar with the general requirements of managing wireless LANs, but must implement wireless-focused solutions for fault diagnostics, configuration management, accounting for network usage, performance monitoring, and policy enforcement.

Managing a small wireless LAN deployment of 5 or 10 access points can be easily accomplished with the builtin functionality of access points. However, managing a larger wireless LAN deployment of dozens or hundreds of access points in a corporate campus or in multiple locations across the country, requires add-on solutions that scale to support the distributed nature of the network.

These wireless LAN management requirements can be satisfied with a combination of 24x7, real-time monitoring of the airwaves and wired-side solutions offered by WLAN infrastructure providers, such as Cisco Systems and Symbol Technologies [1]. Numerous start-up enterprises, such as Aruba

Networks and Trapeze Networks, have introduced wireless LAN switches for an integrated approach to managing all access points in a network [1]. However, most WLAN management systems are often limited by their ability to only manage access points manufactured by the vendor of the WLAN system.

5.1 Configuration

Managing a wireless LAN's configuration across all access points and stations often provides a major challenge to network managers. At the most difficult level, each device must be touched to ensure proper settings for security, performance, and policy compliance. WLAN management offerings, such as Cisco's Wireless LAN Solution Engine (WLSE) or Symbol's Wireless Switch System, can remotely manage access point configuration and apply multiple configuration templates to various segments of a wireless LAN [1].

Managing the user configurations provides a bigger challenge, because network managers may not have direct access to all stations, and touching each station can be a time-consuming project. Real-time monitoring of the airwaves complements wired-side configuration to ensure that access points and stations remain in their defined configurations. Power surges or outages can reset access points to default settings. Employees can alter device settings to allow for open network access. Analysis of the WLAN traffic while in the air identifies these network misconfigurations.

5.2 Fault Diagnostics

Employees and users can benefit from the wireless LAN only when it is up and running. Responding to support calls can be an overwhelming task for an IT department responsible for supporting wireless LANs in remote locations.

WLAN management offerings, such as provided by Cisco and Symbol, can poll network devices from the wire to observe device characteristics and attributes and alert operational staff to some issues [1]. For a higher level of fault diagnostics, real-time monitoring of the airwaves continuously surveys WLAN devices to analyze traffic patterns and alert network managers of device failures and excessive noise in the air that cripples a WLAN. With 24x7, real-time vigilance, wireless monitors alert network managers to network failures the minute they arise.

5.3 Performance Monitoring

After first ensuring that the network is up and running, network managers must then analyze the performance of a wireless LAN to guarantee a maximum return on investment. WLAN management tools, such as Cisco WLSE, can provide some performance information about specific access points by polling information from the wire [1]. In addition, real-time monitoring of the airwaves identifies performance issues that can only be seen from the air, such as signal degradation from channel overlap, frequency interference from non-802.11 devices, and excessive overloading of an access point.

5.4 Accounting: Network Usage

Much like fault diagnostics and performance monitoring, accounting for network usage is accomplished with a combined approach that includes a WLAN management platform and 24x7 monitoring of the airwaves to protect private information. Network management platforms from the likes of Cisco and Symbol track WLAN stations connecting to various applications on the wired side of the network for inhouse accounting purposes [1].

Finally, monitoring of wireless LAN traffic across the airwaves allows network managers to track the network usage based on the peak capacity of each access point and the highest bandwidth-consuming stations and access points. This allows network managers to plan for additional capacity as needed and deal with individual users who abuse the WLAN by downloading large, non-business related files, such as MP3s.

6. SUMMARY AND CONCLUSIONS

So, with the preceding in mind, policy compliance across the wireless LAN touches almost every aspect of network management and security of private information. Network policies govern wireless LAN configuration, usage, security settings, and performance thresholds. However, security of private information and management policies are useless, unless the network is monitored for policy compliance and the enterprise takes active steps to enforce the policy. Finally, real-time, 24x7 monitoring of WLAN traffic identifies policy violations for:

- Rogue wireless LANs – including Soft APs
- Unencrypted or unauthenticated traffic

- Unauthorized stations
- Ad hoc networks
- Default or improper SSIDs
- Access points and stations operating on unauthorized channels
- Insecure stations with default Windows XP settings
- Off-hours traffic
- Unauthorized vendor hardware
- Unauthorized data rates
- Performance thresholds that indicate the overall health of the wireless LAN [1].

7. REFERENCES

[1] "Understanding the Layers of Wireless LAN Security & Management," Copyright © 2003 AirDefense, Inc. All rights reserved. AirDefense, Inc., 4800 Northpoint Parkway, Suite 100, Alpharetta, GA 30022, 2003.

[2] John R. Vacca, *Firewalls: Jumpstart for Network and Systems Administrators,* Digital Press, 2004.

[3] John R. Vacca, *Net Privacy: A Guide to Developing and Implementing an Ironclad ebusiness Privacy Plan,* McGraw-Hill, 2001.

[4] John R. Vacca, *Identity Theft,* Prentice Hall, 2002.

PART II

DESIGNING WIRELESS NETWORK SECURITY

Chapter 6

WIRELESS NETWORK SECURITY DESIGN ISSUES

1. INTRODUCTION

The world is changing dramatically and forcing every enterprise to address new concerns. In particular, enterprises that want to ensure complete satisfaction of their customers' needs must prevent security threats while guaranteeing ready access to information for authorized users. These requirements are integral parts of any systematic approach to network security.

2. WIRELESS NETWORK SECURITY: A PHYSICAL ANALOGY

When it comes to securing physical assets, the typical model is to place them in a central location encircled by layers of defense mechanisms with separate nonaligned entry points. Such a structure guarantees that no single act allows entry. However, in today's environment where technology applications provide the foundation for customer-focused enterprise operations, it is impractical to secure digital information in a similar manner. The very nature of modern organizational wireless network design is to disseminate and disperse data for optimal communication.

A successful wireless network design requires that information be stored in disparate physical locations, with multiple network pathways and access points—both fixed and wireless. The result is that all of these elements are available as means for unsolicited network entry. This in turn makes policing a network an imprecise, costly, and time-consuming operation [1].

3. COUNTING THE COST OF SECURITY

Many wireless network administrators believe their enterprise is unlikely to suffer from technical security lapses. Whether administrators base their assumption on their enterprise's small size, low enterprise profile, simple technology infrastructure, or a seeming lack of critical information, one thing that can be assured is that to make such a judgement is a mistake. In a 2005 FBI survey, 93% of respondents stated their enterprises had detected computer security breaches within the last twelve months. What's more they calculated the average cost of each breach was approximately $78,000. Hence security should not be viewed as a mere enhancement or a luxury limited to enterprises with technologically advanced networks. It should be considered a utility application—a practical necessity that has become a reality for today's wireless networks [1].

4. POLICY FIRST

A vital first step in providing a secure and uninterrupted flow of information is to implement a holistic security strategy. Good security requires three basic elements:

1. A policy that determines the technology choices
2. Technology that uniformly enforces policy
3. People who will manage and weave security into the fabric of the enterprise [1].

5. THE ROLE OF RADIUS IN THE COMPLETE NETWORK SECURITY SOLUTION

Because security risks and threats vary greatly from enterprise to enterprise, technology solutions must be tailored to the specific situation—which can mean higher costs. However, there is a basic level of security that every enterprise should implement as a matter of course. For example, the safety of a system is only as good as the security of its passwords; so as many barriers as possible should be put in place to prevent internal attacks and individuals' use of subterfuge to glean access to information that allows them to enter a wireless network. There are other simple steps an enterprise can take to bridge traditional wireless network weak points and deliver powerful fundamental wireless network security, while limiting the impact upon vital technology budgets [1].

6. THE 802.1X STANDARD BASICS

The 802.1X standard describes wireless network edge access between a supplicant and a network access device. The standard identifies Remote Authentication Dial-in User Service (RADIUS) as a generic means of performing supplicant access validity inspection. RADIUS is not an altogether new technology concept; indeed a great number of remote Internet dial-up and VPN tunneling technologies have been using RADIUS as a security tool for a number of years. It is perhaps not surprising then that the RADIUS standard is fully supported by major software firms, including Microsoft, Funk, and Open, and that its tried and tested format delivers proven high-value wireless network edge security that is also easy to implement and manage. Renewed interest in RADIUS arose from the need to secure networks more effectively at their edge. The vulnerability of wireless network edge security first became apparent as wireless infrastructure access points proliferated in LAN configurations, but other high profile cases of unauthorized access have further emphasized the problem.

RADIUS differs fundamentally from prior application layer security systems in that it operates at the lowest network protocol level to prevent security loopholes. The reason for this move to a lower protocol level is that individuals started to "burrow" beneath application-layer security mechanisms—in the same way a burglar digs underneath a security fence. This inherent weakness of application-layer security came to light with wireless devices, which access the network infrastructure at a protocol level generically lower than the application layer and so allow circumvention of existing security precautions.

> **Note:** Remote Authentication Dial-In User Service (RADIUS) is a centrally administered wireless network access, control, and accounting system that tips the balance of power at the edge of the LAN in the wireless network security manager's favor. When used in conjunction with 802.1X wireless network edge control, RADIUS surpasses other wireless network edge security methods. It is unique in offering full scalability, ease of installation and use, adherence to standards, proven reliability, and low maintenance and cost.

7. HOW DOES RADIUS WORK?

Fundamentally, RADIUS is the industry standard for centralized authentication, authorization, and accounting (AAA) of users connecting to a network. Very brief definitions follow:

- Authentication is the process of deciding whether the user is who he or she claims to be, and whether that user is allowed on the wireless network—as determined by username and password credentials.
- Authorization, also called service delivery, is the system of controlling the wireless network resources such as wireless network privileges and time limits that the user can access once authenticated.
- Accounting is the process of generating log files that record statistics describing each connection session; it is used for billing, system diagnosis, and usage planning [1].

8. DISTRIBUTED SECURITY MODEL

All of the wireless network user security information is centralized in AAA servers, hence the RADIUS security model is organically diffuse and capable of supporting multi-site, multi-domain usage and user roaming. RADIUS is therefore one of the only systems that effectively supports secure client wireless access, a factor that future-proofs RADIUS when compared to other current methods for building walls among different parts of a network [1].

9. AUTHENTICATION

Viewed from the user's perspective the interaction with RADIUS is simple: whenever a user wishes to access a wireless network he or she will be asked to enter a user ID and password, the client will then be either granted or denied network access. Beneath the user interaction, the Network Access Device (NAD) (typically a switch, or a wireless access point) and the client have engaged in a request-challenge-response dialog (illustrated in Fig. 6-1 [1]) that is then verified through a predetermined algorithm by an authentication server. Wireless network access denial will lead to the user being repolled for valid credentials at regular adjustable intervals up to a predefined number of failures.

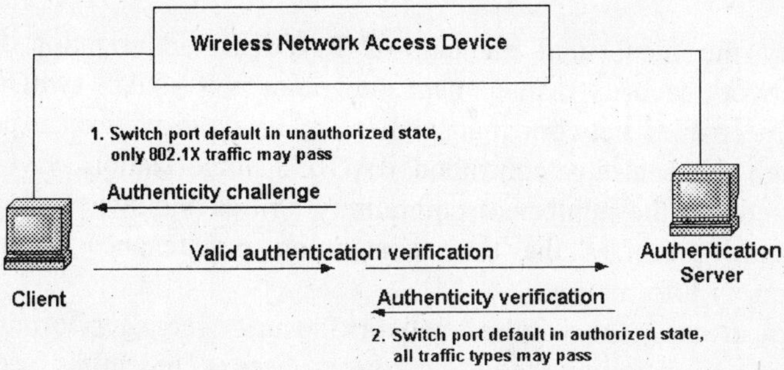

Figure 6-1. Wireless Network Login with RADIUS Client Authentication Process.

RADIUS is not tied to any particular wireless networking scheme. Since it is a foundation technology, it functions regardless of a wireless network's protocol. Such functionality is possible because RADIUS uses the Extensible Authentication Protocol (EAP) that stems from the Point to Point Protocol (PPP), which itself currently underpins the majority of dial-up connections. The NAD ports that are RADIUS aware, begin in an unauthorized state, and only traffic explicitly permitted by RADIUS can pass through. Typically this traffic is limited to authentication details, while all other higher-level protocols are blocked. Therefore, even fundamental wireless network configuration information such as IP address allocation cannot be transmitted until wireless network access has been granted. Only when the client has fully complied with 802.1X authentication is data exchanged freely.

10. AUTHORIZATION

RADIUS provides even further security functionality via its authorization capabilities, which allow the wireless network security manager to apply a number of added network policies in conjunction with basic authentication. RADIUS has the potential to control access to wireless network services and applications, and can enable profile provisioning to be applied on a per user or group basis. It allows access request attributes to be checked for existence or alternatively for specific values. In addition, RADIUS makes it possible to define elementary logic value attributes, such as the time of day login is permitted, duration of login, or number of active sessions with the same user ID. The policy decision outcomes are sent back to the NAD as Access Reply attributes and can be used for client provisioning.

Currently the most vital element of RADIUS authorization from a wireless network security management viewpoint is RADIUS switch login. Switch login enables network management to be controlled on a per-user basis through the centrally controlled RADIUS authorization server. This drastically reduces the number of community strings and switch passwords required, and minimizes the time consuming maintenance of wireless network security information.

In a typical case, the wireless network security manager formulates a user/policy matrix detailing the network users' access credentials (see Table 6-1 [1]). This establishes a portfolio of accessible wireless network utilities for each wireless network user based on their individual requirements. RADIUS authorization also provides a depth of potential functionality that ensures future proofing in enterprise wireless network security.

Table 6-1. A Sample RADIUS User/Policy Matrix

User ID	Password	Policy Profile	Standard Applications	Network Administration	Management Server	Admin Server
David Network	Blue1	Network Administrator	Yes	Yes	No	Yes
Bernard Manager	Yellow2	Management	Yes	No	Yes	No
Samuel Intern	Red3	Admin	Yes	No	No	Yes
Shana Employee	Purple4	Standard	Yes	No	No	No

11. ACCOUNTING

The NAD collects usage data for each session and transmits it to the RADIUS server. Simple harvested usage data may include user identities, individual session durations, and wireless network traffic utilization based on the number of packets and bytes transmitted.

This accounting data is extremely useful for enterprise management. It provides arguably the most accurate data obtainable for infrastructure billing and auditing, while it makes the objectives of wireless network security and trend analysis transparent and easily achievable. The accounting data is also congruent with many predictive management tool requirements, furnishing them with the information they need to perform accurate capacity planning algorithms. RADIUS can also update accounting databases in real time via Structured Query Language (SQL); however, support for generic communication with accounting databases depends on the version of RADIUS software being used.

Fig. 6-2 [1] illustrates how a RADIUS accounting server collates data from Switch units in each department of an enterprise, including one in a remote branch office. The enterprise can use harvested data to track each department's use of the wireless network infrastructure and then allocate wireless network support costs accordingly. Depending on the nature of the data harvested, it is possible to monitor the efficiency of different areas as well as carry out billing on an annual, monthly, weekly, or daily basis.

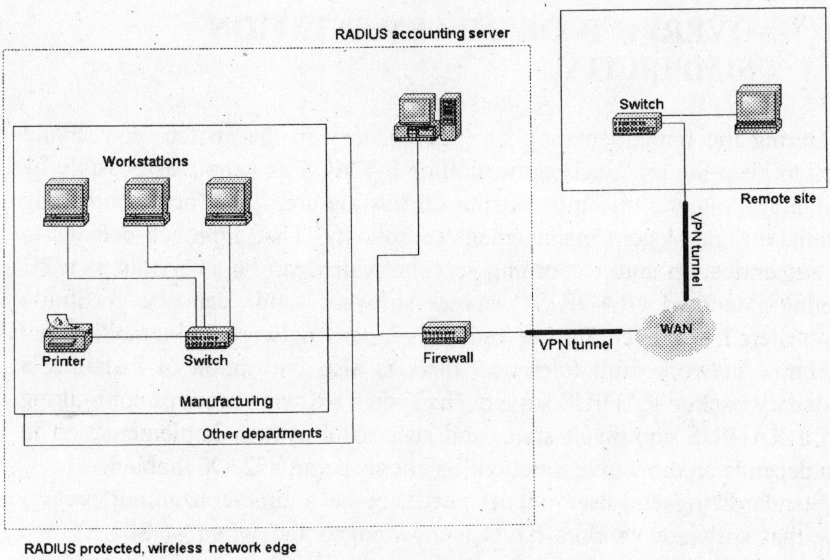

Figure 6-2. Multisite Organizational Accounting Augmented by Wireless Network Login and RADIUS.

12. EXAMINING WIRELESS LAN SECURITY

As mentioned earlier, the 802.1X standard used in conjunction with RADIUS has a positive effect on security for every network, wired and wireless infrastructures or edge access points. However, enterprises must exercise great care when using wireless technology in any secure environment, because wireless is by its nature insecure. For instance, documented implementation flaws have recently been found in the standard wireless security mechanism referred to as wired equivalent privacy (WEP). Though WEP was designed to provide a baseline wired-like protection and scale to a large variety of applications, it is itself vulnerable to a variety of passive and active traffic interception and decryption attacks—which hackers can perpetrate with inexpensive off-the shelf equipment and Internet access. However, now that wireless systems support the 802.1X standard and RADIUS implementation, network user misappropriation becomes much more difficult. That is, because a higher level of security is delivered when combining MAC layer authentication with upper layer authentication [1].

13. OVERVIEW OF IMPLEMENTATION SIMPLICITY

During the implementation of edge security in the switch, you should work to keep the practical application of RADIUS as simple as possible by facilitating integration into existing infrastructure, therefore minimizing installation and system management complexity. That approach extends to the authentication and accounting servers, which can be as simple as a PC running standard RADIUS server software and can be centrally administered. To even further reduce wireless network vulnerability and maximize network fault tolerance, there is also the option of installing a secondary backup RADIUS server. The switch software requires configuring into a RADIUS compliant state; and successful system implementation in turn depends on the wireless network's clients being 802.1X enabled.

Standardizing end-user clients need not be a time-consuming process now that software vendors have acknowledged the value of 802.1X and recognized that the standard is key to integrated local area network security. As such, the 802.1X client is now incorporated into most vendors' new operating systems. For example, it is an automated function in Microsoft Windows Xp and Server 2003; and, Microsoft has also committed to software updates that integrate the 802.1X standard into its earlier operating systems [1].

Note: RADIUS is compatible with enterprises' existing application layer authentication systems that typically use Open Lightweight Directory Access Protocol (LDAP) or similar protocols. The compatibility depends on the RADIUS server selected, but most are able to operate in a man-in-the-middle mode between the NAD and existing user database to interrogate it for authentication details.

14. ASSOCIATED ASPECTS OF SECURITY INCORPORATING RADIUS WIRELESS NETWORK DESIGN

Wireless network login should be enabled on all relevant switch ports. Failure to enable authentication even on a single port could compromise the security of the entire wireless network. However, there are instances when a dumb device, such as printer or scanner, is connected to a switch port—a common scenario in many enterprises. The dumb device requires wireless network access, but is incapable of authenticating itself under the RADIUS requirement. Advocating unrestricted access to the wireless network defeats the object of having the wireless network login capability. To a lesser extent, unlocking the port when a client is successfully authenticated on the port can also result into a security loophole where non-authenticated clients can now access wireless network resources through the authenticated port (see Fig. 6-3 [1]).

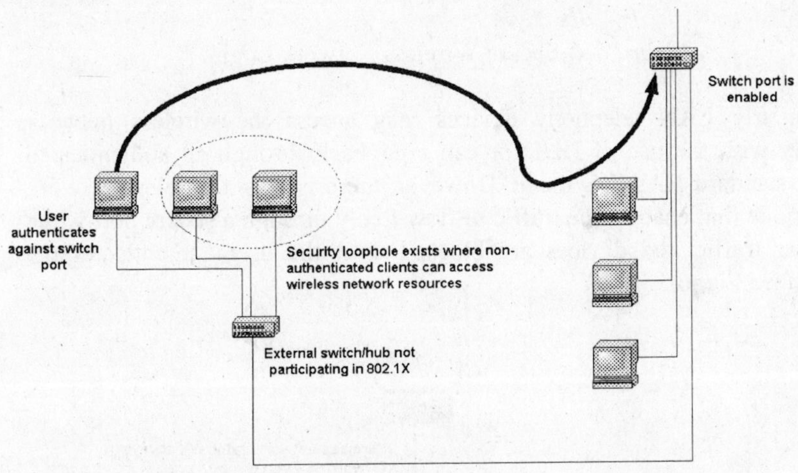

Figure 6-3. Standard Wireless Network Login.

The switch combats this problem via a technique know as Disconnect Unauthorized Device (DUD), which allows the wireless network administrator to instruct ports to pass network traffic only to devices with specific MAC address, while keeping them fully compatible with RADIUS (see Fig. 6-4 [1]). In secure wireless network login mode, the switch will record the MAC address of the client that authenticates, and the switch will then filter traffic based on that MAC address.

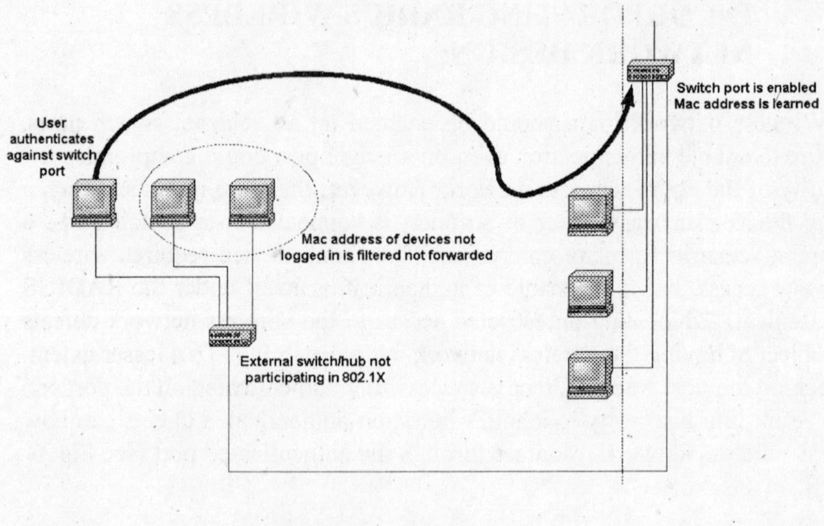

Figure 6-4. Secure Wireless Network Login.

Similarly LAN telephony devices may access the wireless network securely with the use of DUD or can piggyback through an authenticated port in standard RADIUS mode. However, there is also a wireless network login mode that enables the traffic to flow freely through a secure port while all other traffic and devices are blocked until the user authenticates the device (see Fig. 6-5 [1]).

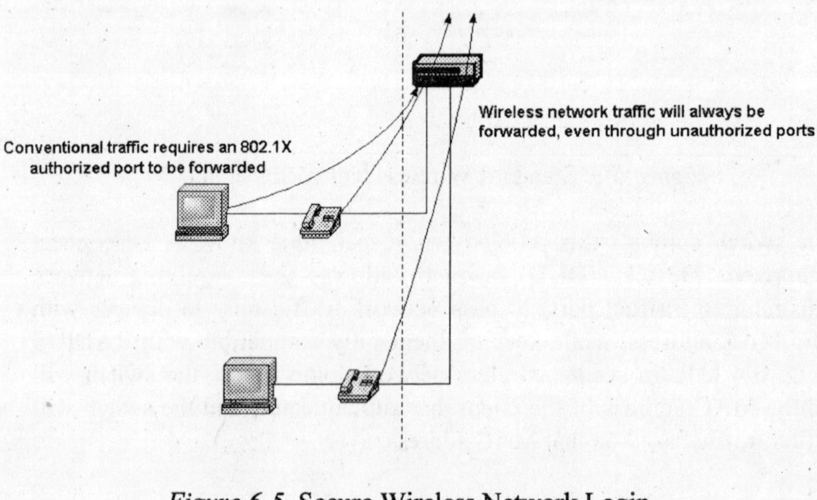

Figure 6-5. Secure Wireless Network Login.

15. FURTHERING WIRELESS NETWORK SECURITY AT THE EDGE AND BEYOND

Although this chapter has focused thus far on the wireless network edge protection provided by RADIUS in conjunction with standard 802.1X, wireless network login and the fundamental network security requirements it satisfies, RADIUS should be complemented by other technology security measures. For instance, every wireless network should have assured firewall protection [2]. It is essential to have a primary firewall at the LAN perimeter in the demilitarized zone (DMZ) between the wireless local area network (WLAN) and wireless wide area networks (WWAN) as shown in Fig. 6-6 [1].

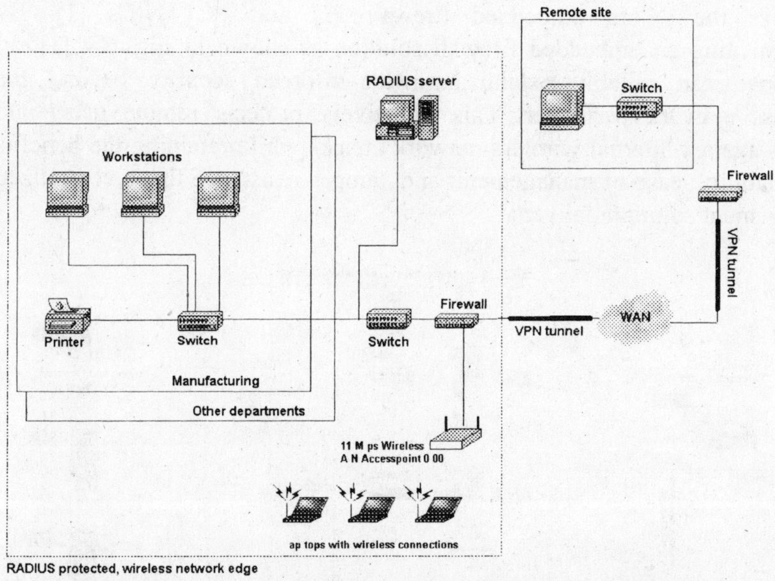

Figure 6-6. The Extent of a RADIUS Protected Wireless Network.

Enterprises with a strong security policy will incorporate many other wireless network security technologies as standard. intrusion prevention systems, intrusion detection systems, encryption, anti-virus software, digital signatures, and virtual private networks are just a few of the safeguards that should be considered seriously when weighing true security requirements and available budget.

16. THE BENEFITS OF WIRELESS NETWORK INTERFACE EMBEDDED TECHNOLOGIES

Pioneering technologies are constantly emerging to expand wireless network edge protection and combat ever-changing security threats. For example, in cases where hard security of end users is a requirement, there should be some form of firewall on the users' network interface cards (NICs). The need is especially critical for unprotected remote end-points and high-security network solutions where information stored in servers and at the wireless network edge must be protected against attack from within the network.

Finally, high levels of protection are now achievable with new products such as the 3Com embedded firewall PC and PCI NICs [1]. By implementing an embedded firewall solution as shown in Fig. 6-7 [1], an enterprise can reliably extend hardware-enforced security beyond the wireless network perimeter. This effectively protects remote users and guards against internal wireless network attacks while retaining the benefits of scalability; ease of management; and, tamper-resistance that a centralized management solution delivers.

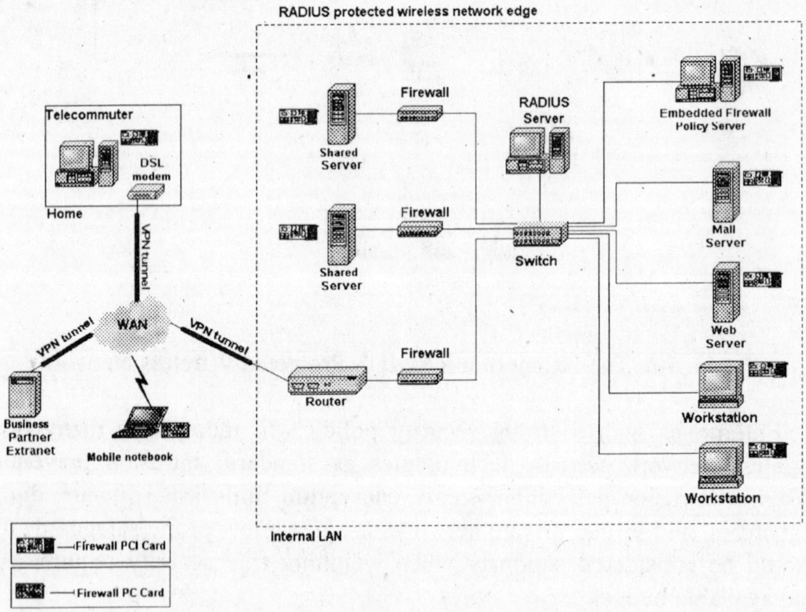

Figure 6-7. A Typical Wireless Network Protected by Embedded Firewall NICs.

17. SUMMARY AND CONCLUSIONS

The theme of this chapter was enterprise critical systems security, by first illustrating the importance of and potential difficulties in protecting information that traverses networks; and, then examining wireless network security as a holistic concept before focusing specifically on the IEEE 802.1X enterprise edge security standard. In addition, the chapter demonstrated how RADIUS used in conjunction with 802.1X provides a long-term, secure, and low-cost system for wireless network authentication, authorization, and accounting.

Finally, there are many positive precautions for an enterprise to consider when determining how best to provide timely and cost-effective wireless network security. Most important is a holistic security policy, which is the essential foundation for any shield against technology-related crimes. And, since the wireless- LAN edge and LAN/WAN perimeter are primary targets for forced entry and latent wireless network cybercrime, they are the obvious location to implement elementary security measures. The system that delivers the greatest wireless network value and flexibility via a distributed security model is the combination of RADIUS and the IEEE 802.1X industry standard. So, it is highly recommended as the first technical security activity in which an enterprise should invest.

18. REFERENCES

[1] "3Com®802.1X Security–Designing a Secure Network," Copyright © 2003 3Com Corporation. All rights reserved. 3Com Corporation, Corporate Headquarters, 5500 Great America Parkway, P.O. Box 58145, Santa Clara, CA 95052-8145, , 2003.

[2] John R. Vacca, *Firewalls: Jumpstart for Network and Systems Administrators*, Digital Press, 2004.

Chapter 7

COST JUSTIFICATION AND CONSIDERATION

1. INTRODUCTION

With the recent viruses making the headlines, you may be thinking more seriously about the cost of not being adequately protected. Investment in security technologies is all about mitigating the risks facing your IT systems. Over the last few years, IT infrastructures have become more complex with an increased use of various applications, which can pose more threats to wireless network security.

This chapter will assess the security risks and identify vulnerabilities to the wireless network. The chapter will also help you, construct the cost arguments needed to define the wireless network security investment.

2. HITTING WHERE IT HURTS

Security attacks disrupt enterprise operations and can incur large recovery costs. Industry analysts have looked at the economic impact of security attacks and how these can be lessened by security investment [1]. They found that security attacks result in five key disturbances to networks:

- Internal disruption: Intentional or unauthorized intrusion from employee or validated user
- Passive code-level intrusion: Worms and viruses
- External disruption: Wireless network hackers and denial of service interruption

- Authentication forgery: Forged identity credentials, to gain access to information or systems
- Extraction: Intentional or unintentional export or deletion of information [1]

According to industry analysts, the most frequent security incidents are external disruption and passive code-level intrusion [1]. External disruption, when the network or information are disturbed by an outsider (a hacker), is also the most costly to resolve. Industry analysts found that external disruption costs a business, on average, $87,613 (USD) [1]. Passive code-level intrusion, when worms or viruses get access to the network, and extraction are the next most costly. When auditing your wireless network, make sure you prioritize the mitigation of those risks that will be most likely to occur and most costly to resolve.

3. ASSESS YOUR VULNERABLE POINTS

As access points to networks have become more complex and staff have become increasingly mobile and reliant on wireless networks, IT security has to deal with more exposed areas. Consider the vulnerable areas and the technologies you need for protection:

- The desktop
- The wireless network perimeter
- Servers
- Wireless network communications [1]

3.1 The Desktop

The desktop can often be the weakest link in a defense system. Without adequate protection, desktops can be vulnerable to viruses, Web attacks and e-mail intrusions. Anti-virus protection is a must.

3.2 The Wireless Network Perimeter

The perimeter of your network could be visible to the outside. If you have embraced any level of e-commerce in your organization, then you will have had to expose parts of your systems to the public. A firewall [3] can protect these open access zones by creating a secure perimeter around your wireless network, and defining exactly what, and who, can get in or out of your network at any time.

3.3 Servers

Servers can have thousands of settings and it could only take one critical subset to be changed without your knowledge to really mess up your day. All too often the source of these unhelpful changes is one of your employees, or even one of your IT staff. Therefore, when assessing your wireless network security risks, also assess the policies and IT training processes in place within your organization.

3.4 Wireless Network Communications

Wireless network communications can be brought to their knees through a Denial of Service attack. As many DoS attacks exploit limitations on the TCP/IP protocols, it's important to protect your network communications with encryption and Virtual Private Networks (VPNs).

4. THE COST OF IMPACT

Building an argument for technology investment can be a challenge for the IT decision-maker. Financial decision-makers tend to apply the Return on Investment (ROI) model to technology, requiring justification of how a purchase will add value to the business. But security should be valued in terms of expenditure prevention.

Security enables the enterprise to more reliably estimate costs and avoid loss of productivity and potentially, a massive clean-up bill if something goes wrong. By investing, you can lower your exposure to losses. Industry analysts found that for each dollar spent investing in security technologies, an enterprise can lower its risk quotient by an amount equal to a multiplier of 33.4. For example, this means that an enterprise spending $40,000 (USD) on a wireless network security investment can effectively lower exposure to losses associated with network risk by an amount equal to or greater than $1,336,000 (USD) [1].

Security managers can build a cost argument by considering the wireless network's level of vulnerability, and prioritizing where spending should be focused. The level of spending should be associated with the cost avoided by preventing security attacks.

To estimate saved costs and thus create an argument for level of investment, it's necessary to take into account the areas where security threatens the economic performance of the enterprise:

- Your data: Research, records and other vital information are expensive (and sometimes impossible) to reproduce or recover

- Your time: When your wireless network administrators are researching a damage trail, they're not working on other critical projects. What are your IT staff costs on an hourly basis?
- Your legal budget: Do you really know what it will cost if you violate BIS or HIPAA regulations, or are you involved in a shareholder or vendor lawsuit?
- Your business: The ability of your enterprise to produce goods or services at an expected rate after a security attack would be impaired – if not ceased temporarily
- Your assets: Do you know what proprietary information has been exposed to the outside world by employees who aren't aware of security gaps [1]?

5. WIRELESS NETWORK SECURITY AS INSURANCE

Every year, enterprises make significant investments in comprehensive insurance policies to protect assets. Enterprises of all sizes, from large enterprises to sole proprietorships, spend billions collectively to guard against liability and property damage, and to keep employees safe and in good health [2].

But, what about the wireless network infrastructure? What about your wireless network? It's the nerve center of your enterprise, through which you communicate with customers, store and maintain data and intellectual property, conduct sales, manage employees, and engage with strategic partners on a 24/7 basis.

Your wireless network is a business-critical entity that needs the same level of protection you would extend to your physical assets. You need a strong wireless network security system working hand-in-hand with your enterprise insurance policies to protect you from loss and liability, and keep you in the enterprise of doing business.

Think about it: a catastrophic wireless network security breach can cause as much serious damage as any fire-or burglary-related loss. The primary sources of computer-related crime and financial losses are:

- Viruses
- Wireless network abuse
- DoS/DDoS Attacks
- Wireless network intrusion [2]

5.1 Viruses

According to the research firm Computer Economics, corporate losses related to viruses rose nearly 50% worldwide in 2005, reaching over US$28 billion.

5.2 Wireless Network Abuse

Up to 81% of employees who have high-speed Internet access at work spend as much as 2 hours a day involved in non-work related activities. At least 7% of employees use the Internet more than 7 hours per week for non-work related activities [2]. This indiscriminate surfing makes wireless networks vulnerable to hacker attacks, spyware/adware exposure, trojans, and virus downloads from rogue Web sites.

5.3 DoS/DDoS Attacks

In a Denial of Service (DoS) attack, targeted systems or wireless networks are rendered unusable, often by monopolizing system resources. A Distributed Denial of Service (DDoS) involves many computer systems, possibly hundreds, all sending traffic to a few choice targets. Denial of Service (DoS) and Distributed Denial of Service (DDoS) attacks were second only to viruses in accounting for corporate financial losses due to computer crime in 2005 [2].

5.4 Wireless Network Intrusion

Over 61% of enterprises surveyed indicated that they do not report their intrusion incidents because the negative publicity would damage their image and/or stock price [2]. Dollar value is one thing; enterprise value is another. Knowing the actual enterprise value of your IT infrastructure as it relates to your enterprise operations and enterprise mission is crucial to understanding the financial consequences of any wireless network security breach. This knowledge is important information to have, and should be top of mind, when developing an effective security plan and a realistic security budget.

To determine the enterprise value of a wireless network security solution, it's imperative to have your existing wireless network security implementation audited by a professional to determine what vulnerabilities there are within your system, how they might be exploited, and what you need to do to protect your enterprise. An auditor will evaluate:

- Enterprise procedures: Are they poor or inadequate?
- Strategic partners: How do they access shared information?

- Vendors: To which systems do they have access?
- Web site: Is your Web server protected?
- IT staff: Are they under-trained and/or overworked?
- Employees: Are they untrained in wireless network security policy, and what are they allowed to do on the enterprise wireless network [2]?

Looking at the elements of your system that are visible to third parties and considering the types of data and resources they could lead to if compromised, an auditor will determine the methods an attacker might use to gain control of your wireless network resources. Servers, modems, routers, and a host of other enterprise-critical architectural components, in the wrong hands, are an open door to your wireless network and your valued resources. Once you have this information, you can get a sense of the bottom-line damage an attack or intrusion will cost you.

The auditor will locate and document the flaws in your current security setup and recommend the best way to systematically address and resolve them, in order of severity. Once you have a plan, you can proactively begin to protect your interests.

6. CONSEQUENCES OF A SECURITY BREACH

Think of the danger to your enterprise's wireless network security in terms of Threat, Vulnerability, and Risk:

- Threat: The degree of potential damage
- Vulnerability: How readily the threat can be exploited on your wireless network
- Risk: The likelihood that the vulnerability will be exploited by the threat [2]

When a wireless network attack is executed against your enterprise, systems and departments across the entire enterprise are affected. By understanding the scope of the consequences of a security breach, you'll have a better idea of the value of your wireless network security insurance policy.

6.1 Loss of Revenue and Damage to Customer Confidence

If your customers cannot access your Web site, or have their e-mails to you bounce back undeliverable, the competition is just a click away — and no doubt more than ready to serve them. With identity theft [5] on the rise at an alarming rate, your customers' confidence plummets every time they learn of a wireless network compromise. They need to know that personal data and other information they entrust to your enterprise will be safe, secure, and untouched by unauthorized and malicious third parties.

6.2 Risk to Proprietary Business Assets

In many industries, an enterprise's most valuable assets include intangibles. Your intellectual capital and other proprietary information is vulnerable to theft, unauthorized distribution, or manipulation whenever your wireless network is breached. The process of replacing or repairing damaged or lost research, archival material, data, and other enterprise-critical information can be at best expensive or at worst, impossible. Are you willing to take that risk?

6.3 Damage to Business Partnerships

Your strategic partnerships are based upon trust and common objectives, and they are one of your most important enterprise tools. When your wireless network is breached, your reliability and reputation are called into question. You'll lose perceived value as a potential enterprise partner, resulting in lost opportunities.

6.4 Potential for Legal Penalties and Litigation

Today, strict regulatory mandates regarding accounting practices, privacy [4], and other industry-specific issues have created challenges for executive management and IT administrators alike. Penalties for violation are significant, and if you are found to be in the wrong, you could be subjected to both civil and criminal prosecution, with fines ranging into six-figure amounts. If you are found to be in violation due to a wireless network attack, are you financially prepared to defend yourself, pay fines, and mount a campaign to repair your damaged reputation with customers, partners, and shareholders?

6.5 Increased Overhead and Productivity Losses

Cleaning up after a wireless network attack is a time-intensive, complex process. You must allocate time and resources toward gathering documentation for the authorities and law enforcement, shutting down your system for assessment and repair, duplicating victimized hard drives, examining forensics, establishing a timeline of attack events, locating suspect files, and of course, revisiting your security strategy.

Executing an incident response effort means that some or all of your ongoing projects will have to be delayed (or abandoned completely) while damage control measures are being taken. This process can take months, and the additional overhead in staff hours can be debilitating.

7. SUMMARY AND CONCLUSIONS

By now, the rationale behind incorporating a strong wireless network security solution into your enterprise insurance plan is clear. In today's digital marketplace, with new and more volatile wireless network threats emerging every week, it's protection rather than reaction that will preserve your bottom line. If strong protection is the answer, the question must then be: what kind?

7.1 Integration and Scalability

Successful enterprises have learned that the most cost-effective wireless network security solutions are integrated, expandable systems that enable them to purchase and use fewer appliances over time. Instead of cobbling together a variety of difficult-to-manage hardware solutions and software services from various sources, they are opting for single-appliance solutions that include firewall, VPN, multi-layered security, antivirus, intrusion prevention, spam blocking, Web content filtering, and a centralized monitoring and management function.

7.2 Upgradeability

In evaluating a security solution, a significant consideration for small-to-medium enterprises is always future growth. As an enterprise matures, growth and change go along with the territory, and this directly affects a wireless network infrastructure. As enterprise objectives change, or products and services are added, the wireless network grows more complex to accommodate those changes. The security solution you invested in two years ago may not be able to keep up with your needs today. If you purchased a static system, your only option may be a fork lift upgrade, sacrificing a significant technology investment and necessitating a new one. By selecting an upgradeable solution from the outset, or replacing an outdated solution with an upgradeable option, you can easily and affordably get more performance whenever your needs change.

7.3 Expandability

The ability to add additional security services on an as-needed basis can extend the functionality of an integrated security system long beyond that of a piecemeal solution. Optional security services such as antivirus for the e-mail gateway, intrusion prevention, spam filtering, and Web content filtering, add value and allow you to customize your security as your requirements become more complex.

8. REFERENCES

[1] "Counting The Cost – What Is Security Really Worth To Your Business?" OnWatch, Copyright © 2005 WatchGuard Technologies, Inc. All rights reserved. WatchGuard Technologies, Inc., 505 Fifth Avenue South, Suite 500, Seattle, WA 98104, October 8, 2003.

[2] "Network Security 101: The Value Of A Protected Network," Copyright © 2005 WatchGuard Technologies, Inc. All rights reserved. WatchGuard Technologies, Inc., 505 Fifth Avenue South, Suite 500, Seattle, WA 98104, July, 2005.

[3] John R. Vacca, *Firewalls: Jumpstart for Network and Systems Administrators,* Digital Press, 2004.

[4] John R. Vacca, *Net Privacy: A Guide to Developing and Implementing an Ironclad ebusiness Privacy Plan,* McGraw-Hill, 2001.

[5] John R. Vacca, *Identity Theft,* Prentice Hall, 2002.

Chapter 8

STANDARDS DESIGN ISSUES

1. INTRODUCTION

Because it is rarely possible to secure every system in an enterprise's wireless network, some form of segmentation is required to defend critical systems from both internal and external attacks. There are various approaches to creating this segmentation, including router filtering, virtual local area networks (VLANs), and hardware firewalls. Some enterprises are using their existing infrastructure to create isolated segments for potentially high-risk systems, effectively creating a wireless network quarantine area. Isolating hosts from the main wireless network until they have passed a security screening process protects these systems from hostile attacks and the rest of the network from these insecure hosts. Creating a wireless network quarantine area is also effective for remote access and wireless users. A sample wireless security architecture standards design diagram is shown in Fig. 8-1 to give an overview of what is involved [1].

> **Note:** The segmentation of different portions of the wireless network using hardware firewalls and VLANs, and the widespread presence of antivirus and software firewalls is illustrated in Fig. 8-1. Interaction between these isolated segments is controlled using VPNs. Additionally, note that intrusion detection systems and other network monitoring systems are an integral part of the wireless network security architecture standards design.

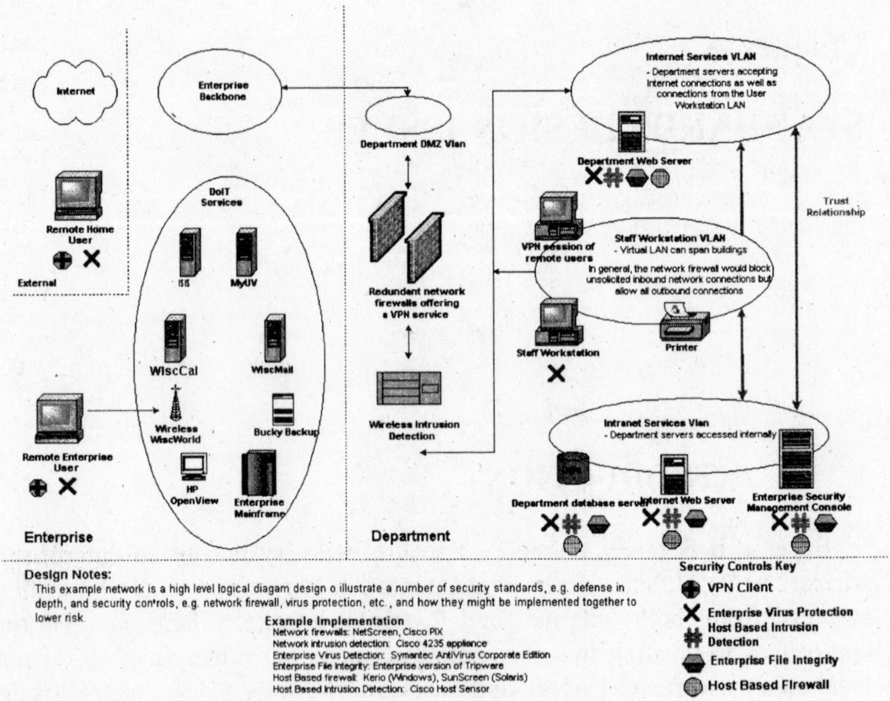

Figure 8-1. Example of a secure wireless network Architecture standards design.

So, with the preceding in mind, this chapter discusses particular aspects of wireless security architecture standards design issues in detail, that can be used for enterprise wireless networks standards design. Given the variety of options for developing a security architecture, some of the more commonly used alternatives are presented here with pros and cons.

2. SWITCHES

Using switches instead of hubs is an effective approach to improving the security and manageability of the wireless network. In addition to making eavesdropping more difficult, switches make it easier to track and/or disconnect problematic hosts. Some switches have port protection/restriction features that can be used to prevent unauthorized individuals from connecting hosts to the wireless network [1].

3. FLEXIBLE IP ADDRESS ASSIGNMENT

Assigning IP addresses using Dynamic Host Configuration Protocol (DHCP) or Bootstrap Protocol (BOOTP) offers a number of benefits that have security implications. For instance, you can require individuals to register their systems prior to getting an IP address, thus giving you a contact person for each system on the wireless network.

With the increasing number of vulnerabilities, some enterprises require computers to pass a security scan before being assigned an IP address. The ability to dynamically reassign an IP address, also gives you the flexibility to make major changes to the network with minimal technical support and little or no inconvenience to users. For instance, you could use DHCP or BOOTP to change the IP addresses of certain hosts, logically moving them to a VLAN that is protected from the Internet and other hosts on your wireless network [1].

4. ROUTER FILTERING

Many enterprises do not have perimeter firewalls and instead perform router filtering at the Internet border. For example, it is standard practice to perform ingress and egress filtering to block incoming traffic that does not have a destination address on your wireless network and outgoing traffic, that does not have a source address on your network (see RFC 2827). Some guidelines are more specific about the types of traffic to filter, recommending that routers be configured to discard packets with any loopback address (127.0.0.0), or link-local DHCP default address (169.254.0.0) or reserved private address (RFC 1918) in the source field [1].

In addition to filtering traffic using their core routers, some enterprises filter traffic on their edge routers to provide internal departments and groups with additional protection. This approach can be used to restrict access to specific systems or block certain protocols without the additional cost of installing a hardware firewall.

Router filtering can function up to a point, but large access control lists are difficult to manage and can impact performance and cause routers to malfunction. Additionally, blocking protocols without providing secure alternatives can create new security problems. For instance, preventing

NetBIOS traffic from entering the wireless network without providing a secure alternative for individuals to access data on their work computers remotely, can drive people to install programs like PCAnywhere that may introduce another serious risk if they are not configured security [1].

5. BANDWIDTH MANAGEMENT

Bandwidth management is a major issue in higher education environments, particularly with the increasing popularity of peer-to-peer (P2P) file sharing applications. Some P2P protocols can be limited using devices to restrict the amount of bandwidth that these applications can consume. However, this approach does not discourage the behavior behind excessive bandwidth usage.

An effective approach to reducing excessive bandwidth use is to temporarily disable the responsible party's wireless network connections, warning individuals that they are hogging network capacity, and to enforce fair penalties for continued misuse. An intrusion detection system can be used to detect certain P2P protocols, triggering further investigation as needed. For example, The University of Florida developed a system called ICARUS (Integrated Control Application for Restricting User Services) to detect excessive bandwidth usage and automatically disable the associated user's wireless network connection [1]. After three violations, the individual associated with the problematic IP address is brought before a disciplinary body.

Notably, the approach of disabling an individual's wireless network connection depends on the ability to identify the person, as previously discussed in the flexible IP address assignment part of the chapter.

6. FIREWALLS AND NETWORK ADDRESS TRANSLATION

Perimeter security is a major challenge on some enterprise wireless networks. Installing a firewall at the Internet border can provide a strong layer of security, but is not feasible in all cases. A recent industry analyst survey shows that nearly 83 percent of the enterprises surveyed have a perimeter firewall. However, this statistic does not measure the effectiveness of this approach and the associated costs.

Installing and managing a firewall is not a trivial task; it requires technical knowledge and significant planning to minimize negative impacts.

Poorly managed or unmanaged firewalls can become a wireless network liability, disrupting legitimate network activity and creating a dangerous single point of failure on the network. This risk has led some enterprises to remove firewalls after the individual responsible for maintaining the device leaves or if the rules become unmanageable. Having to remove a firewall under such circumstances negates the protection that it was intended to provide and wastes the resources that went into purchasing, installing, and maintaining it.

An added challenge is creating a single firewall policy that meets all the needs of every group can be difficult and can result in so many holes in the firewall as to defeat its purpose. One approach to addressing this issue is to offer an opt-in firewall service, using different interfaces or VLANs to separate groups of machines from each other.

Although some enterprises have successfully implemented perimeter firewalls, it is important to realize that they can create a barrier for high-performance wireless networking. Notably, installing a firewall is not a complete solution; there are often other entry points into the wireless network (for example, modems, VPN concentrators, and wireless access points), and intruders may be able to bypass a perimeter firewall in other ways (for example, a Trojan horse or vulnerable Web server). Therefore, it is necessary to take additional steps to harden systems behind the firewall. Given the disadvantages of installing a perimeter firewall, developing a persuasive cost-benefit analysis for a firewall at the Internet border can be a challenge, often requiring data showing a number of attacks and incidents that would have been prevented [1].

6.1 Firewalling Critical Systems

Rather than, or in addition to installing a firewall at the Internet border, some enterprises have deployed firewalls internally to provide a secure sanctuary for critical systems, creating internal defenses by segmenting the network on physical boundaries. A recent industry analyst survey indicates that 80 percent of respondents have interior firewalls and another 20 percent are in the process of implementing them. Creating this sort of safe haven protects core systems from computers on the Internet; as well as, other hosts on your own wireless network [1].

6.2 Departmental Firewall/Network Address Translation (NAT)

Some internal departments reduce their exposure to external attacks by installing and managing their own hardware firewall, restricting access to specific computers, and blocking certain protocols. Also, Network Address Translation (NAT) can be effective for some departments, reducing their exposure to one publicly accessible IP address. However, blocking traffic from the Internet as well as internal system can hinder vulnerability scanning and prevent the central wireless network management staff from administering switches and routers behind the NAT device. Also, unless some form of logging is enabled on the NAT device, it is difficult to attribute activities to a particular host.

As noted earlier, a firewall can also become a liability for a department, if the individual responsible for maintaining the device leaves or if the rules become too complex. To enable departments to protect themselves using a hardware firewall while minimizing the potential negative impact, some enterprises offer managed firewall services or help departments choose and install firewalls that are best suited to their environment [1].

7. VIRTUAL LOCAL AREA NETWORK (VLAN)

Some enterprises have started using VLANs to create a separation between logical groups of computers on their wireless network. In this way, different firewall rules and levels of wireless network monitoring can be applied to different groups. For instance, all administrative systems can be placed on a highly secure VLAN that permits outbound traffic, but denies all inbound connection attempts from the rest of the wireless network. Unknown computers connecting through wireless access points can be placed on an isolated VLAN that does not permit outbound or inbound traffic, requiring individuals to connect to a VPN server and authenticate themselves using their username and password before they can access the rest of the wireless network.

Although VLANs require some administrative overhead to maintain, they offer an inexpensive, flexible, and powerful way to improve wireless network security architecture standards design in enterprises [1].

8. VIRTUAL PRIVATE NETWORK (VPN)

According to recent industry analyst survey, a growing number of enterprises are using VPNs for various purposes, including remote access and wireless network security. Two of the primary functions of a VPN are to protect data in transit and to authenticate individuals or computers. For instance, remote users with a valid username and password can establish an encrypted connection to an enterprise's wireless network via a VPN concentrator, thus protecting all data being transferred over the Internet between their personal computers and the enterprise's wireless network. Similarly, computer systems can be configured to automatically establish IPSec connections that encrypt all traffic traveling between them. Windows Server 2003 and above have the capability to automatically vend machine certificates to computers in a domain and automatically establish IPSec connections between systems.

Notably, VPNs can introduce insecurities in a wireless network, because they can turn a remote user's system into a dual-homed host, creating an avenue of attack from the Internet. For instance, if an individual connects from home via a broadband Internet connection with a VPN that client does not include a software firewall, an intruder could break into the user's home machine from the Internet, and then connect to the enterprise's wireless network via the user's VPN connection. This security weakness can be mitigated by requiring remote users to install software firewalls. Home users can also be encouraged to create layers of security by setting security standards for their home computers and by using inexpensive Small Office Home Office (SOHO) devices that have some firewall/NAT functionality [1].

9. REMOTE ACCESS SECURITY

The security of each entry point into a wireless network deserves some attention. Because remote access VPN concentrators create entry points into a network, they may require independent firewall rules and monitoring systems. Also, enterprises that provide dial-up access have found that intruders connect through their modem banks using stolen usernames and passwords. However, many enterprises cannot afford to install a firewall or intrusion detection system to detect problems on their dial-ups. As a result, stolen dial-up accounts can go undetected for long periods until this becomes a major problem. Additionally, although caller-ID can be useful for investigating these types of incidents, some enterprises cannot afford the additional expense [1].

10. WIRELESS LOCAL AREA NETWORK (WLAN) SECURITY

More enterprises are deploying WLANs as more wireless devices are being used, as the cost of wireless access points decreases, and because of the resulting reduction in cabling costs. Although emerging wireless standards have enhanced security features such as port-based authentication in 802.1x and VLAN tagging for wireless users in 802.11q, most enterprises are deploying wireless networks based on 802.11b standard, despite its security problems. In addition to the configuration of wireless Access Points, the three primary issues relating to wireless security are authentication, encryption, and security management. Encryption of wireless traffic is critical given the increasing number of tools available for eavesdropping on network traffic. Wired Equivalent Protocol (WEP) is not effective because of security weaknesses and lack of end-to-end security; it only encrypts between the client and access point. Advanced Encryption Standard (AES) is stronger but also does not provide end-to-end security between client and network resources. Furthermore, both only support limited roaming. Also, given the distributed nature of wireless networks, some form of centralized management of devices and user accounts/rights is desirable.

Security based on MAC addresses is not flexible enough for most higher enterprise environments, because it does not accommodate the need for individuals to use their own various wireless devices. Additionally, MAC address spoofing enables unauthorized access [1].

11. USING VPN TO IMPROVE WIRELESS SECURITY

A Virtual Private Network (VPN) can be used to provide a low-cost solution for wireless security that enables authentication, network segmentation, and encryption between clients and a VPN concentrator. Although this approach does not provide end-to-end encryption, it is sufficient for most purposes because the traffic is protected while it is transmitted through the air and from the wireless access point to the VPN concentrator. For added protection, when accessing secure resources from wireless networks, it is advisable to require use of cryptographic protocols such as IPSec, SSH, or TLS to provide end-to-end encryption.

Using a VPN does not necessarily prevent malicious individuals from gaining unauthorized access to a wireless network and eavesdropping on traffic. However, important systems and data can be defended from unauthorized wireless users by combining VPN with VLANs and firewalls to segment the network. For instance, allowing anyone to connect to a wireless network that has access to very limited network resources enables guests and student laptops to connect with ease while limiting the risk of abuse. If someone needs to access other resources from the enterprise or the

Internet via wireless, require them to connect through a VPN concentrator, authenticating against existing authentication systems (for example, RADIUS and Kerberos).

Notably, this approach to wireless security does not address the need for roaming, misuse detection, or centralized management of devices and user-based access rights. These features are available in third-party solutions [1].

12. THIRD-PARTY SOLUTIONS

WLAN management solutions provide some features that are not available in many homegrown approaches such as centralized management, access control, and roaming. For this reason, some enterprises have adopted commercial WLAN management systems, but few are using the data encryption capabilities. Third-party systems for securing WLANs have the following advantages:

- Compatible with a variety of access points
- Use existing authentication system (for example, RADIUS)
- Provide guest accounts with limited wireless network privileges for visitors
- Specify access control (which users can access which resources and when)
- Have MAC spoofing detection (for example, detects duplicate wireless network addresses); not clear how effective this is
- Offer session management while moving between access points (roaming)
- Provide centralized, Web-based management [1]

13. DIRECTORY SERVICES

Every enterprise needs to restricting access to systems and the data they contain. Although most operating systems can be configured to require a username and password, and to only permit certain users to access specific data on the system, maintaining this information on each individual system can become unmanageable. To manage authentication and authorization more effectively, some enterprises use directories (LDAP; Windows Server 2003 Active Directory) to store information about users in a central repository. Directory services can be used to support a wide range of security applications, including Internet middleware initiatives in enterprises [1].

14. WEB SERVER SECURITY

Important Web servers such as the enterprise's main server and other publicly accessible Web servers are major targets of attack from the Internet. Additional measures to protect these systems against malicious or accidental harm include making all updates on a staging server, running Common Gateway Interface (CGI) on a separate server, and having a hot backup server. Using a staging server that is separate from the main server enables you to restrict direct access to the main server and make updates to the main server in a more controlled fashion. Placing CGI on a separate server prevents intruders from gaining access to the main server via insecure programs. A backup server that is regularly synchronized with the main server enables you to recover from an incident quickly by replacing the primary server with the backup. Additional recommendations for securing public Web servers are available from CERT (http://www.cert.org/security-improvement/modules/m11.html) [1].

15. APPLICATIONS AND DATABASES

It is not feasible to require a thorough wireless network security standards design in all applications and databases deployed on an enterprise's network. Some application and database guidelines are simple to require such as SSL/HTTPS for Web-based services and logging facilities for auditing in the case of a problem. However, in many instances, it is necessary to accept commercial products that have serious vulnerabilities (for example, MSSQL). Additionally, it may be too costly to implement security in some commercial systems (for example, Oracle) or in-house applications.

Therefore, it is necessary to defend these systems at the wireless network level as discussed at the beginning of this chapter. Using tiered database/application architecture and other middleware solutions when available, can make it easier to defend these systems and protect the data they contain. More advanced requirements such as peer review of source code are difficult to enforce. Participating in the PKI [4] Lite/Federal Bridge project may be feasible in some enterprises, providing support for IPSec, S/MIME, and other security features such as access control. Other technologies such as VoIP and IP security cameras can be difficult to secure and require special attention [1].

16. MONITORING

Intrusion detection systems and other wireless network level monitoring are a critical part of wireless security architecture standards design in enterprises. Your lack of control over the wireless network necessitates a higher degree of awareness of what is occurring to effectively detect and address problems. Open source tools are favored for wireless network and log monitoring within enterprises because they are free, flexible, powerful, and often have broad user bases that can equate to free technical support. However, some enterprises use commercial tools for monitoring wireless network traffic, processing log files, and locating devices. The choice of tools depends on the level of monitoring that is desired and permitted – some enterprises can monitor content and restricting access to certain Web sites while others are prohibited from doing this by their privacy policies [3].

Notably, it is not sufficient to be aware of activities on your network, it is also necessary to locate and deal with problematic hosts. Programs like NEO (http://www.ktools.org/) are useful for locating and disconnecting problematic systems [1]. Some enterprises regularly preserve ARP caches on wireless network devices to help associate suspicious activities with particular hosts even after the computer in question has been disconnected from the network.

Several approaches to monitoring wireless network activities common in enterprise environments are presented here. Importantly, logs of wireless network activities can be used against you to prove involvement in a denial of service attack or P2P file sharing infraction. Therefore, it is important to have a data retention policy governing the logs produced by wireless network monitoring systems.

17. INTRUSION DETECTION SYSTEMS

A common approach to gaining awareness of activities on a wireless network is to install an intrusion detection system. Other enterprises might install multiple intrusion detection systems at several strategic locations on their wireless network. In all cases, these systems are invaluable for detecting and responding to wireless security incidents [1].

18. CENTRALIZED SYSTEM LOGGING

Sending system logs to a centralized server makes it more difficult for intruders to modify these data and enables you to analyze activities from many systems for unusual activities such as multiple systems being targeted by a single intruder. Sending UNIX syslogs to a central logging server is described at http://www.cert.org/security-improvement/implementations/i041.08.html. Windows Event Logs maintained on domain controllers can also be used to gain more awareness of what is occurring on your wireless network [1].

19. METRICS

Finally, the purpose of metrics in wireless network security is to measure progress in implementing the wireless security architecture standards design plan. Such metrics are often closely related to risk assessment metrics, for instance:

- Number of hubs replaced with switches and number of remaining hubs
- Percentage of critical servers placed behind firewall
- Percentage of hosts placed on secure VLAN
- Number of VPN users
- Percentage of servers communicating with IPSec
- Number of secured protocols/passwords (for example, SSL-enabled Web servers, e-mail servers requiring IMAP over SSL or Kerberized POP)
- Number of people using PGP or S/MIME to encrypt e-mail [1]

> **Note:** Metrics from other areas such as vulnerability assessment, security implementation, and incident response may show trends that require changes in the wireless security architecture standards design.

20. SUMMARY AND CONCLUSIONS

Rather than grafting security onto existing systems, it is more effective to redesign systems to make security an integral part of them. However, developing a wireless security architecture standards design for enterprises is complex because of the needs of different groups sharing the network (for example, departments within the enterprise). Many enterprise wireless networks must be able to accommodate unknown devices, including handheld devices and being connected by visitors, internal users, and remote users.

One of the overarching principles when designing a secure system is defense in depth. By combining multiple security measures to create a layered protection, you can protect systems even when one layer of security fails. This concept can be applied to an individual computer system and to an entire wireless network. For instance, at the host level, the practice of disabling unnecessary services can be bolstered by a software firewall [2] that denies access to those services in case they are accidentally enabled. At the wireless network level, systems protected by a firewall can be further protected by requiring users to connect through a virtual private network (VPN) using a username and password.

Scalable host security is another key objective that may require some consideration while developing the wireless security architecture standards design, such as designing a Windows Server 2003 domain or an enterprise wide antivirus solution. Although centralized host management can be costly, in the long term the resources required to install and maintain host management systems can be less than those required to repair large numbers of unmanaged computers that are infected or compromised.

21. REFERENCES

[1] "Security Architecture Design," © Copyright 1999-2005 EDUCAUSE. All rights reserved. EDUCAUSE, 4772 Walnut Street, Suite 206, Boulder, CO 80301-2538, October 8, 2004.

[2] John R. Vacca, *Firewalls: Jumpstart for Network and Systems Administrators,* Digital Press, 2004.

[3] John R. Vacca, *Net Privacy: A Guide to Developing and Implementing an Ironclad ebusiness Privacy Plan,* McGraw-Hill, 2001.

[4] John R. Vacca, *Public Key Infrastructure: Building Trusted Applications and Web Servicess,* Auerbach Publications, 2004.

Chapter 9

AUTHENTICATING ARCHITECTURAL DESIGN CONSIDERATIONS

1. INTRODUCTION

The advances in communication infrastructure and computing devices have led to a paradigm shift in communication services. There is a strong demand for the enterprise to provide advanced communication services. New services demand new multimedia content with multimodal interactions. Both content and interactions are increasingly diversified even within one application. There is an increasing need for enterprise applications to provide multimedia and multimodal services.

On the other hand, the communication infrastructure itself has evolved. One notable trend is that mobile device has become more suitable for integration with components of enterprise application. Increased device functionality coupled with decreased costs has created a flood of devices into the enterprise. Meanwhile, the rapid advance of wireless technology and protocols and popularity of wireless application offers enterprises an opportunity to provide advanced services to mobile user by integrating these technologies with their applications.

Two conventional issues still exist at this new multimedia wireless paradigm, security and authenticating architectural design, to provide advanced communication services to mobile users. Security is a make or break proposition in the realm of wireless applications. Not only do the same risks of cracks and viruses affect wireless systems just as they do in the wired world, but multiple levels of vulnerabilities also come into play when designing and deploying wireless applications, from interoperability issues down to securing the device itself [1]. Wireless also makes it easier to intercept than taping into a wire-line. Approaches based on remote login often lead to a direct data pipe that goes through the firewall [2] to user's mobile device. In a mobile environment, such access methods have become inadequate and insecure, especially for advanced enterprise communication services with multimodal interaction and multimedia content. It is, therefore, important to balance the convenience of wireless applications against the security risk and design such applications with these issues in mind.

Another challenge is how to design a suitable architecture so that requirements for security is satisfied and new technologies and mechanisms can be incorporated into the current system quickly and easily without modifying the current infrastructure. In fact, one of the challenges in enterprise multimodal/multimedia communication is the need to connect and deliver services to a converged communication environment that can consist of multiple network bearers, including traditional PSTN telephone, wireless, LAN, WLAN, VoIP, etc. Many services and applications developed in the past are tied to particular communication bearer [1], and it has become increasingly difficult for an enterprise to port, maintain, and manage applications for each network environment, given the fast pace of the communication infrastructure expansion.

With the preceding in mind, this chapter presents the architecture and the underlying mechanism of the WMTIP. It also describes the experimental authentication of architectural design considerations and testing procedures. Then, it reports the experimental results and their analysis. Finally, the chapter concludes that WMTIP can effectively deliver multimedia content with enhanced user experience and high security.

2. WIRELESS MULTIMEDIA TECHNOLOGY INTEGRATION PLATFORM (WMTIP) ARCHITECTURAL DESIGN

To overcome the limitations outlined earlier for wireless communication services, an event-triggered approach is adopted for wireless enabled secure service notification, multimodal service access and multimedia service content delivery [1]. Also, based upon the principle of separation of concerns, the approach here effectively separates service control from service delivery and content generation from content presentation. This approach leads to the implementation of a loosely coupled system in which the subsystem components communicate with one another purely through mutual exchange of messages.

Fig. 9-1 illustrates the overall architecture of the prototype system [1]. There are two major subsystems within the architecture: the wireless secure broker and the enterprise application. Each one has distinct and well-defined responsibilities. The enterprise application handles all the application logic and control, and content generation. A dialog server is integrated with the enterprise application for providing dialog-based user interfaces. The wireless secure broker, on the other hand, takes care of content presentation and delivery. Collaboration between the wireless secure broker and the enterprise application is achieved through a purely asynchronous message driven interface, with all application level control residing with the enterprise application.

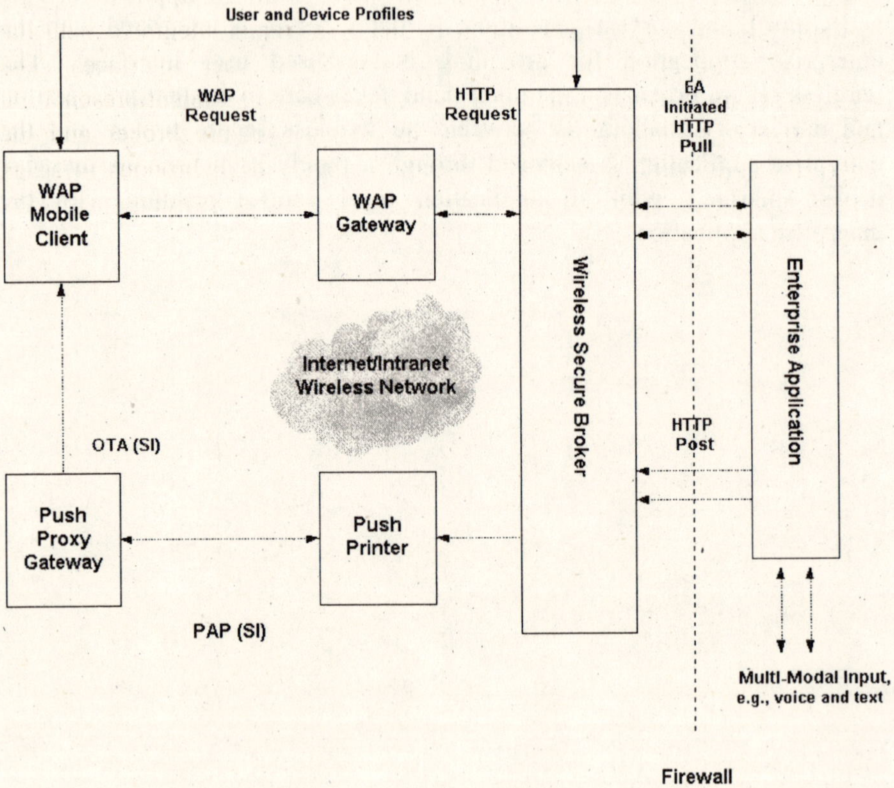

Figure 9-1. Overall architectural flows and protocols of WMTIP. Notations: EA, Enterprise Application; WAP, Wireless Application Protocol; OTA, Over-The-Air protocol; PAP, Push Access Protocol; and SI, Service Indication.

For the implementation of the prototype system, voice-based dialog events are treated as synchronous events within a dialog server, which is almost natural for a dialog-based user interface. From the point of view of the entire system, however, handling events synchronously was considered not to be the best implementation when computing resources are scattered across a network. This is especially true for applications that work across heterogeneous networks (wired and wireless) where the quality of service often varies significantly and unexpectedly. It is thus highly desirable that service content can be generated, transmitted, and processed in an asynchronous manner. Equally important, however, is that these engineering considerations should maintain and enhance, rather than alter or compromise, the underlying functionality.

These objectives have been achieved through careful design of an architectural asynchronous mode of interaction and loosely coupled interfaces between the enterprise application and the wireless secure broker; as well as, between the various components within these subsystems. Any communication between the enterprise application and the wireless secure broker has to be initiated by the enterprise application, thus denying any external system the ability to control the enterprise application. All forms of application control, therefore, lie entirely with the enterprise application.

In order to protect the enterprise application against insecure remote access, the entry process to the enterprise application server is toughened up by putting it behind a firewall. No real sensitive data is stored persistently outside the enterprise application server.

Since wireless devices are far more likely to fall into the hands of unauthorized users, providing strong authentication and authorization approaches is extremely critical. An important principle that has been adopted in order to address this problem is to perform all the authentication and authorization checks at the application level rather than at the web server. An application-driven authentication routine goes and checks a database prior to authenticating the user and allowing the user access to services.

By using a file-based approach to storing registration records, any new registration information would be stored in a text file on the file system of the wireless secure broker. When the enterprise application requested the pending registration records, the same file would be read and used to generate the response document to be sent back to the enterprise application. Once sent, those records would have to be deleted from the file, in order to avoid resending them during the next request from the enterprise application. Considering that a wireless (or even wired) web-based application could possibly be accessed by hundreds or thousands of users at the same time, it is not difficult to imagine a situation where a large number of users are trying to register at the same time as the registration information is requested by the enterprise application. Proper and reliable functioning in such situations would require complex and low-level file locking and text processing mechanisms to be implemented within the system. In addition, this approach would not be secure enough, since the registration information is stored on the server in plain text files on the file system, which offers neither password protection nor some form of encryption or encoding of the data.

In addition, you can also build the system to a hybrid model-view-controller architecture, where most of the application logic is encapsulated in reusable components independent of the presentation technology used. These components are shared by modules within the system, and could also be used across systems. This approach provides a much cleaner and adaptable system architecture [1].

3. EXPERIMENTS

As previously mentioned, the infrastructure here provides a number of security features. Among them is the Wireless Secure Server (WSS), which acts as a secure broker between the end-user space and application. This is one of the most important features, which provides security in service delivery and flexibility in content generation. In this part of the chapter, a measurement is taken of how the introduction of WSS could affect the overall performance by studying two applications: Airlines Services and Credit Card Services [1].

3.1 Test Environment

The approach here is implemented as a prototype of an advanced enterprise communication server. It is a wireless multimedia technology integration platform (WMTIP) to support various applications in enterprise customer relation management (CRM) services for mobile users. The enterprise application is based on a multimedia dialogue server, MTIP, which has a distributed dialogue system architecture. The dialogue system and web convergence in MTIP is achieved through an approach based on Hybrid-VoiceXML which has VoiceXML as its voice modality [1]. WMTIP is a step further beyond MTIP, with convergence of a wireless web for mobile users.

The prototype system of WSS is developed and deployed using Java Web Service Developer Pack (JWSDP) with tomcat. The service is hosted by a machine sitting outside the enterprise firewall. WSS and the enterprise application infrastructure (in this case, the Multimedia Technology Integration Platform [1]) are connected through an Ethernet with a speed of 100 Mbps. AT&T Wireless is chosen as the wireless service provider and the mobile phone used in the test case is a Sony Ericsson with GPRS and WAP-push services [1]. WSS will deliver the content to the phone in a XHTML format [1].

The two applications of Airlines Services and Credit Card Services, are studied to evaluate the performance. The Airlines Services provides customers information with regards to flight, luggage claim, and limo pick-up services. Besides a text instruction and map, a telephone link to reach the airlines customer service center or limo driver is also provided, which is embedded in a web page. The customer just needs to click the OK key, so that a call will be made automatically. The Credit Card Service offers customers the ability to view their statements on a mobile device and transfer the balance over the phone. Using this service, the customer is able to obtain a detailed current statement including account number, user name, previous balance, current balance, closing date, and balance due. The balance transfer service gives the customer an option to transfer the current balance from one credit card to another one [1].

3.2 Understanding Various Belays

Before you can collect data, some terms should be clarified to better understand the various delays. For example, Push Request Time (PRT) is defined as the time that MTIP sends push-requests to WSS via HTTP. This is the delay caused by the introduction of WSS because MTIP would invoke Push Initiator directly with the absence of WSS.

And, Push Initiation Time (PIT) is the time that the Push Initiator is launched, whereby WSS sends a push request to PPG. In this case, the push process is constructed to include two operations, push submission and status query. The push submission is to deliver a push message from a Push Initiator to a PPG, which should deliver the message to a user agent in a device on the wireless network. The status query is the operation that the Push Initiator requests, regarding the current statue of a message that has been previously submitted [1]. Therefore, in this case, the push initiation consists of one pair of push/response messages, and at least one pair of status-request/status-message, as shown in Fig. 9-2 [1].

> **Note:** The Push Initiator queries the status at most three times or exists upon receiving the delivered status message.

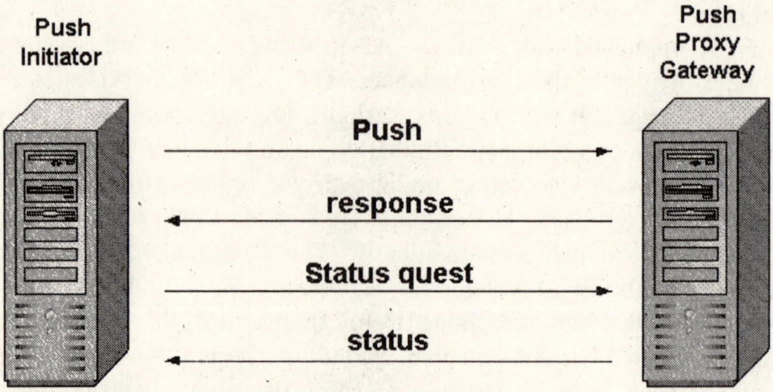

Figure 9-2. Push Initiation.

Now, the Service Provider Time (SPT) includes the airtime from PPG to mobile phone, the time that the phone proceeds message alert, the airtime from the mobile phone to WAP gateway for pulling content, and the time used by the gateway to proceed the WAP request. On the other hand, the WSS Service Time (WST) is the time that WSS delivers the application content in the XHTML format through HTTP to the WAP gateway.

Fig. 9-3 depicts the clarification for the terms defined in the preceding [1]. Please refer to Fig. 9-3 for more details on protocols and authenticated architecture design.

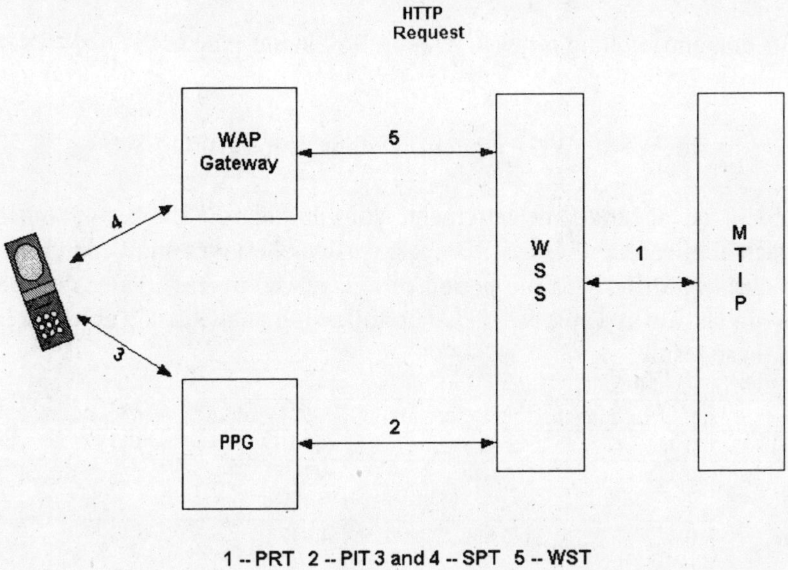

Figure 9-3. Time Defintion.

4. PERFORMANCE EVALUATION

After setting up the testing environment, you should run tests to collect performance data. This part of the chapter briefly describes the testing procedure, report and analyses of the results.

4.1 Test Procedure

You should use a software tool [1] for capturing and analysis of packets transmitted on a network, to record the traffic on the MTIP side and WSS side, respectively. The time stamp for each packet should be recorded by the tool. PRT, PIT, SPT and WST can be calculated by using the following formula, whereas T is annotated as the packet time stamp [1]:

PRT = response packet WSS->MTIP – request packet MTIP->WSS

PIT = status packet PPG->WSS - push request packet WSS->PPG

SPT = content-fetching request WG->WSS - status packet PPG->WSS

WST = packet WSS->WG - content-fetching request WG->WSS

To obtain an accurate measurement, you should run the test 5 times against each application of Other Services, Airlines Services and Credit Card Services during a different time period of a day. The average values of the five runs are shown in Table 9-1 [1]. The following analysis in Table 9-1 is based on these results.

Table 9-1. Experimental Result

Services	PRT (seconds)	PIT (seconds)	SPT (seconds)	WST (seconds)
Other Services	0.365	0.685	8.701	0.115
Airlines Services	0.321	0.674	6.310	4.681
Credit Card Services	0.327	0.699	4.942	0.182
Average	0.338	0.686	6.651	
Standard Deviation	0.024	0.012	1.902	2.617

4.2 Result Analysis

Before you analyze the results, you should clarify that the PIT and SPT are not related to the applications, and the application has very little impact on PRT while WST is application dependent. More detail explanations are follow.

During the process, MTIP sends push-requests to WSS, and MTIP sends data in an XML format through HTTP. The data consists of two parts: control entity and content entity, similar to PAP. The control entity is the information related to the subscriber, such as subscriber id, phone number, gateway identification, phone screen size, etc; therefore its size is fixed. The content part carries the application specific data. For example, for the credit card services, it is a statement in XML format. The size of the content entity varies and could affect the PRT, but slightly, because it only carries data in plain text format and the image files are stored in the local machine of WSS for efficiency purposes. In the future, you should provide WAP push services for any users based on SOAP. In such a situation, PRT may vary dramatically since the SOAP message could carry large size files.

From Table 9-1, you can see that PIT is very stable with a standard deviation of 12 milliseconds. The reason for this is that PAP will define the push-message, push-response, status-query-message, and status-query-response using DTD. Furthermore, in most test cases, the Push Initiator has received a delivered status, upon the first time status query. Therefore, only one pair of status-query-message/response is transmitted.

> **Note:** The average value of PIT doubles the value of PRT, because PAP is transmitted over HTTP, and the push initiation process consists of two rounds of HTTP request/response while PRT just has one.

The test results should show that for direction services, SPT is much longer than the other two applications. However, this difference is not caused by the application, but reflects the wireless network situation during the test. On the other hand, WST is determined by applications. For example, the service time for airlines services is much longer than the other two, because this application delivers two colorful pictures to the user agent of the mobile phone.

As previously mentioned, PRT is the delay caused by the introduction of WSS; however, compared to the delay on the service provider side, it is minimal. It accounts for only 5% of SPT. One of the issues in measuring SPT, is that you need to send Service Indication, instead of Service Loading to the user agent in the mobile phone. This is because SI gives users options to load immediately or store the message for possible postponed viewing and does not disrupt the user's current browser activities. The side affect for the experiment is that you have to load the content manually by clicking the OK key, which causes the recorded SPT to be longer than the actual value. However, even if you use one second to click the OK key, PRT is still only 6% of SPT [1].

Finally, by analyzing your experimental results, you could draw a conclusion that the delay caused by the introduction of WSS is minimal, compared to the delay on the service provider side. However, with such a small overhead, your infrastructure provides a secure mechanism to delivery services to mobile users.

5. SUMMARY AND CONCLUSIONS

This chapter presented an experimental report of a multimedia wireless system with secure delivery of multimodal services to end-user over voice and data converged networks. The design should lead to an easily adaptable and authenticated extensible system architecture, in which system components interact with each other through loosely-coupled interfaces based on asynchronous exchange of messages. It effectively separates service logic from service delivery, through the introduction of a secure broker between the end user devices and the enterprise level application that contains the service logic. This separation also provides additional levels of security and reliability at the application level. This chapter also describes a system implementation and its performance results.

Finally, this chapter has presented an approach for secure and reliable delivery of multimedia services and content for multimodal applications over converged networks. The approach here leads to the architecture and design of a system with a very clear and well-defined separation of concerns between various subsystems and components. This results in the architectural design of flexible and adaptable systems that can be easily upgraded or extended without inconvenience to end users of the system. The chapter also showed how to implement a prototype system that has been designed and developed in order to validate the proposed approach. In addition, the chapter showed how to run a set of experiments to validate that the additional flexibility in the authenticated architectural design, does not compromise the performance of the overall system. In fact, the additional delay is minimal, as compared to the wireless network delay caused by the service provider.

6. REFERENCES

[1] J. Jenny Li and Eric Wong, "Case Study of a Multimedia Wireless System," [Avaya Labs Research, (formerly part of Bell Labs) Basking Ridge, NJ 07920 and Department of Computer Science, University of Texas at Dallas, Richardson, TX 75083] © 2004 Avaya Inc.. All rights reserved. Avaya Labs Research - Headquarters, 233 Mt. Airy Road, Basking Ridge, NJ 07920, 2002.

[2] John R. Vacca, *Firewalls: Jumpstart for Network and Systems Administrators*, Digital Press, 2004.

PART III

PLANNING FOR WIRELESS NETWORK SECURITY

Chapter 10

IMPLEMENTATION PLAN DEVELOPMENT

1. INTRODUCTION

This chapter describes the overall security implementation plan development for wireless networks. It aims to give you a thorough understanding of the security planning for the wireless network solution and the reasons for designing it that way. It should also give you enough information to help you adapt the plan to the particular needs of your enterprise.

The chapter begins with a description of how 802.1X and Protected Extensible Authentication Protocol (PEAP) work to secure access to the wireless network. This is followed by a description of the target enterprise for the implementation of the wireless network security plan development and explores some of their key requirements.

The middle portion of the chapter describes the security planning for the wireless network solution, including: the network design; Internet Authentication Service (IAS) server placement; selection of hardware and software; obtaining certificates; and client configuration. How to migrate from an unsecured WLAN to 802.1X and PEAP is also outlined.

The end of this chapter deals with variations on the basic implementation of the wireless network security plan development. The most important of these planning variations is how to scale the implementation of the wireless network security plan development for use in larger enterprises, which is discussed at some length. Other planning options covered are:

- Reusing the IAS infrastructure for wired LAN security
- Using IAS for remote access authentication

- Deploying WLANs in SOHO environments [1]

As part of planning for your secure wireless network implementation you need to ensure that you have the right skill sets available in your enterprise and that you involve the right people in the decisions affecting the deployment. To get the best out of this chapter, it would be helpful if you were familiar with the following topics:

- Networking concepts particularly Wireless LANs.
- Certificate services and public key infrastructure (PKI) concepts.
- General security concepts such as authentication, authorization and encryption.
- Security features such as users, groups, auditing, and access control lists (ACL).
- Application of security settings [1].

2. HOW WIRELESS LAN SECURITY WORKS

The original 802.11 WLAN security scheme, known as Wired Equivalent Privacy (WEP), has serious security deficiencies that allow an attacker to discover the network key and break on to the network. This scheme is known as Static WEP in this, because of its use of a fixed network access and encryption key shared between all members of the WLAN.

Use of the IEEE 802.1X gives a strong access control mechanism for the WLAN. This must be coupled with a secure Extensible Authentication Protocol (EAP) method. The choice of EAP method defines the type of credentials that can be used to authenticate users and computers to the WLAN.

PEAP is a means of protecting another EAP method (such as MS-CHAP) within a secure channel. The use of PEAP is essential to prevent attacks on password–based EAP methods.

Strong data protection of the WLAN traffic can be provided by either dynamic WEP or WPA. Master encryption keys for data protection are generated as part of the 802.1X authentication process (although dynamic WEP and WPA use these keys differently).

The distinction between static WEP and dynamic WEP is crucial. Dynamic WEP uses the same encryption algorithms as static WEP, but continually refreshes the encryption keys, therefore defeating known attacks on static WEP. Dynamic WEP only refers to the network data protection mechanism; network authentication is handled separately by 802.1X.

2.1 How 802.1X with PEAP and Passwords Works

Fig. 10-1 illustrates how 802.1X with PEAP works [1]. The figure shows the following four main components:

- Wireless Client
- Wireless access point (AP)
- The RADIUS server and directory
- The internal network [1]

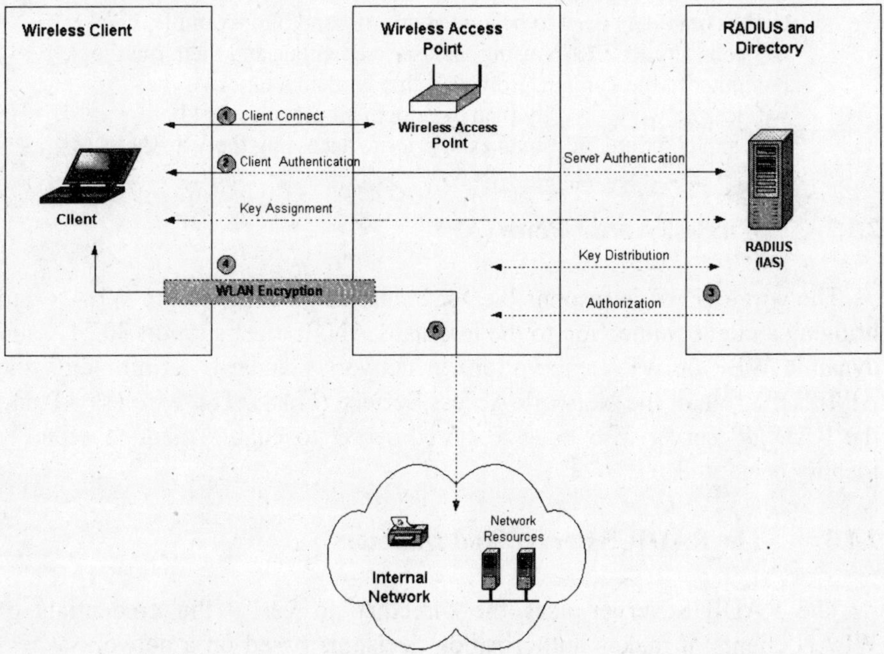

Figure 10-1. 802.1x and PEAP authentication to the wireless LAN.

2.1.1 Wireless Client

This is a computer or device running an application that requires access to network resources. The owner of the credentials that are used to authenticate the client to the network can be a user or a computer. The client must have a WLAN network adapter that supports 802.1X and dynamic WEP or WPA encryption. The client is also referred to as the station (STA) in many network standards documents.

Before the client can be granted access to the WLAN, it must agree on a set of credentials with the authentication service (the RADIUS server and directory) in some out–of–band operation. In this case, the domain accounts of the user and computer are created prior to connecting to the WLAN. The client knows its password and the domain controller (the directory) is able to verify the password. The client must also be preconfigured with the correct WLAN settings, which include the WLAN name and the authentication method to use.

> **Note:** Strictly speaking, only one set of credentials (either the user or the computer) need to be agreed out-of-band. For example, you can connect the WLAN using the user credentials and then join the computer to the domain. However, this implementation of the wireless network security plan development assumes that both user and computer accounts exist prior to accessing the WLAN.

2.1.2 Wireless Access Point (AP)

The wireless AP is responsible for controlling access to the WLAN and bridging a client connection to the internal LAN. It must support 802.1X and dynamic WEP or WPA encryption. In network standards terminology, the AP fills the role of the Network Access Service (NAS). The wireless AP and the RADIUS server also have a shared secret to enable them to securely identify each other.

2.1.3 The RADIUS Server And Directory

The RADIUS server uses the directory to verify the credentials of WLAN clients. It makes authorization decisions based on a network access policy. It may also collect accounting and audit information about client access to the network. This is referred to as the authentication service (AS) in network standards terminology.

2.1.4 The Internal Network

The Internal Network is a secure network to which the wireless client application needs to gain access. The following steps describe how the client makes a request and is granted access to the WLAN and thus the internal network. These step numbers correspond to the numbers in Fig. 10-1 [1].

2.1.4.1 Step 1

When the client computer is in range of the wireless AP, it tries to connect to the WLAN that is active on the wireless AP and identified by its Service Set Identifier (SSID). The SSID is the name of the WLAN and is used by the client to identify the correct settings and credential type to use for this WLAN.

2.1.4.2 Step 2

The wireless AP is configured to allow only secured (802.1X authenticated) connections. When the client tries to connect to it, the AP issues a challenge to the client. The AP then sets up a restricted channel, which allows the client to communicate only with the RADIUS server (blocking access to the rest of network). The RADIUS server will only accept a connection from a trusted wireless AP; that is, one which has been configured as a RADIUS client on the IAS server and provides the shared secret for that RADIUS client.

The client attempts to authenticate to the RADIUS server over the restricted channel using 802.1X. As part of the PEAP negotiation the client establishes a Transport Layer Security (TLS) session with the RADIUS server. Using a TLS session as part of PEAP serves a number of purposes:

- It allows the client to authenticate the RADIUS server; this means that the client will only establish the session with a server holding a certificate that is trusted by the client.
- It protects the authentication protocol against packet snooping.
- The negotiation of the TLS session generates a key that can be used by the client and RADIUS server to establish common master keys. These keys are used to derive the keys used to encrypt the WLAN traffic [1].

Secured within the PEAP channel, the client authenticates itself to the RADIUS server using the MS-CHAP v2 EAP protocol. During this exchange, the traffic within the TLS tunnel is only visible to the client and RADIUS server and is never exposed to the wireless AP.

2.1.4.3 Step 3

The RADIUS server checks the client credentials against the directory. If the client is successfully authenticated, the RADIUS server assembles information that allows it to decide whether to authorize the client to use the WLAN. It uses information from the directory (such as group membership) together with constraints defined in its access policy (for example, the times of day that WLAN access is allowed) to either grant or deny access to the client. The RADIUS relays the access decision to the AP.

If the client is granted access, the RADIUS server transmits the client master key to the wireless AP. The client and AP now share common key material that they can use to encrypt and decrypt the WLAN traffic passing between them.

When using dynamic WEP to encrypt the traffic, the master keys are directly used as the encryption keys. These keys need to be changed periodically to thwart WEP key recovery attacks. The RADIUS server does this by regularly forcing the client to re–authenticate and generate a new key set.

If WPA is used to secure the communication, the master key material is used to derive the data encryption keys, which are changed for each packet transmitted. WPA does not need to force frequent re–authentication to ensure key security.

2.1.4.4 Step 4

The AP then bridges the client WLAN connection to the internal LAN, allowing the client to talk freely to systems on the internal network. The traffic sent between the client and AP is now encrypted.

2.1.4.5 Step 5

If the client requires an IP address, it can now request a Dynamic Host Configuration Protocol (DHCP) lease from a server on the LAN. Once the IP address has been assigned, the client can begin communicating normally with systems on the rest of the network.

2.2 Computer and User Authentication to the WLAN

The process just described illustrates how a client (a user or a computer) successfully connects to the WLAN. For example, let's say you are using Windows XP--it authenticates both the user and the computer independently. So, when the computer first starts up, it uses its domain account and password to authenticate to the WLAN. The authorization of the computer to the WLAN follows all of the steps outlined described earlier. Having the computer connect to the WLAN, using its own credentials, permits it to be managed even when no user is logged on. For example, group policy settings can be applied and software and patches distributed to the computer.

When a user logs on to the computer, the same authentication and authorization process is repeated, but this time with the user's name and password. The user's session replaces the computer WLAN session; this means that the two are not active simultaneously. It also means that an unauthorized user cannot use an authorized computer to access the WLAN.

> **Note:** Windows XP allows you to override this behavior and specify that either only the computer or user credentials are used.

These are not recommended configurations. The former allows users to connect to the WLAN without authorization. The latter prevents the computer from connecting to the WLAN until a user logs on, which will interfere several computer management functions.

3. TARGET ENTERPRISE PROFILE

This preceding solution is designed for a small enterprise of 100–200 people. Although the enterprise is fictitious, its characteristics and requirements are derived from extensive real–world research. These real–world requirements have helped shape the style and scope of this chapter; as well as, the choices made in the design.

It is important to understand that the implementation of the wireless network security plan development is not restricted to enterprises of this size. The simplicity of the design and the scalability of the components used mean that the same PEAP–based WLAN implementation of the wireless network security plan development can be easily scaled for much larger (with thousands of users) and much smaller enterprises. By understanding the characteristics of the target enterprise, you will be better placed to understand the assumptions of the design and adapt them to your own enterprise.

Using the implementation of the wireless network security plan development in larger enterprises is covered in the "Scaling for Larger Enterprises" part of this chapter. For very smaller enterprises, all of the required components can be installed on a single existing server.

3.1 Enterprise Layout

The physical and information technology (IT) layout of the enterprise is shown in Fig. 10-2 [1]. There is a single large head office, which houses most of the IT systems and the majority of the users. All Active Directory domain controllers are located here. The head office has a connection to the Internet through a firewall server [2]. There are a number of WLAN clients and wireless APs connected to the internal network.

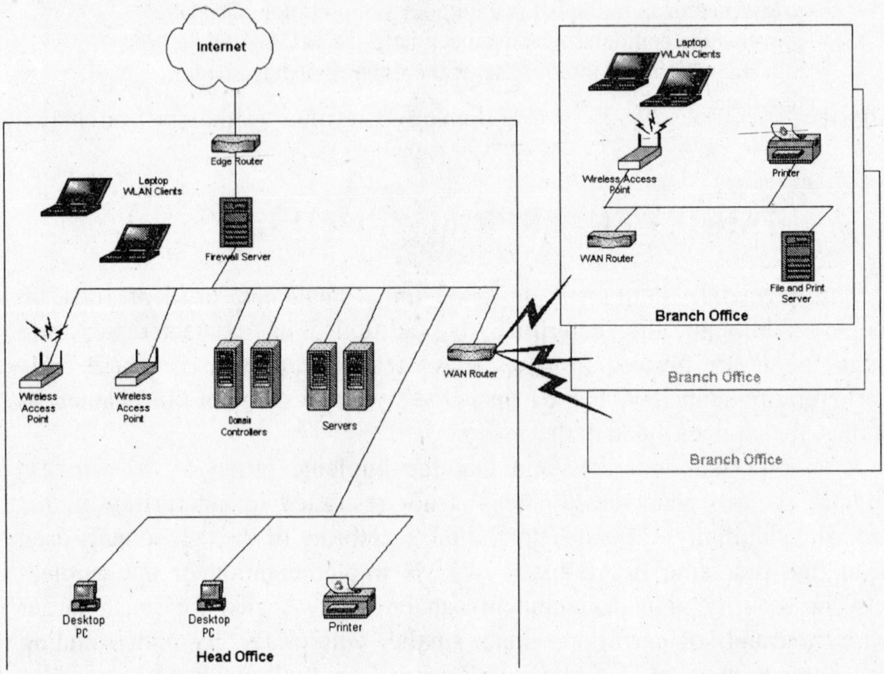

Figure 10-2. The physical and IT layout of the target enterprise

There are also one or more outlying offices with very few local IT services beyond network connectivity to the head office. There are a small number of clients (possibly all wireless) at this office and it frequently receives visitors from head office who bring their WLAN clients to be able to connect back to their applications and data at the head office.

The wide area network (WAN) connectivity between offices is provided either by private lines (for example a T1—1.5Mbps) or by DSL Internet connections and a virtual private network (VPN) router–to–router link across the Internet. The WAN connection is typically not resilient to failure.

> **Note:** If the WAN connection between offices is provided by a VPN connection across the Internet, each office will typically have a firewall protecting it from threats on the Internet. The presence of this firewall is not relevant to the WLAN discussion and has been omitted to aid clarity.

4. DESIGN CRITERIA

The enterprise described previously, will typically have the following types of criteria for a WLAN implementation of the wireless network security plan development, as shown in Table 10-1 [1]. These criteria have been extended to cover a broad category of enterprises. The design presented in the rest of the chapter explicitly uses these criteria.

Table 10-1. WLAN implementation of the wireless network security plan development Design Criteria

Design Factor	Criteria
Security Requirements	–Robust authentication and authorization of wireless clients. –Robust access control to permit network access to authorized clients and to deny any unauthorized access. –High strength encryption (128–bit) of wireless network traffic. –Secure management of encryption keys.
Scalability—Min/Max users supported	25 to 5,000 or more WLAN users. See Table 10-2 for authentication loads for different sizes of WLAN.
Scalability—Number of sites supported	Basic: Single large site with local domain controllers and IT services; one or more small sites with no domain controllers. Minimum users required are 25. High end: Single central site with multiple domain controllers; large offices with single domain controller and/or resilient WAN connectivity to head office; multiple small offices with no domain controller, probably no WAN resilience. Maximum users allowed are 5000.
Availability Requirements	Use of multiple wireless APs, IAS, or domain controllers gives WLAN resilience to single component failure for larger offices. Small office WLANs are vulnerable to WAN failure unless redundant connectivity is installed.
Platform Support	Server platforms: Windows Server 2003, Standard Edition or Enterprise Edition (for IAS and Certification Authority (CA) installation). Standard edition supports

	maximum of 50 wireless APs (RADIUS clients) per server. Client platforms: Windows XP Professional or Tablet Edition; Pocket PC 2003.
Extensibility (reuse of the implementation of the wireless network security plan development components for other applications)	Other network access applications (remote access VPN, 802.1X wired network access, and firewall authentication) can be supported by same authentication infrastructure.
IT Enterprise Requirements	Installation and management of the implementation of the wireless network security plan development requires IT professional with latest MSCE certification or equivalent knowledge and 2 to 3 years experience in IT industry.
Manageability Requirements	The implementation of the wireless network security plan development will require minimal management to maintain healthy operation. Alerts are sent through e-mail and/or Windows event log (or modified to trigger other alter types). IAS component can be monitored by Windows monitoring solution (using event logs and performance counters), by RADIUS logging and by Simple Network Management Protocol (SNMP) management system.
Standards Conformance	The implementation of the wireless network security plan development supports the following standards: –IEEE 802.11 (a, b, or g) network standards. –IEEE 802.1X authentication with PEAP. It can be used with other EAP methods such as certificate-based EAP-TLS and PEAP-EAP-TLS. –Dynamically keyed WEP and WPA WLAN protection. Future capabilities and standards (for example, 802.11i). –RADIUS support for RFC 2865 and 2866.

Table 10-2 gives an indication of the WLAN authentication requirements for different sizes of enterprise [1]. The New Authentications per second is part of the steady load and assumes an average of four new full authentications per user per day as users rove between wireless APs. The Peak New Authentications per second column indicates the type of load expected when all users authenticate over a 30 minute period (for example at the start of the day). The Re-authentications per second column shows the number of periodic re-authentications to force the renewal of dynamic WEP keys. The figures/amounts shown in Table 10-2, are referred to later in the chapter when discussing IAS server sizing.

Table 10-2. WLAN Authentication Requirements

Number of WLAN Users	New Authentications per Second	Peak New Authentications per Second	Re–authentications per Second
100	> 0.1	0.1	0.1
1000	0.1	0.6	1.1
10,000	1.4	5.6	11.1

5. WLAN ARCHITECTURE

This part of the chapter covers the architecture of implementing the wireless network security plan development. Fig. 10-3 illustrates the basic network layout for the head office [1].

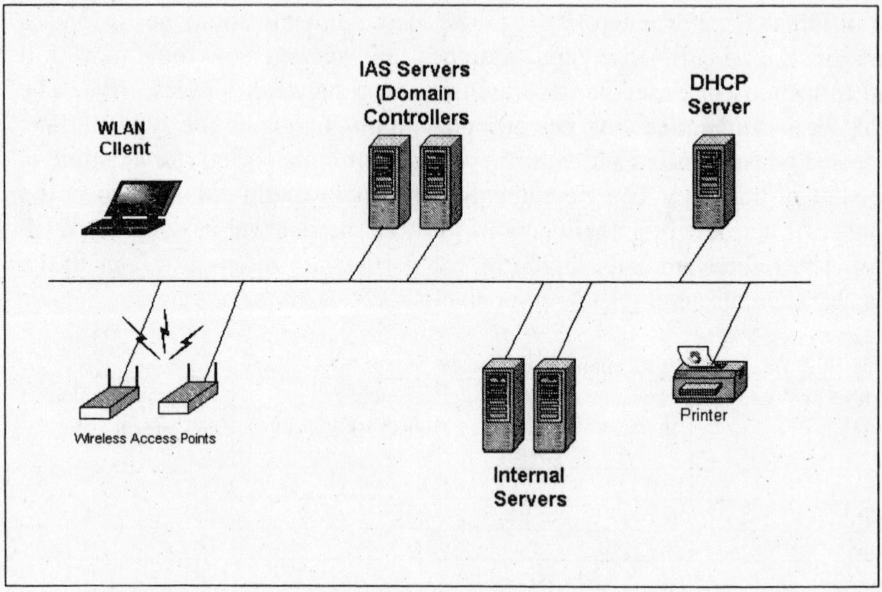

Figure 10-3. The network layout for the head office.

Fig. 10-3 shows wireless clients, two or more wireless APs, two IAS servers running on Active Directory domain controllers, a DHCP server, and other network connected servers, clients, and devices. With the exception of the WLAN clients, all items are connected on a single LAN using one or more layer 2 switches. A single subnet is used for the entire internal network at this site. There are routed connections (not shown in Fig. 10-3) through the firewall to the Internet and other offices.

5.1 Network Design

Larger enterprises are more likely to have a routed environment within a single site. Although this does not make any difference to the authentication infrastructure, it may have an impact on how the wireless APs are connected to the rest of the network. To make it easier for users to rove between multiple APs across a site, it is a normal practice to place all the wireless APs and WLAN clients on the same IP subnet. This allows users to rove between wireless APs but still keep the same IP address. A detailed discussion of this topic is beyond the scope of this chapter. In your network design, you must ensure the following: First, the wireless APs have connectivity to both primary and secondary IAS servers. If the APs are on a different VLAN/subnet from the IAS servers, traffic must be routable between these subnets.

Second, the WLAN clients must have connectivity to DHCP servers. If the servers are not on the same subnet, you need to have DHCP/BOOTP relay agents to forward the client DHCP request to a DHCP with a scope defined for that subnet. The clients will, of course, also need connectivity to their normal network services such as domain controllers, file servers, and so on.

5.2 Selection of Wireless Network Hardware

You should ensure that your wireless APs and wireless network adapters support the following:

- 128-bit WEP encryption (if using dynamic WEP) or TKIP encryption (RC4) or AES encryption (if using WPA).
- 802.1X Authentication.
- Dynamic Key Update (WEP encryption only).
- WPA Support (even if you are using dynamic WEP, you should have a clear commitment from vendor to provide firmware update to provide WPA support.) [1]

You should have sufficient wireless APs to provide coverage for WLAN clients across the physical locations that you need to support. You should also plan the wireless AP placement so that there is adequate backup coverage in all locations in case a wireless AP fails.

5.3 IAS Server Placement

The goal of IAS server placement is to achieve a resilient WLAN service while keeping implementation and management costs low. A WLAN service that is resilient to a single component failure has the following characteristics:

- All physical zones where you need WLAN coverage must have two or more wireless APs in range.
- Each wireless AP must be able to communicate with a backup IAS server in the event that its primary IAS server fails or the network connection to this server fails.
- The services on which IAS and the WLAN clients are dependent (for example, Active Directory, DHCP, and DNS) must also provide resilience [1].

The second of these is the most significant for planning IAS server placement. During the implementation of the wireless network security plan development, IAS is on existing domain controllers. This provides the highest performance configuration together with relatively low implementation and management costs. As a general recommendation for enterprises of all sizes, you should deploy IAS to every site that has a domain controller (although you may not need to install IAS on every domain controller).

Fig. 10-4 illustrates the placement of IAS servers in the enterprise [1]. IAS is deployed on two existing domain controllers in the head office. The network CA (refer to the part "Obtaining Certificates for IAS Servers" later in this chapter) is also installed on one of these domain controllers. All APs in the head office are configured to use these IAS servers.

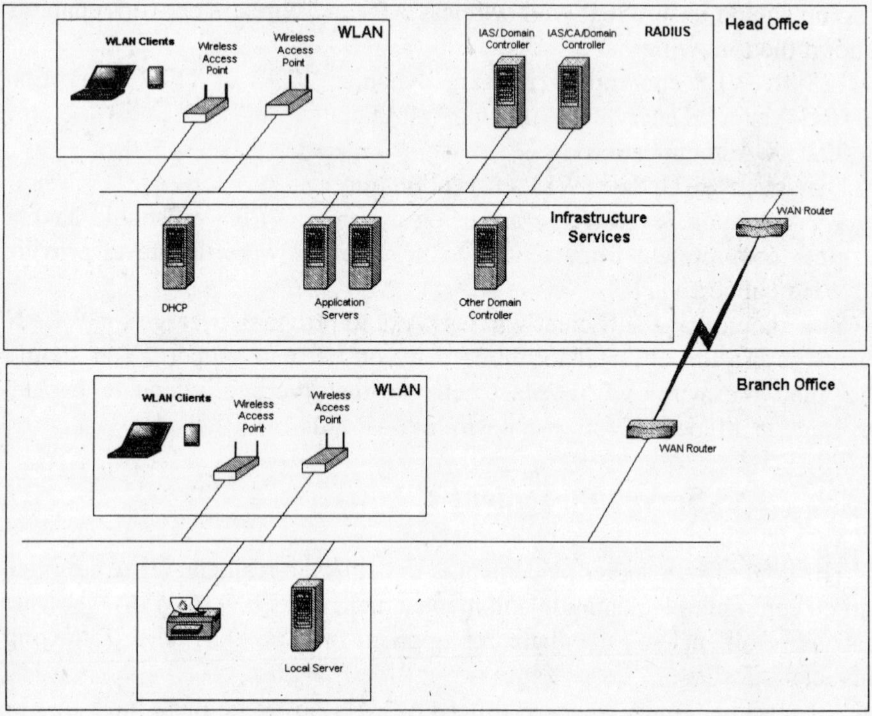

Figure 10-4. The head office and branch office infrastructure.

The enterprise has a small branch office with no local domain controller. The wireless APs in this site use the two IAS servers in the head office for all authentication requests. This means that users will be unable to authenticate to the WLAN if the WAN connection to head office fails. This may be an unacceptable risk for many enterprises.

To resolve this, you should either install redundant WAN connectivity or install a local IAS and domain controller. Although this might seem like an unreasonable cost for this type of office, WAN failure will also cause most other network services to fail (for example, local file servers) without access to a domain controller. Rectifying this will therefore benefit the reliability of these services as well as the local WLAN. Deploying domain controllers to branch offices is discussed in the part "Scaling for Larger Enterprises", later in the chapter.

For small offices where WAN connectivity is highly unreliable and deploying a local domain controller is not feasible, you may decide to deploy a standalone WLAN. For more information on this, you should read the part "SOHO Environments" later in the chapter.

5.4 Assignment of APs to RADIUS Servers

You must assign all your wireless APs to IAS servers. Each wireless AP requires a primary and a secondary RADIUS server. This allows the wireless AP to use the secondary RADIUS server in the event that the primary server has failed or is not contactable. This arrangement is shown in Fig. 10-5 [1].

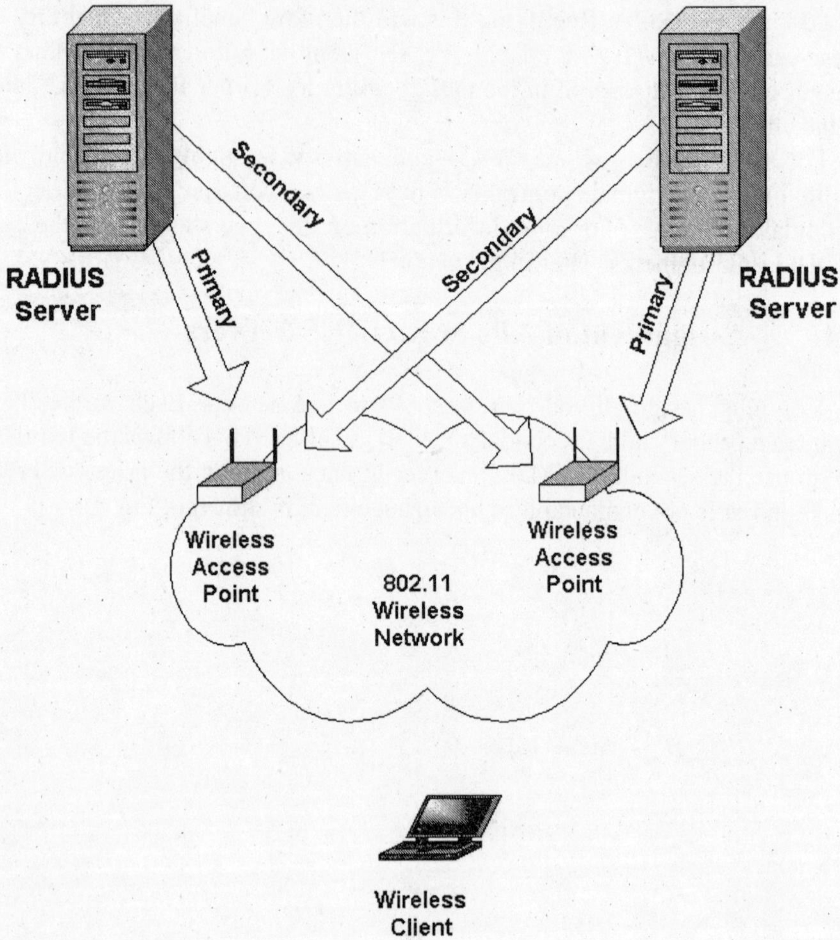

Figure 10-5. Balancing AP between primary and secondary IAS servers.

Figure 10-5 shows how each of the wireless APs is configured with different primary and secondary RADIUS servers. This allows load balancing between the servers. Wireless APs in sites with no local IAS server will follow the same pattern using thc IAS servers in the head office as primary and secondary RADIUS servers.

For wireless APs in sites with only one local IAS server, the local server should always be the primary server and the server in the head office (or other suitable location where there is reliable connectivity to an IAS server) should be the secondary server. This is illustrated in Fig. 10-6 [1].

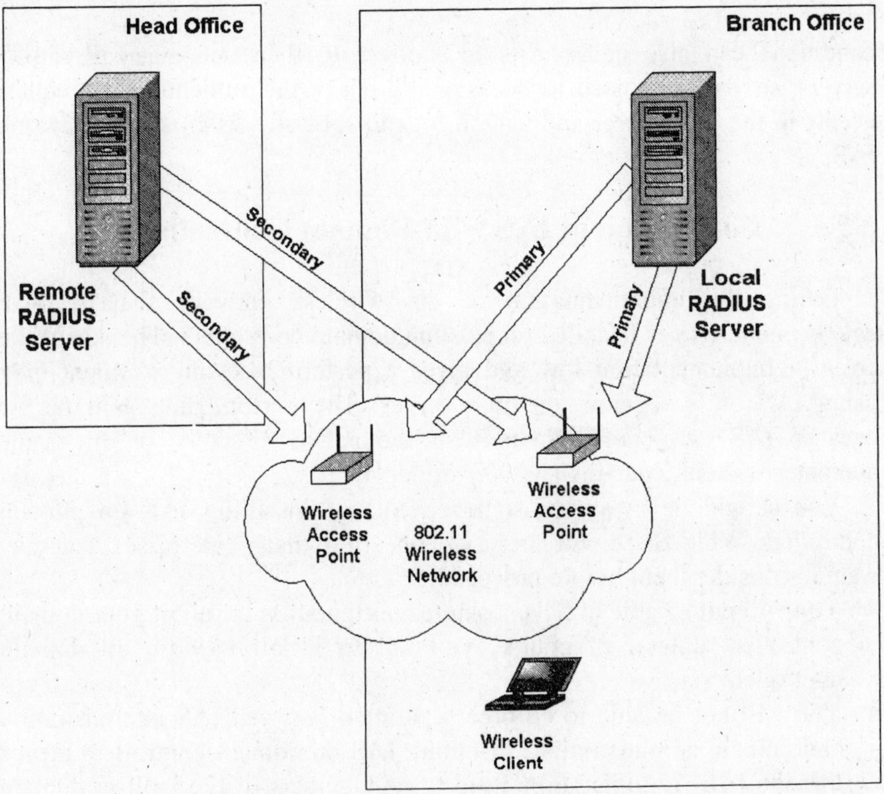

Figure 10-6. Configuring APs to use local and remote IAS servers

If you have many APs, you should carefully document the assignment of APs to IAS servers. You can use this record to ensure that every AP has been assigned a primary and a secondary server and that the load from APs is evenly balanced across the available servers.

> **Note:** All wireless APs will failover to the secondary IAS server when the primary IAS server is not available. However, most APs do not automatically revert to using the primary when it becomes available again (they will only fall back if the secondary subsequently fails). This is not a major problem where both IAS servers are at the same location; it will simply make load across the servers uneven. However, where the secondary IAS is remote, a temporary failure of the primary may leave all APs authenticating to the secondary over a non–optimal WAN link.

If your APs do not automatically revert to their designated primary server, you may need to manually reset the APs so that they start using the local IAS server when it recovers after a failure. Transient network conditions can also cause APs to failover to their secondary RADIUS servers, so you may need to occasionally check the authentication request events in the IAS server application logs to spot any APs using the wrong IAS.

5.5 Co-location of IAS with Domain Controllers

During the implementation of the wireless network security plan development, IAS is installed on existing domain controllers. This keeps the costs of implementation low and gives a performance improvement over using IAS on a separate member server. The performance gain occurs because IAS can communicate with the active directory on the same computer without incurring any network delay.

You should be aware of some caveats of installing IAS on domain controllers. While these will not be of concern to many enterprises, you may want to consider them before proceeding:

- You will not be able to have a single configuration for all of your domain controllers unless, of course, you opt to install IAS on all domain controllers.
- You will not be able to enforce separation between IAS administration and domain administration. Installing IAS on domain controllers means that the IAS Administrators need to be members of the built-in domain Administrators group as well.
- High load on the domain controller functions will adversely affect the performance of IAS and vice versa. You may want to put these on separate servers to have more control over their individual performance and operation of these services [1].

5.6 IAS Software and Hardware Requirements

For a target enterprise of 100 to 200 users, it is unlikely that IAS load on servers will ever be an issue as long as you are using the recommended hardware specification. However, for larger enterprises, this may be a consideration, particularly if they are running IAS on existing domain controllers. The load on the IAS will be affected by:

- Number of users and devices requiring RADIUS authentication.
- Choice of authentication options such as EAP type and the re-authentication frequency.
- Whether RADIUS logging is enabled [1].

You can use the figures/amounts given in Table 10-2, in the earlier "Design Criteria" part, to estimate the number of authentications per second that you can expect from a given user population. You should consider the steady state load when users are authenticating normally and the worst case load during peak times. Extrapolating the figures from this table, 200 users generate a steady state load of less than one full authentication every 50 seconds and one fast re–authentication every ten seconds. These are such small numbers that the only really significant figure is how long it takes to authenticate all users following an outage — when all users need to connect back to the WLAN immediately. This is a much more extreme peak than the start-of-day logon, which will tend to spread over 30 minutes or more.

Authentication options have a significant effect on IAS server load. Protocols such as PEAP perform a CPU–intensive public key operation upon initial log on; although for subsequent re-authentications, cached session information is used, allowing what is known as a fast reconnect. If you are using dynamic WEP, the clients will re-authenticate every 15–60 minutes to generate new encryption keys. However, with WPA you need to enforce re-authentication much less frequently, typically every 8 hours. Table 10-3 shows the approximate number of authentications per second for IAS on an Intel Pentium 4 2 GHz server running Windows Server 2003 with Active Directory on a separate server [1].

Table 10-3. Authentications per Second.

Authentication Type	New Authentications	Fast Reconnect Authentications
PEAP authentications per second	36	166
Time to authenticate 200 users	6 sec	2 sec
Time to authenticate 1000 users	30 sec	7 sec

The figures in Table 10-3 were calculated with RADIUS logging enabled and with Active Directory running on a separate server; both of these factors reduce the performance of IAS, so these figures can be considered a pessimistic estimate. As the figures in Table 10-3 show, this type of server will allow 200 WLAN users to authenticate to the network in six seconds and 1000 users in 30 seconds.

5.7 IAS Configuration

IAS settings can be broken down into four major categories:

- IAS Server settings
- RADIUS Logging configuration
- Remote Access Policies
- Connection Request Policies [1]

These categories are described in detail next. All of these settings are common to all of the IAS servers used in the implementation of the wireless network security plan development; this allows the settings to be configured on one IAS server and copied to the others.

In addition, each IAS server will also have one or more wireless APs configured as RADIUS clients. RADIUS clients were described in the earlier part "Assignment of APs to RADIUS Servers." The set of RADIUS clients is usually different for each server and therefore they are not replicated between the servers in the same way as the other settings.

5.8 IAS Server Settings

IAS server settings includes the logging of authentication requests to the event log. Logging of both successes and failures is enabled in during the implementation of the wireless network security plan development.

> **Note:** Request logging is covered in the "RADIUS Logging" part later in this chapter.

The User Datagram Protocol (UDP) ports is what the IAS server listens on for RADIUS authentication and accounting requests. The implementation of the wireless network security plan development uses the default RADIUS ports 1812 and 1813 for authentication and accounting, respectively.

5.9 RADIUS Policies

IAS policies control the authentication and authorization of accounts to the network. There are two types of policies: Remote Access Policy (RAP) and Connection Request Policy (CRP).

The RAP controls how or whether a connection is authorized to the network. A RAP contains a set of filter conditions that determine whether that policy applies to a given connection request. Some examples of filter conditions are: specifying the Windows security group of which a client must be a member, specifying the connection type (such as Wireless or VPN) of the requesting client, and specifying the time of day that the client is attempting to connect. Each RAP has a policy action, which is set either to allow or deny a connection request. Connection requests matching the RAP condition filter will be allowed or denied access to the network according to the policy action setting.

A RAP also contains a set of parameters that apply to an allowed connection, known as the RAP profile. These parameters include the authentication methods that are acceptable for this connection, how an IP address will be assigned to the client, and the amount time for which the client can remain connected before re–authentication is required. There can be multiple RAPs on an IAS; each connection request is evaluated against them (in order of priority) until a matching RAP either allows or denies the request. The RAP used in during the implementation of the wireless network security plan development is configured as shown in Table 10-4 [1].

Table 10-4. Remote Access Policy Configuration.

Configuration Item	Setting
--Policy Name	Allow Wireless LAN Access
--Policy Type	Allow
RAP Conditions:	
--NAS-Port-Type matches	IEEE 802.11 Wireless Other Wireless
--Windows Group matches	Wireless LAN Access
RAP Profile:	
--Dial-in constraints — Client Timeout	60 minutes (dynamic WEP) 8 hours (WPA)
--IP Address assignment	Server settings determine IP assignment
--IP Filtering	None
--Authentication	All disabled apart from EAP
--Authentication—EAP Type used	Protected EAP (PEAP)
--Authentication—PEAP EAP type used	EAP
--Authentication—Fast Reconnect	Enabled
--RADIUS Attributes	Ignore-User-Dialin-Properties = "True" Termination Action = "RADIUS-Request"

The condition filter matches all wireless clients and all members of the domain group Wireless LAN Access. For more details on the use of security groups with the RAP, refer to the "WLAN User and Computer Administration Model" part later in this chapter.

The Dial-in constraints — Client Timeout setting can have an impact on the security and the reliability of the implementation of the wireless network security plan development. Reasons for using different values to those given in the table are discussed in the "Security Options for Dynamic WEP" part later in the chapter.

The RADIUS attribute "Ignore-User-Dialin-Properties" is used to bypass per user control of network access permissions. See the "WLAN User and Computer Administration Model" part for an explanation of per user and per group access control.

A connection request policy controls whether to process the request at a particular RADIUS server or send it to another RADIUS server (called

RADIUS proxy). RADIUS proxies are typically used where the RADIUS server does not have the information to process the request itself and must forward it to an authoritative RADIUS server, for example, to a server in another active directory forest. RADIUS proxies are not used during the implementation of the wireless network security plan development and are outside the scope of this chapter.

5.10 RADIUS Logging

You can configure IAS servers to log two types of optional information: Successful and rejected authentication events; and, RADIUS authentication and accounting information. Successful and rejected authentication events generated from WLAN devices and users can be recorded on the IAS server. Authentication events information is most useful for troubleshooting authentication issues, although this information may also be used for security auditing and alerting purposes.

Initially, you should keep success and reject event logging enabled but you may consider disabling success events once the system has stabilized. Successful WLAN access events will bloat the system event log, but may be needed for audit purposes.

IAS also has the ability to record authentication and network access session information in the form of RADIUS request logs. You normally use RADIUS logging either when you need to charge for network usage (for example, as an Internet service provider (ISP), you need to charge based on connection time); or, where you need specialized security audit information (although, for most purposes this is already covered by the authentication events logged to the event log).

5.11 IAS Security

You should treat IAS with similar security precautions as you use with a domain controller. Secure control of your network is dependent on the security of your IAS infrastructure. There are several simple measures, which you can implement to improve the security of IAS:

- Use strong passwords for your RADIUS clients (wireless APs). The implementation of the wireless network security plan development includes scripts to generate true random passwords to make dictionary attacks difficult.
- Enable RADIUS Message Authenticator for all RADIUS clients to prevent the IP addresses of wireless APs from being spoofed. It is

enabled during the implementation of the wireless network security plan development.

- Ensure that your server security settings are appropriate.
- Ensure that your servers are patched with the latest security hotfixes and that you maintain up-to-date patches on an ongoing basis.
- Ensure that you are using strong domain account settings. In particular, you should ensure that strong passwords and regular password changes are enforced. You may also consider enabling domain account lockout to block password guessing attacks. However, you should only enable account lockout if you have the support resources to unlock accounts for users in a timely fashion.
- Consider using IPSec to secure the RADIUS traffic and strengthen mutual authentication between your wireless APs and IAS servers. However, not all wireless APs support the use of IPSec for this [1].

5.12 WLAN User and Computer Administration Model

Access to the WLAN during the implementation of the wireless network security plan development, is controlled using domain security groups. Although it is possible to use the dial–in properties of domain user objects to allow and deny access to individuals, this is tedious to administer for many users.

The implementation of the wireless network security plan development, uses a very simple scheme of allowing all domain users and computers access to the WLAN. For many enterprises, controlling access through domain membership is a strong enough control and minimizes additional management overhead associated with the WLAN. However, to give more control to enterprises that require it, security groups can be used to define who is allowed to access the WLAN.

As described in the part "RADIUS Policies," the IAS remote access policy uses a filter condition that grants WLAN access to all members of the Wireless LAN Access group. Table 10-5 shows the membership of the Wireless LAN Access group [1].

Table 10-5. Wireless Access Groups to Allow All Users and Computers.

Top Level Universal Group (Granted Access in RAP)	First Level Members(Domain Global Groups)	Second Level Members(Domain Global Groups)
Wireless LAN Access	Wireless LAN Users	Domain Users
Wireless LAN Access	Wireless LAN Computers	Domain Computers

The group in the first column of Table 10-5, Wireless LAN Access, has two members listed in the second column namely, Wireless LAN Users and

Wireless LAN Computers. These "First Level" groups themselves have members (shown in the third column — Second Level Members") namely, the Domain Users and Domain Computers groups respectively. This arrangement of nested groups allows all users and computers in the domain to connect to the WLAN.

If allowing all users and computers to access the WLAN is overly permissive for your enterprise, you can remove either or both Domain Users and Domain Computers from these groups. You will then need to add the specific user and computer accounts or groups to the Wireless LAN groups. Table 10-6 illustrates how to use the Wireless LAN Access group structure in this manner [1]. For more information on the use of these security groups in a multidomain forest, refer to the "Scaling for Larger Enterprises" part later in this chapter.

Table 10-6. Wireless Access Groups to Allow Selected Users and Computers

Top Level Universal Group (Granted Access in RAP)	First Level Members (Domain Global Groups)	Second Level Members(Domain Global Groups)
Wireless LAN Access	Wireless LAN Users	User1 User2 User3
Wireless LAN Access	Wireless LAN Computers	Computer1 Computer2 Computer3

5.13 Obtaining Certificates for IAS Servers

The IAS servers need to have certificates to authenticate the WLAN clients. Server certificates are required to create the TLS encrypted tunnel between IAS servers and clients. TLS helps protect the authentication exchange between the server and clients.

> **Note:** TLS is an RFC standard based on the similar Secure Sockets Layer version 3.0 (SSL 3.0). Both are commonly used to secure Web traffic as part of the Hypertext Transfer Protocol, Secure (HTTPS).

5.14 Embedded CA Versus Commercial CA

To provide these certificates, you have the choice to either install a CA yourself or buy the certificates from a commercial certificate provider. Both options are valid and choosing one over the other creates no real technical difference to the WLAN implementation of the wireless network security plan development. The major pros and cons to using in–house CA compared to buying certificates from a commercial provider are summarized in Table 10-7 [1].

Table 10-7. Pros and Cons of Using Your Own CA versus Commercial Certificates

In-house CA	Commercial CA
No per certificate cost	Per certificate cost
CA software to be installed and managed	No server software
Automatic enrollment and renewal	More complex enrollment process, manual installation of certificates

The balance of the argument depends on how complex and costly it is to manage your own CA. If the cost of setting up a local CA is low and the management is simple, it is often a more attractive proposition than purchasing external certificates.

The implementation of the wireless network security plan development uses a simple internal CA to provide the certificates. The terms "embedded CA" and "network CA" have been used in this chapter to indicate that it is a special purpose CA, which is essentially invisible to users and administrators and which issues certificates of a single type. The limited functionality of the CA during the implementation of the wireless network security plan development means that it can be installed and used with little or no user or management intervention. For example, in the implementation of the wireless network security plan development, the CA can issue a certificate with a lifetime of 25 years; therefore, you will not have to renew it during the lifetime of the implementation of the wireless network security plan development. Automatic enrollment and renewal of the IAS server certificates means that there is no manual certificate distribution to perform.

Contrast this with using external certificates. You must remember to renew the certificates of all IAS servers every year or two years. Each time you have to manually create the certificate request on each IAS server, send the request to the commercial CA then retrieve and manually install the issued certificate. Failure to do this will prevent users from connecting the WLAN. For many enterprises, the management overhead of this is far more onerous than the simple internal CA described in the implementation of the wireless network security plan development.

5.15 Limitations Of The Implementation Of The Wireless Network Security Plan Development CA

The implementation of the wireless network security plan development uses a special CA configuration to issue certificates to the IAS servers. It was only designed to meet this specific need and is not suitable as a general purpose certification authority.

Digital certificates are used in many applications, such as secure e-mail and Web browsing, IP Security (IPSec), Virtual Private Networks (VPN), Encrypting File System (EFS), and others. Each of these applications has its own security requirements. Your enterprise will have its own unique security requirements with respect to these applications. For these reasons, it is strongly recommended that you do not attempt to use the implementation of the wireless network security plan development CA for any other purpose.

If you plan to use these or other certificate applications, design your certificate infrastructure around their requirements. Things that you need to consider include:

- The implementation of the wireless network security plan development CA is a self-signed root CA, so you cannot revoke it (you have to revoke the issued certificates in the case of CA compromise).
- Industry or country-specific regulations may require you to use a multitier CA hierarchy for some or all certificate types.
- Do not install a CA on a domain controller for high-security certificate uses [1].

If you do not have clear requirements for other types of certificates at present, however, you can deploy the CA described in the implementation of the wireless network security plan development without closing your options in the future. At a later stage, if you identify other requirements for certificates, you can deploy a more sophisticated PKI [3] alongside the implementation of the wireless network security plan development. You are then free to either run them side by side or migrate to issuing all certificates from the new PKI.

5.16 WLAN Clients

The WLAN implementation of the wireless network security plan development supports several, different types of WLAN client either explicitly or implicitly. This chapter does not cover using other types of clients that support 802.1X with PEAP.

Computer authentication to the WLAN is used when no user is logged on to the computer. This allows the computer to obtain Group Policy object

(GPO) settings, run startup scripts, and have patches downloaded to it. It is also required during the initial stages of user log on. The user cannot start to authenticate to the WLAN until after the user profile is loaded; therefore log on scripts, other GPO settings, and roaming profiles will fail if the computer does not have an existing connection to the network before the user logs on.

5.17 Other 802.1X Clients

The WLAN implementation of the wireless network security plan development described here supports the use of WPA WLAN protection in place of dynamic WEP. WPA is always preferred over WEP because it provides better key management and implements a stronger network encryption algorithm. WPA also supports the use of AES encryption, if the hardware (wireless APs and network adapters) provide the necessary support.

So, it may not be possible to upgrade the existing WLAN equipment (wireless APs and client network adapters) to support WPA. The cost of buying and deploying new hardware may also be prohibitive. Thus, WEP coupled with 802.1X authentication continues to provide a very high level of protection for WLANs and is the default choice for the implementation of the wireless network security plan development.

5.18 Migration From An Existing WLAN

If you have an existing wireless network in place, you should plan a migration strategy upfront to ensure minimal disruption to users and the environment. Many enterprises have 802.11–based WLANs operating without network authentication or encryption. Other enterprises have implemented static WEP using shared key encryption often combined with Media Access Control (MAC) address filtering. The process of migrating from either of these scenarios to an 802.1X secured WLAN involves the following steps:

1. Deploy certificates to IAS Servers.
2. Configure the wireless network remote access policies on the IAS servers.
3. Deploy a WLAN configuration for the new WLAN to client computers: The new 802.1X – enabled network needs a new network Service Set Identifier (SSID). WLAN group policy should be deployed well in advance of wireless AP reconfiguration to ensure that mobile computers with only occasional LAN access receive the configuration.
4. Configuration of wireless APs to require 802.1X security: This is usually best done site by site, (for example, by building or campus) and it is

either performed out of hours or with adequate warning to users about possible WLAN outage. You should create RADIUS client entries for the IAS for all APs on the site, configure the APs with the addresses of the IAS servers for the AP's RADIUS entries, and finally, switch the AP over to allow only 802.1X authenticated clients. To facilitate rollback, you may want to back up the wireless AP settings before starting this step, so that they can be restored in an emergency [1].

This type of rollout minimizes the disruption to users and allows you to rollback a site easily, if anything goes wrong. There will inevitably be some problems experienced by users during the switch over, so you should keep the users informed about the migration and be prepared to handle more support calls than normal.

As with all migration strategies, careful planning and testing is essential. The steps involved in configuring client computers and wireless APs can cause disruption to the environment if they are not tested thoroughly to iron out nascent problems.

Detailed planning of migration from unsecured and static WEP WLANs or from proprietary WLAN security schemes are not included in this chapter. All these are similar in principle and follow the pattern previously described.

6. SCALING FOR LARGER ENTERPRISES

This part of the chapter describes some of the key considerations for using the implementation of the wireless network security plan development in a large enterprise. For example, this would involve one with several thousand users.

6.1 IAS Server Placement

As you increase the number of locations where you support WLANs, you need to decide how these wireless APs will be serviced by IAS servers. There are essentially two approaches: First, use a small number of central IAS servers: Use a small number of central IAS servers to handle all WLAN authentication traffic (two are likely to be sufficient). You also need to ensure that the WAN connections between your outlying offices and the IAS servers are resilient.

Second, distribute IAS servers to each office. There is a lower limit to the size of an office where this makes economic sense, but, as a rule of thumb, any office that is large enough to have its own domain controller can have IAS locally (usually installed on the domain controller).

Although the investment in network resilience may seem an expensive option, you need to weigh this against the administration cost of managing many distributed IAS servers. Even if IAS is installed on the same physical server as an existing domain controller, there will be some cost of managing each IAS instance. In practice, most large enterprises will use a hybrid of the two in the following ways:

- Centralize IAS servers and invest in WAN resilience wherever possible.
- Distribute IAS to offices where WAN resilience is not achievable or is prohibitively expensive.
- Use pre–shared key (PSK) WPA WLANs for very small offices, with poor connectivity or for employees' home offices [1].

The centralized IAS strategy was illustrated in the "IAS Server Placement" part, earlier in this chapter. The use of a local domain controller and IAS in a branch office is shown in Fig. 10-7 [1]. This shows a larger outlying office that is linked by a WAN to the Head Office shown in Fig. 10-4 earlier.

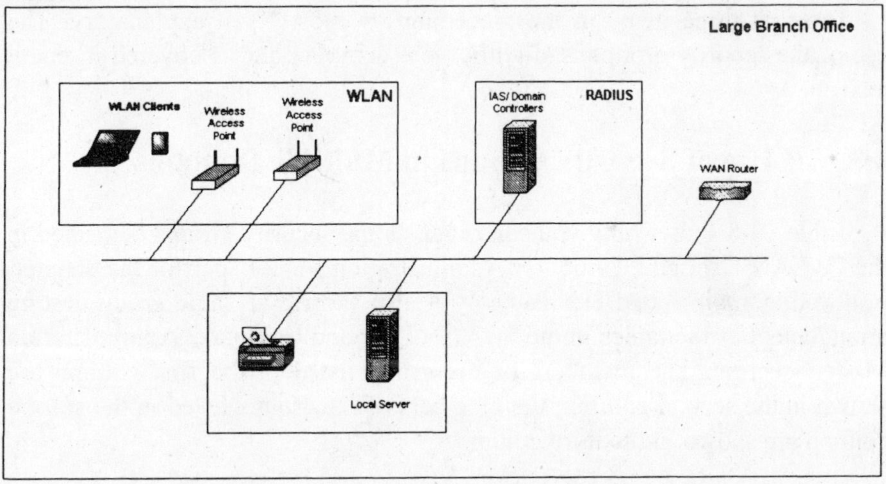

Figure 10-7. Larger branch office with local domain controller and IAS

In this site, the APs are configured to use their local IAS server as the primary RADIUS server and one of the IAS servers in the head office as the secondary RADIUS server. This means that WLAN clients can be authenticated, even if the local IAS server or WAN connectivity fails.

However, if you have resilient WAN connectivity (for example, multiple WAN links with different providers), there is little to be gained by deploying additional servers at branch offices; it only adds complexity and additional management overhead.

6.2 Multiple Domains

The basic implementation of the wireless network security plan development scales transparently to multi–domain forests. The key points to consider while using the implementation of the wireless network security plan development with multiple domains are as follows:

1. IAS servers must be registered in each domain that has users or computers that will be accessing the WLAN.
2. The GPOs for server settings and automatic certificate request settings must be imported into every domain in which IAS servers are installed.
3. The GPO that controls client computer WLAN settings must be created in each domain where there are client computers that will access the WLAN.
4. The security groups used by IAS to filter remote access policies need to be configured to support multiple domains [1].

The first three items in the preceding are mostly self-explanatory. The use of the security groups is slightly more complex and is covered in detail next.

6.3 Use of Security Groups in Multiple Domains

Table 10-8 shows how you can organize the security groups described in the "WLAN User and Computer Administration Model" part of the chapter, in a multidomain forest [1]. Table 10-8 also shows the same group nesting arrangement as the tables in the "WLAN User and Computer Administration Model" section. The members of the group listed in the first column are shown in the second column; the members of the groups listed in the second column are shown in the third column.

Table 10-8. Wireless Access Groups to Allow All Users and Computers.

Top Level Universal Group (Granted Access in RAP)	First Level Members(Domain Global Groups)	Second Level Members(Domain Global Groups)
RootDom\Wireless LAN Access	UserDom1\Wireless LAN Users	UserDom1\Domain Users
RootDom\Wireless LAN Access	UserDom2\Wireless LAN Users	UserDom2\Domain Users
RootDom\Wireless LAN Access	UserDom3\Wireless LAN Users	UserDom3\User1 UserDom3\User2 UserDom3\User2
RootDom\Wireless LAN Access	UserDom1\Wireless LAN Computers	UserDom1\Domain Computers
RootDom\Wireless	UserDom2\Wireless	UserDom2\Domain Computers

LAN Access	LAN Computers	
RootDom\Wireless LAN Access	UserDom3\Wireless LAN Computers	UserDom3\HR Computers UserDom3\Finance Computers

The example in Table 10-8 uses fictitious names. For example, RootDom is the name of the domain where IAS servers are installed and UserDom1 and UserDom2 are other domains containing users and computers to be granted WLAN access.

In the example shown, all users and computers from UserDom1 and UserDom2 are implicitly granted access to the WLAN because the Domain Users and Domain Computers groups from those domains are members of the Wireless LAN Users and Wireless LAN Computers groups for the same domain. However, the users from UserDom3 are individually added to the Wireless LAN Users group of UserDom3. The computers are granted access by using enterprise unit security groups (for example, all computers in the HR department).

The global groups for each domain, namely, Wireless LAN Users and Wireless LAN Computers, are then added as members of the universal group Wireless LAN Access. All members of this latter group are granted access to the WLAN in the IAS remote access policy.

6.4 PKI Architecture

As mentioned in the earlier part "Obtaining Certificates for IAS Servers," many applications can utilize certificates. It is important to note that while appropriate for the implementation of the wireless network security plan development, a standalone CA is unlikely to meet the more varied needs of larger enterprises. Before implementing the CA described in this chapter, ensure to carefully consider other uses for certificates that you might have in the future, as well as alternate PKI architectures more appropriate for these scenarios.

7. VARIATIONS ON THE IMPLEMENTATION OF THE WIRELESS NETWORK SECURITY PLAN DEVELOPMENT ARCHITECTURE

This part of the chapter discusses variations to the core design. The following parts of this chapter look at alternatives to the default security settings of the implementation of the wireless network security plan development, by using the IAS servers for wired and remote access authentication; creating guest WLANs for visitors; and, deploying WLANs to very small environments such as home offices.

7.1 Security Options for Dynamic WEP

The "How 802.1X with PEAP and Passwords Works" part earlier in this chapter discussed the use of dynamic WEP encryption in the implementation of the wireless network security plan development. The security of dynamic WEP relies on its ability to renew the encryption keys at regular intervals to thwart known attacks on the WEP protocol. IAS ensures that the keys for each wireless client are renewed at a set interval by using a client session time-out, which forces the client to re-authenticate to the WLAN.

Reducing the session time-out value increases security, but may reduce reliability and performance. A 60 minute time-out gives adequate security for most circumstances and certainly for 11 Mbps 802.11b networks. Normally, wireless clients will never transmit enough data in 60 minutes to allow a WEP key to be recovered by an attacker. The latest research indicates that static WEP keys can be recovered by capturing between 1 and 5 million network packets encrypted with the same key [1].

> **Note:** The figure of 1 million packets was obtained from testing static WEP WLANs using relatively weak keys (a "user–memorable passphrase") and is not directly applicable to dynamic WEP WLANs. Unlike typical static WEP WLANs, dynamic WEP uses strong random encryption keys and renders one of the key optimizations used by the authors much less effective. Nevertheless, it is good security habit to err on the side of caution and use the pessimistic figure of 1 million packets to assess the security threat to dynamic WEP WLANs.

One million packets typically equates to around 500 MB of data (assuming an average packet size of 500 bytes). For the encrypted data to be secure, the session time-out needs to be set so that it forces each client's key to be renewed before the client sends more than this amount of data.

For typical client network usage, such as e-mail, Web browsing, and file sharing, average data transfer rates are 160 Kbps or less. At this transfer rate, assuming 500 bytes packet size, it would take approximately 7 hours for an attacker to accumulate enough data to recover the client's current encryption key.

> **Note:** In laboratory conditions, 500 Mb could be transmitted in much less than 7 hours; around 10 minutes on a11 Mbps WLAN or less than 3 minutes on a 54 Mbps WLAN. However, this assumes a single client having exclusive use of the WLAN and

streaming unacknowledged UDP packets in one direction. This scenario, or anything near it, is extremely unlikely for a real world WLAN.

A 60 minute session time-out is more than adequate for most enterprises. This means that an average client would transmit around 150,000 packets before each key refresh; nearly an order of magnitude less than the 1 million packet threshold required for WEP key cracking. However, you may want to use a shorter time-out value for one or more of the following reasons:

- If you have wireless clients that send or receive large amounts of data over the WLAN in relatively short time periods, you should set the time-out to a shorter duration than the time it takes a single client to send and receive 75 MB (this is less than 20 percent of the amount of data required to recover a WEP key, so has a large safety margin).
- If you use 802.11a or 802.11g 54 Mbps WLANs, it is easier to transmit larger numbers of packets in a given time. You may want to reduce the session time–out to 15 minutes on these faster WLANs.
- If the capabilities of WEP key cracking techniques improve significantly, less data will be needed to recover the WEP keys. For example, if a new cryptanalytic technique is discovered that allows keys to be recovered with only 100,000 packets, you would need to lower the session time-out to prevent wireless clients reaching this limit before their encryption keys are renewed.
- If you have specialist security needs, you may wish to reduce the time-out below the threshold at which even theoretical attacks on WEP would be successful (10 minutes or 3 minutes as described in the previous note). However, you need to weigh this decision against the caveats described later in this part of the chapter. If the data is sensitive enough to require this level of precaution, you should seriously consider using only WPA data protection on the WLAN and using IP security to help protect the data when traveling over wired LANs [1].

There are two main disadvantages to reducing session time-out: First, lower WLAN reliability: Very short WLAN session time-outs could cause clients to fail re-authentication and disconnect from the WLAN if communication with a domain controller or IAS server was lost temporarily. Second, increased load on IAS servers: The shorter the time-out, the more often the client needs to re-authenticate with an IAS server and domain controller. As a result, the load on IAS servers and domain controllers will increase. Because IAS caches client authentication sessions, the load increase will normally be significant only for enterprises with a very large number of wireless clients or when using extremely short session time-out values.

7.2 Other Network Access Services

The RADIUS design used in the implementation of the wireless network security plan development can provide authentication, authorization, and accounting services for other network access servers. such as 802.1X wired network authentication and VPN and remote access authentication.

7.3 802.1X Wired Network Authentication

802.1X wired network authentication is the simplest such application requiring no modification of the basic RADIUS design. Enterprises that have a widely-distributed wired network infrastructure may find it difficult to control unauthorized use of the enterprise network. For example, it is often difficult to prevent visitors plugging in laptops or employees adding unauthorized computers to the network. Some parts of the network, such as data centers, may be designated high security zones and only authorized devices should be allowed on these networks; this could even mean the exclusion of employees with enterprise computers. How a wired network access solution would integrate with the design is illustrated in Fig. 10-8 [1].

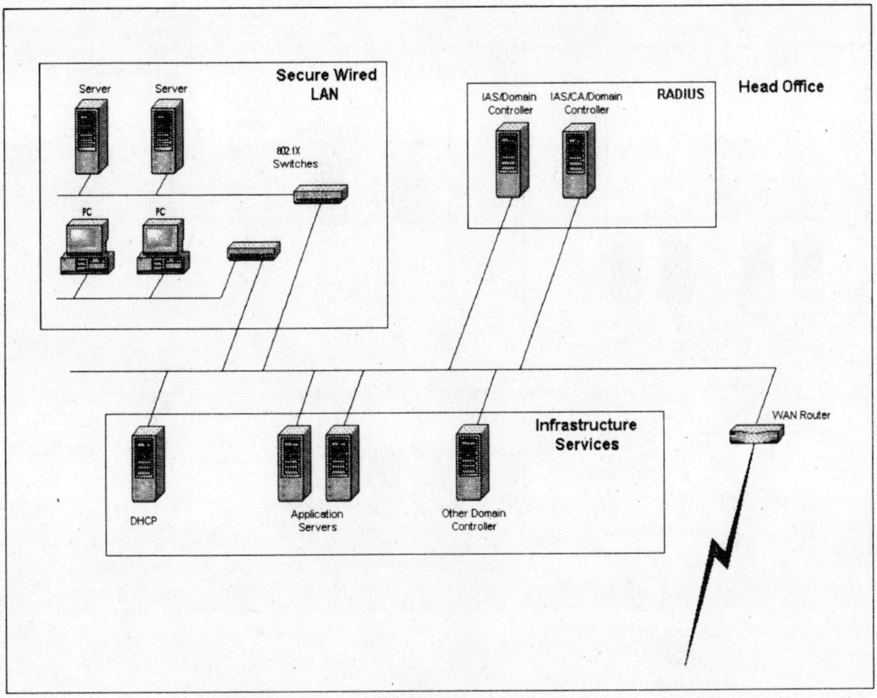

Figure 10-8. Using 802.1X wired authentication.

The bold–edged box shown in Figure 10-8 represents the 802.1X wired components and the other boxes contain the relevant services. Compare Figure 10-8 to Figure 10-4. Only the head office is shown in Figure 10-8.

802.1X – capable network switches play an identical role to the wireless APs in the core implementation of the wireless network security plan development and can take advantage of the same RADIUS infrastructure to authenticate clients and selectively authorize access to the appropriate network segment. This has the obvious advantages of centralizing account management in the enterprise directory while still retaining network access policies under the control of the network security administrator.

7.4 VPN and Remote Dial–Up Authentication

Another network access service that could use the RADIUS components is VPN and remote dial-up. Particularly in larger enterprises, it is likely that some additions would need to be made to the design as it stands, such as the addition of RADIUS proxies. The extended implementation of the wireless network security plan development is shown in Fig. 10-9 [1].

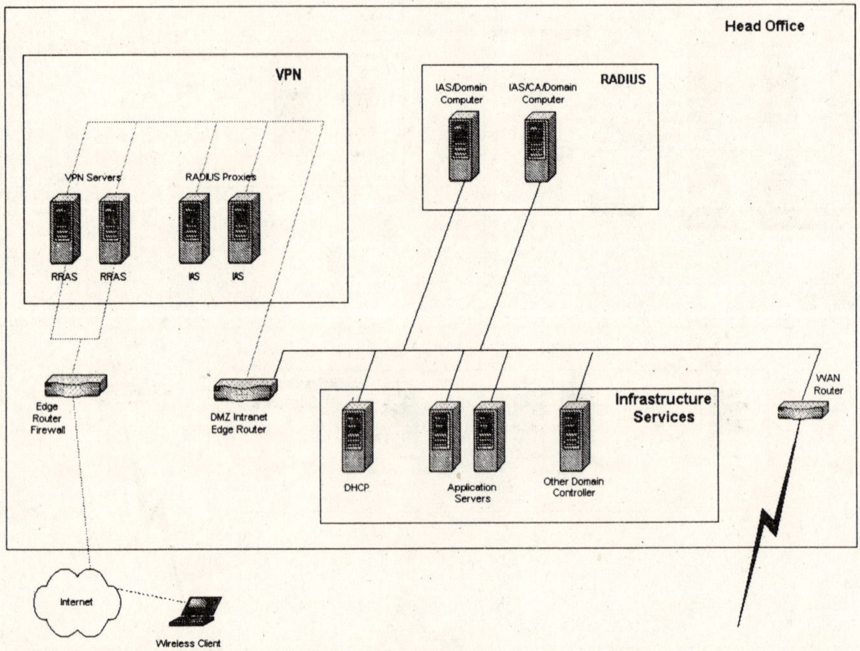

Figure 10-9. Extension of the RADIUS component to support VPN

The VPN servers, in this variant, play the same functional role as the wireless APs in the core design. They pass the clients' authentication requests to the RADIUS infrastructure. It is possible to have the RADIUS requests passed directly to the internal IAS servers. However, many enterprises like to add an additional layer of RADIUS proxies providing an extra security layer and routing the requests to the internal IAS servers.

As with wired network security, the implementation of the wireless network security plan development brings the same advantages of reusing existing infrastructure and centralizing account management. Further enhancements are possible, such as using smart card-based user authentication.

A major advantage that using IAS brings to both VPN and remote dial-up is the ability to perform checks on the client at connect-time; for example, ensuring that the client has up-to-date antivirus software, or that it is running a enterprise-approved operating system build. If the client fails these checks, the RADIUS server denies it access to the network. Therefore, even a properly authenticated user and computer may be denied access if they present a possible security threat to the enterprise network.

7.5 Bootstrapping Client Computers

Most wireless capable computers also have a wired network interface. This makes it relatively easy for the clients to join the domain and download WLAN settings prior to connecting to the WLAN. However, this may not always be the case. Already, handheld devices have only wireless, and do not have wired network adapters. This presents the problem of bootstrapping a client prior to connecting to the WLAN, because it does not have the settings and credentials that it needs to connect to the WLAN. This problem becomes even more difficult if an enterprise decides to use both wired and wireless 802.1X security, because a client cannot connect to a wired LAN without first having the correct credentials and settings.

There are two main approaches to bootstrapping a client computer if you cannot use a wired LAN to retrieve settings and credentials; these are: using a guest LAN and using an alternative authenticated connection (for example a VPN connection) to obtain credentials and settings; and, manually configuring clients. Manual configuration is a simple way to achieve this. To configure the client computer settings and join it to the domain, the person configuring the computer needs to be a member of the local Administrators group on the computer. To bootstrap a computer using manual configuration:

1. Manually configure WLAN settings for the correct WLAN SSID.
2. Connect to WLAN using a valid user domain credentials. You will not be able to connect using the computer account until the computer is joined to the domain.
3. Join the computer to the domain and then restart it.
4. After rebooting, the client will be able to connect to the WLAN using the computer account and will download the WLAN GPO settings. These settings will simply overwrite the manually configured settings.
5. Both the user and the computer can now connect to the WLAN.

7.6 SOHO Environments

You may need to deploy WLANs to locations where it is not possible or practical to authenticate users using your IAS infrastructure. For example, home offices for users who regularly work from home or small offices with very unreliable or low bandwidth connectivity to the main enterprise network.

Finally, previously, the only solution to this was to configure static WEP security and hope that no one was determined enough to bother to attack your WLAN. A far better solution is to use WPA in PSK mode. All Wi-Fi certified wireless APs now ship with WPA security, although older APs may not support this. You should ensure that your APs support WPA PSK, because of the additional security value that it brings. Unlike static WEP, the WPA authentication key is not recoverable from the encrypted traffic; therefore, it is much more difficult for an attacker to break on to the network. You should also ensure that your users are trained to use strong WPA keys and to change them regularly; and, understand the security implications of not doing so. To implement WPA PSK, you need a wireless AP, wireless network adapters, and a client operating system that supports WPA. You do not need a RADIUS server or other server infrastructure.

8. SUMMARY AND CONCLUSIONS

This chapter began with a description of how 802.1X wireless LAN security works. To provide focus for the design, a picture of the target enterprise for the implementation of the wireless network security plan development was given along with the enterprise's key design criteria for the WLAN solution. Following this, the main aspects of the chosen WLAN design were discussed. The design covered the network; IAS server placement and IAS configuration; the use of certificates; and, the different types of wireless clients. The key points on migration from an existing WLAN were also outlined.

Finally, the two parts at the end of the chapter discussed important variations to the basic design. Firstly, how to scale the implementation of the wireless network security plan development for larger enterprises was described, along with instructions on how to deal with the main points of divergence from the core implementation of the wireless network security plan development. This was followed by illustrations of how to use the same basic authentication infrastructure to support other network services such as remote access, VPN, and wired network security; and, how to deal with the sticky problems of bootstrapping clients and deploying WLANs to SOHO environments.

9. REFERENCES

[1] "Securing Wireless LANs with PEAP and Passwords: Planning a Wireless LAN Security Implementation," Microsoft TechNet, © 2005 Microsoft Corporation. All rights reserved. Microsoft Corporation, One Microsoft Way, Redmond, WA 98052-6399, April 3, 2004.

[2] John R. Vacca, *Firewalls: Jumpstart for Network and Systems Administrators*, Digital Press, 2004.

[3] John R. Vacca, *Public Key Infrastructure: Building Trusted Applications and Web Servicess*, Auerbach Publications, 2004.

Chapter 11

WIRELESS NETWORK SECURITY PLANNING TECHNIQUES

1. INTRODUCTION

Motivated by the need to reduce IT costs while increasing employee productivity, enterprise-wide wireless local area network (LAN) and wireless wide area networks (WWANs) solutions are becoming increasingly viable. Proliferation of mobile computing devices has boosted employee demand for access to their enterprise's wireless network beyond the tether of their office workstation. Meanwhile, accelerated wireless transmission rates and increasing vendor adherence to standards-based interoperability are enhancing the practicality of wireless LANs. And WWANs.

Yet, the same wireless technologies that can erase the physical limitations of wired communications to increase user flexibility, boost employee productivity, and lower cost of wireless network ownership, also expose wireless network-based assets to considerable risks. The security embedded in wireless LAN and WWANs technologies falls short of providing adequate protection. Early-adopting enterprises have found that evaluating, and where possible, mitigating these risks before deploying a wireless LAN is beneficial. This chapter briefly discusses wireless network security planning techniques by providing an overview of the security risks and technical challenges in this area, as well as summarizing key recommendations for secure wireless LANs and WWANs.

2. SECURITY RISKS AND TECHNICAL CHALLENGES

Security is a principal consideration when planning, designing, implementing, and managing a wireless network infrastructure. This is especially true for wireless LANs and WWANs, which present a unique set of challenges to IT and security professionals. In addition to the typical problems that new wireless network and device technologies engender, including incompatibilities and ongoing support issues, non-secure wireless LANs and WWANs can expose an enterprise's wireless network traffic and resources to unauthorized outsiders. Such individuals may capture data and exploit wireless network-based resources, including Internet access, fax servers, and disk storage [3]. More importantly, wireless access to a network can represent the entry point for various types of attacks, which can crash an entire network, render services unavailable, and potentially subject the enterprise to legal liabilities [1].

2.1 Leaky Buildings

Wireless LAN radio signals can extend beyond the intended perimeter and "leak" through the physical boundaries of a floor or building. As these transmissions seep into common, public, or private areas such as roads, parking lots, and other buildings, they may fall prey to war driving or a drive-by hacking attack. Using off-the shelf hardware and freely available Internet software, unscrupulous individuals can defeat WEP encryption capabilities and access enterprise wireless data [1].

2.2 Unapproved Deployments

Insiders, including employees and contractors, may choose not to wait for the IT Department. They succumb to the low price and easy installation of WiFi starter kits (two wireless NICs and a WiFi Access Point), which can be purchased for about $260 and set up with minimal technical know-how in under ten minutes. When unapproved technology is plugged into an enterprise network, a number of challenges ensue, including end user and equipment support difficulties as well as potential disruptions to existing services.

Malicious outsiders who gain office physical access could quickly place an unobtrusive wireless AP in a conference room or lobby area. Such devices are easy to hide and simple to implement; history is replete with stories of such bugs even in supposedly secure foreign embassies. Operating from a nearby location, malicious outsiders can capture data, access enterprise resources, and interrupt services [1].

2.3 Exposure Of Wireless Devices

Many of today's laptops ship with embedded WiFi capabilities. Hackers can access a device's data and the enterprise's wireless LAN and WWAN, even if that particular device has never been used to send or receive wireless transmissions.

Most new machines, including gateway servers, do not ship with optimal security settings. The default settings are intended for easy installation and deployment, not for protecting assets [1].

2.4 Signal Interference

Walls, columns, and other building features can reduce signal strength between a wireless NIC and an AP, severely limiting a wireless LAN's range and connection quality. These problems may be mitigated with additional equipment. Other wireless technologies sharing the same public spectrum (such as Bluetooth, cordless phones, and other wireless equipment) can also adversely impact transmission range and quality [1].

2.5 Evolving IEEE Standards

Enterprises contemplating a wireless LAN deployment can choose to implement an 802.11b-based wireless LAN today, or wait for upcoming variations, which are intended to address performance and security issues. IEEE and its workgroups are continually defining and refining standards in light of emerging needs and perceived weaknesses in existing technologies. To the extent that vendors' 802.11 implementations deviate from the various IEEE standards, their equipment can create interoperability challenges [1].

3. RECOMMENDATIONS

Even as new 802.11 vulnerabilities are identified and exploited, enterprises can mitigate or eliminate many of wireless LAN's security risks with careful education, planning, implementation, and management. The following steps aid this process:

1. Establish wireless LAN security policies and practices
2. Design for security
3. Logically separate internal networks

4. Enable VPN access only
5. Remove unnecessary protocols
6. Restrict AP connections
7. Protect wireless devices [1]

3.1 Establish Wireless LAN Security Policies And Practices

The cornerstone of an effective wireless LAN strategy involves planning, defining, standardizing, documenting, disseminating, and enforcing wireless LAN security policies and practices. These include specifying the make, model, configuration (see Fig. 11-1 [1], and settings of the wireless LAN equipment authorized for use, as well as documenting and managing the APs and connected network infrastructure.

Employee education increases awareness of security risks. Some employees may not realize that deploying an unauthorized wireless LAN or using a WiFi product out of the box may increase security risks. Clear and frequently conveyed guidelines usually promote active cooperation [1].

3.2 Design For Security

When placing wireless APs for strategic coverage, installers should consider signal bleed into uncontrolled areas where transmissions can be intercepted. Wireless coverage should be implemented only where needed [1].

3.3 Logically Separate Internal Networks

The LAN segments that connect to wireless APs should connect to a corporate Virtual Private Network (VPN) gateway, but not directly to the production network. Eliminating APs from the production network minimizes the risk of attack techniques such as packet sniffing.

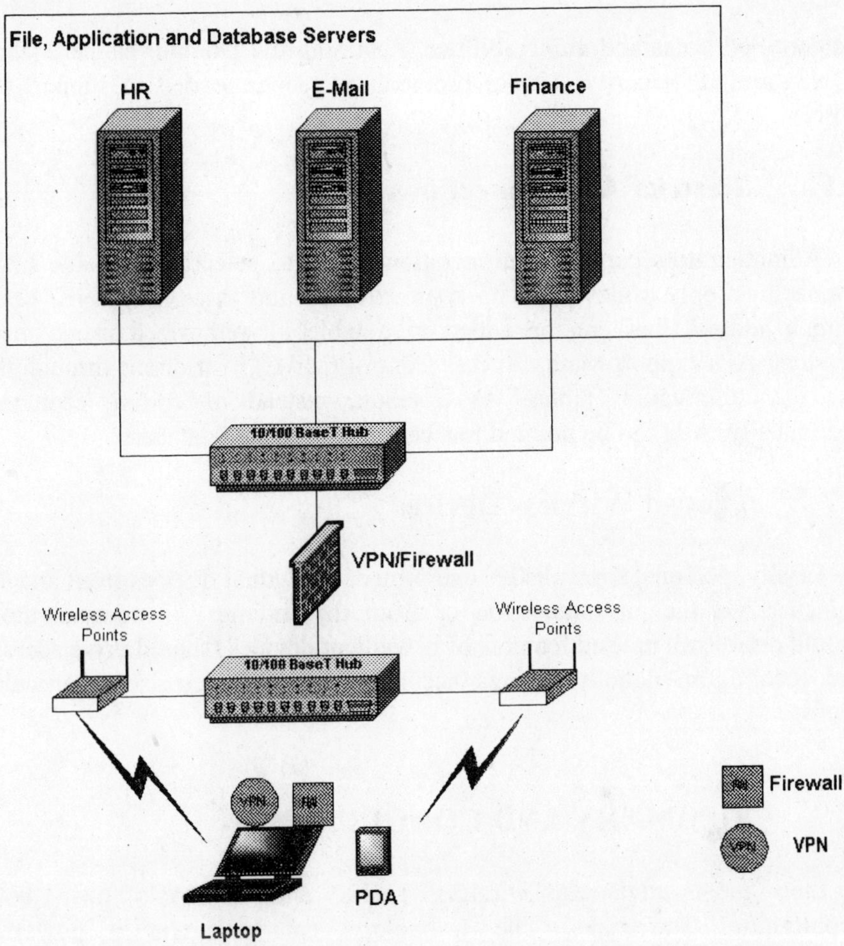

Figure 11-1. Recommended configuration of wireless LAN

3.4 Enable VPN Access Only

Requiring users to connect to the wireless LAN via a VPN is recommended. Once authenticated, authorized users communicate using an encrypted tunnel between the connecting device and the LAN, reducing the risk that a transmission will be captured.

3.5 Restrict Unnecessary Protocols

Restricting unnecessary or redundant protocols from the LAN segments that connect the APs to the VPN gateway reduces the possibility of unidentified holes and vulnerabilities. Retaining the Domain Name System (DNS) and IP Security (IPSec) protocols is recommended to support the VPN.

3.6 Restrict AP Connections

Administrators can use authorization tables to selectively enable LAN connections only to devices with approved NIC addresses. Each NIC has a unique address that can be added to a table of authorized users; most vendors' APs support Media Access Control (MAC) restrictions through the use of authorization tables. As a result, instead of editing each AP individually, APs can be pointed to a centrally managed database.

3.7 Protect Wireless Devices

Finally, personal firewalls [2] can protect individual devices from attacks launched via the air connection or from the Internet. IT administrators should disable all unused features of new client devices (shared drive access) and reconfigure default settings according to the enterprise's particular needs.

4. SUMMARY AND CONCLUSIONS

Like most advances, wireless LANs and WWANs pose both opportunities and risks. The technology can represent a powerful complement to an enterprise's networking capabilities, enabling increased employee productivity and reducing IT costs. To minimize the attendant risks, IT administrators can implement a range of measures, including establishment of wireless security policies and practices, as well as implementation of various LAN and WAN design and implementation measures. Finally, achieving this balance of opportunity and risk allows enterprises to confidently implement wireless LANs and realize the benefits this increasingly viable technology offers.

5. REFERENCES

[1] "Wireless LAN Security: Enabling and Protecting the Enterprise," Copyright © 2002 Symantec Corporation. All rights reserved. Symantec Corporation, 20330 Stevens Creek Blvd., Cupertino, CA 95014, 2002.

[2] John R. Vacca, *Firewalls: Jumpstart for Network and Systems Administrators,* Digital Press, 2004.
[3] John R. Vacca, *The Essential Guide To Storage Area Networks,* Prentice Hall, 2002.

PART IV

INSTALLING AND DEPLOYING WIRELESS NETWORK SECURITY

Chapter 12

TESTING TECHNIQUES

1. INTRODUCTION

There are several techniques to performing penetration testing on your wireless network, the objective of all of them being to improve the security and integrity of the network itself. What wireless lacks in the security of the physical layer and medium must be compensated for in protections on other layers of the stack. As you'll recall from earlier chapters, there are many different attacks that a nefarious individual can carry out on your wireless network.

In this chapter, let's focus more on a cracker attempting to penetrate your network and hack one of the servers held therein. The three phases of this hypothetical, but entirely realistic attack, consist of:

1. Gaining access to the wireless network, even though it is protected by WEP2. Finding available servers on the network
2. Determining the services on those servers available for connection (and exploit)
3. Taking advantage of a well-known vulnerability to gain unauthorized access to a machine [1]

2. PHASE 1: WEP KEY CRACKING

The first task is to figure out how to gain access to the WEP-protected wireless network. Using AirSnort, named after the venerable intrusion detection system Snort, you can passively monitor transmissions across a wireless network and, from that monitoring, derive the encryption key for a WEP-protected network once you have an adequate base of packets. The number of packets required is somewhere between 5 and 10 million packets, but once this foundation of packets for reference has been gathered, it takes less than one second to identify the key. AirSnort is a Linux-based application and requires two things of your network card: it must support RF monitor mode; and, it must have the ability to pass these RF-monitor mode packets to the PF_PACKAGE interface [1].

> **Caution:** All 802.11b networks with 40/128 bit WEP encryption are vulnerable. Furthermore, since using AirSnort constitutes a passive attack, nothing can be done to detect the program being run, either.

> **Note:** There is now a Windows release of AirSnort. From the latest information on the porting effort, it appears to require considerable experience from developers in order to get it working on Windows, however.

Having a wireless card that operates in monitor mode, allows the card to capture packets without associating with an access point or ad-hoc network. This way, you can sniff packets from a specific channel without ever needing to transmit any packets. In monitor mode, you can also discover access points that might not otherwise be available to you and attempt to crack the WEP key associated with that AP that way.

AirSnort also allows you to use promiscuous mode, which is similar to monitor mode, but requires you to associate with a nearby access point--meaning you can effectively only sniff networks that trust you. If you are in promiscuous mode, you will not be able to sniff packets until after you have associated with a AP.

The suggested cards from the developers of the product include Cisco Aironet cards, any Prism-based cards using wlan-ng drivers, and Orinoco-based cards using the newer, patched set of orinoco_cs drivers. You will also need an up-to-date version of libpcap and gtk+-2.2 and gtk+-devel, since the AirSnort interface runs inside a graphical user interface (GUI). Once all of the prerequisites have been filled, download the latest version of AirSnort, and then, from a command line, execute the following [1]:

```
tar -xzf airsnort-0.2.3a.tar.gz
cd airsnort-0.2.3a
./autogen.sh
make
```

Once those commands have completed, AirSnort is installed on your machine. Run AirSnort within your favorite GUI environment on Linux [1].

Once you have your wireless card active, press Start and begin collecting packets. The most common question now is, how long should you wait Here's how AirSnort cracks a WEP key. A weak IV, can assist in exposing only one key byte. AirSnort collects these weak IVs and sorts them according to which key byte each assists in exposing. When a sufficient number of weak IVs have been gathered for a particular key byte, AirSnort computes the probable value for that key byte using some advanced statistical methods. Once these probable values have been generated, AirSnort makes a guess at the key based on the highest ranking values found in the statistical analysis. Typically, there is approximately a 95% chance that a weak IV will reveal nothing at all about a key byte. It may require only a few packets before a key byte is revealed, or it may require many times more. Thus, some keys will be generated and tried fairly quickly, whereas others will crack much more slowly. Regardless, after a while, AirSnort will calculate the key and present it to you. At that point, associate it with the wireless network, and you're now effectively connected to that LAN as if you were plugged in via a standard wire port [1].

Another tool much like AirSnort is WEPCrack, a utility that simply cracks 802.11 WEP encryption keys via much the same method as AirSnort. WEPCrack, however, requires a lot more manual intervention, sniffing packets out of the air and then logging them to a file in a specific format so that a Perl script can be run to glean weak IVs from the packets. Another Perl script is then run to generate the WEP key from that list of IVs. AirSnort has quite a bit more polish, and allows you to complete all of the tasks associated with the key breaking process from within one application. It is recommend that AirSnort be used over WEPCrack; however, if you are working in a forced text-only environment with no GUI, then give WEPCrack a look [1].

3. PHASE II: PORT SCANNING

Next, let's discover some of the things running within your newly-cracked network. Port scanning continues to be an effective and simple way to detect anomalies and openings from your internal network to the outside world. Using the Linux-based NMAP utility, which is the gold standard of port scanning software, helps you to determine both the operating system and version of a system detected through the firewall [2], but also what ports are open and what usual applications are behind those open ports. The port scanning process essentially knocks on the doors of your computer, asking who is there and why.

NMAP and port scanning is a good place to start when beginning your penetration testing process. It provides an overview of your networks, with pointers into where to look for weaknesses [1].

4. PHASES III AND IV: IDENTIFYING VULNERABLE APPLICATIONS AND EXPLOITING THEM

The next step, once you've gained access to the wireless network and discovered open ports on systems connected to the network therein, is to probe the applications behind those ports, determine their identities and any possible vulnerabilities. Again on Linux, there are a couple of tools you can use to try to bring down applications and target their weaknesses.

The first application is NetCat, which is a multipurpose tool for the TCP/IP protocol-it's been dubbed the "swiss army knife" of network administrators everywhere [1]. It is a simple Unix utility which reads and writes data across network connections, using the TCP or UDP protocols. It is designed to be a reliable "back-end" tool that can be used directly, or easily driven by other programs and scripts. At the same time, it is a feature-rich network debugging and exploration tool, able to create almost any kind of connection imaginable with several built-in features. Some of the activities you can perform with NetCat include:

- Create outbound or inbound connections, TCP or UDP, to or from any ports
- Implement full DNS forward/reverse checking, with appropriate warnings
- Use any local source port
- Use any locally-configured network source address
- Use integrated port-scanning capabilities, with randomizer
- Use integrated loose source-routing capability
- Read command line arguments from standard input
- Use Slow-send mode, one line every N seconds
- Create a hex dump of transmitted and received data
- Let another program service established connections
- Use the optional telnet-options responder [1]

One of the primary uses a hacker would have with NetCat is to send improperly crafted, illegitimate, or handily cloaked TCP/IP packets to a certain host and port and watch the reaction of the application behind that port. For example, one can construct a packet with enough HTTP code to request the web server at port 80 to identify itself and its platform. But, let's start a bit earlier than that. To conduct a simple port scan with NetCat -- for instance, of the server at 192.168.0.2 using ports 1-500, you'd execute and receive in return the following [1]:

```
linux:~ # netcat -v -w2 -z 192.168.0.2 1-500
mercury.hasselltech.local [192.168.0.2] 464 (kpasswd) open
mercury.hasselltech.local [192.168.0.2] 445 (microsoft-ds) open
mercury.hasselltech.local [192.168.0.2] 444 (snpp) open
mercury.hasselltech.local [192.168.0.2] 443 (https) open
mercury.hasselltech.local [192.168.0.2] 389 (ldap) open
mercury.hasselltech.local [192.168.0.2] 139 (netbios-ssn) open
mercury.hasselltech.local [192.168.0.2] 135 (epmap) open
mercury.hasselltech.local [192.168.0.2] 88 (kerberos) open
mercury.hasselltech.local [192.168.0.2] 82 (xfer) open
mercury.hasselltech.local [192.168.0.2] 80 (http) open
mercury.hasselltech.local [192.168.0.2] 53 (domain) open
mercury.hasselltech.local [192.168.0.2] 42 (name) open
mercury.hasselltech.local [192.168.0.2] 25 (smtp) open
```

From this list, you'll see that port 80, the usual web server port, is open. You can use NetCat to identify the banner of the service, a fairly accurate way to determine the software used for that service. Using NetCat, displays the following [1]:

```
linux:~ # netcat -v -n 192.168.0.2 80
(UNKNOWN) [192.168.0.2] 80 (?) open
GET HTTP
HTTP/1.1 400 Bad Request
Server: Microsoft-IIS/7.0
Date: Tue, 26 Sep 2005 22:56:11 GMT
Content-Type: text/html
Content-Length: 87
Connection: close
Content-Length: 34

<html><head><title>Error</title></head><body>The parameter?
</html>
```

Tip: Clever administrators can change this, but most don't bother unless they're creating honeypot systems.

The banner indicates that you've found what is probably a Windows 2003 server running IIS 7.0. Clever hackers will then research valid exploits on unpatched Windows 2003 servers and try them. For instance, issuing the GET statement above with an address `http://192.168.1.90/scripts/..%255c../winnt/system32/cmd.exe?/c+dir+c:` to an unpatched IIS 5 server will result in a listing of the contents of the root directory in C: in your NetCat window. That's a problem for the administrator, but you just accomplish the hacker's objective (finding a way into your system) using NetCat [1].

NetCat can also run as a backdoor once installed on a machine, listening on a specific port and waiting for connections. Most hackers would use NetCat on both ends of a connection for a full-featured communications session. For instance, on a remote machine at IP address 192.168.0.2, running nc -L -p 7896 -d -e cmd.exe would instruct NetCat to stay open listening for a connection (-L), listen on port 7896 (-p), run cmd.exe upon connection (-e), and detach itself from that process (-d). Then, on a local machine, run nc -v -n 192.168.0.2 7890 and watch as NetCat connects to the remote machine and a shell on the remote machine opens, ready for your commands. It's a powerful tool for penetrating networks [1].

Finally, if web server exploitation particularly concerns you, you might want to have a look at Whisker. It's another Linux tool that uses documented vulnerabilities in web server software to test running instances of those programs. Take, for instance, the machine in the earlier example detected by NMAP and NetCat to be running Microsoft Internet Information Services (IIS). Whisker takes common vectors of attacks-buffer overflows, FrontPage extensions, IIS password administration, and Unicode URL attacks that have been well documented-and uses them against a running instance of IIS. It's a great way to tell if a system is hardened.

5. SUMMARY AND CONCLUSIONS

In this chapter, you've discovered how to find wireless networks, gain access to them even if they're using WEP encryption, find vulnerable systems and ports, and use various utilities to assist you in those endeavors. This is mostly what a hacker would do to gain access to your systems. Finally, by knowing your hacker, you can then successfully implement strategies to circumvent these methods.

6. REFERENCES

[1] Jonathan Hassell, "Wireless Attacks and Penetration Testing," Copyright © 2005 SecurityFocus. All rights reserved. SecurityFocus, SecurityFocus Symantec Corporation, Suite # 1000, 100 4th Avenue S.W., Calgary, AB T2P 3N2, Canada [Symantec Corporation, 20330 Stevens Creek Blvd., Cupertino, CA 95014], June 14, 2002.

[2] John R. Vacca, *Firewalls: Jumpstart for Network and Systems Administrators,* Digital Press, 2004.

Chapter 13

INTERNETWORKING WIRELESS SECURITY

1. INTRODUCTION

With the exponential growth of the Internet and expected huge success of future wireless networks, the convergence of Internet and wireless multimedia applications and services is of great interest. It will further lead to wireless Internet access with high business penetration in wireless communications. With ongoing industry activities in third-generation (3G) wireless networking, industry analysts envisage the importance of IP-based services over 3G networks. TCP is tuned to perform successfully in wired networks due to transmitting packets over a stable network, where the major cause of packet losses is congestion in the network. But, random bit errors on the wireless link are not properly handled by TCP and cause degradation of TCP performance with respect to throughput. In this case TCP will respond to losses caused by link errors, by invoking unnecessary congestion control and avoidance algorithms, resulting in reduction of window size and leading to reduction of overall throughput.

Regarding the problem of degradation of TCP/IP performance when operating over wire less links, there are many proposals to increase IP stack performance. This chapter focuses on the concept of performance enhancing proxies (PEPs), which were introduced in a working group of the Internet Engineering Task Force (IETF) [1]. A conventional IP-based wireless network in Fig. 13-1 is shown to explain the implementation approach of PEPs in it [1]. Fig. 13-1 illustrates a typical 3G Universal Mobile Telecom Telecommunications System (UMTS) network with an IP-based core network connected to many radio access networks (RANs), where PEPs are implemented. The wireless network is connected to the IP-based backbone through gateways, such as a serving of the General Packet Radio Service (GPRS) support node (SGSN). TCP connections can be established between the core wireless network and mobile hosts/user equipment (UE), whereas the performance of the TCP connection is much improved due to PEP deployment in the communication chain. The remainder of this chapter discusses the centralized and distributed modes of PEP implementation. In addition, the observed security concerns and their potential solutions are discussed. After introducing the schematic models of different PEP approaches, the chapter compares the performance of these approaches and gives an adaptive solution to utilizing them. The problem of wireless security compatibility and its solution are given; as well as, conclusions.

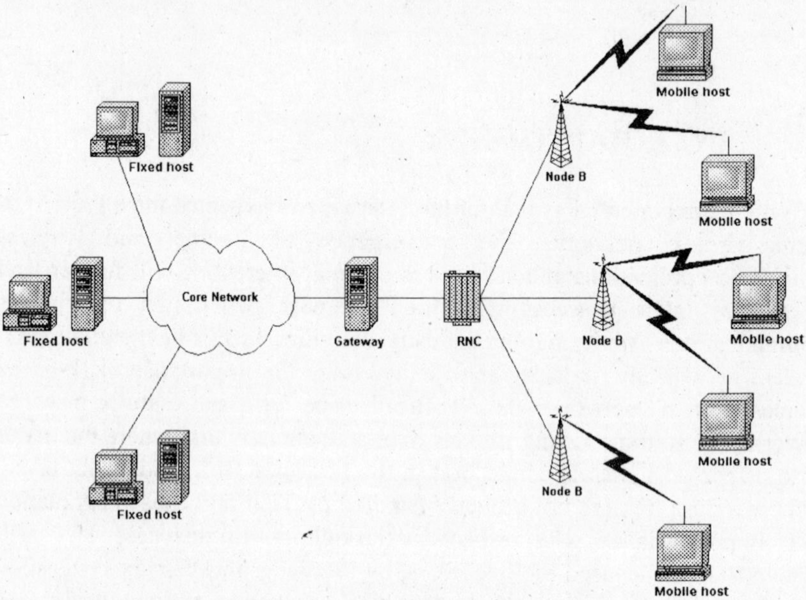

Figure 13-1. Wireless Network Security Topology.

2. OPERATION MODES OF PEP

Different approaches are proposed to improve the protocol performance over wireless networks, which can be classified by implementation mode (centralized or distributed), as shown in Fig. 13-2 [1]. In the distributed mode, the TCP connection is implemented end to end between the fixed host and the mobile host without the intervention of a PEP module in the intermediate node, such as a radio network controller (RNC) at the transport layer for performance enhancement.

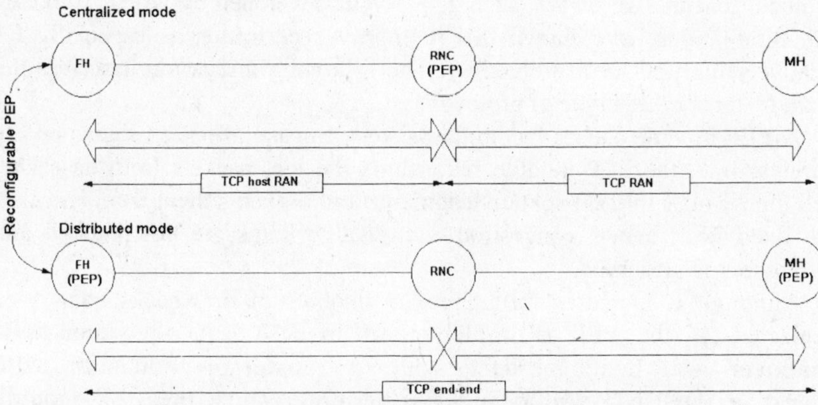

Comparison between the modes	Centralized mode	Distributed mode
Advantages	• High throughput • Works with legacy fixed hosts	• Better security • Less signaling during inter-RNC handover
Disadvantages	• Signaling overhead during inter-RNC handover (shifting context) • Context is required to transfer for inter-RNC handover • Security is compromised	• Fixed hosts need to understand PEP protocols

Figure 13-2. A comparison of centralized and distributed modes.

This mode can be applied where channel conditions are acceptable and do not greatly affect protocol performance. In the centralized mode, the TCP connection goes through an intermediate node equipped with PEP functionality for performance enhancement. This mode can be configured when channel conditions are more severe (a system is highly loaded and a mobile host is badly shadowed). The two modes are based on different approaches with different protocol stack penetrations improving TCP performance in lossy systems (in wireless networks) [2].

In the centralized mode, the PEP is implemented in the RNC. The PEP proxy hides any non-congestion-related losses from the TCP sender and therefore requires no changes to the protocol implementations of existing senders. The intuition behind this approach is that since the problem is local, it should be solved locally (on the wireless link). In the distributed mode, losses on the wireless link are handled through the use of selective acknowledgment (SACK) or explicit loss notification (ELN), as explained later. The table in Fig. 13-2 shows the advantages and disadvantages of both modes.

One example of a centralized mode approach is a reliable link layer protocol that is TCP aware, such as SNOOP developed by UC of Berkeley [1]. Simulations have shown that it improves performance, especially for relatively high bit error probability. The following discussion describes the main features of this type of protocol.

A PEP module caches incoming packets. In case of packet losses on the wireless link, the PEP module retransmits the lost packets from its cache, and blocks all duplicate acknowledgments and prevents them from reaching the fixed host; hence, congestion control algorithms are not invoked and throughput is preserved.

Although it increases throughput, a number of drawbacks cannot be neglected. If the PEP is implemented in RNC, in the inter-RNC handover case, handover delays due to transfer of connection state stored in the RNC will occur. Furthermore, since the PEP module must decode the protocol headers, it is incompatible with the use of IPsec, since it cannot access the encrypted TCP header.

The distributed mode adopts end-to-end protocols that attempt to make the TCP sender (fixed host) handle losses by the use of two techniques. First they use some sort of SACKs [1] to allow the sender to recover from multiple packet losses in a window without resorting to a coarse timeout. The SACK RFC [1] proposes that each acknowledgment contain information about up to three noncontiguous blocks of data that have been received successfully by the receiver. Each block of data is described by its starting and ending sequence numbers. Due to the limited number of blocks, it is best to inform the sender about the most recent blocks received. Second, they attempt to have the sender distinguish between congestion and other forms of losses using ELN. The sender can then avoid invoking congestion control algorithms when non-congestion-related losses occur. In this method, the corruption of a packet at the link layer is passed to the transport layer, which sends an ELN message with the duplicate acknowledgments for the lost packet. The end-to-end approach with SACK shows good performance at a medium bit error rate; it is simple to implement and compatible with the use of IPsec.

3. ADAPTIVE USAGE OF PEPS OVER A RAN

The RNC of the wireless network has the ability to choose connection modes (select centralized or distributed mode). The switch is between the centralized and distributed mode, as *intermode handover* in this chapter. The RNC works as follows to dynamically utilize the PEP algorithms. While in the distributed mode, the RNC continuously monitors channel conditions such as fading and interference levels. Based on the available information, the RNC decides whether to move to the centralized mode or maintain the current mode. As long as the channel conditions are good and interference is low, the radio link control (RLC) can handle the losses, and there is no need to switch to the centralized mode.

3.1.1 Distributed to Centralized Mode

The steps of the intermode handover from distributed to centralized mode can be carried out as follows:

1. An intermediate node (RNC) monitors the channel of the current wireless connection.
2. If the channel condition is good and the data load is low, repeat Step 1; otherwise. the RNC triggers the mobile host to get into the switching procedure.
3. The PEP function for the connection will be activated/integrated in the RNC.
4. The mobile host has the option to accept the new settings and benefit from the performance enhancement or reject it due to some constraints. [1].

3.1.1.1 Security Constraints

The security requirement might be violated if the connection is switched back to the centralized PEP mode, with regards to Step 4 in the preceding. Since the PEP needs to lie part of the wireless network security association between the mobile and the fixed host, it must be able to break the end-to-end semantics of the connections by disabling end-to-end use of IPsec in order to act on the encrypted TCP header. Otherwise, the centralized PEP cannot be implemented.

3.1.1.2 Protocol Constraints

Due to the reality of mobility, the handover process is inevitable. Handover procedures add considerable overhead to connection, due to the transfer of the connection state between the intermediate nodes. If the mobile host is compatible with the changes, it will he reconfigured accordingly.

Once the negotiation is terminated, the connections that accept the performance enhancement and the involved mobile host that agrees to the preceding constrains will switch to centralized mode.

3.1.2 Centralized to Distributed Mode

Intermode handover from centralized to distributed can be carried out by the following steps:

1. The RNC estimates the improvement of the wireless link.
2. The RNC prompts the mobile host to switch back to distributed mode.
3. If the mobile host accepts, it will be reconfigured accordingly and will download the end-to-end PEP functionality, and so will the fixed host [1].

Therefore, the connections that experience improved channel quality will switch back to the distributed mode. One realization of a message sequence chart is shown in Fig. 13-3 [1].

As can be seen from the simulation results, the implementation of PEP in the intermediate node (base station) conserves high throughput with high bit error rates in the channel, compared to other approaches. But, a problem arises when the connection uses IPsec. In such a situation the user has to sacrifice one service for the sake of the other, because centralized PEP and IPsec are mutually exclusive, as explained later.

4. PROBLEMS OF PEP WITH IPSEC: THREATS

To motivate the necessity for security in communications in wireless networks, some attacks are outlined that could be performed by malicious core software if no protection mechanisms are in place.

4.1 Attacks Against the User

Mobile hosts could be used to compromise the user's private sphere. Data stored in the terminal, such as address book, date book, and called numbers, could be changed or made available to third parties. The effectiveness of functions for secure e-commerce, such as payment or computation of digital signatures for a user, could be affected.

Furthermore, the mobile host could be changed into a surveillance device by installation of manipulated core software that transmits environmental sounds to an attacker. Since the core software controls the user interface, the terminal could appear to be switched off or in standby mode, and the attack would therefore not be noticed by the user. The user could also be harmed by being billed for telephone calls set up by malicious software (especially calls to premium-rate numbers).

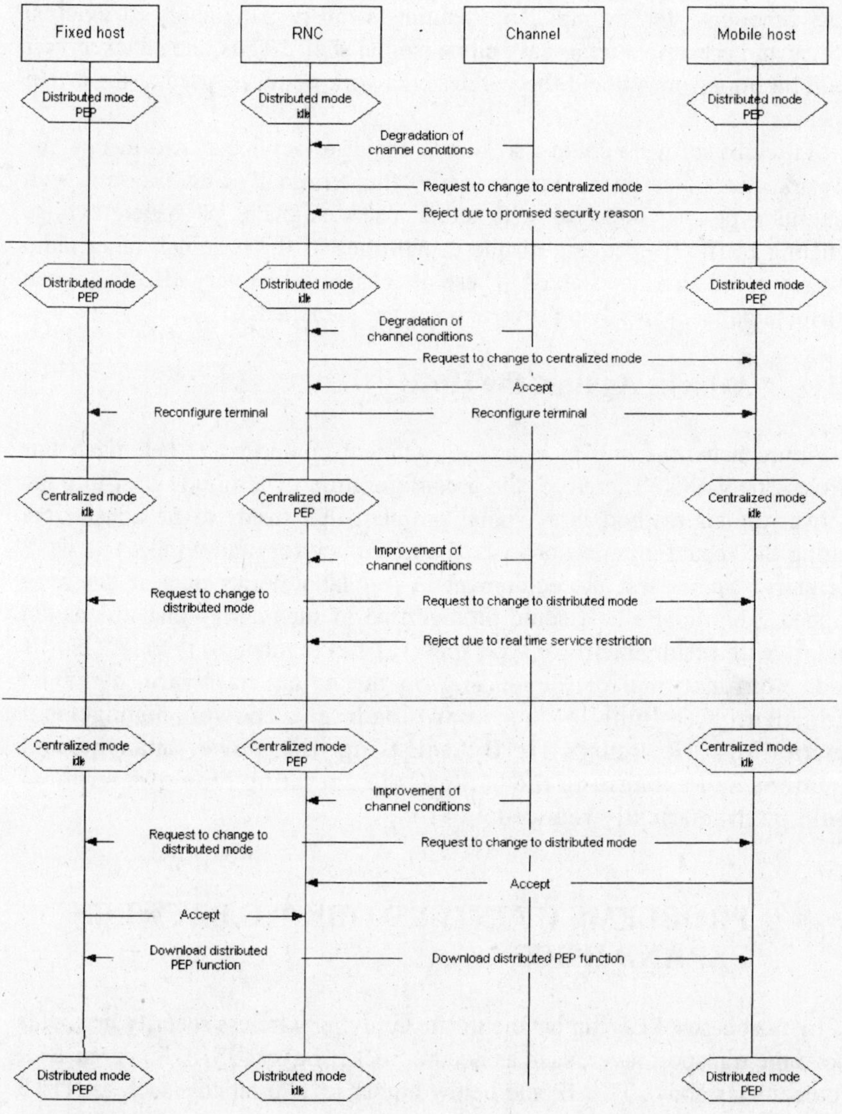

Figure 13-3. Message sequence chart of the adaptive usage of PEP.

4.2 Attacks Against the Wireless Network

The mobile host contains security functions that protect the communication at the air interface. Malicious software could circumvent these functions, for example, by sending security parameters to external entities and creating a temporary clone (rogue shell). Thus, the attacked user could be impersonated and the wireless network could be used at his or her expense.

Malicious software could also launch denial-of-service attacks against the network and other users. For example, the terminal could transmit with maximum power, interfere with other users' signals, or generate huge signaling traffic. Depending on the capabilities of the terminal, other radio systems could also be affected. These attacks could be very effective when distributed attacks involving several hosts are launched.

4.3 Attacks Against the Host

Future networks will meet an environment composed of heterogeneous wireless networks. Therefore, the reconfiguration of a mobile host through an over-the-air method is a crucial problem that needs to be considered. During the reconfiguration process, the manufacturer (and wireless network operators) cannot test the equipment in test laboratories once it has been shipped. The mobile host could be modified in such a way that it does not comply with requirements for type approval and regulation. It could also be made nonfunctional or, depending on the actual hardware, damaged permanently (a mobile host downloading a power management function by IP transport). By misusing the power management functions and exhausting the battery, the operating time of a terminal could be dramatically reduced.

5. PROBLEMS OF INTERWORKING BETWEEN PEP AND IPSEC

In most cases, PEP can handle traffic applying wireless security protocols above the transport layer, such as Secure Socket Layer (SSL). Since the SSL protocol runs above TCP/IP and below higher level protocols such as HTTP or IMAP, the TCP header is therefore not encrypted, and PEP can manipulate acknowledgments. However, only a limited number of applications include support for the use of transport layer security nowadays. On the other hand, wireless network (IP) layer security, such as IPsec, can generally be used by any application, transparent to the application, offering access control, data integrity, authentication, and confidentiality. Providing all these services guarantees protection for IP and upper layer protocols.

In fact, IPsec not only offers "information" data integrity, but also "control" data integrity, such as acknowledgments on the transport layer, so attackers cannot use information such as wireless network statistics (flow of acknowledgments) to make an attack. An example could be that hackers attack the TCP header and change the window size to zero, forcing the host to go into persist mode and stop sending data. Another is to regenerate the same acknowledgment number, making the host interpret it as duplicate acknowledgments, causing reduction in window size and consequently in throughput. Also, the connection could be made to shut down.

IPsec works in two modes, transport and tunnel, explained briefly here.

Transport mode is typically used in peer-to-peer communications to provide Intranet security. The data packet is encrypted, but the IP header isn't. Therefore, it can be read and used by any standards-based device or software, as depicted in Fig. 13-4 [1].

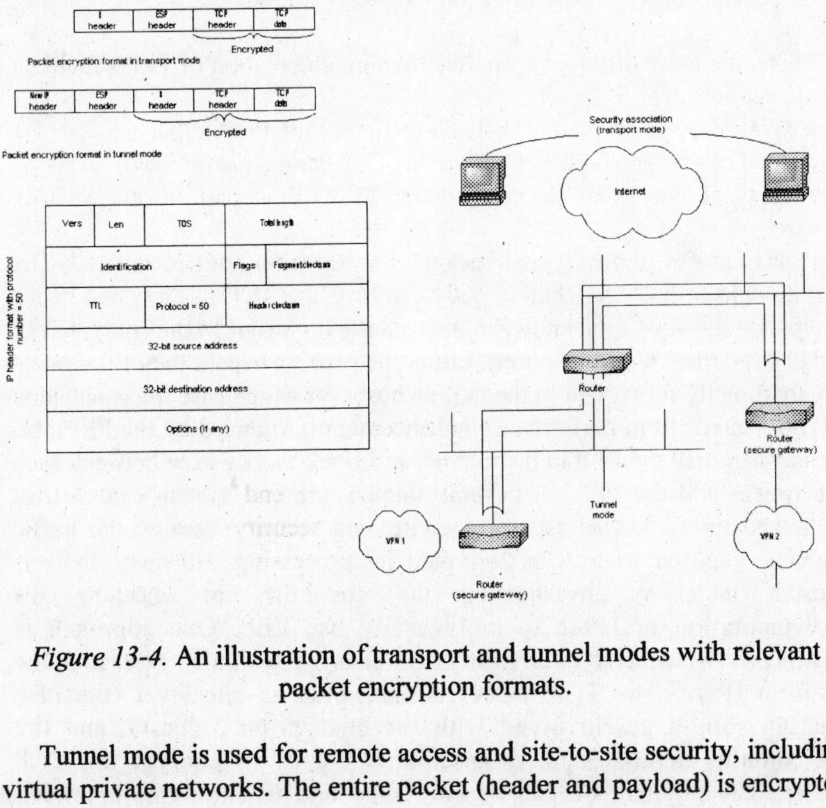

Figure 13-4. An illustration of transport and tunnel modes with relevant packet encryption formats.

Tunnel mode is used for remote access and site-to-site security, including virtual private networks. The entire packet (header and payload) is encrypted and given a new IPsec header that is divided into an encapsulating security payload (ESP) header field and an IP header field with an authentication trailer, thus hiding the topology of the protected sites (see Fig. 13-4).

In these circumstances, PEP employed in the centralized mode (implemented in an intermediate node) cannot act on traffic protected by the use of Ipsec, because it cannot examine the TCP headers of IP packets. Due to encryption of IP packets via IPsec's ESP header in either transport or tunnel mode, a TCP header is rendered and the payload is unintelligible to the PEP [1].

Therefore, the user or administrator should decide whether to protect data using IPsec or activate the PEP for throughput enhancement, but not both at the same time. From this, arises the need to find a way to compromise between data security, integrity, and high throughput.

6. SOLUTIONS TO THE PROBLEMS

There are some different scenarios the implementation of PEP should be able to support:

- Arrival of nonencrypted packets that require high throughput
- Arrival of encrypted packets where the PEP cannot act on them
- Arrival of encrypted packets where the PEP is part of the security association [1]

In the case of nonencrypted packets, no problems are encountered. The PEP module in the intermediate node can read the TCP headers and hence manipulate the flow of acknowledgments autonomously. In the encrypted IP packet case, there are two options. Either the packets bypass the PEP module and are directly forwarded to the mobile hosts (in which case the connection will not benefit from performance enhancement introduced by the PEP); or, the user can trust the PEP in the middle, and IPsec can be used between each end system and the PEP. In general, though, the end system cannot trust PEPs, and this is not as secure as end-to-end security because the traffic might be exposed when it is decrypted for processing. However, there is research underway investigating the possibility of changing the implementation of IPsec to more easily use PEP. One approach is multilayer IP security over protocol stacks by partially releasing the freeform IPsec. The TCP header is encrypted as one layer (the PEP function should get involved with the encryption process), and the PEP should include security associations used to encrypt the TCP header. The payload of TCP is encrypted end to end. This approach still requires trust of PEP, but to a much lesser extent. However, it of course increases the complexity of IPsec.

From the user viewpoint, the option of using IPsec with different levels of security requirements is needed. If an application requires a high level of security, IPsec cannot be compromised. If an application does not require high security, but high throughput is the greatest demand, security can be compromised to improve the significance of the PEP in the connection. Of course, the terminal should be configured properly to support both encrypted and nonencrypted packet types.

All in all, there must be a smart decision making module in the intermediate node. This would forward IP packets to the PEP in case IPsec is not applied or the PEP is trusted and can decrypt and reencrypt the packets, and let the encrypted packets with IPsec bypass the PEP if it cannot take part in the wireless security association. In order to conceive this, note that in the IP protocol, there is a number in the header field. It is valid for both a normal IP header or a new IPsec header, where the header determines the protocol to be used. With this number, an application can easily determine if a packet is protected with IPsec or not and thus make a decision. When the protocol used on top of IP is ESP, the number in this field will have a value of 50 (decimal), as shown in Fig. 13-5.

Finally, if PEP is trusted (it has the security association key), the packets will be decrypted, passed to the PEP module, and then re-encrypted and sent to their destination. If the PEP is not part of the security association, the packets should bypass the module. Fig.13-5 illustrates the traffic of IP packets from one terminal to the other through the RNC [1]. It also illustrates the functionality of the decision making module according to security. Moreover, it is possible in most IPsec implementations to have security associations between two different IP addresses and switch this security association on or off according to the approach. This will give flexibility to choose to send some packets protected by IPsec and others freely without the use of any protection.

7. SUMMARY AND CONCLUSIONS

This chapter proposed the interworking between performance enhancing proxies (PEPs) and IPsec in wireless networks. The low-throughput problem due to TCP/IP in a radio access network was illustrated. Performance comparison among different PEPs implemented in the RAN was carried out in order to optimize spectrum efficiency. By using PEP, end-to-end security is compromised, and a concept was proposed to circumvent this problem. Furthermore, a way to utilize PEP for different wireless network loads was also proposed. In this chapter, it was also suggested that a scheme that allows the coexistence of IPsec and PEP over wireless networks be installed and deployed through the adding of an intelligent module in the node where the PEP is implemented.

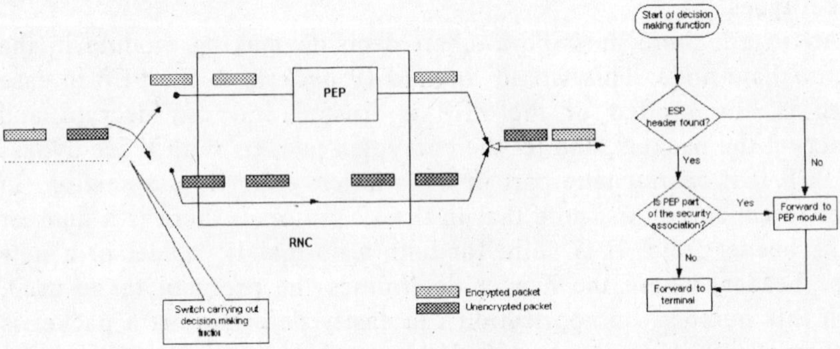

Figure 13-5. A switch in the intermediate node and the flow chart of its decision function.

Finally, this chapter explained the adaptive usage of PEPs mainly for security reasons in IPsec connections. An intermediate node in the radio access network is to decide whether a TCP connection should be established directly to the terminal or run over a PEP module in order to benefit from the performance enhancement added by a PEP. Dynamic PEP utilization depends not only on load information, but also on security limitation. If end-to-end security is demanded, any PEP in the communication chain will compromise this requirement. To decide, you must consider three communication modes: transmission with nonencrypted packets, with encrypted packets where the PEP does not have the security association key, and with encrypted packets where the PEP has the security association key. Potential security concerns were also discussed with regards to using PEP and IPsec concurrently. It was also proposed that an intelligent switch for internetworking between IPsec and PEP be devised. Depending on the chosen communication modes for security, there are many different approaches to internetworking between IPsec and PEP, which is subject to operator policy and user wireless security demands.

8. REFERENCES

[1] Nadim Assaf, Jijun Luo, Markus Dillinger and Luis Menendez, "Internetworking Between IP Security and Performance Enhancing Proxies for Mobile Networks," Copyright © Copyright 2005, IEEE. All rights reserved. IEEE Corporate Office, 3 Park Avenue, 17th Floor, New York, New York, 10016-5997 U.S.A., May, 2002.

Chapter 14

INSTALLATION AND DEPLOYMENT

1. INTRODUCTION

Wireless application security encompasses a broad range of elements, namely just about anything that negatively impacts the application. When developing wireless application software, you should be thinking about potential security risks and applicable countermeasures. This often goes well beyond what the software was intended to do, but the extra time and expense will be well worth it.

Wireless networks can also increase productivity and reduce infrastructure costs, but how do you balance the needs of users with the requirements of security? What are the pitfalls of wireless installations and deployments, and how can you avoid them? The goal of this chapter is to ease your concerns about the security of wireless networks installations and deployments, by increasing your knowledge on the subject.

2. INSTALLATION

Wireless devices are small, and users can lose or misplace them easily. Or, they could end up stolen. Most IT enterprises have fairly good authentication for servers and databases residing on the enterprise network, but the wireless device itself is still vulnerable to attack.

In order to guard against unauthorized access to data and applications on the wireless device make sure the device's keypad automatically locks after a number of minutes of non-use, and require a form of authentication (such as a PIN or password) to unlock the keypad [1].

But don't overlook the fact that a hacker might try to figure out the PIN through brute force methods, such as systematically entering different PIN combinations. As a result, incorporate a cool-down period to temporarily and completely disable access to the device if the authentication is not successful after several attempts. In addition, consider performing memory and application wipe functions to restore the device to factory default settings if the thief types in the wrong PIN or password numerous times.

Another way a hacker can gain access to user information is by monitoring a user's password over an unsecured connection, such as an unencrypted POP3 email account over a public WiFi network. Because many users use the same passwords for multiple services, the hacker will likely try the password to unlock access to a secured account, such as the user's enterprise email or applications. In order to counter this, encourage users to not use the same password for different accounts, or automatically generate unique and appropriate passwords for users [1].

Hackers are very good at finding ways into wireless devices. For example, a hacker may attempt to synchronize the device to a PC or laptop in order to view and copy data stored on the device. Wireless adapters are especially vulnerable because a hacker can exploit them without physically stealing the device. For a knowledgeable hacker, Bluetooth, Wi-Fi, IrDA and cellular wireless interfaces allow easy entry into the device. For example, a user associated with a public Wi-Fi wireless LAN is part of the same network as other users. Thus, a hacker associated with the same network can access files on the wireless device unless the user takes precautions.

Wireless devices have many interface ports, so be sure to incorporate a personal firewall that locks them down [3]. Firewalls do a good job of limiting which ports are active and which protocols and users can use them. Also, consider whether users can install applications, and possibly disallow that ability. Some applications may install software that unknowingly will open a wireless port that may allow hackers in [1].

Because storage cards are portable, they also offer a security risk, similar to removable storage disks in PCs [4]. A thief taking possession of a user's Palm Pilot, for instance, can eject the storage card and read the data into another device. This can be prevented by encrypting the data on the cards or disabling their use.

It's difficult to fully secure wireless devices from the smarter hackers. That's why some enterprises establish the means for remote destruction of the wireless device data and applications [1].

An issue with remote destruction, however, is that it can only happen if there is a connection available to the device. The thief may somehow disable the radio, which precludes the destruction from taking place. Of course the connection will be made (and the destruction possible) if the thief tries to access an enterprise application.

Also, it can be inconvenient to perform the destruction for a misplaced device that might be found within a few days. If the user is located in a remote overseas location, then the replacement of the data and applications will be difficult [1].

So, make use of remote destruction if possible, but don't depend on it. Because radio waves propagate outside the physically controlled area of the user, there is a potential for unauthorized monitoring of wireless network application data. Most enterprises require the use of encryption for wireless connections, but it's important to use effective mechanisms. Hackers can monitor the transmission of 802.11 wired equivalent privacy (WEP)-encrypted data packets and eventually decipher the data. Better choices for wireless encryption in enterprises include the Wi-Fi Protected Access (WPA) and the 802.11i standards.

When operating from public WiFi networks, there is no encryption on the wireless network between the wireless device and the access point. This isn't a huge problem if the user is using a credit card over the network, assuming that the e-commerce site employs secure socket layer (SSL) encryption [1].

Virtual private network (VPN) software is much safer when accessing corporate systems from public networks. A VPN client on the wireless device communicates with the remote application, with encryption taking place from end-to-end. This is a well understood technology—enterprises have been requiring travelers to utilize VPNs for years when using dial-up services.

In fact, some enterprises even require VPNs when connecting wireless devices directly to the enterprise wireless network. The access points in this case reside on the un-trusted side of the enterprise firewall. This is a safe solution, but it incorporates higher costs for hardware in order to support a greater number of VPN connections. You're all accustomed to using anti-virus software on PCs and laptops, but many smart phones go unprotected. Traditionally, virus writing culprits haven't targeted smart phones, because there aren't enough in circulation to make it worthwhile. Authors of viruses generally stick to writing them for platforms that are very common—they strive for the biggest bang for the virus [1].

The proliferation of smart phones, however, will eventually grab the attention of virus writers. As a result, be certain to include anti-virus utilities to your smart phone installations.

With wireless networks, denial of service (DoS) can happen if a hacker continually sends disassociation frames addressed to specific users to keep them from having access to a WiFi wireless network. This can also be done by transmitting a relatively powerful interfering signal within the area of the wireless network. WiFi protocols are very polite and let radio frequency interference use the frequency spectrum and force wireless users to wait until the interference goes away. Unfortunately, a mischievous person can take advantage of this situation [1].

Also, after disabling access to a legitimate access point, a hacker can surface a rogue access point, and the user's radio card will automatically associate with it. When this occurs, it's possible for the hacker to intercept and even take over sessions between the user and the application server. The rogue access point can be configured to look just like the legitimate one did to the enterprise wireless network. These types of man-in-the-middle attacks can be very damaging to information security.

Some enterprises deploy effective authentication systems using digital certificates and require access points to verify their identities to the enterprise wireless network. This prevents the hacker from getting anywhere when attempting the man-in-the-middle attack [1].

Before investing a great deal of time and money in guarding against DoS attacks, evaluate the potential for such an event. If such an attack will impose hazards to people, substantial loss of revenue or embarrassment to the enterprise, then strongly consider deploying anti-intrusion systems that monitor the presence of DoS attacks and proceed with countermeasures automatically.

So, with the preceding in mind, wireless networks can increase productivity and reduce infrastructure costs, but how do you balance the needs of users with the requirements of security? What are the pitfalls of wireless deployments, and how can you avoid them [1]?

The goal of this part of the chapter is to ease your concerns about the security of the deployment of wireless networks by increasing your knowledge on the subject. In fact, after you understand and install a few of these guidelines, you will be able to reduce your level of anxiety, embrace these new tools, and usher in a new era of wireless freedom for your company.

3. DEPLOYMENT

Here's the golden rule of wireless network security: You don't need to be faster than the bear. You just need to be faster than you.

Security for networks and wireless LANs is no different then any other form of security. For example, if you want to protect your expensive car stereo, there are several steps you can take. You can lock your door. You can install a car alarm system. You can even hire a 24-hour armed guard. However, you know no security system is 100 percent safe. If somebody wants your car stereo or data) badly enough, they'll find a way to get it. But chances are, if you take measures to make the task difficult, the thief (like the bear) will move on to attack easier prey [2].

Ultimately, your choice of security measures depends on the perceived value of the assets you're protecting and your willingness to spend money to protect them. It's a classic cost-benefit consideration. Although it's true, many wireless security measures can be defeated in theory and in practice, it's nevertheless worthwhile to employ their protection, especially if the cost is negligible.

The following list shows some of the measures you can take, from least to most expensive and time consuming.

3.1 Service Set Identifier (SSID) Obfuscation

Service Set Identifier (SSID) obfuscation is one of the first and easiest steps you can take to secure a wireless LAN. SSID is a unique 32-character ID used to connect wireless devices. To associate with a wireless network, the client must use the same SSID as the access point (AP). This information is passed in clear text and sniffable even if you have WEP enabled for data encryption. Also, because most access points, by default, broadcast their SSID to the world, AP discovery tools such as Netstumbler can pick these up with ease [2].

At a minimum, you should rename your SSID using something other than its default value. The name you choose should clearly identify your wireless network as private, without divulging your enterprise name or physical location. You want to identify your wireless network as private, so it's clear to people you don't want them associating with your network.

By default, most APs turn on management beacons to advertise the existence of the AP. In a perfect world, this would be a good feature. However, if your AP has the option, you should turn off management beacons because they can lead intruders to your SSID [2].

3.2 MAC Address

Each Ethernet client card has a globally unique MAC address hardcoded at the time of production. This address consists of two components: The manufacturer's unique ID as assigned by the IEEE, and the card's unique ID as assigned by the manufacturer. Combined, this value represents the client card's MAC address. Some APs let you restrict access based on the MAC address of the client card.

For very small environments (just a handful of users), you may consider implementing MAC-based restrictions because the population of MAC addresses is small and manageable. If you have a large population of users, the administrative burden of keeping track of all those addresses is likely to be prohibitive [2].

The other problem is that even with WEP encryption, MAC addresses travel in clear text. Further, many wireless client cards can define (spoof) their own MAC address in their driver configuration. An attacker can simply sniff the traffic, choose a valid MAC address, and assign it to his or her own wireless client card.

The press has had a lot to say about the security (or lack of it) for WEP encryption. The problem, however, is more complex than is commonly portrayed. First, there are known security weaknesses in the implementation of the RC4 cipher that make WEP traffic vulnerable to cracking. Tools such as Airsnort, WepCrack, and dWepUtils can break the encryption after accumulating enough "interesting" traffic. The passive listening attack could take hours, days, or even weeks, but once an adequate amount of interesting traffic is captured, cracking WEP is a trivial, non-CPU intensive task [2].

> **Note:** Interesting traffic means that certain WEP packets contain weak cryptographic keys. Only a small percentage of total traffic contains these weak keys; however, the more weak keys that are accumulated over time, the greater the likelihood of cracking WEP.

The other issue is the challenge of key management. Even if WEP vulnerabilities didn't exist, system administrators would still have to consider the significant administrative challenges involved in widespread shared secret key deployment. If your client base is relatively small, it may be simple enough to tell everybody the secret key and change it when necessary However, if your enterprise requires a large number of clients to have wireless access, key management becomes an issue. What happens if somebody leaves the enterprise, or if a laptop is stolen? You must operate under the assumption that the key has been compromised and you need to re-key every client machine.

3.2.1 802.1 X Potential

One potential solution to the key management and encryption problem is the use of 802.1x. 802.1x will allow wireless enterprises to scale to the hundreds or thousands of users by providing a centralized security administration process. After authenticating with a Remote Authentication Dial In User Service (RADIUS) or Kerberos server, 802.1x provides dynamically generated session keys which, if compromised, would be useless for decrypting other sessions. Think of it as the ability to change your secret key on the fly for every session.

Because 802.1x is port-based, you can use it to secure both wired and wireless networks. Furthermore, you can use it for authentication, encryption, or both. The challenge with 802.1x is that only a small number of products currently support it. If you're using Microsoft Windows for your client operating systems, you must be running Windows 2003 to natively support 802.1x in your environment.

3.2.1.1 What Can You Do?

By now, you might be wondering if there's any way to go wireless without endangering your enterprise's wireless network. The answer is simple: Treat your wireless segments just like you would a T-1 or DSL connection from a service provider. That is, treat it as an untrusted wireless network. You don't give the public Internet free reign on your wireless network, so why would you give that privilege to the wireless world?

Wireless Ethernet signals don't stop at your enterprise walls. In fact, with the proper antenna, amplification and line of-sight, your 802.11 signal can be picked up many miles away. The bottom line is this: Use a firewall and encrypt your network traffic. Operate under the assumption that your wireless networks will be compromised. Remember, if no enterprise assets are available on the unauthenticated, unencrypted wireless segment, there's no risk of damage from an unauthorized wireless client or the occasional "war driver."

3.2.2 802.1 X Revisited

Perhaps the best hope for a standardized security solution for 802.11 lies in the adoption of the 802.1 x standard. 802.I x is now supported by products from Linksys, Dlink, SMC, Netgear, and other lower cost manufacturers. 802.1x authentication is built on three components:

1. The supplicant: The client requesting access to the wireless network.
2. The authenticator: The "enforcer" that permits or denies access to the wireless network. This is typically an Access Point.
3. The authentication server: The server that maintains authentication information, checks the supplicant's credentials, and reports the results to the authenticator. This is typically a RADIUS server.

To put it in nightclub terms, the supplicant is the person trying to get into the club, the authenticator is the bouncer keeping people out or letting people enter, and the authentication server is the VIP list of who's allowed in. Each port, whether physical in the case of a switch or virtual in the case of a WLAN, begins in an "uncontrolled" state. In this state, a supplicant can't directly access the network because all traffic except 802.1x is denied. When a request (or WLAN association) is detected, the authenticator sends an identity request to the supplicant. The supplicant responds to the authenticator with its identity.

Next, the authenticator forwards the identity to the authentication server. The authentication server requests a specific type of credential (password challenge) to authenticate the supplicant. This request is once again proxied by the authenticator, because the supplicant and authentication server still can't communicate directly with each other at this point.

After successfully validating the supplicant's credentials, the authentication server provides an encrypted authentication key. The supplicant uses this authentication key to unlock the port and transition from an uncontrolled state to a controlled state, and access to the wireless network.

You can configure 802.1x to request periodic re-authentication, thereby creating a new key as often as every five or ten minutes without user intervention. This reduces the risks posed by the WEP vulnerabilities described earlier.

So, if you are operating in a non-Microsoft environment, the basic building for 802.lx deployments remain the same. Open source alternative's for each component are readily available.

3.2.2.1 The Reality Of Wireless

It's often said that the most secure computer is one that's turned off. As a security professional, the key to implementing technology is to find ways to support the needs of your user community, while protecting the information assets of the enterprise. This balance requires a healthy appreciation for technology as a business enabler, while respecting the delicate security risk considerations. Wireless networks, however, are here to stay, and no amount of wishful thinking will make the problem go away.

According to industry analysts, at least 2 0 percent of large enterprises already have rogue WLANs in place that were installed by users, not the IT department. The rogue wireless networks were implemented, because users grew impatient with their IT department's unwillingness to support new technologies, so they just did it themselves. Have you performed a wireless security assessment recently to identify if you already have unauthorized APs?

Finally, a proactive approach to wireless network security is your best bet. Don't drive your users to impulsively deploy their own wireless APs. Anticipate your WLAN requirements and start a wireless pilot project right away. Wireless Ethernet deployments can be a bumpy road, but the enormous productivity gains and workflow benefits are well worth the trip.

4. SUMMARY AND CONCLUSIONS

Wireless software applications can realize a level of security that satisfies the needs of most enterprises, assuming you plan effective solutions and have security policies in place that enforce conformance to security mechanisms. Keep in mind, however, that the strongest authentication and data encryption won't keep a hacker from merely looking over the shoulder of someone using a wireless device in a public place. So don't forget to educate users on this simple hacker tactic as well as the more sophisticated ones.

5. REFERENCES

[1] Jim Geier, "Implementing Mobile Security," Mobilized Software, Copyright (c) CMP Media LLC. All rights reserved. CMP Media LLC, 600 Community Drive, Manhasset, NY 11030, February 1, 2005.

[2] Lee Barken, "Wireless Security Step by Step," copyright ©1983-2005 Advisor Media, Inc. All Rights Reserved. ADVISOR MEDIA, Inc. 4849 Viewridge Avenue, San Diego, CA 92123-1643, April, 2002.

[3] John R. Vacca, *Firewalls: Jumpstart for Network and Systems Administrators,* Digital Press, 2004.

[4] John R. Vacca, *The Essential Guide To Storage Area Networks,* Prentice Hall, 2002.

Chapter 15

SECURING YOUR WIRELESS E-COMMERCE STOREFRONT

1. INTRODUCTION

There is a common perception that wireless network environments are inherently less secure than wired environments. Reports of phone masquerading and phone call tapping in mobile wireless environments have led many to believe that this is not an environment conducive for e-commerce [1]. While this was certainly true in the past, the wireless industry has been working hard at providing security protections strong enough for real mobile-device based e-commerce. Well, maybe (See sidebar, "Risky Business With 50 Million MasterCard Files").

Risky Business With 50 Million MasterCard Files

MasterCard International recently reported that more than 50 million credit card accounts of all brands might have been exposed to fraud through a computer security breach at a payment processing company--perhaps the largest case of stolen consumer data to date. MasterCard said its analysts and law enforcement officials had identified a pattern of fraudulent charges that were traced to an intrusion at CardSystems Solutions of Tucson- Ariz., which processes more than $26 billion in payments for small to midsize merchants and financial institutions each year. About 31 million Visa and 19 million MasterCard accounts were compromised: the other accounts belonged to American Express or Discover cardholders. The accounts affected included credit cards and certain kinds of debit cards. The F.B.I. is still investigating.

An infiltrator had managed to place a computer code or script on the CardSystems network that made it possible to extract information. MasterCard found that CardSystems, in violation of MasterCard's rules, was storing cardholders' account numbers and security codes on its computer systems. That information, supposed to be transferred to the bank handling the merchants' transactions, but not retained by CardSvstems.

There were conflicting accounts on how the investigation began. CardSystems identified a potential problem and contacted the F.B.I, then the Visa and MasterCard associations. It took steps to immediately ensure all systems were secure.

MasterCard began the investigation when it was notified by several banks that they had detected atypical levels of fraudulent charges. In turn, MasterCard began monitoring information from the accounts for common purchasing points. Using data-analysis systems, it was able to focus on an unspecified bank receiving spending data from merchants.

Visa had been aware of the data breach, but kept quiet at the request of the authorities. Discover and American Express just recently learned of the breach and had been closely monitoring accounts.

According to MasterCard, an unauthorized person was able to exploit the security vulnerability and gain access to CardSystems' network, exposing cardholders' names, account numbers and expiration dates; as well as, the security code. Typically, three or four digits were also printed on the credit card.

The processing companies are hubs for millions of payment records, which could be infiltrated as information passes through. It is the juiciest target for an individual who wants account numbers. It is a honeypot for identity thieves.

Other personal data that might contribute to identity theft [8], like Social Security numbers and dates of birth, was not stored on the cards and therefore not at risk. In this case, credit card holders would not be liable for any fraudulent charges to their accounts.

The credit card industry is organized in a complicated way so that a consumer's transaction makes several stops before a shop owner gets paid. When a customer swipes a credit card, the information travels from the merchant's terminal to the merchant's bank along an electronic network, like Visa or MasterCard. In the process, a third-party processor, like CardSystems, serves as a router, recording the payment at the merchant's terminal before sending it along to the merchant's bank so the shop gets paid.

Customers to be likely at risk, are now being notified. The banks are now closely monitoring any accounts that might have been affected.

Although 50 million accounts were reported to have been put at risk, is not clear whether data on all of those accounts or only some, was actually stolen. Nor would MasterCard and investigators detail the number of individuals affected or dollar amounts involved in any of the fraud.

The breach is by far the largest in a relentless string of recent security failures. For example, the financial giant Citigroup announced that nearly five million consumer records, stored on magnetic computer tapes, had been lost during a shipment by United Parcel Service to a credit reporting agency. The tapes were not encrypted and they have not been found.

The growing concern over many of these breaches has been that information like Social Security numbers, names, addresses and dates of birth can be used to open new lines of credit, secure loans and otherwise engage in identity theft. But, the account numbers exposed in the most recent incident are the real lingua franca of cybercriminals, who either use them to purchase stolen goods, secure cash advances or sell the numbers in bulk at underground sites on the Internet.

> Three of the most notorious online sites engaged in credit card fraud and peddling, known as ShadowCrew, DarkProfits and Carder Planet, were taken down in an extensive F.B.I. investigation. Other sites (often based in Russia and other parts of the former Soviet Union) continue to thrive, and "dumps" of card numbers are routinely advertised, bought and sold.
>
> It is far from clear where the CardSystems data was being siphoned to. The breach appeared to be particularly savvy. What's unique about this is that it appears to be a very targeted attack, which makes it sound very clever and insidious [3].

This chapter focuses on two "flavors" of wireless service, as provided by the WAP Forum and by NTT DoCoMo of Japan [2]. WAP is the Wireless Application Protocol, an open, global specification that empowers mobile users with wireless devices to easily access and interact with information and services instantly [1]. iMode is a proprietary mobile ISP and portal service from NTT DoCoMo, Japan [2].

The WAP Forum is an industry association of over 900 members that has developed the de-facto world standard for wireless information and telephony services on digital mobile phones and other wireless terminals [1]. The primary goal of the WAP Forum is to bring together enterprises from all segments of the wireless industry value chain to ensure product interoperability and growth of the wireless market [1]. The focus of the current and recent past work by the WAP Forum has been to ensure mobile devices are sufficiently secure to allow e-commerce transactions of real value to occur. The iMode effort from NTT DoCoMo in Japan [2] focused first on market penetration with insecure, or vanilla type, services and handsets. Rather than adopt or develop a new approach to wireless transactions, DoCoMo adopted the Internet model and protocols. Security was added after the business case for wireless transactions had been conclusively demonstrated.

Both the WAP Forum and iMode are adopting security in a "staged" approach, although the WAP Forum has been more aggressive in their specification and adoption of security functionality and requirements. In the latest class of secure wireless protocols from the WAP Forum, client-side certificates are specified and used as part of client-side authentication and non-repudiation services. The current release of iMode allows for end-to-end Secure Sockets Layers (SSL) [2] with server-side authentication; client-side certificates are identified as future work by DoCoMo [2].

Fig. 15-1 shows the environment that this chapter is concerned with [1]. Users for some time have been able to access e-commerce (web sites) from traditional wired browsers (Netscape, Internet Explorer, and so on). Users are also able to access e-commerce sites from new wireless devices such as PDAs and mobile phones. This chapter examines the security services that are provided at the user's browsing device, be it a traditional wired browser or a newer wireless device. The chapter also focuses on security of the data in transmission from the user's browser to the e-commerce web site (see sidebar, "Outsourcing Leads To Data Theft In India" [4]). Security issues related to the web site (such as storage of credit cards [6]) are outside of the scope of this chapter.

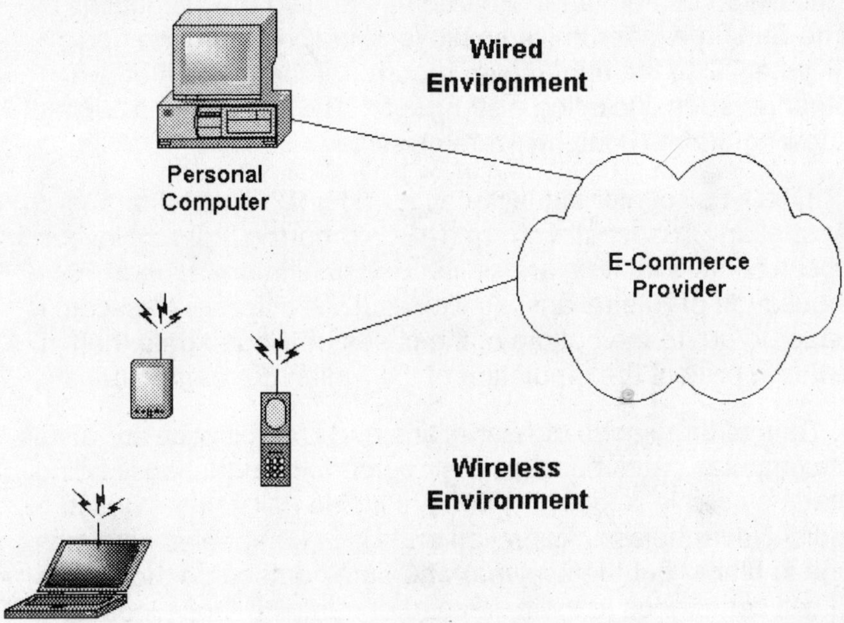

Figure 15-1. E-Commerce Security.

Outsourcing Leads To Data Theft In India

Although it has offered to help investigate recent allegations of data theft, India's National Association of Software and Service Companies (Nasscom) says that its hands are tied, as complaints have not been filed against anyone alleged to have been involved in thefts from India's call center companies. There is no formal complaint with police in India against Karan

Bahree, the person alleged to have sold information on U.K. bank accounts to a reporter of the London tabloid, The Sun.

The Sun's reporter, operating undercover, was sold secret information on 2,000 bank accounts by Bahree in Delhi, who said he obtained the data from contacts at call centers in and around Delhi. London Police confirmed that they were investigating the allegations by The Sun, but did not have jurisdiction in the matter. Bahree has denied the allegations by The Sun, in a letter to his employers, indicated that he had given a CD to the undercover reporter at the request of two other persons, including a stringer for The Sun in India, without knowing that its contents were classified.

India's call center industry came under fire again after the Australian Broadcasting Corp. (ABC) reported that employees of call centers in India are selling personal information of thousands of Australians; as well as, U.S. citizens. Nasscom is eager to get to the bottom of the cases of alleged data theft in order to protect the reputation of the Indian call center industry.

Tens of thousands of Australians and U.S. citizens are at risk of computer fraud, because their personal information is being made available illegally by workers inside call centers based in India. Nevertheless, Australian and U. S. companies have no one to blame but themselves, when they outsource thousands of call center jobs to India, instead of retaining those jobs for their own workers, who are now unemployed.

Nasscom will work with legal authorities in Australia, U.S. and India to ensure that those responsible for any criminal breaches are prosecuted and face the maximum penalty, the organization reveled in a statement. Nasscom is concerned that such reports emanate from "entrapment operations," and no person has reported any harm yet. In the absence of a formal complaint, the enforcement officials cannot launch formal investigations and apprehend the criminals. This is a real whitewash, that protects criminals and call centers, but leaves consumers of both countries, in a vulnerable position of having their personal data sold for purposes of identity theft; as well as, to terrorists, who could use the stolen information to disguise their entry into both countries, for the purpose of carrying out catastrophic attacks [4].

In order to provide security for e-commerce transactions in both the wired and wireless world, it is necessary to provide at least the following services [1]:

- User authentication: Provides the system proof that a user is who they claim to be.
- Data authentication: Consists of two sub-services: data integrity and data origin authentication. With data integrity, the receiver of data can be convinced that the data was not changed in transit. Data origin authentication proves to the receiver that data actually did come (originate) from the stated sender.
- Data confidentiality: Data confidentiality protects against disclosure of any data while in transit and is provided by encryption of data.
- Authorization: Authorization is the act of determining whether an (authenticated) entity has the right to execute an action. This is the responsibility of the system providing the e-commerce transactions/services.
- Audit: An auditing service provides a history of actions that can be used to determine what (if anything) went wrong, when it went wrong, and what caused it to go wrong. Audit services can also be used to pinpoint the last known "good" state of information [1].

In addition to these well-known services, it is increasingly common to expect that e-commerce transactions provide *non-repudiation*. Non-repudiation is proof that the user did in fact initiate a transaction. Non-repudiation is usually implemented by requiring a user to *digitally sign* a transaction. Digital signatures are unique to users and are used to provide proof that a user was involved in a given transaction. If the process of binding a user's name to the signing key (to create a digital certificate) meets certain security and legal requirements, a digital signature can be "strong enough" to provide non-repudiation of the user's actions at a later stage.

Non-repudiation is required for those transactions that are considered to be "out-of-economy" transactions, such as bill payments from a user's account to an account owned by a different entity. An "in-economy" transaction, such as a transfer from a given user's checking to their savings account, is fairly easy to unwind (if necessary). That is, recovering from a mistaken or fraudulent in-economy transaction is easily handled. Unwinding an out of-economy transaction is a much more difficult process. Rather than provide a mechanism to unwind such transactions, most enterprises would choose to rely on non-repudiation proofs as evidence that a user did in fact agree to a transaction (so that unwinding is not necessary).

This chapter examines the security of traditional Internet (wired) environments and new mobile wireless environments based on WAP and iMode in the context of their suitability for e-commerce. Later on, the chapter provides an overview of the networking environment for wired and wireless protocols (WAP and iMode). Next, the chapter gives a high-level introduction to the WAP specification and iMode. The security functions implemented by WAP and iMode are discussed next. Finally, an analysis of WAP and iMode security, and how it relates to wired security is presented. The chapter finishes with a summary and conclusions.

2. WIRELESS NETWORKING ENVIRONMENTS

This part of the chapter briefly describes the networking environments common to the wired and wireless worlds. A description of the typical protocols and security requirements seen in the traditional wired world are also presented here. This will provide a basis for comparison with the wireless world capabilities presented later in this chapter.

2.1 Traditional Wired Environment

Fig. 15-2 shows the traditional wired environment for a "home" user accessing an e-commerce provider (also known as the back-end or the Enterprise) from a browser [1]. The user connects to their local Internet Service Provider (ISP). The connection can happen over telephone lines, a cable television network, etc. The ISP provides the user access to the Internet and routes the user's requests. The networking protocol used from the user's browser to the e-commerce provider is TCP/IP [1].

2.2 WAP Wireless Networking Environment

WAP, or Wireless Application Protocol, is an industry initiated world standard [1] that allows the presentation and delivery of information and services to wireless devices such as mobile telephones or handheld computers. The major players in the WAP space are the Wireless Service Provider (WSP) and the Enterprise. The Wireless Service Provider is the wireless equivalent of an Internet Service Provider (ISP). The role of the WSP is to provide access to back-end resources for wireless users. The WSP provides additional services because wireless users must transition from the wireless to wired environments (unlike an Internet environment where the user is already "on" the Internet). The WSP's space contains a Modem Bank,

Remote Access Service (RAS) server, Router, and potentially a WAP Gateway.

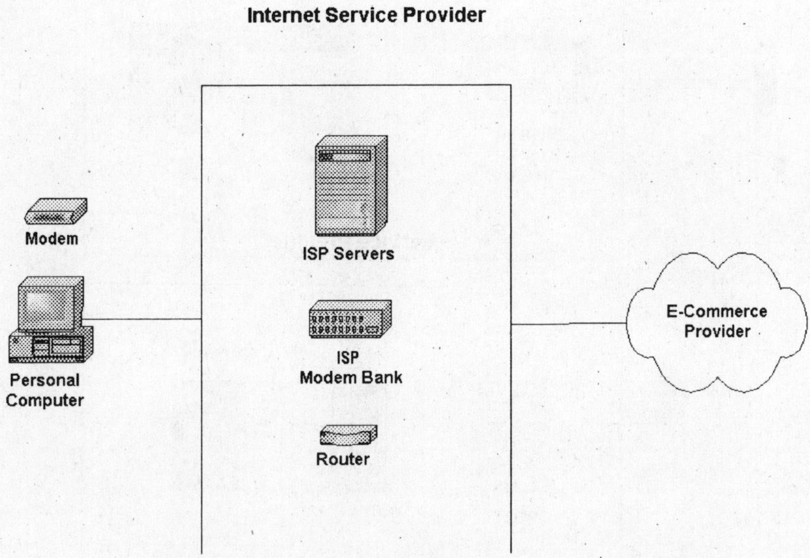

Figure 15-2. Traditional Wired Environment.

Fig. 15-3 illustrates what may be considered the "traditional" WAP wireless networking environment [1]. This environment is analog to the wired environment, where all "connection-type" services are provided by the Wireless Service Provider. Much of this functionality overlaps with functionality currently provided by the telecommunications industry. It is anticipated that the majority of this functionality will be implemented and managed by Telecommunication Companies such as Wireless Service Providers.

The WSP handles the processing associated with the incoming WAP communications, including the translation of the wireless communication from the WAP device through the transmission towers to a Modem Bank and Remote Access Server (RAS) and on to the WAP Gateway. The Modem Bank receives incoming phone calls from the user's mobile device, the RAS server translates the incoming calls from a wireless packet format to a wired packet format, and the Router routes these packets to the correct destination.

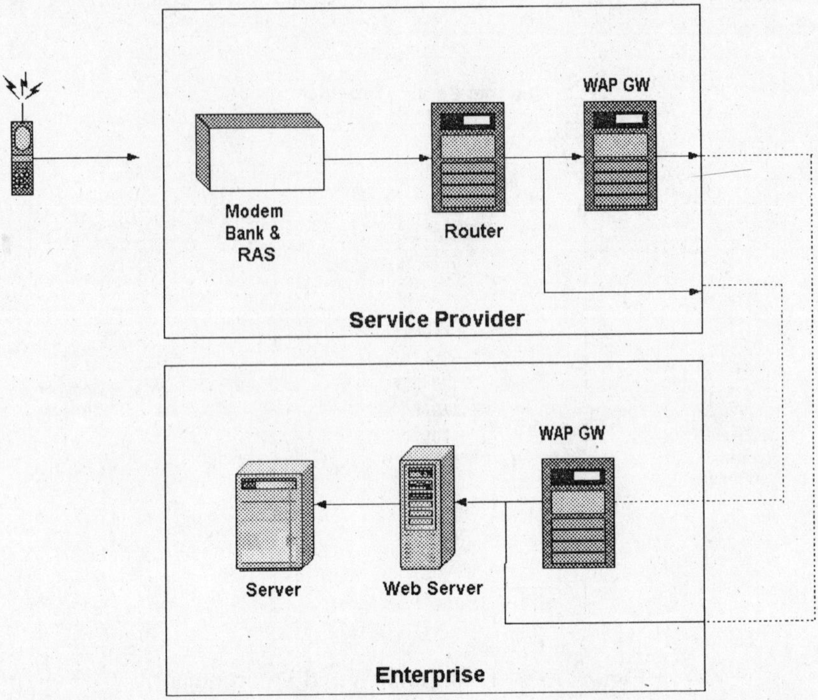

Figure 15-3. "Traditional" WAP Wireless Networking Environment.

The WAP Gateway is used to translate the WAP protocols (protocols that have been optimized for low bandwidth, low power consumption, limited screen size, and low storage) into the traditional Internet protocols (TCP/IP). The WAP Gateway is based on proxy technology. Typical WAP Gateways provide the following functionality:

- Provide DNS services, for example to resolve domain names used in URLs.
- Provide a control point for management of fraud and service utilization.
- Act as a proxy, translating the WAP protocol stack to the Internet protocol stack [1].

Many Gateways also include a "transcoding" function that will translate an HyperText Markup Language (HTML) page into a Wireless Markup Language (WML) page that is suited to the particular device type (such as a Nokia or Motorola Timeport mobile phone). The Enterprise space contains the back-end Web and application servers that provide the Enterprise's transactions.

While it seems "natural" for the Wireless Service Provider to maintain and manage the WAP Gateway, there are circumstances under which this is not desirable. This is due to the presence of an encryption "gap", caused by the ending of the Wireless Transport Layer Security (WTLS) [1] session at the Gateway. The data is temporarily in clear text on the Gateway until it is re-encrypted under the SSL session established with the Enterprise's web server. This problem is discussed in detail later in the chapter.

In such cases, the WAP Gateway should be maintained at the Enterprise, as shown in Fig.15-4 [1]. Maintaining a WAP Gateway does not require any telecommunications skills; the Gateway receives regular UDP packets. The problem with this solution remains the absence of the DNS client at the mobile device, which would require the storage of profiles for every target on the mobile device. This also requires that the Enterprise set up a relationship with the Service Provider whereby all incoming packets destined for the Enterprise (identified by IP address) are immediately routed by the WSP directly to the Enterprise and are never sent to the WSP's Gateway.

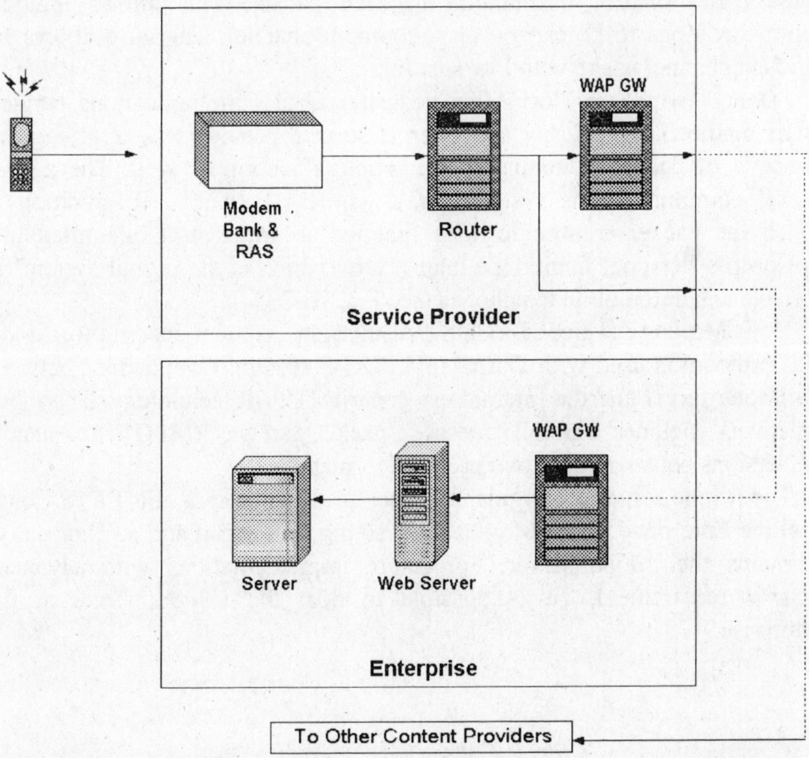

Figure 15-4. e-Commerce Specific WAP Networking Environment.

Note: This environment is referred to as an "e-commerce specific" environment as it is within the realm of e-commerce transactions that this type of architecture will be required.

2.3 iMode Networking Environment

iMode is the proprietary protocol of NTT DoCoMo of Japan. iMode provides Internet service using Personal Digital Cellular-Packet (PDC-P) and a subset of HTML 3.0 for content description [2]. In addition, iMode also allows application/content providers to distribute software (Java applets) to cellular phones and also allows users to download applets (games). Furthermore, iMode uses packet-switched technology for the wireless part of the communication and is carried over TCP/IP for the wired part of the communication.

Packet switching systems send and receive information by dividing messages into small blocks called packets and adding headers containing address and control information to each packet. This allows multiple communications to be carried on a common channel. This also allows for efficient channel usage with low cost [2].

"DoPa," which is DoCoMo's dedicated data communications service, offers connections to LAN and Internet service providers by applying this principle of packet switching to the wireless section as well. The mobile packet communications system has a wireless network configuration in which the packet communications function is added and integrated into DoCoMo's Personal Digital Cellular (PDC) which is the digital system for portable and automobile telephones (see Fig. 15-5 [1]).

DoCoMo has developed a data transmission protocol specific to iMode. This protocol is used with DoCoMo's PDC-P system. Connections between the iMode server and the Internet use generic TCP/IP technology. The PDC-P network includes a mobile message packet gateway (MPGW) to handle conversions between these two protocol formats.

The iMode server is a regular Web server. It can reside at NTT DoCoMo or at the Enterprise. DoCoMo has been acting as a portal and so "normally" maintains the iMode server. For future implementations with advanced security requirements, it is possible to host the iMode server at the Enterprise.

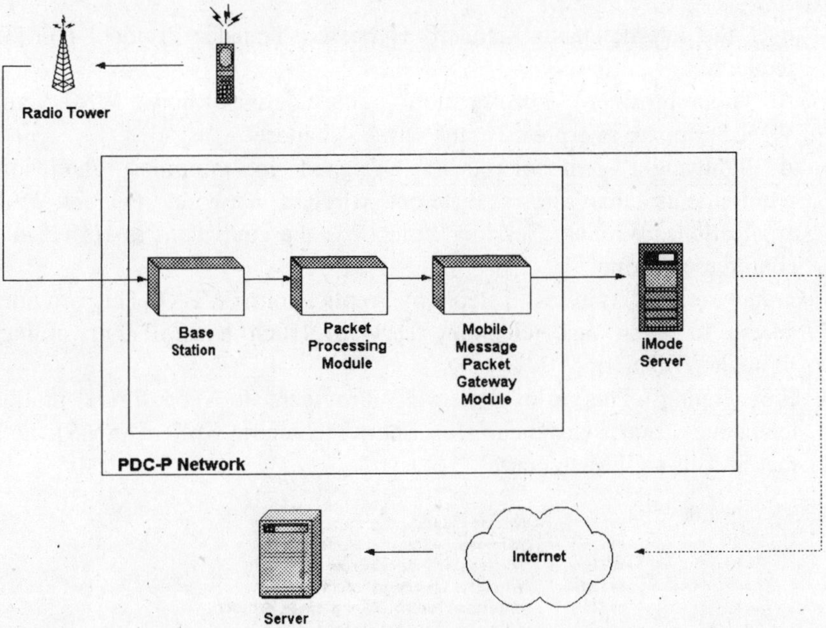

Figure 15-5. DoCoMo iMode Wireless Networking Environment.

3. THE WIRED AND WIRELESS PROTOCOLS

This part of the chapter provides a brief introduction to the two wireless protocols, WAP and iMode. There is also a discussion of the similarities and differences in these protocols. There is however, a discussion of the wired protocols, SSL/TLS [1], TCP/IP [1] and so on, for which there are plenty of excellent references available.

3.1 The WAP Specification

The WAP specification defines an open, standard architecture and a set of protocols for the implementation of wireless access to the Internet. Fig. 15-6 shows the relationship between the traditional Internet protocols and the WAP protocols [1]. The WAP specification includes among others [1]:

- An XML-type markup language, Wireless Markup Language (WML): WML and WMLScript provide a set of markup tags appropriate for wireless devices. WML content is accessed on a (traditional) web server over the Internet using standard Hypertext Transfer Protocol (HTTP) requests.

- A "microbrowser" specification: This defines how WML and WMLScript are interpreted at the wireless handset.
- A lightweight protocol stack: Designed to minimize bandwidth requirements, this allows different wireless networks to run WAP applications. Wireless Session Protocol is the equivalent to HTTP in a compressed format.
- Framework for Wireless Telephony Applications (WTA): This provides access to traditional telephony services (such as Call Forwarding) through WMLScript.
- Provisioning: This allows Service Providers to reconfigure mobile telephones from a distance using Short Messaging System (SMS) (note that SMS is a GSM standard) [1].

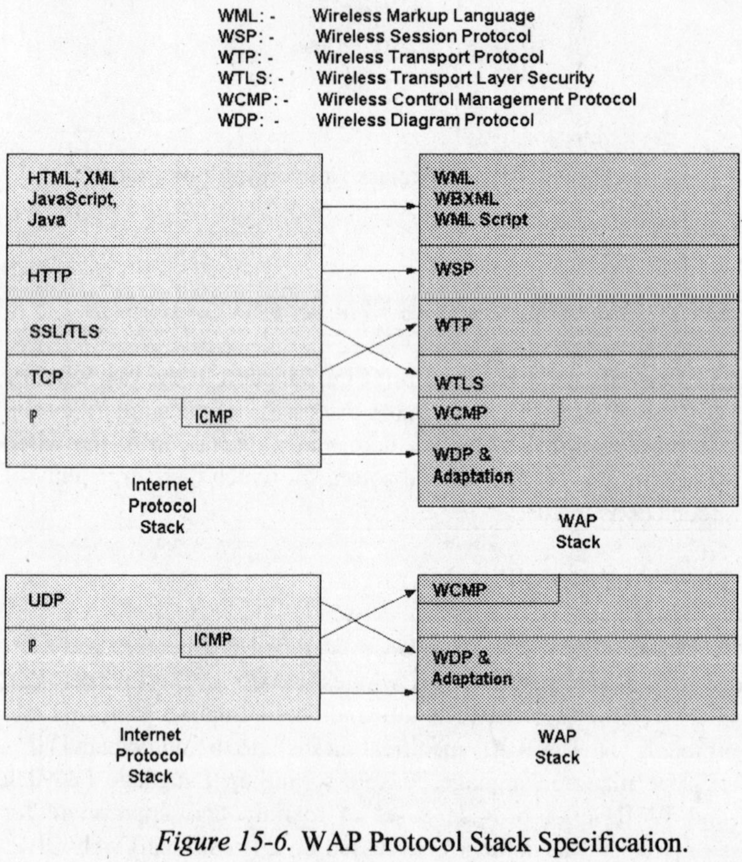

Figure 15-6. WAP Protocol Stack Specification.

3.2 The iMode Specification

iMode is a proprietary service offered in Japan [2]. The iMode specification is a proprietary protocol of NTT DoCoMo of Japan. The protocol stacks used by iMode have been reported in public forums (at RSA and at IETF [2]).

Fig. 15-7 shows the iMode protocol stacks [1]. iMode Security is provided at the transport layer using SSL/TLS and is based on the security provided by these Internet protocols.

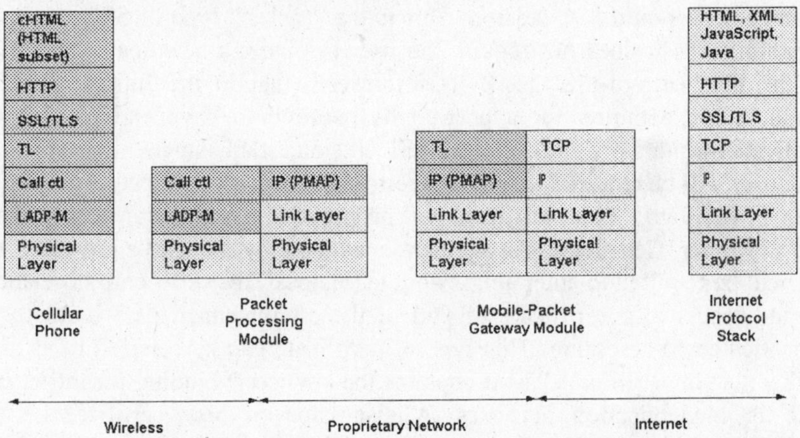

Figure 15-7. iMode Protocol Stacks.

The TL and LAPD-M protocols are standards of the Association of Radio Industries and Business (ARIB) [2]. iMode uses "compact HTML", or "cHTML" for representing on-line (on-air) content. The structure of cHTML means that a user can also view "traditional" HTML pages (although cHTML pages look better). This is in contrast to WAP, where HTML pages must be translated to WML.

4. WIRED AND WIRELESS SECURITY FEATURES

This part of the chapter provides a brief overview of the security features of each of the wired and wireless protocols. The majority of e-commerce applications run over TCP/IP protocols. Typically, resources accessed by a user are HTML pages or Java-based resources. These resources are accessed using the HTTP. If the user requires access to a sensitive resource, they will be required to establish an SSL session with the back-end. This session establishment will be prompted by the back-end and may be "server-side" or "mutually authenticated" SSL.

A "server-side" SSL session means that the server presents a digital certificate to the user's browser as proof of the server's identity (providing a binding of the public key presented with the name contained in the certificate). A "mutually authenticated" SSL session means that both the user's browser and the server present to each other a digital certificate to prove the authenticity of their identities (as bound to the public keys presented) to the other party. Most back-end systems will (at a minimum) require a server-side SSL session. This is the "easiest" type of SSL session to establish, as it does not require the user to obtain a certificate (thereby passing the "ease-of-use" test). It is foreseen, that in the future, it will become more common for a user to be required to have and present a certificate for mutually authenticated SSL session establishment.

A user will be required to obtain a certificate from a Certificate Authority (such as VeriSign). This certificate will be stored in a certificate store at the user's browser. Certificate management is handled by the user's browser. A user will be required to enter a password to "unlock" the certificate store and provide access to a certificate as part of the establishment of a mutually-authenticated SSL session. This type of certificate can be referred to as an "authentication certificate," as it contains the key-name binding for the keys used for authentication purposes. A discussion of how certificates are managed and distributed in the Internet world is beyond the scope of this chapter.

> **Note:** Certificates may also be used to provide a binding of user name to signing key. It is recommended that a user have separate signing and authentication keys.

4.1 WAP Security

There are several components to the security features available with the WAP specification. These include the WTLS protocol for securing communications, WAP Identity Module (WIM) smartcards for storing user certificates, and functions such as Crypto.signText() to allow for signing of WAP transactions. This part of the chapter briefly describes these features and the implications of their use.

4.1.1 The WTLS Protocol

WAP communications are protected using the WTLS protocol [1]. WTLS provides entity authentication, data confidentiality and data integrity.

It is based on the IETF SSL/TLS [1] protocols. WTLS is used to secure communications between the WAP device and the WAP Gateway. There are three different classes of WTLS:

- Class 1: This type implements unauthenticated Diffie-Hellman key exchange to establish the session key.
- Class 2: This enforces server side authentication using public key certificates similar to the SSL/TLS protocol. The WAP Gateway uses a WTLS certificate (a particular form of X.509 certificate compressed to save on bandwidth).
- Class 3: Clients implementing this level are able to authenticate using client side certificates. These certificates are regular X.509 format and can be stored either on the client or on a publicly accessible server (in this case a pointer to the certificate will be stored on the mobile device) [1].

Early WAP devices only implement WTLS Class 1. This level of security is insufficient [1] and should not be used for e-commerce transactions. Devices supporting WTLS Class 2 are currently available. These devices are being used in several read-only access and in-economy banking applications in Europe and the UK (see sidebar, "Breaking The Bank"[5]. WTLS Class 3 devices are now available [1].

Breaking The Bank

Recently, Finland called on its citizens to take more care securing their Wi-Fi networks after it emerged that about €200,000 ($245,400) had been stolen from a local bank using an unprotected home network. The Helsinki branch of global financing company GE Money called on police to investigate the theft. The money, which has since been recovered, was stolen from one of GE Money's accounts at a local bank, said Jukkapekka Risu, investigating officer for the Helsinki police.

Police now believe that the company's 26-year-old head of data security in Helsinki stole banking software from the company along with passwords for its bank account. Accomplices then accessed the account from a laptop computer using an unprotected network at a nearby apartment building in Helsinki's Kallio district.

They used the passwords to transfer the money to a different corporate account that they had set up six months earlier. The thieves apparently thought that using someone else's Wi-Fi network would help cover their tracks.

Suspicion initially fell on the owner of the Wi-Fi network until police searched his apartment and determined he was not involved. They then deduced from the laptop's MAC address that it belonged to GE Money, and fingers started to point toward the bank's security officer. The MAC address had been saved in the wireless LAN's ADSL box.

Police are still completing their investigation and the security officer, along with three other suspects, have not yet been charged. The case came to light after the Finnish newspaper Helsingin Sanomat obtained documents related to the case and published a story.

The case will likely be sent to prosecutors and charges will follow. Despite not having been charged yet, the security officer was immediately dismissed. GE Money is a subsidiary of General Electric.

The security officer and the other suspects were held by police for several weeks. One suspect remains in custody, and the security officer is temporarily barred from travel.

Police declined to name the suspects or identify the bank from which the money was stolen. The case was picked up by television news programs in Finland and caused something of a buzz. It also prompted the Finnish Communications Regulatory Authority to remind citizens about the dangers of not securing their wireless networks.

Wi-Fi is starting to become popular in Finland, particularly among home users. Probably less than half of the networks in use have been secured. The agency advised people to employ at least the standard WEP encryption [5].

4.1.2 The WAP Identity Module

To facilitate client side authentication, new generation WAP phones will provide a WIM [1]. The WIM will implement WTLS Class 3 functionality.

The WIM has embedded support for public key cryptography–RSA [1] is mandatory and Elliptic Curve Cryptography [1] is optional. An example of a WIM implementation is a smart card. In a wireless phone, it could be part of the Subscriber Identity Module (SIM) card (in the case of GSM [1]) or an external smart card (referred to as a WIM card). In the case of a combined SIM-WIM card this is typically called a SWIM card.

> **Note:** The manufacturer's certificates bind the manufacturer's name with the public keys configured on the WIM. Thus *all* WTLS sessions established through a WIM and a WAP Gateway will use the same public keys for the initial session negotiations. Each session will (potentially) include a different certificate for this key. Manufacturer's certificates are used by default until a user has registered additional certificates with back-end Enterprises.

A WIM will be configured (at the manufacturer) with two sets of private-public key pairs (one for signing and one for authentication) and two *manufacturer's* certificates. The WIM is also able to store some number of user certificates or user certificate references, such as a URL-based reference

> **Tip:** Certificate references, such as a URL that are used to access a back-end certificate store, are the preferred means of storing certificate information. Because the user certificates are X.509 based, storing full X.509 certificates on the WIM will very quickly use up all available storage space.

A user will be required to "enroll" or otherwise register a certificate at each Enterprise (such as Bank A and Insurance Company Z). These certificates will bind the user's public key (hardcoded on the WIM) with their local Enterprise name (hence the requirement of one certificate per Enterprise). The process of enrollment may require "proof-of-identity" information from the user; a discussion of the enrollment process is beyond the scope of this chapter.

A basic requirement for WIMs is that they are tamper-resistant. This means that certain physical hardware protection is used, which makes it not feasible to extract or modify information in the module (volatile, non-volatile memory and other parts). This is a strong requirement due to the presence of the user's private keys on the WIM.

> **Note:** These private keys *never* leave the WIM.

4.2 iMode Security

The iMode protocols are based on Internet protocols. The HTTP and SSL/TLS protocols are used end-to-end by iMode. Lower level protocols are proprietary NTT DoCoMo protocols. There are five security issues with iMode. These issues must be addressed separately:

1. Security of the radio link between iMode handset and the cellular base station (this link uses proprietary protocols and encoding controlled by NTT DoCoMo).
2. Security of the transparent public Internet connection between iMode sites and the handset in the cHTML layer.
3. Security of private networks on iMode.
4. Security of private network links between the iMode center and special service providers such as banks.
5. Password security [2].

Some of the preceding points are no longer valid, given iMode's adoption of SSL. As such, points 1, 2, and 3, in the preceding, are arguably handled by the use of SSL (although it is still not possible to make any statements about the strength of the lower level protocols and their use).

4.2.1 Security Protocols

iMode security relies on "standard" Internet security as provided by SSL. The security protocols used within the DoCoMo network and over the air, are proprietary protocols that run over SSL (using the packet switched capabilities of the DoCoMo network).

iMode has adopted SSL. Before SSL, iMode provided 'only air interface security' [2]. This required that the PDCP network be secure and trusted. With the adoption of SSL, however, iMode now provides end-to-end security in the entire wireless network (security within a carriers wireless network and security between a carrier's network [2].

4.2.2 Certificate Management

iMode does have the ability to handle server-side authenticated SSL sessions. iMode phones are pre-configured with root CA keys from PKI [7] vendors Baltimore and VeriSign [2]. This will allow for the establishment of a server-side authenticated SSL session between the iMode device and the Enterprise. This SSL session is established between the iMode device and the iMode server. For most e-commerce applications, the iMode server will be hosted by the Enterprise.

iMode does not yet have the capability of handling client-side certificates, and as such, there are no requirements for management of client-side certificates. This also implies that non-repudiation is not possible with current implementations of iMode.

Technical presentations on iMode have listed smart cards as a new, future direction. These smart cards will be similar to the WAP WIM smart cards, with all of their advantages.

4.2.3 Downloadable Applications

iMode allows users to download Java applications. iMode devices have a Java Application Manager (JAM) that controls download and management of Java applications. Java applets cannot control the JAM, nor can they launch new applets, nor access "traditional handset resources" (such as a user's telephone book). Thus, iMode has provided a version of the Java sandbox on the iMode device to contain these downloadable applications.

5. WIRED AND WIRELESS SECURITY SERVICES

This part of the chapter provides a brief discussion of security services such as user identification, authentication, and non-repudiation. There is also a discussion of the so-called "security gap," as it exists in the WAP environment.

5.1 User Identification and Authentication

User identification and authentication can take several forms in the wired world, the most common being username and password. Username and password authentication can be accomplished using "Basic Authentication" in HTTP or through a "forms-based" authentication process (a user fills in their username and password in an HTML form). Additional forms of authentication are provided through the use of token (SecurID, for example), or digital certificates.

Depending on the class of WTLS service, the type of user identification and authentication possible with WAP differs. In all classes of WTLS, it is possible to do a username/password identification and authentication using WML forms sent between the server and the mobile device.

With WTLS Class 3 service, it is possible to have client-side identification and authentication based on the public/private key pair that is hardcoded on to the WIM card and bound with the user's name in their certificate.

User authentication in iMode is the same process as in the Internet. HTTP Basic Authentication is supported by iMode. It is not clear if forms-based authentication is supported.

5.2 Non-Repudiation

Non-repudiation is one of the security services required by "advanced" e-commerce transactions. Non-repudiation requires client-side certificates that bind the user's signing key with their name. The process of generating the signing keys and of generating the user's signing certificate must be "secure" (requirements are often legislated and vary from country to country), so that the user cannot repudiate a transaction based on the keys, certificate, or signature used.

To support the requirement for non-repudiation, the WAP browser (on the WAP device) provides a WMLScript function, Crypto.signText() [1]. It is somewhat similar to JavaScript/Java in the Internet environment. This function will require a user to sign a string of text. A call to the signText() method displays the text to be signed and asks the user for confirmation. Some implementations may require the user to enter data and simultaneously sign it while other implementations will require users to enter information, send it to a server, where the server will return information with a signText call.

After the data has been signed and both the signature and the data have been sent across the network, the server can extract the digital signature and validate it, and possibly store it for accountability purposes.

Current implementations of iMode do not facilitate non-repudiation. Once DoCoMo and iMode allow for client-side certificates and clientside authentication, non-repudiation should follow.

5.3 The Security GAP

Security "gaps" appear when a secure session is terminated "prematurely". For example, consider an intended end-to-end secure application using SSL between components A and B, passing through component Z (see Fig. 15-8) [1]. A security gap would result if the SSL session (supposedly between A and B via Z) was broken at component Z, and reestablished between Z and B. This would result in two secure sessions, between A and Z and between Z and B. The message would be secure while on the wireless network, but would be in an insecure state on component Z.

The risks associated with this situation can be mitigated if components Z and B are placed within a secure system (so that there is no security gap while on the Internet, for example). Another means of mitigating the risks associated with a security gap is to provide integrity protection on information (for example using a digital signatures). This will allow the back-end system (component B) to determine if the information has been modified by Component Z while in the security gap.

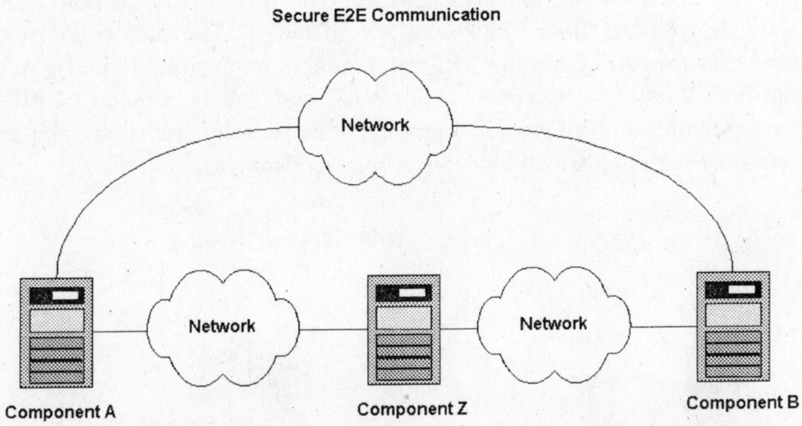

Figure 15-8. Security "Gap"

A WAP Gateway introduces an encryption "gap," since the WTLS session exists only between the WAP device and the Gateway (only between components A and Z as shown in Fig. 15-8). In a secure architecture, an SSL session will then be established between the Gateway and the back-end. One way to mitigate this "problem" is to place a WAP Gateway within the premises of the Enterprise.

Finally, the WAP Forum has adopted the Transport Layer E2E (end-to-end) Security Specification [1] to address this issue. The Transport Layer E2E specification works as follows (see Fig. 15-9) [1]:

- The WAP client tries to send a request through its default Gateway to a secure domain.
- The secure content server determines that the request must arrive through the WAP Gateway in its domain and returns a HTTP redirect message.
- The default Gateway validates the redirect and transmits it to the client.
- The client then caches the new connection and transmits transactions destined for the secure domain to the subordinate WAP Gateway.
- After the connection is terminated, the default Gateway is re-selected [1].

6. SUMMARY AND CONCLUSIONS

The perceived lack of security in the wireless environment has delayed many initiatives in providing access to e-commerce applications from wireless devices. Many enterprises are still skeptical that the same kind of security protections that they are used to in the current Internet (wired) e-commerce environment, are also available for wireless transactions. This chapter showed that these perceptions are misplaced. The chapter described the security properties and mechanisms available for Internet (wired), WAP based and iMode e-commerce. Both WAP and iMode provide excellent security features and are geared to provide other security provisions over and above those commonly available in a wired environment.

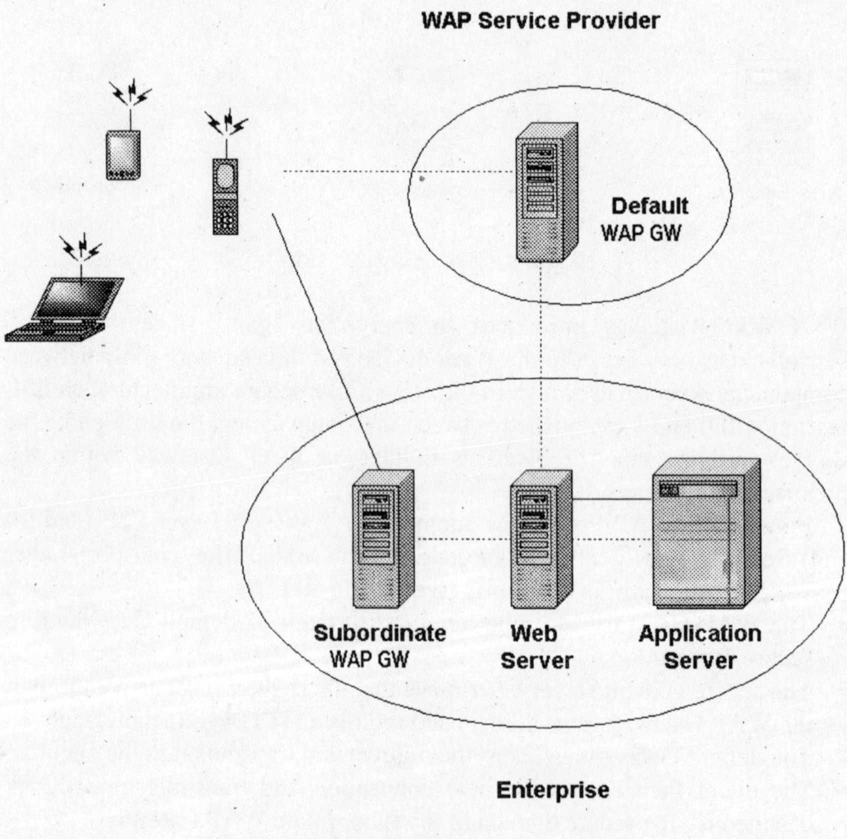

Figure 15-9. WAP End-to-End Security.

Note: This specification enforces the type of architecture described in Fig. 15-4.

The security that can be achieved for WAP or iMode enabled transactions, is at least equal to the one that can be achieved in current

Internet browser based transactions. Furthermore, WAP seems to be a step ahead with regards to specifications, using the public private key pairs hardcoded on the WIM, the ease of- use of client-side certificates and the built-in `Crypto.signText` function. All client-side public-private key pairs are generated in a known, secure manner. Through the adoption of the WIM module it is possible to achieve a more reliable form of authentication (public key and smart card). Because the process of key and certificate generation is controlled and known, WAP can claim non-repudiation at a level not generally possible in the Internet world. Whether WAP or iMode will be a great success or not, therefore should not depend on security issues. The remaining issues, such as usability and openness, should determine in how far WAP and/or iMode can be globally adopted [1].

7. REFERENCES

[1] Paul Ashley, Heather Hinton and Mark Vandenwauver, "Wired versus Wireless Security: The Internet, WAP and iMode for E-Commerce," IBM Software Group – Tivoli, Copyright (c) 2002 IBM Corporation. All rights reserved. IBM Corporation, 1133 Westchester Avenue, White Plains, New York 10604, 2002.

[2] John R. Vacca, *i-mode Crash Course*, McGraw-Hill Professional Book Group, 2001.

[3] Eric Dash And Tom Zeller Jr., "MasterCard Says 40 Million Files Are Put at Risk," The New York Times, Copyright 2005 The New York Times Company. All rights reserved. The New York Times Company, 229 West 43rd Street, New York, NY 10036, June 18, 2005.

[4] John Ribeiro, "No Complaints Filed Over Data Theft In India," Computerworld, Copyright © 2005 Computerworld Inc. All rights reserved. Computerworld, One Speen Street, Framingham, MA 01701, August 18, 2005.

[5] James Niccolai, "Finns Urge Better Wi-Fi Security After Bank Break-In," Networkworld, copyright 1995-2005 Network World, Inc., Network World, Inc., 118 Turnpike Road, Southborough, MA 01772-9108, August 18, 2005.

[6] John R. Vacca, *The Essential Guide To Storage Area Networks*, Prentice Hall, 2002.

[7] John R. Vacca, *Public Key Infrastructure: Building Trusted Applications and Web Servicess*, Auerbach Publications, 2004.

[8] John R. Vacca, *Identity Theft*, Prentice Hall, 2002.

Chapter 16

CERTIFICATION OF WIRELESS NETWORK SECURITY PERFORMANCE

1. INTRODUCTION

When wireless LANs first became available in the early 1990s, primary applications were wireless bar code solutions for needs like inventory control and retail price marking. Data transfers for these types of applications don't demand very high performance. In fact, 1Mbps data rates are generally sufficient to handle the transfer of relatively small bar codes for a limited number of users.

Today, enterprises are deploying wireless LANs for larger numbers of users with needs for enterprise applications that involve e-mail, Web browsing, and access to various server-based databases. The need for higher data rates and techniques to improve performance of wireless LANs is becoming crucial to support these types of applications. To get that extra performance, you have a lot to consider.

1.1 Choose the Right 802.11 Physical Layer

An important element that impacts the performance of a wireless LAN is the selection of the appropriate Physical (PHY) Layer (802.11a, 802.11b, or 802.11g). 802.11a offers the highest capacity at 54Mbps for each of twelve (maximum) non-overlapping channels and freedom from most potential RF interference. 802.11b provides 11Mbps data rates, with only three non-overlapping channels. 802.11g will eventually extend 802.11b networks to have 54Mbps operation, but the three non-overlapping channels limitation will still exist. Of course requirements dictate needs for performance, which will point you toward a particular PHY. If you need maximum performance, then 802.11a is the way to go, but you may need more access points because of the weaker range it has compared to 802.11b [1].

1.2 Properly Set Access Point Channels

The 802.11b standard defines 14 channels (11 in the U.S.) that overlap considerably, leaving only three channels that don't overlap with each other. For access points that are within range of each other, set them to different channels (1, 6, and 11) in order to avoid inter-access point interference. You can also take advantage of the automatic channel selection features that some access points offer. All too often, enterprises are setting their access points all to the same channel. The problem with this is that sometimes roaming will not work as users move about the facility, and the transmission of a single access point blocks all others that are within range. As a result, performance degrades significantly. With 802.11a, this is not an issue, because the 802.11a standard defines separate, non-overlapping channels [1].

1.3 Provide Adequate RF Coverage

If access points are too far apart, then some users will be associating with the wireless LAN at something less than the maximum data rate. For example, users close to an 802.11b access point may be operating at 11Mbps; whereas, a user at a greater distance may only have 2Mbps capability. In order to maximize performance, ensure that RF coverage is as spread out as possible in all user areas, especially the locations where the bulk of users reside. The completion of an effective RF site survey will aid tremendously with this exercise. The proper setting of transmit power and selection of antennas will also aid in positioning access points for optimum performance [1].

1.4 Avoid RF Interference

Cordless phones and other nearby wireless LANs can offer significant interfering signals that degrade the operation of an 802.11b wireless LAN. These external sources of RF energy in the 2.4GHz band periodically block users and access points from accessing the shared air medium. As a result, the performance of your wireless LAN will suffer when RF interference is present. So obviously, you should strive to minimize sources of RF interference and possibly set the access point channels to avoid the interfering signals. Again, an RF site survey will help you discover interference problems before designing and installing the wireless LAN. If it's not possible to reduce potential interference to an acceptable level, then consider deploying 5GHz, 802.11a networks [1].

1.5 Consider RTS/CTS

The optional request to send/clear to send (RTS / CTS) protocol of the 802.11 standard requires a particular station to refrain from sending a data frame until the station completes a RTS / CTS handshake with another station, such as an access point. RTS / CTS reduces collisions associated with hidden nodes and may improve performance. Collisions can occur when hidden nodes blindly transmit when another station (blocked by some obstruction or significant range) is already transmitting. This causes a collision and results in each station needing to retransmit their frames, doomed again by a possible collision due to the hidden node scenario. The outcome is lower throughput. If you suspect hidden nodes are causing collisions / retransmissions, then try setting the RTS / CTS threshold lower through a trial and error process while checking the impacts on throughput [1].

1.6 Fragmentation

An 802.11 station can use the optional fragmentation protocol (see sidebar, "Wireless Network Security Fragmentation") to divide 802.11 data frames into smaller pieces (fragments) that are sent separately to the destination. Each fragment consists of a MAC Layer header, frame check sequence (FCS), and a fragment number indicating its ordered position within the frame.

Wireless Network Security Fragmentation

In order to make the best use of wireless LAN bandwidth, you need to be aware of various 802.11 configuration parameters. As an optional user feature, the 802.11 standard includes the ability for radio-based network interface cards (NICs) and access points to fragment packets for improving performance in the presence of RF interference and marginal coverage areas. Let's take a closer look at how fragmentation works and cover some implementation tips.

Fragmentation in Action

A source (NIC or access point) uses fragmentation to divide 802.11 frames into smaller pieces (fragments) that are sent separately to the destination. Each fragment consists of a MAC Layer header, frame check sequence (FCS), and a fragment number indicating its ordered position within the frame. Because the source station transmits each fragment independently, the receiving station replies with a separate acknowledgement for each fragment.

An 802.11 station only applies fragmentation to frames having a unicast receiver address. This includes any frame (data frame) directed toward a specific station. In order to minimize overhead on the network, 802.11 doesn't fragment broadcast (beacons) and multicast frames.

The Sequence Control field of each fragment header includes a Fragment Number subfield, indicating the fragment number of the frame. The number is zero for the first fragment, then increments by one for each successive fragment of a particular frame. The single-bit More Fragment field in the fragment header indicates whether or not a frame is the last of a series of fragments. The More Fragment field is set to "1" if the source station will be sending additional fragments of the same frame. It's set to zero if no more fragments will follow.

The destination station reassembles the fragments back into the original frame using fragment numbers found in the header of each frame. After ensuring the frame is complete, the station hands the frame up to higher layers for processing. Even though fragmentation involves more overhead, its use can result in better performance if you tune it properly.

Fragmentation Implementation Tips

The use of fragmentation can increase the reliability of frame transmissions. Because, if you send smaller frames, collisions are much less likely to occur. The fragment size value can typically be set between 256 and 2,048 bytes. This value is user controllable. In fact, you activate fragmentation by setting a particular frame size threshold (in bytes). If the frame that the access point is transmitting is larger than the threshold, it will trigger the fragmentation function. If the packet size is equal to or less than the threshold, the access point will not use fragmentation. Of course, setting the threshold to the largest value (2,048 bytes) effectively disables fragmentation.

Similar to RTS/CTS, a good method to find out if you should activate fragmentation is to monitor the wireless LAN for collisions. If you find a relatively large number of collisions, then try using fragmentation. This can improve throughput if the fragmentation threshold is set just right.

If very few collisions (less than 5 percent) are occurring, then don't bother. The additional headers applied to each fragment would dramatically increase the overhead on the network, reducing throughput, that, you want to avoid.

If significant numbers of collisions are occurring, try setting the fragmentation threshold to around 1,000 bytes first, then tweak it until you find the best results. After invoking fragmentation, follow-up with some testing to determine if the number of collisions is less and the resulting throughput is better. You should try a different setting or discontinue using it altogether if the throughput drops (even if you have fewer collisions).

The use of 802.11 simulation tools can aid in determining optimum fragmentation threshold sizes; but, you'll need to develop a simulation model that mimics your wireless network. It will be difficult to accurately portray the RF interference on the actual network. As a result, fine-tune the actual WLAN in conjunction with real testing results [2].

Note: Keep in mind that the use of RTS/CTS could be a better way to reduce collisions if hidden nodes are present. It's best to jointly consider the use of RTS/CTS and fragmentation before settling on which one to use. As with any 802.11 tuning mechanisms, the goal is to improve performance. If what you do improves throughput, then you're doing the right thing [2].

Finally, with thresholds properly set, fragmentation can reduce the amount of data that needs retransmission. RF interference often causes only a small number of bit errors to occur. Instead of resending the entire data frame, the station implementing fragmentation only needs to retransmit the fragment containing the bit errors. The key to making fragmentation improve throughput is to set the thresholds properly. A threshold too low, will result in smaller fragments (making retransmissions efficient), but the greater number of fragments requires substantial overhead because of the additional headers and checksums. As with RTS / CTS, use a trial and error process to set the threshold while keeping an eye on consequential throughput. If there is no appreciable RF interference, then it's best to deactivate fragmentation [1].

2. SUMMARY AND CONCLUSIONS

In IEEE 802.11 networks, the Service Set Identifier (SSID) is viewed by some security professionals as an unneeded advertisement of the wireless network to attackers and these professionals assert that all measures should be taken to hide the SSID. But this advertisement is the essential role that SSIDs are designed to play. The broadcast of the SSID improves the performance of a wireless network and the SSID cannot be hidden without degrading proper WLAN operations. Finally, efforts to hide the SSID are at best half-measures which lead to a false sense of security and to a degradation of wireless network performance, particularly in a roaming situation."

3. REFERENCES

1] Jim Geier, "Maximizing Wireless LAN Performance," Wi-Fi Planet, Copyright 2005 Jupitermedia Corporation All Rights Reserved. Jupitermedia, 23 Old Kings Highway South, Darien, Connecticut 06820, December 20, 2002.

[2] Jim Geier, "Improving WLAN Performance with Fragmentation," Wi-Fi Planet, Copyright 2005 Jupitermedia Corporation All Rights Reserved. Jupitermedia, 23 Old Kings Highway South, Darien, Connecticut 06820, September 23, 2002.

PART V

MAINTAINING WIRELESS NETWORK SECURITY

Chapter 17

CONFIGURING SECURE ACCESS

1. INTRODUCTION

Greater numbers of users today are accessing enterprise IT resources remotely — either via wireline broadband such as DSL or via wireless such as 802.11 public wireless LANs (PWLAN). Remote access users are using an Internet service provider and traversing the public Internet to reach their enterprise resources. While this method provides a valuable benefit to the remote worker, it also presents a severe security problem for the enterprise, as valuable intellectual property is potentially exposed to malicious behavior. In particular, PWLANs have the added security risk of an over-the-air transmission that is relatively unsecured due to the immaturity of the technology and the lack of educated users.

IPSec is the standard for the wireless networking layer across IP networks and, as such, is the mechanism of choice for securing remote access sessions. Virtually all enterprises today manage their own remote access sessions. The user will typically have a client (often proprietary) running on his or her laptop that has been downloaded and configured by the enterprise's IT department. The user will use this client to create an IPSec session that traverses the public Internet and terminates on proprietary IPSec devices. The downside of this implementation is that it has been burdensome for enterprises to have to purchase, maintain, configure, and monitor expensive IPSec termination devices at multiple locations. This is not only an ongoing capital expense due to the initial purchases and the ongoing hardware and software upgrades, but is also a tremendous operational expense requiring dedicated IT staff to manage the platforms and the end user clients. Furthermore, the enterprise typically has to manage a RADIUS server to provide authentication, authorization, and accounting.

2. THE SECURE REMOTE ACCESS OPPORTUNITY

While virtually all enterprises desire to provide secured remote access with IPSec, not all want to take on the burden of implementing the solution internally. Outsourcing secure remote access management to the service provider reduces enterprise capital outlay and administrative overhead, allowing the enterprise to focus on their core business. At the same time, this additional service creates a new revenue stream for service providers resulting in a sizeable revenue opportunity, by taking complexity out of the enterprise and creating value in the network. Routers provide a single platform supporting IPSec- based secure remote access, which is just one of many services in a rich edge toolkit for maximizing revenue and minimizing operations costs [1].

3. WIRELESS NETWORKS SECURE REMOTE ACCESS SOLUTION

With the introduction of secure remote access on the router platform, service providers who are offering broadband aggregation and services via the router can layer on secure remote access as a new service offering to their enterprise customers. Now, a remote access user can simply use the IPSec client installed in various versions of Microsoft Windows to encrypt their traffic and terminate that IPSec session on the router. Specifically, the client PC runs IPSec and L2TP Access Concentrator (LAC) contained within Microsoft Windows, while the router runs IPSec and L2TP Network Server (LNS) to terminate the IPSec session. The provider is then able to use the router to map the user traffic into a number of Virtual Private Network (VPN) types, including Layer 3 2547bis VPNs, Virtual Router (VR) VPNs, and IPSec session aggregating multiple user flows or L2TP tunnels. Therefore, the operation is a simple, two-step process consisting of creating an IP connection and authenticating the client; and, establishing a VPN connection with the E-series router and assigning the client to the proper VR instance (see Fig. 17-1) [1].

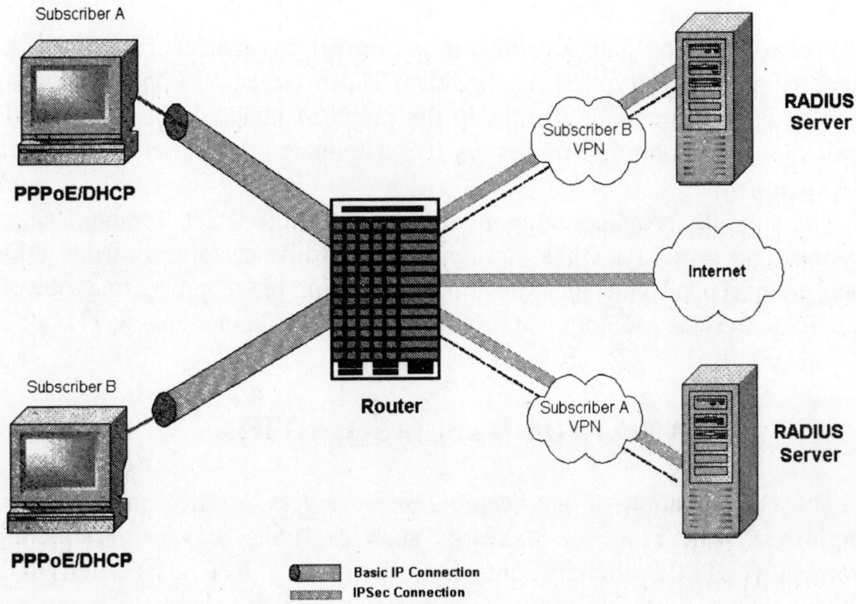

Figure 17-1. Remote VPN Access

4. THE IPSEC SERVICE MODULE

The secure remote access solution leverages the IPSec Service Module to provide hardware-based IPSec functionality at highly scalable throughput rates. The IPSec Service Module does not have a physical interface, but provides encryption for other line cards in the router chassis. IPSec solutions for the router platforms accelerate two critical aspects of IPSec: Key calculation and packet encoding and decoding [1].

4.1 Key Calculation

Key calculation is performed by dedicated hardware. This is critical at boot time when many tunnels must be established before packet forwarding can begin and because key calculations can occur at any time independent of network load [1].

4.2 Packet Encoding And Decoding

Packet encoding and decoding is performed by dedicated hardware to support line-rate encryption functionality. Hardware-based implementations perform packet encoding at rates in the range of hundreds of Mbps (half-duplex) while supporting numerous IPSec tunnels or Security Association (SA) pairs.

The module provides support for up to 5,000 IPSec remote access sessions and up to a 1 Gbps throughput rate, while performing triple DES encryption. To provide incremental increases in IPSec processing power, two IPSec Service Modules can be supported in each router chassis [1].

5. A STANDARDS-BASED SOLUTION

The configuration of the secure remote access solution in the router employs a wide range of standards such as IPSec, Layer 2 Tunneling Protocol (L2TP), voluntary tunneling defined in RFC 3193, RADIUS authentication and the Internet Key Exchange (IKE) to simplify operations. The solution supports both encapsulation and the Authentication Header (AH) and Encapsulation Security Payload (ESP) protocols as specified by the IPSec standard. It also supports the two encryption protocols specified by the standard, Data Encryption Standard (DES) and triple DES [1].

6. ALL ACCESS METHODS SUPPORTED

IPSec-based secure remote access on the router also affords service providers with dexterity in deployment due to its access-agnostic nature. Secure remote access services can be offered over all leading access methods, including:

- Point to Point Protocol over Ethernet (PPPoE)
- Layer 2 Tunneling Protocol (L2TP)
- Dynamic Host Configuration Protocol (DHCP)
- Leased or Private Lines
- Wireless LAN connections or "Hot Spots [1]"

Service providers can have a mix of access protocols and still offer this service to all users. For example, there is an opportunity for providers to support both DSL and 802.11 PWLAN services from the same router platform and then to layer secure remote access on top of both of these access types, thus maximizing revenue from the platform.

7. COMPREHENSIVE VPN PORTFOLIO

With the remote user's IPSec session terminated at the router at the edge of the service provider's wireless network, the user's traffic can be mapped to any VPN supported by the router. This includes Layer 3 (L3) 2547 VPNs, Virtual Router, another IPSec session, or L2TP. There are particular benefits with each approach as follows next [1]

7.1 L3 2547 VPNs

In the case of 2547 VPNs, there is a full mesh between all members of the VPN. Therefore, the user's traffic can be directed to a branch office without first traversing the headquarters site, reducing transaction latency and increasing wireless network reliability with no single point of failure [1].

7.2 IPSec Session

The router can map many IPSec sessions from remote access users into an aggregated IPSec session and then connect that IPSec session to the enterprise. Now, the enterprise can purchase less powerful IPSec termination equipment and it does not have to manage a large number of user sessions. The flexibility afforded by this comprehensive VPN portfolio enables providers to maximize their addressable market with different VPN types suited to different customer needs [1].

8. STRAIGHTFORWARD CONFIGURATION

Finally, configuring the router platform to support secure remote access adds only a few steps to normal router configuration. IPSec can be set up on the router to use either pre-shared keys or digital certificates for authentication and encryption. Using preshared keys makes initial configuration slightly easier, but using digital certificates provides a higher level of automation and security [1].

9. SUMMARY AND CONCLUSIONS

With the proliferation of broadband access worldwide, remote access is becoming more prevalent in increasingly mission critical applications for the competitive enterprise. Cable modems and xDSL access are making remote office/telecommuting an increasingly viable and useful scenario for individual subscribers. Similarly, the dramatic increase in the availability of public wireless LANs (802.11) is enabling mobile workers to stay connected to the enterprise network at broadband speeds. Securing these client sessions and protecting intellectual property is clearly a key enterprise requirement, and IPSec defines the architecture for this security. However, equipment costs, scaling issues, and administrative overhead have made deployment and ongoing management of remote access by the enterprise problematic. This enterprise pain point presents service providers with an opportunity — to offer a managed secure remote access service.

Service providers can now use IPSec secure remote access services as a foundation for a value-added services portfolio. Finally, these providers can now offer IPSec secure remote access to customers who desire this critical service, but do not want to provide the equipment and staffing requirements themselves.

10. REFERENCES

[1] "Secure Remote Access," Copyright © 2004, Juniper Networks, Inc. All rights reserved. Juniper Networks, Inc., 1194 North Mathilda Avenue, Sunnyvale, CA 94089, 2004.

Chapter 18

MANAGEMENT OF WIRELESS NETWORK SECURITY

1. INTRODUCTION

Because of Sarbanes-Oxley, HIPAA, Visa's CISP program and other regulations, wireless network security is increasingly becoming an executive-level issue in enterprises and other organizations. Wireless LAN security has been an especially hotly-debated subject for the past several years. IT security experts have focused primarily on three WiFi security issues:

1. Data encryption
2. Authentication and access control
3. Intrusion detection [1]

While this debate has driven the WiFi industry to rapidly improve security standards in each of these areas, security specialists have largely failed to address a critically important fourth component of WLAN security: wireless network management. Wireless network management and wireless network security are not separate issues that can be addressed in isolation. They are two sides of the same coin: if a WiFi network is not managed, it cannot be secure.

2. WIFI SECURITY: THE BASIC REQUIREMENTS

Any enterprise planning a wireless network must have a strategy for addressing each of these critical areas of wireless LAN security:

1. Data Encryption: In a secure wireless network, the most obvious requirement is that all data transmitted between the wireless client devices and the wireless access points must be encrypted to prevent intruders from viewing this data.
2. Authentication and Access Control: The second absolute requirement of a secure wireless network is that only authorized, authenticated users are able to connect to the network.
3. Intrusion Detection: Because WiFi access points are inexpensive and easily available to the general public, it is essential that the enterprises have the ability to detect when any unauthorized, unsecured wireless access points have been connected to their networks.
4. Wireless Network Management: To implement a secure wireless network, the enterprise must define security policies and ensure that these policies are implemented uniformly across the entire network infrastructure. The enterprise must continuously audit the infrastructure to ensure that these policies remain in effect at all times. If this is not done, the enterprise must assume that its wireless network is not secure [1].

3. WIRELESS DATA ENCRYPTION

Early WiFi networks relied on the Wired Equivalent Privacy (WEP) standard, which used a flawed security algorithm that used a single security "key" for all users. This proved relatively easy to crack. As a result, early enterprise adopters of WiFi technology often required wireless users to connect to the WLAN using a VPN or proprietary client for encryption.

To address the shortcomings of WEP, the WiFi industry introduced a new standard known as WiFi Protected Access (WPA). WPA uses longer 128-bit encryption keys and utilizes Temporal Key Integrity Protocol (TKIP) to provide unique encryption keys for each user and each session. WPA provides much stronger data encryption than WEP and is being adopted rapidly by many enterprises. The primary barriers to WPA adoption are that it requires all client devices and access points to support the WPA standard and for the appropriate settings to be enabled on each of those devices. While most devices sold in recent years are WPA-compliant, many legacy devices remain today.

To address this timing issue, an increasing number of enterprises are using multiple VLANs on their wireless networks. One VLAN is for users with WPA-enabled devices while other users are segmented on an "untrusted" VLAN which has limited wireless network access privileges and often forces users through a VPN or firewall [2].

To provide even more secure data encryption, the WiFi industry has ratified the WPA2 (802.11i) standard, which uses an even stronger encryption algorithm known as AES. While WPA2-compliant products are currently on the market, few enterprises have yet adopted WPA2 since it also requires older client devices and access points to be updated or replaced. In many cases, legacy WiFi hardware may have to be replaced entirely to convert to WPA2, since the computational requirements of AES encryption are relatively high. As a result, enterprise conversion to WPA2 will likely be a very gradual process. Still, most enterprises are planning to implement WPA2 at some point in the future and are starting to specify that all new APs and client devices be WPA-2 compliant to facilitate this transition [1].

4. AUTHENTICATION AND ACCESS CONTROL

Most enterprises today are migrating to 802.1x port-based authentication and access control solutions for their wireless networks. In an 802.1x authentication framework, every port is protected and each network user is verified through an authentication server, usually RADIUS. 802.1x was originally developed with wired network security in mind, but has been costly to implement on wired network because enterprises would have to replace many existing switches with 802.1x compliant switches. Ironically, since all enterprise-grade wireless APs support the 802.1x framework, 802.1x is being adopted much more rapidly for wireless network security than on the wired network. For example, both WPA and WPA2 rely on 802.1x and Extensible Authentication Protocol (EAP) for authentication.

Most enterprise IT organizations are familiar and comfortable with the 802.1x framework. The primary barrier to adoption of the 802.1x framework is the requirement that all client devices and access points have the appropriate settings software and firmware. There are many different EAP variants (PEAP, LEAP, WPA, WPA2, etc.). Enterprises using the 802.1x framework must set a policy and ensure that the EAP variant they select is uniformly supported on all their clients and APs. As enterprises transition, many are using two or more VLANs on their wireless networks, a secure one for 802.1x/EAP-enabled devices and a separate VLAN for other devices. The enterprises experiencing the greatest difficulty migrating to the 802.1x framework are:

- Retailers and manufacturers using legacy or non- Windows-based handheld devices that do not support 802.1x.

- Colleges and universities with a wide variety of client devices with non-standard configurations.
- Wireless internet service providers (WISPs) who must support a wide variety of users' client devices [1].

Where it has not been feasible to implement 802.1x port-based authentication, enterprises have adopted other means to secure their wireless networks, including wireless gateways, access control lists (ACLs), and various remote access control solutions like VPNs and firewalls. Because wireless gateways provide access control without requiring specific client configurations, they are particularly useful in environments where there are many different types of client devices or where the IT enterprise does not have full control over clients (such as in the university and WISP markets). However, proprietary gateways can be expensive and difficult to manage when deployed in large numbers, making them a costly access control solution for most enterprises.

MAC-based access control lists are often used in environments with large numbers of legacy client devices that do not have sufficient 'intelligence' for robust encryption and where cost is a major factor. In these instances, the access point is configured to permit only a predefined list of end devices (identified by MAC address) to connect to the network. While inexpensive to deploy, ACLs are susceptible to MAC spoofing and can be hacked without great difficulty. As a result, they are typically not used in high-security environments.

In other environments, VPNs and firewalls are required to provide extremely high levels of security. Since most enterprises have these solutions in place, they can leverage their existing investment in security on the wireless network. However, since most VPNs were designed to support remote users, as the wireless network expands and the user base grows, these systems can be stretched beyond their limits, forcing IT to consider costly updates as the WLAN scales.

While enterprises today may adopt different approaches to access control and authentication, there is growing momentum to adopt 802.1x-based solutions for secure, scalable wireless LAN roll-outs. No matter what policy is adopted, however, it is critical for the enterprise to define those policies centrally and ensure they are enforced uniformly across the wireless network [1].

5. INTRUSION DETECTION

Unauthorized, unsecured rogue access points connected to the enterprise network without IT's knowledge or permission are a clear and common security threat. With prices of home access points falling below $100, APs are inexpensive and easy for anyone to install in the enterprise environment.

Whether deployed by well-meaning employees or malicious intruders, rogue access points provide an open window into the enterprise wireless network. Every enterprise must have a strategy to quickly detect and remove unauthorized rogue access points. There are two primary methods through which rogue APs can be detected on the wireless network: Wired network scans designed to identify rogues from the wired side of the wireless network to which they are connected; and, wireless scans using authorized wireless devices to detect the RF signal being broadcast by the rogue access points.

Ideally, enterprises should use a combination of wired and wireless network scans to gather as much information as possible about rogue access points and to locate them on the wireless network. As always, the appropriate intrusion detection solution for any enterprise will be determined by a combination of requirements, time, and budget [1].

5.1 Wired Network Scans

Because few (if any) enterprises have wall-to-wall wireless coverage via authorized APs or sensors, wired network scans represent the baseline intrusion detection that must be used by every enterprise. Enterprise-grade wireless intrusion detection systems conduct wireline scans in two ways: Fingerprinting and interrogating routers and switches [1].

5.1.1 Fingerprinting

Fingerprinting allows the makes and models of wireless access points to be identified, so that they can be detected automatically as soon as they are connected to the wireless network. It is particularly important to detect the SOHO access points that are rarely used in enterprise deployments and are the most common rogue APs [1].

5.1.2 Interrogating Routers And Switches

Interrogating routers and switches, allows the identification of every device connected to the wireless network and assigns each device a score based on the likelihood of it being an unauthorized AP. Wireline scans provide a great deal of information about the rogue device (such as the switch port to which the device is connected, the IP address of the device, etc.) that can be used to pinpoint the location of the device on the wireless network [1].

5.2 Wireless Network Scans

Wireless network scans are a particularly effective means to detect rogue APs, because any functioning access point (including rogues) broadcasts an RF signal that can be detected wirelessly (see Fig. 18-1) [1]. Because wireless scans detect the RF signal of the rogue device, they are highly accurate and have a very low rate of false positive identifications. The greatest limitation of wireless scans is that a sensor device (an authorized AP, client device, or dedicated hardware probe) must be within a few hundred feet of the rogue in order to detect it.

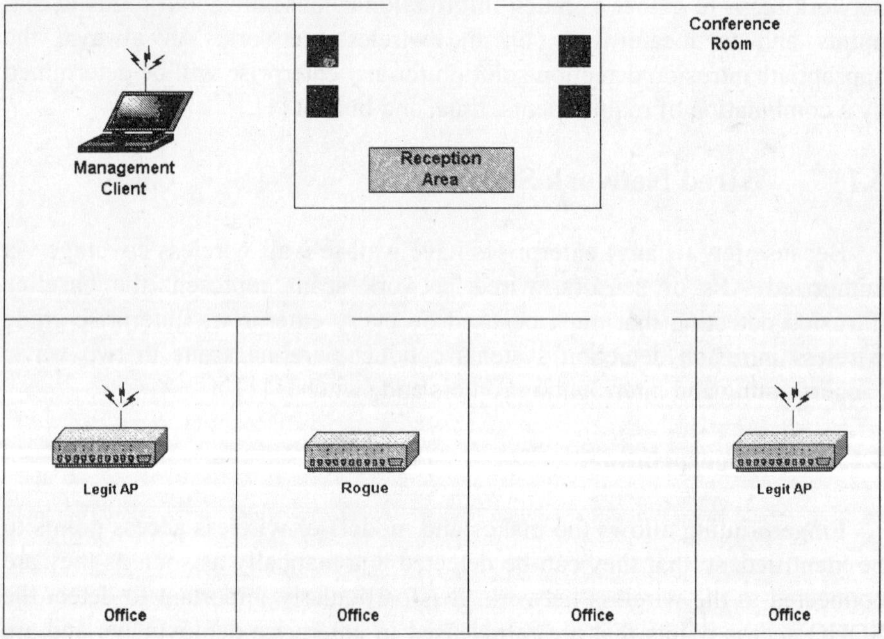

Figure 18-1. How wireless network scans work.

The most cost-effective way to conduct wireless scans is typically to use existing, authorized access points to listen for other unknown APs and ad hoc wireless networks within range. The access points can instruct most enterprise grade APs to conduct this type of wireless scan and to report the results. The list of APs discovered wirelessly are compared to a database of authorized access points to identify potential rogues. RF scanning via existing access points is very cost-effective, because it requires no new hardware. The primary limitation of this methodology is that only rogue APs within range of authorized APs can be discovered. The combination of AP-based wireless scans with wireline detection techniques, helps address this limitation.

Similarly, with appropriate software, wireless-enabled client devices can be used to report RF activity within range. Access point management software, for example, enables Windows-based client devices to report all the APs. The access point software then determines which of the detected APs are authorized devices and which are potential rogues. In this way, client devices can help fill coverage gaps where there is no wall-to-wall RF signal from existing APs.

Enterprises with stringent security requirements and large capital budgets, may consider using wireless IDS systems based on proprietary RF probes in addition to wireline scans and wireless detection via authorized APs and clients. While requiring costly hardware to be installed in all covered locations, these wireless IDS systems gather robust RF data that enables them to detect rogue APs, unauthorized client devices, man-in-the-middle-attacks, denial of service attacks, and more [1].

6. WIRELESS NETWORK MANAGEMENT

A robust, comprehensive WiFi management solution is the final element of a secure wireless security architecture. To safeguard a wireless network, a management solution must give administrators total control over the network infrastructure [1].

6.1 Wireless Network Discovery

A wireless LAN management solution must first automatically discover all wireless access points and other WiFi devices connected to the wireless LAN infrastructure to ensure that wireless network administrators have authorized each of these devices. To ensure discovery of all APs, the management solution must use a combination of Layer 2 discovery protocols (such as CDP, OSU NMS, WNMP, etc) as well as SNMP and HTTP wireless network scans. Once all APs have been discovered, the management solution must then generate accurate inventory reports for the IT staff to use to ensure that no unknown devices have been connected to the wireless network; and, they can account for all previously installed devices [1].

6.2 Automated Configuration

Implementing secure encryption and access control on a wireless network requires that the enterprise define the centralized configuration policies; and, that these policies are applied uniformly to all wireless APs and other devices. For example, if an enterprise specifies a security policy based on WPA with PSK, then every AP must have all the appropriate WPA settings enabled on each access point. If this enterprise is also using separate VLANs and/or SSIDs for different classes of wireless network users (employees versus guests), then these specific settings must also be applied correctly to each AP on the network.

Configuring these settings manually exposes numerous opportunities for human error that could jeopardize wireless security. Industry analysts estimate that up to 81% of wireless LAN attacks will be the result of misconfigured APs and devices. The only way to provide true WLAN security is to use a wireless network management solution that automatically configures wireless network hardware from all leading vendors.

Good wireless network security also requires that passwords, SSIDs and other key security settings be rotated on a frequent basis. Without a centralized wireless network security management solution that makes configuration changes quickly and efficiently, best wireless security practices are unlikely to be followed because of the labor required to manage settings on thousands of devices.

An efficient configuration solution is also essential when a wireless network is under attack. With a centralized wireless network security management solution, remote administrators can immediately shut down entire segments of the wireless network when it is under attack. They can even schedule the entire wireless network to shut down after hours, when there are no legitimate enterprise users of the wireless network--and schedule the network to turn on once again at the start of business the next day [1].

6.3 Audit Management

It is not enough to configure wireless APs and infrastructure devices correctly once, at the time of installation. Especially in large enterprises with multiple IT staff members, it is extremely common for an access point to become misconfigured during trouble-shooting or due to human error. Worst case, if a malicious intruder connects to an AP's interface, he or she may be able to alter the configuration in a way that undermines all security policies put in place.

To combat misconfigured APs, wireless network management solutions must provide a detailed audit trail and system log to track each configuration change by user. By tracing the source of all configuration changes and errors, a wireless network security management solution ensures accountability and helps IT ensure that any staff members who cause configuration errors are better trained in the future.

In addition to providing accountability and training, enterprises must conduct frequent audits of the configuration of each AP to ensure that its actual configuration always conforms to security policies. These audits simply cannot be conducted manually, since there can be hundreds of settings that need to be checked on each wireless access point. Only a centralized wireless network security management solution can efficiently compare actual AP configurations to pre-defined policies, and automatically report any discrepancies. A centralized wireless network security management solution uses a highly efficient process for polling devices and assessing their current configurations, thus enabling it to conduct audits on a continuous basis. Administrators can even auto-repair any configuration mismatches to ensure that all APs conform to wireless network policies at all times.

In this way, a wireless network security management solution can eliminate the risk that misconfigured access points will expose the network to attack. A wireless network security management solution also provides a detailed system log that tracks which user implemented configuration changes, to ensure accountability and trace any configuration errors to their source [1].

6.4 Firmware Management

An efficient solution for updating the software on access points and other wireless network devices is a requirement for WLAN security. First, as hardware vendors identify and release patches to address security vulnerabilities in their devices, it is essential for the enterprise to be able to distribute these updates efficiently to hundreds or even thousands of devices. Second, as enterprises migrate to new security standards like WPA and WPA2, many of their legacy access points and devices will need to receive firmware updates to support these standards [1].

6.5 Wireless Network Monitoring And Alerting

Finally, a real-time monitoring solution that tracks each user by username is an essential component of wireless network security. Using real-time monitoring views, administrators can determine exactly who is connected to the wireless network, where they are connected, whether they have been authenticated, and more. With this information, it is possible to quickly identify any unauthenticated users and determine where they are connected to the wireless network [1].

7. SUMMARY AND CONCLUSIONS

Without a wireless network security management solution, it is impossible to secure a WLAN. Without management, other steps to take to secure the WiFi network are incomplete and ineffective, for example:

- Protecting wireless data requires that encryption settings be applied uniformly on all wireless network devices. This is a management task.
- Implementing access control policies require that wireless devices, RADIUS servers, and other wireless network infrastructure are configured properly.
- Intrusion detection solutions require that all authorized devices and users be discovered and monitored in order to determine whether any unauthorized devices are connected to the network [1].

Finally, all of the preceding are management tasks. The information security team at any enterprise deploying a wireless LAN must insist on a comprehensive wireless network security management solution. Without it, they cannot do their jobs and cannot guarantee the security of the enterprise's secure wireless network resources.

8. REFERENCES

[1] "Managing Secure Wireless LANs: The Inextricable Link Between Wireless Network Security and Network Management," © 2005, AirWave Wireless, Inc. All rights reserved. AirWave Wireless, Inc., 1700 South El Camino Real, Suite 500, San Mateo, CA 94041, 2005.

[2] John R. Vacca, *Firewalls: Jumpstart for Network and Systems Administrators*, Digital Press, 2004.

Chapter 19

ONGOING MAINTENANCE

1. INTRODUCTION

IEEE 802.11-based wireless LANs, also called WiFi networks, are quickly expanding into mainstream areas of business from their traditional niche applications in warehouses and on retail floors. As a result, it is becoming equally as important for network engineers and technicians to have the necessary tools to troubleshoot and/or maintain and secure their wireless networks, as it is their wired networks.

Especially useful are portable, integrated wireless/wired analyzers. Having a single device for troubleshooting and/or maintaining both network segments allows technicians to quickly determine whether the sources of problems are wireless or wired issues (or non-wireless network issues altogether) so they can maximize wireless network availability for users, who are growing increasingly mobile.

2. THE WIRELESS NETWORK ENVIRONMENT

There are several modes of WiFi configurations, and visibility into all devices, RF channels, and protocol types in the various modes is critical for quick problem resolution. For example, it is important that ad-hoc peer-to-peer wireless networks, as well as, bridged, switched, and mesh infrastructure networks can all be analyzed by device category, interface, and switch port using a single device [1].

2.1 Wireless Network Architectures

Ad-hoc wireless networks consist of client devices communicating directly with one another in a peer-to-peer workgroup fashion. Ad-hoc wireless networks can pose a threat if an unauthorized client(s) should automatically associate with a legitimate client that contains sensitive data or if they piggyback onto that client's connection to gain access to wired network resources.

Wireless infrastructures are comprised of access points (APs) which are either connected directly to the wired network, or to wireless switches. They provide the RF environment for client devices, and can be configured to create point-to-point networks for bridging wireless networks between buildings, such as across a parking lot.

Yet another infrastructure type is mesh networking. A mesh network consist of APs that communicate with one another using wireless routing protocols. Mesh networks enable communications with the wired network through a minimal number of access points that are connected to the wired network. Mesh networks are often considered in order to provide flexibility in access point placement and to reduce the costs and complexity of running cable from wiring closets to each AP [1].

2.2 Multimode Channel Scanning

In the *radio access network* of wireless clients and APs, it is becoming common that the full suite of 802.11 types (802.11b and 802.11g, which operate in the 2.4GHz band, and 802.11a, which operates in the 5GHz band) will be in use in a given enterprise environment. The reason is that enterprises desire to take advantage of the maximum number of non-interfering channels, avoid RF interference, and optimize WLAN capacity.

Even if an enterprise is using just one 802.11 mode, having a wireless analyzer that can scan all the channels in the 802.11b, a, and g bands, is recommended as a best practice. Otherwise, your enterprise risks security threats from ad-hoc and rogue APs operating in the other bands as sghown in Fig. 19-1 [1].

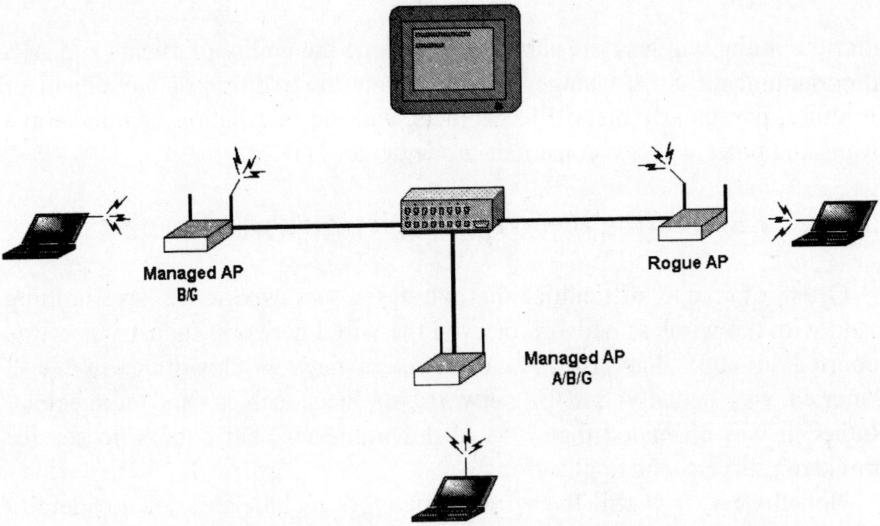

Figure 19-1. Troubleshooting and securing a mixed-mode WiFi environment.

A multimode analyzer scans the 802.11 channels in the 2.4GHz and 5GHz freqencies in a given geography to check for proper configuration, signal-to-noise ratio (SNR), bandwidth utilization levels, and other issues. If utilization on an AP is topping out, for example, it could be because there are temporarily too many wireless clients associated with it. On the other hand, perhaps a particular user or protocol is "hogging" bandwidth. Technicians equipped with wireless analyzers can discover those "top talkers," enabling the enterprise to decide whether MP3 downloads or other greedy traffic should be banned from the wireless environment.

2.3 Possible RF Problems

Unlike the wired network, the performance of the wireless LAN and users' ability to access the wireless network are prone to change as the environment surrounding APs and clients changes. Because users connecting to wireless APs are often mobile, it can be challenging to predict how many will be using a given AP at one time. In addition, intermittent coverage holes, or dead zones, may materialize when an AP becomes temporarily overloaded or when clients roam to areas where the RF signal strength is too weak to maintain association.

Dead zones in out-of-the-way areas where APs have not been installed can become a problem when new wireless applications, such as wireless voice over IP, are deployed. Also, changes to the physical environment made after the initial wireless site survey can impede the ability of clients and APs to communicate. Such changes might include the addition or movement of furniture, particularly metal file cabinets, and the installation of microwave ovens and other wireless consumer-grade devices [1].

2.4 Eliminating the Wireless Network as a Suspect

Often, of course, difficulties that wireless users experience have nothing to do with the wireless network or even the wired network. Industry analysts reported in 2005 that just 23% of wireless network downtime in North America was actually due to network products, cables, and connectors. Rather, it was estimated that 70% of downtime was attributable to service providers, servers, and applications.

Nonetheless, it is still the wireless network technician's job to identify whatever is causing perceived wireless network problems. In many enterprises, application support teams require that wireless network issues be eliminated as possible culprits before they will troubleshoot their applications [1].

3. THE TROUBLESHOOTING PROCESS

When users encounter problems with their WiFi connections, they typically call an internal help desk. When simple troubleshooting over the phone is not sufficient, the help desk dispatches a technician to the client's location.

If a wireless user is having trouble logging in, the first thing the technician will want to determine is exactly where the problem is occurring. Using a portable test and measurement device that tests both the wireless and wired network is generally the quickest means to this end.

If the technician can use the wireless analyzer in client mode to successfully authenticate and associate from the problem location, then the problem may lie in the user's client device configuration or in that client's access rights. If the analyzer cannot reach the authentication server, the problem could lie in either wireless or wired physical layer. Not enough bandwidth, falling out of range, or interference, for example, could be at the root of the problem.

The technician can use a wireless analyzer to scan the wireless environment to measure signal strength and AP capacity from the problem location. Scanning in this manner is often referred to as *passive mode,* as the analyzer is not actually associated with an access point while performing these tests. In *passive mode,* the analyzer's wireless NIC is only receiving wireless data and is not transmitting. If RF quality is satisfactory, then the technician will use the analyzer to link to the wireless network, in *client mode,* to conduct other tests such as authentication tests, ping, and throughput tests.

Often, technicians must verify that the client configuration conforms to the enterprise's security policies for packet encryption and authentication method (such as Extensible Authentication Protocol, or EAP, type). A mismatched security parameter would prevent successful authentication and authorization.

A well-designed portable wireless/wired analyzer should be able to monitor and troubleshoot every step of the authentication process to see if and where it breaks down. If the authentication server is denying the user access, for example, the issue might lie in the authentication server itself, the user's security configuration, or the user's access rights. Supervising the EAP authentication process from a wireless analyzer will eliminate a number of possibilities [1].

4. BOLSTERING WIRELESS SECURITY AND PERFORMANCE

As mentioned earlier, wireless networks are dynamic. Once deployed, the wireless network environment continues to change. This happens in part through human error and sometimes through the addition of unauthorized devices to the network by employees seeking to improve their wireless access. In some cases, because wireless connectivity is three-dimensional in nature, outsiders beyond the physical walls of an organization can also use unauthorized APs to gain access, either by happenstance or by design [1].

4.1 Finding Rogue APs and Ad-Hoc Wireless Networks

Not all enterprises can justify the expense of deploying an overlay sensor network, such as an intrusion detection system (IDS), to seek out unauthorized, or *rogue* APs. In most cases, the process of locating unauthorized rogue APs and ad-hoc wireless networks can be managed effectively by performing walk-around wireless network audits that test for vulnerabilities. This involves configuring a portable test device so that production APs are designated as "authorized" in the test system software. The test device will then be able to quickly and clearly identify unauthorized APs and ad-hoc wireless networks in real-time during periodic audits [1].

4.2 Conducting Wireless Network Audits

From a security standpoint, industry analysts predict that through 2007, 81% of successful WiFi attacks will occur due to the misconfiguration of APs and client software. Wireless analysis tools can help prevent this by contributing to the best practice of conducting regular wireless audits to make sure APs and clients are configured in accordance with enterprise policy.

It is recommended here that enterprises regularly check each AP's configuration and make sure it accurately reflects the enterprise's internal security policies. For example, if an enterprise has adopted WPA and has selected, say, Protected Extensible Authentication Protocol (PEAP), one of several available authentication methods, wireless network administrators should regularly check that all APs are indeed configured for PEAP.

Periodically, after the initial wireless site survey, wireless network technicians can use their portable analyzers to analyze the RF environment and look for changes that might cause performance degradation. They can also watch for user trends (such as finding where wireless users congregate) which may indicate areas where additional APs should be installed [1].

5. FORM FACTOR CONSIDERATIONS

There are several types of analyzers available for troubleshooting and securing your wireless network. At this juncture, the most useful type will likely be a portable device that is designed to troubleshoot both the wireless and wired enterprise network segments [1].

5.1 Portable Systems

Ruggedized, integrated wireless network analyzers have several advantages over laptop computers and handheld, personal digital assistant (PDA)-style devices, as well as centralized systems (see detail discussion later in the chapter). Laptops, for example, are limited in performance by the Windows Network Driver Interface Specification (NDIS) drivers, which specify how communications protocols, such as TCP/IP, communicate with the laptop NIC. NDIS limitations often cut performance in half. From a usability perspective, laptops are also less desirable as technicians hesitate to loan their laptops to others to conduct tests; and, they may not want to leave their laptop somewhere to conduct long-term test and analysis.

For their part, PDAs lack onboard cardbus support, which is necessary in order to enable (802.11a/b/g) WiFi channel scanning. As noted earlier, this is a critical capability required for doing a thorough job of troubleshooting the wireless enviornement [1].

5.2 Centralized Systems

Systems that support some RF management capabilities in a wiring closet or data center switch or controller are useful; however, they have visibility only into what the distributed infrastructure APs can "see" and are able to report back to the centralized system. If there is a dead zone, for example, due to a change in the physical environment, a centralized RF management system may not be able to discover it.

Similarly, a centralized system may be able to indicate the general location of a rogue AP, but to the technician dispatched to disable it, nearby APs visually look the same. Portable analyzers, on the other hand, serve as a complement to the centralized systems by providing audible and visual signal strength indicators that lead technicians directly to the rogue AP.

Finally, many enterprises today support legacy WiFi infrastructures with traditional APs. They simply have not had the budget or justification to upgrade to centralized infrastructures or install proprietary Intrusion Detection Systems (IDS). In these environments, frequent audits with a portable wireless network analyzer offers an efficient management and maintainance solution [1].

6. SUMMARY AND CONCLUSIONS

As wireless LAN technology continues to proliferate, wireless LAN users will increasingly call upon help desk resources to report wireless network issues. Fortunately, technicians no longer need to carry several tools in order to test and troubleshoot their networks. Integrated wireless/wired portable analyzers can quickly isolate problems to the wireless or wired network, client device, or application, enabling technicians to accelerate problem resolution.

Wireless analyzers discover network-connected devices and provide information regarding their associated health, signal strength, and security configurations. They also have the ability to operate as a wireless client which helps technicians to immediately determine whether the issue is specific to the given user's device. Finally, portable, integrated network analyzers have performance advantages over laptops, multimode scanning advantages over handhelds, and cost and granularity advantages over centralized systems.

7. REFERENCES

[1] "Troubleshooting Wireless LANs To Improve WiFi Uptime And Security," Copyright ® 2005 Fluke Corporation. All rights reserved. Fluke Corporation., P.O. Box 777, Everett, WA USA 98206-0777, 2005.

Chapter 20

STANDARDS DEVELOPMENT

1. INTRODUCTION

As seen through recent developments on campus, wireless network technology has experienced meteoric growth both in importance and in usage. By allowing users to access the Internet without having to be attached by cumbersome wires, wireless technology has not only cleaned up large messes of wires in homes and offices, it has freed people to take full advantage of the mobility of their laptops. On campus, you can see people connected this way not only in libraries and classrooms, but out in the open as well. In the business world, wireless technology has allowed for offices with less clutter and more flexibility. However, all of this convenience has not come without a price. As of right now, wireless technology has been fraught with various security issues. With packets of information sent over radio waves, information can be intercepted, replicated, and faked much more easily than it can be over wired networks. As of now, the popularity of wireless technology has grown faster than what it was ready for.

The development of the 802.11b wireless standard's security, was originally based mostly on the scarcity of the technology. Nowadays, with more people turning to wireless technology as the next level of connectivity, improvements must be made to bring it a level of security closer to that of its wired counterpart.

2. THE PRESENT STATE OF WIRELESS

Despite the name, wireless technology is not completely free of wires. In order for computers free themselves from wires, they must send and transmit packets through access points. These, in turn, are wired to the rest of the network. Computers with wireless Ethernet cards interact with the APs to send and receive packets of information via microwaves. These packets are governed by carrier sense multiple access / collision avoidance as opposed to the collision detection of wired networks. This makes wireless somewhat less efficient at the moment. Access points usually come equipped with two antennae, which can be adjusted to set the space in which the wireless network can work. As of right now, APs can provide service within about 300 ft of the point. For providing service to long areas, such as down hallways, leaky coax cable can be used. This is basically done by putting holes through the cable linking the access point to the network. The microwaves "leak" in and out of these holes, providing service for the areas close around the holes [1].

2.1 802.11

The current standard used for most wireless networks is the IEEE's 802.11b. This allows for a bandwidth of 11Mbps on a 2.4 GHz band. In practice, though, the throughput is usually closer to 4-6 Mbps. Dropping prices and widespread availability have made this the standard of choice, as most people have turned to it for their wireless needs. However, the demand for greater bandwidth has lead to the development of newer standards. 802.11g, is just now being incorporated into hardware, promises a bandwidth of 54 Mbps over a band of 5.0 GHz. Not only that, but it promises to be backwards compatible with 802.11b products. Thus, people and enterprises can use the new technology along with legacy hardware while upgrading in piecemeal fashion. The 802.11b will just run at its normal slower speed.

In contrast, the 802.11a standard, which also runs at 54 Mbps over a 5.0 GHz band, has been out for over a year. This standard, however, uses orthoganol frequency division multiplexing and operates at a range of only 60 feet. This would lead to a need for many more access points to cover the same amount of area. Also, 802.11a products are not interoperable with 802.11b products, meaning that installing 802.11a networks would require the scrapping of any 802.11b technology [1].

2.2 Security on Wireless

Currently security on 802.11b is run through three different mechanisms. The first is the Service Set Identification, or SSID. This is a wireless network name used by computers to gain access to a wireless network. In order to connect through that access point, the computer would need to have the appropriate SSID. This would help to keep unwanted and unregistered users from gaining access to the Internet through that point. However, most vendors either have their products set with the SSID disabled, or with it set to broadcast. When the SSID is broadcast, any computer could receive the SSID from the AP and thus gain access. Still, as the SSID is plaintext and not encrypted, it is liable to be stolen even if not broadcasted. Also, because many people need to configure their own computers to wireless networks, SSID's are often public knowledge, and very easy to find.

The second layer of security comes from the Media Access Control address filtering. This allows administrators to set the IP addresses of those who are allowed to use a certain AP. While it provides another means of protecting the wireless network, the work of adding and deleting addresses can be tedious and hard to manage. Also, with some sniffing and poking around, outsiders could obtain and spoof the addresses of registered users to get access to the wireless network.

The final piece of the wireless security structure comes from Wired Equivalent Privacy encryption. This is the most controversial aspect of wireless security. Since it uses the first 24 bits as an initiation vector, WEP offers 40 or 104 base encryption protection. Still, according to some sources, both are equally flawed and easily bypassed. In the WEP system, all computers set on the access point have a key for encrypting and reading the data. WEP uses RC4 encryption to prevent people from intercepting and reading packets sent over microwaves (see Fig. 20-1) [1]. However, WEP has come under attack as being easily cracked by readily obtained programs. The static nature of the WEP key makes it even worse, since cracking it once gives the attacker unfettered access until the key is manually changed everywhere [1].

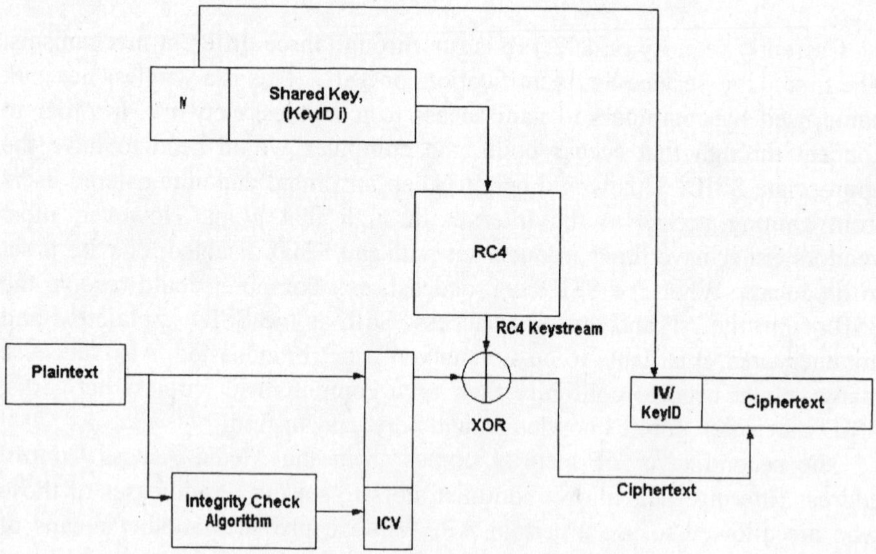

Figure 20-1. Standard WEP Encryption.

2.3 Wireless Security Risks

The open nature of wireless opens it to a variety of different attacks. War driving, or war chalking, involves moving around with wireless equipment, finding open access points. This allows people to find unsecure APs and places where they people can leech off of the shared bandwidth of others. This leads to insertion attacks, where attackers can put themselves into wireless networks that they were not intended to be a part of.

Using sniffers, attackers can also attempt to intercept and even hijack the wireless connections of others. Unlike normal sniffers, which must be attached to the wireless network somehow, wireless sniffers need only to pick up the packets that are floating freely over microwaves within the coverage area. With the weak encryption of WEP, monitoring information is relatively easy. This information could also be used to hijack other sessions, taking control of their accounts. Also, attackers can use their own transmissions to attempt to jam wireless networks, leading to denial of service attacks.

Another security issue created by hackers has been the use of clone access points to steal information. By introducing their own access point, attackers can try to get others to mistakenly connect to their AP, and thus get access to any private information they submit. In a large number of cases, attackers can get by just by using vendor-set SSID's and default configurations. Many wireless networks do not even have SSID or WEP enabled, and even more beyond that have not set new or sufficiently hard to guess passwords [1].

3. POSSIBLE SOLUTIONS AND THE FUTURE OF WIRELESS SECURITY

From the wealth of problems noted in the preceding, it would seem that 802.11b was a standard put up more to get the ball rolling on wireless technology than to establish a lasting standard for it. Currently, the IEEE has a task force working on new standards to fix exiting problems.

Prominent among these solutions is the development of 802.11i. This standard was developed by many of the people who were first to discover and point out the flaws in WEP. The primary new feature of 802.11i is the Temporal Key Integrity Protocol, which replaced much of the maligned WEP. The key for TKIP, while also using RC4 encryption, changes for every 10 Kb of data transmitted. This makes it much harder to get a working key and keep using it in the wireless network. The RC4 encryption has been replaced by the more reliable AES.

Finally, 802.11i was drafted in late January of 2003, its adoption as a standard is now complete. WiFi Protected Access (WPA) has also been developed as a security system that will be forward compatible with 802.11i. This incorporates the 802.1x authentication standard, which allows for key management. In this system, a user connects with a pass phrase. This pass phrase leads to a TKIP-generated key, which is specific to the user and not static [1].

4. SUMMARY AND CONCLUSIONS

It is very likely that the IEEE and other enterprises will continue to find ways to improve the security of wireless technology. Because this technology is becoming so popular and widespread, the technological community cannot afford for it to continue to be as insecure as it presently is. Though new problems may come out in 802.11i, the wireless industry can be confident that security will be continuously improved until it is at least comparable with that of wired networks. Not only that, but it should allow for both security and the convenience and mobility that makes wireless attractive. The world is quite possibly moving towards a wireless state. Before that happens, these serious security issues must be addressed. Finally, as this technology is too important to give up on, you can rest assured that it will be continuously improved until the security reaches an acceptable level.

5. REFERENCES

[1] "Future Forecast -- Wireless Technology And Security," Copyright 2003. The University of North Carolina at Chapel Hill, Chapel Hill, NC 27599, 2005.

Chapter 21

ENSURING SITE INTEGRITY

1. INTRODUCTION

It's becoming increasingly clear that the current model for wireless network security (defend the perimeter and patch, patch, patch) has some serious shortcomings. First of all, relying on signature files and patches doesn't provide the absolute protection that some vendors promise. Even if your perimeter systems are fully up to date, new attacks that signature files don't recognize will still get through. That was the case in January 2003 when the Slammer worm struck, spreading so quickly around the world that it slipped right past signature-based defenses and reached most vulnerable hosts within 18 minutes.

Fast worms such as Slammer and new blended attacks that combine worms and viruses will likely become more common in the years to come. Because only their authors know what forms these attacks will take, IT teams have no way of blocking them with signature files. For all the investment being made in perimeter defenses, enterprise wireless networks remain vulnerable.

Second of all, this maintenance-heavy approach to wireless network security is expensive -- too expensive. Recently, industry analysts found that the largest area of enterprise IT spending, 36%, is allocated to staffing costs. Why are IT enterprises spending so much on staffing? In part, because today's security model is so labor-intensive. IT enterprises need staffers for a growing list of low-level security tasks, such as reading the latest pile of security bulletins, tracking down patches, reprogramming firewalls [3] and so on. When you consider that all this security work still leaves wireless networks vulnerable to fast worms and blended attacks, perhaps it's time to put down the patch CDs, sit back and rethink the approach to wireless network security [1].

For enterprises today, the wireless network is where business takes place. Every department in an enterprise relies on the wireless network for applications and for a growing share of communications, not only e-mail and instant messaging, but soon telephony as well. The mission of wireless network security is to ensure that applications can do their jobs and that applications have the wireless network bandwidth and the availability needed to support the operations of the enterprise.

There's also a broader perspective on wireless network requirements. It's a holistic view that encompasses security as well as availability, bandwidth and control. It's called wireless network security site integrity. This is the real goal behind securing a wireless network. When the wireless network is functioning properly, providing applications with the bandwidth and availability they need, then the wireless network has integrity, and security is doing its job, even when the wireless network is under attack.

Instead of investing primarily at the perimeter, wireless network managers would do well to adopt this broader approach, recognizing the unique vulnerabilities and requirements of each area of the wireless network and deploying a layered security architecture designed to coordinate wireless network operations overall and achieve wireless network integrity. It is recommended here that enterprises make wireless network integrity an essential element of their application security architectures and invest in the following four layers:

1. Perimeter defenses
2. The network integrity systems layer
3. The application gateway layer
4. The host integrity layer [1]

1.1 Perimeter Defenses

Keep your perimeter security, including firewalls, intrusion-detection systems and antivirus filters and use these defenses to keep bad traffic off the wireless network. But don't fool yourself into thinking that a secure perimeter equals a secure wireless network. Make sure you still have resources for the next three layers of security [1].

1.2 The Wireless Network Integrity Systems Layer

The wireless network integrity systems layer is a critical area between your perimeter and your application defense systems. Security here relies on automated, policy-driven traffic management systems that recognize traffic anomalies and react in real time to block, redirect and throttle problematic traffic, ensuring that bandwidth is available for mission-critical applications.

By applying intelligent traffic management in this layer, enterprises can not only minimize the effects of attacks that get through the perimeter; they can also intelligently manage surges of legitimate traffic and surges from problematic applications such as instant messaging and peer-to-peer file-sharing. Vendors delivering network integrity system features include Arbor Networks Inc., Captus Networks Corp., DeepNines Inc., ForeScout Technologies Inc., Lancope Inc., Mazu Networks Inc., NetScreen Technologies Inc., Network Associates Inc., Radware Ltd., Riverhead Networks Inc., Symantec Corp., TippingPoint Technologies Inc. and TopLayer Networks. Enterprise security architects are familiar with the concepts of wireless network integrity and should evaluate the vendors against wireless network integrity site requirements [1].

1.3 The Application Gateway Layer

Security at the application gateway layer focuses on the contents of traffic reaching applications. Web application gateways, e-mail spam filters, XML security systems and Secure Sockets Layer virtual private networks help ensure that application traffic is clean, efficient and secure [1].

1.4 The Host Integrity Layer

The host integrity layer security systems protect configurations on hosts and include host-based antivirus applications, intrusion-prevention software, spyware tools and personal firewalls. As the innermost layer of security, these products provide essential "last-resort" security for applications.

If current trends continue, security attacks will become more frequent and more virulent in the coming years. Investing in signature-based security systems is of limited use. A wiser course is to develop a multilayered security architecture that recognizes the strengths and the limitations of each type of security product. When deployed effectively, this layered approach creates a wireless network that can withstand not only security attacks, but also unpredictable surges of legitimate traffic. By investing in wireless network integrity, you can control the rising labor expenses for IT, while improving the wireless network bandwidth and availability your applications (and your enterprise operations) require [1].

2. CONFIDENTIALITY AND INTEGRITY DURING COMMUNICATIONS

Deploying wireless networks enable freedom of mobility, more communication channels, and limitless access to information. Combined with the ease of implementation and the relatively small amount of infrastructure required , portable wireless networks can greatly benefit emergency-response personnel, consultants, or any other type of field personnel.

It is inherently difficult to have physical control over a wireless network's entire area of coverage, therefore, it must be assumed that the wireless network is being placed in a hostile environment—one where unauthorized outsiders can access information.

Portable wireless networks are inexpensive, ranging between $200 and $1,000. These costs can be further minimized because outdated hardware can often be used. While implementation time can range from a few hours to a few days, preconfiguration of these wireless networks dramatically reduces the deployment time [2].

2.1 Matching the Technical Solution to the Enterprise Objectives

Confidentiality prevents eavesdropping. A malicious attacker could eavesdrop on wireless communications and compromise sensitive data, such as passwords. Confidentiality can be achieved by establishing an encrypted tunnel between the roaming stations and a portal on the wired infrastructure. There are many mechanisms that can be used to achieve this goal depending on the application. IPSecurity (IPSec) is a standards based encryption infrastructure that can provide confidentiality and integrity. IPSec has been proven effective for a number of years and can provide the necessary encrypted tunnels. Other options to provide confidentiality are encrypted port forwarding using a protocol such as SSH or using an encrypted protocol such as Secure Sockets Layer (SSL) or Transport Layer Security (TLS) [2].

2.1.1 Authentication of Users

Authentication is important to verify that legitimate users only use the wireless infrastructure. 802.1x is a solution that provides authentication using Extensible Authentication Protocol (EAP) to authenticate wireless stations to a wired network. 802.1x authentication is implemented using encrypted RADIUS-based authentication to a backend server for station authentication. In some implementations, the WEP key is also generated on a per session basis. In addition to providing a better authentication mechanism than the current 802.11 standard (see sidebar, "802.11 Wireless Security Problems"), 802.1x simplifies WEP key management among the stations [2].

802.11 Wireless Security Problems

Recently, the 802.11 standard has been proven to be susceptible to attack methods and therefore cannot provide adequate wireless network security. Let's look at why:

Confidentiality or Integrity

802.11 provides a recently proven flawed encryption mechanism, based on the RC4 algorithm called Wire Equivalent Protection (WEP). It is based on a shared secret key that resides in the access point and on the mobile stations. All participants and all access points share this shared secret across the wireless network. The integrity mechanism is nothing more than a basic CRC-32 checksum, which is more for error checking than a cryptographic mechanism for providing integrity to information in the communications channel.

Authentication

The 802.11 standard provides a flawed shared-key authentication method that presents a cleartext challenge and a ciphertext challenge to the attacker, providing a trivial means to mount a cryptographic attack on the RC4 stream, which is needed for encryption.

Mobility Management

While the current standard allows a distribution system to enable roaming between access points, it does not define how the distribution system should function or what to do when a client roams off a distribution system or to a different subnet, so when implementing these features you may be required to use a single vendor's product [2].

2.1.2 Mobility Management for Stations

Wireless users are often mobile and need the ability to roam around the network. This involves connection to multiple access points. Mobile IP (RFC-2002) can provide such mobility management. Mobile IP has two agents that allow roaming machines to properly route back to a home wireless network. The first agent is the Home Agent, which is a static server that registers a static address to each mobile station. The static address is used to keep VPN connections alive while roaming. When a mobile station roams from the home wireless network to a foreign network, the mobile station looks for a foreign agent through which the VPN connection can be re-established [2].

2.2 Network Architecture

When the wireless network coverage area is small, such as a conference room, an exposition area at a conference, or a quickly constructed triage center, a single access point (AP) can usually cover the affected area, and secure access is ensured when combined with a traditional VPN solution. This solution manages risk in a prudent manner assuming the stations are fixed, and not moving beyond a defined area. This is known as Scenario 1: which is defined as the area where limited mobility is needed; and, the population is less than 25.

The solution for a second scenario (where there is a large roaming area; and, the population is greater than 25) is similar to the first, but requires the addition of a distribution system. A distribution system allows users to roam between access points without losing connectivity. Wireless network designers should use a single subnet for the distribution system; this preempts the need for users to change IP addresses while roaming or using Mobile IP. The 802.11 does not specify the operation of the distribution system, so in most cases, all access points will need to be the same brand and capable of operating in a single distribution system [2].

2.3 Description of Architecture

In either application, the tactical wireless network traffic will need to be sent off to another wireless network through a gateway of sorts (see Fig. 21-

1) [2]. This gateway may be a wired 10 or 100M/bit Ethernet interface, a telephone line, a satellite terminal, or another terrestrial wireless link. While gateways may simplify wireless integration, they are not critical in every case. For instance, if email is the only application used, then a hardened email server can provide security over SSH or SSL [2].

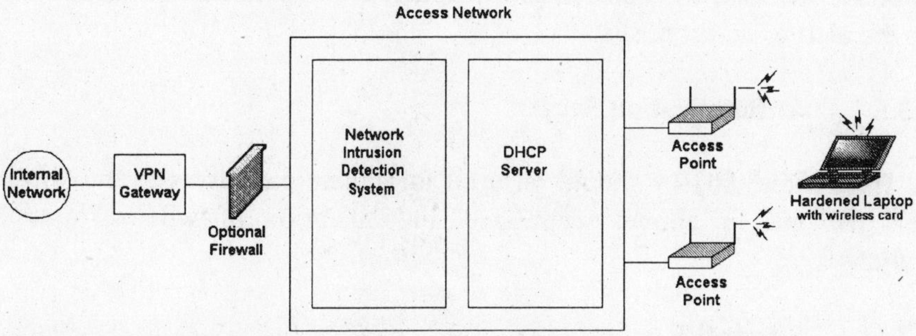

Figure 21-1. The access wireless network.

2.3.1 Access Points

802.11 access points provide connectivity similar to a traditional wired Ethernet. APs continuously send messages to potential wireless users to inform them about the radio network; these messages are called beacons. When configuring access points, the frequency of beacons should be minimized, beacons should be broadcast at the lowest bandwidth that is to be used by wireless users, and broadcasting too much information about the configuration of the AP, such as SSID, should be avoided. Out-of-band management should be used for all APs [2].

2.3.2 Switch

An Ethernet switch can be used to consolidate all access points. The switch should be physically separate from the wireless network—on the other side of the gateway. This will help prevent layer-2 attacks that could be used to circumvent VLAN and other layer-2 access controls.

2.3.3 VPN Gateway

An IPSec or SSH-based VPN Gateway will provide confidentiality and integrity for users over the wireless link. Additional IPSec VPNs may also be used to allow users to access enterprise networks remotely.

2.3.4 Intrusion Detection System

A Wireless Network Intrusion Detection System (WNIDS) should be deployed on all wireless segments. WNIDS signatures should be very easy to create, because the authorized traffic should be known, so WNIDS turns up the alert to detect misuse.

2.3.5 Authentication Server

If possible, 802.1x should be used to authenticate users to the wireless network. 802.1x allows centralized authentication of wireless users or stations.

2.3.6 Firewall/Router

Depending on configuration, a firewall may be optional. Often VPN gateways serve the same function, and can be hardened. Additionally, the IDS system can be used to analyze traffic.

2.3.7 User Workstations

User workstations and support servers should be hardened to prevent host hopping and exploitation of available services. Personal firewalls should be used as an additional countermeasure.

2.3.8 Vulnerabilities

While the recommended architecture in the preceding addresses several of the security issues involved when deploying wireless networks, there are vulnerabilities that the architecture does not take into account. The design does not take into account RF denial of service attacks. It is possible to use RF flooding to disrupt a wireless network, rendering it useless, but without compromising the information. Little can be done to design against RF interference, because the frequencies used for 802.11 are unlicensed. In addition, standard issues with key management are still relevant. Keys must be deployed, maintained, and revoked in a secure fashion to prevent unauthorized users from gaining access to the wireless network. Finally, man-in-the-middle attacks are still possible, for instance, an attacker could deploy an unauthorized AP either in closer proximity to, or with more power than the authorized AP, in order to get users to associate to the unauthorized

AP. Properly employed IPSec will lessen the viability of many of these attacks, but the threat will still exist.

3. SUMMARY AND CONCLUSIONS

Secure wireless networks can be deployed quickly and cost effectively once the architecture and hardware are procured. However, complexity increases as the amount of serviceable users and coverage area grows. When users must "roam" between different access points, new criteria enter into the equation, due to the complexities involved in configuring a wireless network that provides users the mobility they desire. Finally, when these factors are taken into consideration, a wireless network can be designed with appropriate countermeasures, to provide both mobility and security for wireless networks in hostile environments.

4. REFERENCES

[1] Eric Ogren, "Using A Layered Security Approach To Achieve Network Integrity," Computerworld, Copyright © 2005 Computerworld Inc. All rights reserved. Computerworld Inc., One Speen Street, Framingham, MA 01701, February 12, 2004.

[2] David Pollino and Matt Miller, "The Deployment of a Wireless Network in a Hostile Environment," Copyright © 2004 @stake, Inc. All rights reserved. stake, Inc., 196 Broadway, Cambridge, MA 02139-1902, 2004.

[3] John R. Vacca, *Firewalls : Jumpstart for Network and Systems Administrators*, Digital Press, 2004.

PART VI

INFORMATION WARFARE COUNTERMEASURES: THE WIRELESS NETWORK SECURITY SOLUTION

Chapter 22

DEFENSIVE WIRELESS NETWORK SECURITY STRATEGIES FOR GOVERNMENTS AND INDUSTRY GROUPS

1. INTRODUCTION

Information warfare, or sneak electronic assaults, could easily crash power grids, financial wireless networks, transportation systems, and telecommunications, among other vital services. The National Security Agency (NSA) traces the macro threat to hostile or potentially hostile governments as well as drug lords, criminal cartels, and increasingly computer-savvy guerrilla groups. Some of these rogue organizations are doing reconnaissance today on U.S. wireless networks, mapping them, and looking for vulnerabilities.

Cyberblitzes like those that briefly knocked out major Web sites a few yeas ago (including Yahoo! Inc.'s Internet gateway, eBay Inc.'s auction service, and Amazon.com Inc.'s retail site), could easily be copied on a larger scale. Criminals, crackers, foreign governments—when President Bush reads that intelligence briefing, he had better move faster than he did during those infamous 7 minutes of pondering what to do when told of the 9-11 attacks!

Such warnings are not new for the agents at NSA, who have frequently conjured up a "digital Pearl Harbor,"—a reference to the Japanese surprise attack that threw the United States into the Second World War. But the NSA and other U.S. officials seem to be stepping up a public awareness campaign, spurred by the spread of information technology, growing knowledge of malicious computer code, and ever greater U.S. reliance on wireless networked systems.

2. IS THE UNITED STATES GOVERNMENT PREPARED FOR INFORMATION WARFARE?

The answer to the preceding question is a resounding "no." A reasonable question that should be asked is "Why are we vulnerable?" In a recent report by the Defense Science Board Task Force on Information Warfare, the task force unequivocally lays the blame at the U.S. government's own doorstep.

The reality is that the vulnerability of the Department of Defense (and of the nation) to offensive information warfare attack is largely a self-created problem. Program by program, economic sector by economic sector, the U.S. government has based critical functions on inadequately protected telecomputing services. In aggregate, the U.S. government created a target-rich environment and the U.S. industry has sold globally much of the generic technology that can be used to strike these targets. From the standpoint of psychological operations, it's not so much exploited technology as it is that the U.S. government has created a global information system it does not control and does not understand, and this in turn creates additional targets for exploitation. Most recently, this problem is being exacerbated by the growing emergence of "always-on" connections being made to individual homes and small businesses.

Recently, for example, a private security company alerted the FBI that it found a malicious program on some 3,000 computers that could be remotely activated to launch an attack on a site of choice—a Trojan. Many of these computers are privately owned and are on cable-modem or digital line subscriber (DSL) always-on connections. In addition to the technological risk posed by the fact that many of these computers have very limited or no security, the users of these computers often are attractive targets for social engineering efforts for a simple reason. The very thought that they would be targeted for an attack is foreign to the owners.

From an information warfare perspective, there are three primary target audiences for the attacker using psychological operations—psyops. The attacker can focus on the enemy, those who are friendly to his or her cause, or the neutrals; with each target chosen for a specific purpose. If the attacker is simply a hacker/cracker/script-kiddie, it might be for nothing more than to grab a credit card number or prove to friends that he or she could do it. Unfortunately, the dangers the U.S. government faces are not limited to those groups. The government also faces the threat of multinational efforts to subvert their defenses and find an economic, diplomatic, or military advantage. These efforts might be aimed not only at the U.S. defense structure, but also at the U.S. utility infrastructure, transportation, communications, finance, and much more. The U.S. government also cannot discount the potential entrance of organized crime into the equation, not to mention political activists and disgruntled employees.

So, as more individuals (the neutrals) turn to the Internet to help them with tasks that have usually been served by personal service or other traditional means, tasks such as banking, tax filing, shopping, and personal communications, the Internet as a loci for commerce and communication becomes increasingly critical both to the individual and to the business and industries that serve the individual. And although the commercial sector is beginning to realize the importance of security, the information on the virtually unprotected personal machines may very well hold the key to a crippling attack on any sector simply because those sectors exist to allow the personal machines to connect to do business.

From a psyops point of view, however, how is it done? In any attack, finding and exploiting a trust relationship can be a key to success for the attacker. Let's look at how a trust relationship can be exploited. One of the most often cited examples of a physical trust relationship that was exploited successfully is the Mitnick attack. Here Kevin Mitnick discovered a relationship between host A and host B. He was able to determine the probable response that host A would give to host B after receiving the initiating packet in a three-way handshake. He blocked host B with a Denial of Service attack and sent a packet to A crafted to look as if it came from B. He then sent the expected response along with a small payload that contaminated host A's .rhost file and caused host A at that point to completely trust Mitnick's computer. He dropped the attack on host B and simply accessed A as a trusted root user.

So, how might an attacker employ psyops against a trust relationship? One of the more common examples used to explain trust exploitation is that of the overworked call center. Imagine a worker at a large corporate call center. The caller has done some research and discovered that a new personal report has been hired by the CEO. He calls and identifies himself as Bert Jackson who has just been hired by the boss. He tells him he's been working all day researching a project that the CEO wants a report on in the morning and he needs access to the system to put the report together. Unfortunately, he's forgotten his password and it's already 11 p.m. Can he get a new password or should he call the CEO and have him call? In a shop with strong security that would be an easy call, but it's easy to see that, in many cases, the call center worker would simply trust that the caller is who he or she says he or she is and give out a new password. The net result? The attacker gets in and can probably hide his tracks before the real Bert Jackson complains.

If the company is also a prime contractor for the government, a public utility, or even a company whose success or failure can severely impact the stock market, then the attacker has gained a tremendous advantage by simply manipulating information he or she has gained by infiltrating the system. But, let's go back to Bert Jackson.

Assume, for this scenario, that a group wanted to create a deleterious impact on the stock market. That group, perhaps over a period of months, maps IP ranges that are known to belong to public Internet service providers (ISPs) providing high-speed, always-on access to individuals and small-businesses and they map for the Netbios ports. As they map, a second team begins the infiltration process, finding those machines that are unprotected and that contain information, such as passwords to personal investment accounts, banking, and the like. Even though these passwords may be encrypted, with modern cracking tools being what they are, at the end of the mapping period, they very well could have discovered thousands of accounts, including Bert Jackson's, that could be exploited. Choosing the time to strike, they simultaneously use these accounts to issue massive sell orders to the various brokers and close thousands of bank accounts with the money transferred to offshore accounts that they may or may not care about accessing. The distributed nature of this attack would make detection and prevention difficult, if not impossible, and would certainly sow an atmosphere of fear and distrust that would severely affect the general economy.

Again, the question is why? Let us look at the three basic types of attack— strategic, consolidation, and battlefield. If the preceding scenario were executed by organized crime, it would probably fall into the battlefield type because they probably would be looking to cause a drop in stock market prices where they could step in and buy cheaply, thus allowing them to see an impressive gain as confidence rebounded. If a foreign government perpetrated the attack, it might very well fall into one of the other two categories. The attackers might be trying to distract the attention of the current administration away from what they might be attempting elsewhere (strategic) or attempting to bring together the economic resources needed to launch a more serious battlefield attack against us later (consolidation).

But, what is it that causes you, as a whole, to make it easy for those who would want to abuse that trust? In a culture where the phrase "trust is earned" is a familiar maxim, it would seem that you would be more eager to challenge than you really are. However, trust also seems to be a social construct between two or more individuals. In both social and business milieu, as alluded to earlier, a need to trust develops out of the need to foster cooperation to achieve goals and objectives.

If that is, in fact, the case, then how does the U.S. government overcome this tendency and manage to protect their critical resources? Part of the difficulty they face here is that their focus tends to be on strengthening the security of their physical defenses, whether that be through encryption, perimeter-based defenses, host-based defenses, or, preferably, a combination of the three. Unfortunately, the U.S. government still has too few in system administrative positions who are security-aware enough to alter default installations on whatever machine they are setting up (whether it be Microsoft based or Unix based) to give an acceptable initial level of protection to their users. But these are technological trust defenses and likely will always be open to attack. And although hardening those physical defenses is undeniably important, the U.S. government often overlooks the most dangerous vulnerability (their users), and that is where they spend the least amount of time in education. Why do computer viruses such as the "I Love You" virus of few years ago work? Because users, whether corporate, governmental, or private, haven't been taught how to protect themselves and change the paradigm of automatically trusting the e-mail that announces it comes from Aunt Barbara.

The U.S. government must begin focusing on the end user and on those who provide connections to the end users. When virtually all private connections to the Internet were made over modems connecting to a Dynamic Host Configuration Protocol (DHCP) server where each session was served with a different IP address, it was much less likely for a private machine to be compromised and efforts to compromise machines tended to be focused on commercial, governmental, and educational systems. Today, however, that situation is rapidly changing and ISPs must accept the responsibility of advising or requiring their customers to install personal firewalls [6] and give them the advice needed to properly configure and maintain those firewalls. They also must understand the need to properly filter their outgoing traffic to block and detect activity coming from within their wireless networks that can be harmful to the general Internet community.

Educating the end user is going to be the most daunting task. The recent proliferation of e-mail-related viruses has certainly helped to awaken many to the dangers, but there must be a broader effort to educate and assist users in protecting themselves and the U.S. government from the bad guys. To do this, the security community needs to do a better job in educating first the media and then the public through the media. Psyops can work both ways. The difference between the U.S. government and the bad guys is that the government has permission—they have the intent to do what is right. So it is with perception management. The U.S. government can manage perception so that people will realize the risks they actually face and take steps to protect themselves. In helping them to protect themselves, the U.S. government also helps to protect the rest of the users on the Internet who could be attacked by their systems if they are compromised. Trust is wonderful when exercised in an environment where it is reasonable. In a global environment where criminals, unfriendly political forces, and people who just don't care about others have the same rights and access as anyone, trust can be dangerous.

Education, not legislation, is the key component. The U.S. government can pass all the laws it wishes, but it won't affect the traffic that is coming out of countries such as Korea, China, and Singapore. The government needs to be communicating these messages with intelligence. If the U.S. government knows what needs to be done and doesn't communicate it effectively, then whatever else it does is irrelevant. If the government scattershots their communications without filtering them through an understanding of the message they need to pass, then all they are sending out is noise [7].

3. ARE OTHER GOVERNMENTS PREPARED FOR INFORMATION WARFARE?

Are other governments ready to use information-age tricks to use against their adversaries? Yes, to some extent. Case in point is as follows:

At first, the urgent phone call from the U.S. Transportation Department confounded Cheng Wang, a Long Island-based webmaster for Falun Gong, the spiritual movement that has unnerved Chinese authorities. Why did the department think his computers were attacking theirs? The answer turned out to be startling. The electronic blitz hadn't come, as it seemed, from various Falun Gong Internet sites. Rather, someone had lifted their electronic identities. Computer sleuths followed a trail back to the XinAn Information Service Center in Beijing–where an operator identified it as part of the Ministry of Public Security, China's secret police [7].

Web hacking, it seems, isn't just for amateurs anymore. While the recent rash of cybervandalism against some of E-commerce's [1] biggest names has garnered headlines, that's only part of the story. From Beijing to Baku, governments and their surrogates are using the Internet to harrass political opponents and unfriendly neighbors, to go after trade secrets, and to prepare for outright warfare. Burma's military junta, for instance, is blamed for targeting the "Happy 99" e-mail virus at opponents who use the Net to advance their cause. Dissidents describe the attacks as inept–proof, perhaps, that dictatorships are still behind the hacking curve [7].

3.1 Hack Attack

But Burma is not alone in trying. In January 2000, hackers from Azerbaijan with names like "The Green Revenge" and "Hijack" tampered with dozens of Armenian-related Web sites, including host computers in the United States. Experts suspect involvement or support from the Azerbaijani government, which imposes tight controls over Internet use within its borders. Relations are tense between Azerbaijan and Armenia, who fought a war over the disputed territory of Nagorno-Karabakh, so it wasn't long before the Armenians retaliated in kind. It is the first precedent of a physical battle going on-line.

In Cheng Wang's case, his computers in Hauppauge, N.Y., were among Falun Gong sites around the world hit by a barrage of hacking attempts and e-mail "bombs" that coincided with a physical crackdown on the group's practitioners in China. Several of the hacking incidents were traced to the mysterious XinAn office.

It is often difficult to track down who is to blame. But for wireless networked Americans, who own 60% of the world's computing capacity, such electronic conflict should be unsettling. True, the scariest scenarios dreamed up by experts, such as a hostile government disrupting financial markets, haven't come to pass—yet. But more than a dozen countries,among them Russia, China, Iraq, Iran, and Cuba, are developing significant information warfare capabilities. A senior CIA official cited a Russian general who compared the disruptive effects of a cyberattack on a transportation or electrical grid to those of a nuclear weapon. China is considering whether to create a fourth branch of its armed services devoted to information warfare. The Pentagon isn't sitting still either. The U.S. military's offensive cyberwarfare programs are presently being consolidated at the U.S. Space Command in Colorado.

Nearly as worrisome as a cyberattack to experts is electronic espionage. From March 1998 until January 2001, intruders broke into computer systems belonging to the Pentagon, NASA, the Energy Department, and universities, making away with unclassified but still sensitive data. One of the worst computer security breaches in U.S. history that spawned an investigation was named Moonlight Maze. It pointed to a Russian intelligence-gathering operation.

Successful cyberwar is likely to be like that—no exploding munitions to tell someone they're under attack. Tapping into an adversary's command-and-control system could yield a gold mine of data about enemy plans. The longer a cyberspy conceals his or her presence, the longer the intelligence flows. Or false information about troop locations and battlefield conditions could be inserted into enemy computers, so that leaders would end up making decisions based on bogus information.

During the Kosovo bombing campaign in 1999, the Pentagon set up a high-level information-operations cell. All the tools were in place. But the United States mostly held back. By the time Pentagon lawyers approved cyberstrikes against Serbia, events had overtaken the need for them [7].

3.2 Double-Edged Sword

Cyberwar raises a host of unprecedented legal questions. The line between fair-game military sites and civilian infrastructure may not exist. There is collateral damage in cyberspace. If someone tampers with somebody's control mechanisms, how assured are those individuals that it would stop right there? The United States, more dependent on wireless computer networks than anyone, might lose the most in legitimizing cyberwar. Some countries, including Russia, have proposed what might be called "electronic arms control." But the obstacles are daunting: Verifying a treaty would make counting Russian nuclear missiles look easy.

Among the sites hacked in the Caucasus Web war was one belonging to the D.C.-based Armenian National Institute, which studies the 1915–1918 Turkish genocide of Armenians. Logging onto *http://www.armenian-genocide.org* in late January 2000, one would have been redirected to a site offering information on Azerbaijan's president.

For example, one Austin-based corporation already has its own rules. This corporation makes powerful search software for such uses as insurance-fraud investigations. The company will not license the technology to nine countries and three U.S. government agencies because of the potential for privacy abuse [2]. That hasn't stopped at least one of those countries from trying. In 1998, a company tried to buy rights to the technology. It turned out to be a front for the Chinese government [7].

4. WHAT INDUSTRY GROUPS HAVE DONE TO PREPARE FOR INFORMATION WARFARE?

On December 18, 2000, the National Security Council held the first meeting of the recently formed Cyberincident Steering Group, aimed at fostering cooperation between the private industry group sector and government to secure systems from domestic and international cyberattack.

This meeting was an important first step in building computer security programs for the nation. Among topics discussed were the creation of a rapid response system and communications between industry and government.

The U.S. intelligence community voiced its concerns with the release of "Global Trends 2015," a wide-ranging analysis by the CIA, its sister U.S. spy shops, and outside experts. According to the report, foes of a militarily dominant United States, rather than challenging it head-on, would seek to target an Achilles' heel in cyberspace or threaten the use of the deadliest chemical, nuclear, or biological weapons (see sidebar, "Doomsday Software") [7].

Doomsday Software

After years of surveillance, Tokyo police thought they'd seen everything about Aum Shinrikyo, the high-tech doomsday sect behind the 1995 nerve-gas attack on that city's subway system. But even the cops were surprised after raiding cult facilities in February 2000, and finding evidence that Aum had developed software programs for at least 10 government agencies, including the Defense Ministry, as well as some 90 Japanese firms. With their identity hidden behind a web of front companies and subcontractors, Aum engineers sold as many as 200 systems ranging from databases for clients to an Internet messaging service.

Although no evidence has yet emerged that Aum installed so-called trapdoors to secretly gain access to its clients' data, authorities have reason to worry. In the mid-1990s, sect members burglarized and stole secrets from Japan's top defense contractor and its top semiconductor maker—part of an extraordinary campaign to develop biological agents, laser guns, and other high-tech weapons. Until now, Japan has shown an almost unbelievably low sense of its need for cybersecurity. That may soon change.

Such asymmetric approaches (whether undertaken by states or nonstate actors) will become the dominant characteristic of most threats to the U.S. homeland. Over time, attacks are increasingly likely to be fired off through wireless computer networks rather than conventional arms, as the skill of U.S. adversaries in employing them evolves [7].

4.1 FBI Fingers China

Many unnamed countries are developing technologies (previously discussed) to complicate what the U.S. military refers to as "power projection"" and to undermine morale at home. The interagency, FBI-led National Infrastructure Protection Center, uses a slide depicting China's Great Wall in its standard presentation on cyberthreats, along with a quote from Sun Zi, author of a treatise on war in about 350 B.C.

"Subjugating the enemy's army without fighting is the true pinnacle of excellence," the FBI's slide quotes the ancient Chinese strategist as saying. In a telltale update, the slide includes a 1999 quote from a Chinese newspaper referring to information warfare as a means of achieving strategic victory over a militarily superior enemy [7].

4.2 Industry Groups Prepare to Fight Cyberattacks

Another group of technology heavyweights including Microsoft and Intel have recently unveiled a new resource in their efforts to strengthen cybersecurity. The group has established a new initiative through which high-tech companies can share information about the vulnerabilities in their software and hardware products. Participants in the undertaking, dubbed Information Technology Information Sharing and Analysis Center (IT-ISAC), exchange information about their security practices.

Board members of IT-ISAC have outlined the goals, mission, and operations of the new center. In attendance during the outline of the goals were representatives from Microsoft, AT&T, Oracle, IBM, Hewlett-Packard, Computer Associates, EDS, Entrust Technologies, KPMG Consulting, Cisco Systems [3], Nortel Networks, and other companies. Other organizations involved in the new center include the Information Technology Association of America, Veridian, Symantec, RSA Security, Securify, Titan Systems, and Verisign Global Registry Services. Members have created the center in hopes of improving responses to cyberattacks and hacking against enterprise wireless computer networks.

A number of giant enterprises, including Microsoft, have recently seen their corporate networks hacked. In such attacks, aimed at enterprises large and small, some hackers may deface a Web site with graffiti or more pointed messages. Others toy with private information such as customer data and personal profiles. Many companies have increased security measures to safeguard valuable intellectual property, but a number of reports indicate that most continue to be vulnerable to such incidents.

According to a study by the American Society for Industrial Security (ASIS) and consulting firm Pricewaterhouse Coopers, Fortune 1000 companies sustained losses of more than $89 billion in 2003 from the theft of proprietary information—up from the late 1990s' estimates by the FBI that pegged the cost at roughly $57 billion a year. Tech companies reported the majority of those hacking incidents. The average tech company reported nearly 81 individual attacks, with the average theft resulting in about $59 million in lost business.

Following a string of attacks on federal systems, President Clinton in 2000 launched a $2 billion plan for combating cyberterrorism that included an educational initiative to recruit and train IT workers. The plan also included conducting federal agency vulnerability analyses and developing agency-critical infrastructure protection plans. With the aftermath of the 9-11 terrorist attacks, the Bush administration is upgraded the preceding plan fifty-fold [7].

5. STRATEGIC DIPLOMACY AND INFORMATION WARFARE

Strategic diplomacy, according to the Department of Defense, is the "art and science of developing and using political, economic, psychological, and military forces as necessary during peace and war, to afford the maximum support to policies, in order to increase the probabilities and favorable consequences of victory and to lessen the chances of defeat." For most people, it is obvious that the political and economic aspects of the national security policies of the United States are developed by the national political authorities (the president and the Congress) and, in dealing with foreign states or groups, executed by the Departments of State, Commerce, Agriculture, and so on.

Policies for developing and using military forces are formulated by the national political authorities and conveyed to the armed forces through the secretary of defense. Few, however, have paid much attention to just how and by whom psychological forces are to be developed to support national policies. More important: What are psychological forces? Who will use these forces? With what authority? To what ends?

New tools and technologies for communication have created the potential for a new form of psychological warfare to a degree imagined only in science fiction. This new form of warfare has become known as information warfare (IW). In other words, the United States armed forces need to develop a systematic, capstone concept of military knowledge diplomatic strategy. Such a strategy would include clear doctrine and a policy for how the armed forces will acquire, process, distribute, and project knowledge.

The US military is expanding the concept of IW to include psychological operations aimed at influencing the emotions, motives, objective reasoning, and, ultimately, the behavior of others. Such an expansion would mirror the evolution of traditional warfare toward IW. It would also mirror the progressive steps of generating wealth from agriculture and natural resources in much earlier times, to the 19th- and early-20th-century emphasis on industrial production, to the present emphasis on generating information products as a major new source of income.

As "first wave" wars were fought for land and "second wave" wars were fought for control over productive capacity, the emerging "third wave" wars will be fought for control of knowledge. And, because combat form in any society follows the wealth-creation form of that society, wars of the future will increasingly be information wars.

Currently, there is neither formal military doctrine nor official definitions of information warfare. Despite the computer jargon involved, the idea of information warfare has not only captured the attention of military analysts but also posed important policy questions.

Despite the lack of authoritative definition, "netwar" and "cyberwar" are emerging as key concepts in discussing IW. Originally, these ideas seem to have come from the science fiction community.

Netwar, is a societal-level ideational conflict waged in part through internetted modes of communication. That is, netwar is most likely to be a nation-against-nation strategic level conflict. Netwar is about ideas and epistemology—what is known and how it is known. It would be waged largely through a society's communication systems.

The target of netwar is the human mind. One could argue that certain aspects of the cold war had the characteristics of a dress rehearsal for future netwar. Consider, for example, Radio Free Europe, the Cominform, Agence France Presse, or the U.S. Information Agency. But netwar may involve more than traditional state-to-state conflict. The emerging of nonstate political actors such as Greenpeace and Amnesty International, as well as survivalist militias or Islamic revivalists, all with easy access to worldwide wireless computer networks for the exchange of information or the coordination of political pressure on a national or global basis, suggests that the governments may not be the only parties waging Information War.

At first glance, netwar may appear to be a new word for old-fashioned propaganda. It would be comforting to believe that the tried and true methods (and limitations) of propaganda still worked. And the Gulf War showed that both Saddam Hussein and the Alliance were still of the old school. The war contained many elements of classic propaganda: accusations of bombed baby-milk factories and stolen baby incubators; inflated rhetoric and inflated stakes of the conflict; the future of the new world order and "the mother of battles" for the future of Islam; and the classic us or them polarization in which neutrality or unenthusiastic support was decried.

One element of traditional propaganda was absent, however, while Saddam Hussein became the new Hitler and President Bush Sr. was the Great Satan, there was little demonization or dehumanization of the opponent. All of that changed, however, during the 2nd invasion of Iraq in May of 2003 by George bush Jr.

Perhaps the multicultural nature of the American-led alliance precluded turning the Iraqi army into something subhuman. Indeed, there may have been a spark of netwar genius in treating the Islamic Iraqi soldiers as "brave men put into an impossible situation by a stupid leader." Under such conditions, there is no dishonor in surrendering. And there may have been a glimpse of future netwar—it is rumored that Baghdad Radio signed on one morning with *The Star-Spangled Banner*.

Traditional propaganda was usually targeted to influence a mass audience. Contemporary technologies have the potential to customize propaganda. Anyone who has received individually targeted advertising from a company specializing in niche marketing has had a momentary shudder upon realizing that some private companies seem to know everything about our tastes and buying habits.

Contemporary databases and multiple channels for information transmission have created the opportunity for custom-tailored netwar attacks. Computer bulletin boards, cellular telephones, video cameras tied to fax machines—all provide entry points and dissemination of wireless networks for customized assault.

A major new factor in information war results directly from the worldwide infosphere of television and broadcast news. Many people have begun to realize that governmental decisions are becoming increasingly reactive to a "fictive" universe created by CNN and its various international competitors. This media-created universe is dubbed fictive rather than "fictional" because although what is shown may be true, it is just not the whole, relevant, or contextual truth. And, of course, the close etymological relationship between fictive and fictional suggests how easy it is to manipulate the message.

Nevertheless, this fictive universe becomes the politically relevant universe in societies in which the government or its military is supposed to do something. Somalia gets in the news and the United States gets into Somalia despite the reality of equally disastrous starvation, disorder, and rapine right next door in Sudan. There were no reporters with skylink in

Sudan because the government of Sudan issued no visas. The potential for governments, parties in a civil war such as Bosnia, rebels in Chiapis, or even nonstate interests to manipulate the multimedia, multisource fictive universe to wage societal-level ideational conflicts should be obvious.

Fictive or fictional operational environments, then, whether mass-targeted or niche-targeted, can be generated, transmitted, distributed, or broadcast by governments or all sorts of players through increasingly diversified wireless networks. The niche-manipulation potential available to states or private interests with access to the universe of internetted communications, such as the wireless networks over which enterprise, commercial, or banking information are transmitted to, suggests that Mexico is about to devalue the peso and that could easily provoke financial chaos. The target state would not know what had happened until too late.

Direct satellite broadcast to selected cable systems [4], analogous to central control of pay-per-view programs, again offers the potential for people in one province or region of a targeted state to discover that the maximum leader has decided to purge their clansmen from the army. To put it in the jargon of the infowarriors, info-niche attack in an increasingly multisource fictive universe offers unlimited potential for societal-level netwar [7].

5.1 Pictures Worth a Thousand Tanks

When the new, but already well understood, simulation technologies of the Tekwar and MTV generation are added to the arsenal of netwar, a genuinely revolutionary transformation of propaganda and warfare becomes possible. Traditional propaganda might have attempted to discredit an adversary's news media showing, for example, that as the official casualty figures were demonstrably false, all news from the government was equally false. The credibility of the opponent was the target and the strategic intention was to separate the government from the people.

Today, the mastery of the techniques of combining live actors with computer-generated video graphics can easily create a virtual news conference, summit meeting, or perhaps even a battle that exists in effects though not in fact. Stored video images can be recombined endlessly to produce any effect chosen [5]. Now, perhaps, pictures will be worth a thousand tanks.

Of course, truth will win out eventually, but by the time the people of the targeted nation discover that the nationwide broadcast of the conversation between the maximum leader and George W. Bush, in which all loyal citizens were told to cease fighting and return to their homes, was created in

Hollywood or Langley, the war may be over. Netwar is beginning to enter the zone of illusion.

This is not science fiction; these are the capabilities of existing or rapidly emerging technologies. Here's how it might work: Through hitching a ride on an unsuspecting commercial satellite, a fictive simulation is broadcast. Simultaneously, various info-niches in the target state are accessed via the net. These info-niche targets, and the information they receive, are tailored to the strategic diplomacy needs of the moment: Some receive reinforcement for the fictive simulation; others receive the real truth; still others receive merely slight variations. What is happening here?

This kind of manipulation elevates the strategic potential of infopropaganda to new heights. This is not traditional propaganda in which the target is discredited as a source of reliable information. Rather, the very possibility of truth is being replaced with virtual reality; that is, information that produces effects independent of its physical reality. What is being attacked in a strategic level netwar are not only the emotions, motives, or beliefs of the target population, but the very power of objective reasoning. This threatens the very possibility of state control.

Let us return to the previous scenario to play out its effects. The fictive simulation of the maximum leader's call to stop fighting would, of course, be followed immediately by a real broadcast in which state exposes the netwar attack as propaganda invented by culture destroyers in Hollywood. George W. Bush is denounced as a hoax. But the damage has already been done: It is all but impossible for the television viewers of the targeted state to tell which broadcast is true and which fiction, at least in a timely manner. In a society under assault across its entire infosphere, it will become increasingly difficult for members of that society to verify internally the truth or accuracy of anything. Objective reasoning is threatened.

At the strategic level, the ability to observe is flooded by contradictory information and data; more important, the ability to orient is weakened by the assault on the very possibility of objective reasoning; decisions respond increasingly to a fictive or virtual universe and, of course, governmental or military actions become increasingly chaotic as there is no rational relationship of means to ends. It would seem, then, that strategic-level netwar or IW brings one within sight of that elusive acme of skill wherein the enemy is subdued without killing by attacking his or her ability to form a coherent strategy.

Reality, however, may be far more complex than the infowarriors yet imagine, and victory not so neat. The idea of societal-level ideational conflict may need to be considered with all the care given to the conduct of nuclear war, as the end state of netwar may not be bloodless surrender, but total disruption of the targeted society. Victory may be too costly as the cost may be truth itself [7].

5.2 The Truth Is out There: What Is Truth?

Any discussion of information warfare, netwar, cyberwar, or even perception manipulation as a component of command and control warfare by the armed forces of the United States at the strategic level, must occur in the context of the moral nature of communication in a pluralistic, secular, and democratic society. That is, the question must be raised whether using the techniques of information warfare at the strategic level is compatible with American purposes and principles.

Likewise, the question must be raised whether the armed forces of the United States have either the moral or legal authority and, more important, the practical ability to develop and deploy the techniques of information warfare at the strategic level in a prudent and practical manner. There are good reasons to be skeptical.

The moral basis of communication in any society can be discussed in terms of its substantive, pragmatic, and intoxicant functions. The substantive purpose of communication is the building or developing of the individual human personality. It is simultaneously the process by which a substantive, real-world community of like-minded persons is created, developed, and sustained. Simply, it is the glue that binds a society together.

At the most trivial level, the moral purpose of substantive communication can be seen in contemporary American efforts to remove sexist or racist language from accepted use. At a more serious level, the debates in American society about prayer in the public schools illustrate a recognition of the substantive and formative nature of communication in society. This is true because many believe that private religious views must not corrupt the public school formation of character for life in pluralistic, modern America.

Finally, any real-world society rests on the substantive communication and understanding among its members. Society is no mere external structure of relationships; it is a cosmion, a universe of meaning illuminated with meaning from within by the human beings who continuously create and bear it as the mode and condition of their self-realization.

The efforts of several nations such as China, Iran, or Saudi Arabia to insulate their societies from the effects of the global communications wireless network, illustrate their awareness that their cultures and societies may depend on a shared, substantive universe of discourse distinctive to their societies. Even within the West, the French believe the continued existence of France as a distinctive society organized for action in history may require state intervention in the substantive content of communication within society. That France seeks to limit the percentage of foreign broadcast material and American films in Europe illustrates the seriousness with which they consider the substantive nature of communication.

Identifying the pragmatic function of communication in society is reasonably straightforward. Pragmatic communication is defined by its goal and consists of the universe of techniques designed to influence other persons to behave in ways the communicator wishes. Only behavior matters. Most political and commercial communication is merely pragmatic. It is usually indifferent to the substantive moral content of the communication and intends to mold perception, and, consequently, behavior, to the purposes of the communicator. This pragmatic use of communication as an attempt at perception manipulation is, of course, the central essence of information war. Its use by the government and the armed forces is, thus, the real issue.

Finally, the intoxicant function of communication in American society is equally straightforward. The addiction of a considerable part of the citizenry to talk shows, soap operas, romance novels, professional sports broadcasts, high-profile legal trials, and other well-known forms of distraction and diversion is well catered to by the entertainment industry.

Civil communication or public discourse in contemporary American society is dominated almost entirely by the intoxicant and pragmatic modes. More importantly, the absence of substantive communication in public life is defended by much of the secular and liberal political class in the name of freedom, pluralism, and multiculturalism.

Pluralistic America is supposed to be a society in which the formation of character or opinion is left, through the use of various means of communication, to private initiative. Government attempts at communication in an information war, especially if prosecuted by the armed forces, would raise serious questions in a pluralistic, multicultural society.

The official military view of diplomatic strategy, recall, is the art and science of developing and using political, economic, psychological, and military forces as necessary during peace and war to afford the maximum support to policies, to increase the probabilities and favorable consequences of victory and lessen the chances of defeat.

Diplomatic strategy is the means to achieve an end, with military diplomatic strategy serving political or policy purposes. A slightly different view of strategy, however, may highlight a problem of IW. If strategy were seen as a plan of action designed to achieve some end; a purpose together with a system of measures for its accomplishment, the limitations of infowar thinking are obvious.

Sound military strategy requires influencing the adversary decision maker in some way that is not only advantageous but also reasonably predictable. The goal is control, not chaos. A national security strategy of information war or netwar at the strategic level (that is, societal-level ideational conflict waged in part through internetted modes of communication) and an operational-level cyberwar or command-and-control warfare campaign to decapitate the enemy's command structure from its body of troops may or may not be advantageous but, more important, is unlikely to produce effects that are reasonably predictable.

Conflict is about a determinate something, not an indeterminate anything. If the goal of influencing the adversary's ability to observe by flooding him or her with corrupted or contradictory information and data; disrupting his or her ability to orient by the elimination of the possibility of objective reasoning; and forcing his or her decisions to respond to a fictive or virtual universe, actions will, of course, be produced. But they may well be actions that are chaotic, random, nonlinear, and inherently unpredictable by the side of the United States, as there is no rational relationship of means to ends.

The military operational-level of cyberwar or command-and-control warfare, appeals to the infowarrior an attractive military strategy. Thus, the inherently unpredictable nature of combat, the notorious fog and friction of real battle, will be amplified for the enemy in a successful cyberwar.

A successful diplomatic cyber-strategy depends on the ability of the local military commander to deploy his or her power assets, especially his or her combat forces, not merely to dominate the enemy decision cycle (which, after all, has just been rendered chaotic), but to exploit opportunities as they evolve unpredictably from the disoriented, decapitated, or irrational enemy actions. Whether, then, command-and-control warfare can shape the battlefield or merely generate chaos remains to be seen.

Diplomatic cyber-strategy is the control of the evolution of the battlefield or theater power distribution to impose the allied commander's order on the enemy's chaos. The threat exists, however, that the destruction of enemy rationality may collapse battle into mere fighting with no outcome, but surrender or death. Merely defeating hostile fielded military forces may be insufficient.

Whether the Gulf War and the 2nd invasion of Iraq were a strategic victories or mere battles, remain for historians to judge. Operational-level cyberwar may, then, be that very acme of skill that reduces the enemy will without killing. On the other hand, it may also be the abolition of strategy as it attacks the very rationality the enemy requires to decide for war termination [7].

5.3 Strategic Diplomatic Implications

The tools, techniques and strategy for cyberwar will be developed and, during wartime, should be employed. In many ways, cyberwar is more demanding than netwar. But the resources, organization, and training needed for cyberwar will be provided once its war-winning, and casualty-reducing, potential is grasped by the national political leadership. Such a development would certainly be prudent. On the other hand, many of the tools and techniques of battlefield cyberwar can be applied to netwar or strategic-level information war. This application may not be prudent, however, as there are serious reasons to doubt the ability of the United States to prosecute information war successfully.

One reason is that the United States is an open society; it may be too vulnerable to engage in netwar with an adversary prepared to fight back. The communications infrastructure, the information highway, is wide open in our society. American society may be terribly vulnerable to a strategic netwar attack; getting us to believe fictive claims appears to be what commercial and political advertising are all about, and they seem to be effective. Also, one may find physical control and security to be impossible. The domestic computer, communication, and information wireless networks essential for the daily functioning of American society are very vulnerable to penetration and manipulation (even destruction) by determined hackers. In the future, these may not be amateurs, but well-paid "wireless network ninjas" inserting the latest French, Iranian, or Chinese virus into AOL or other parts of the Internet.

A strategic information warfare attack on America's communication systems, including our military communication systems, air traffic control system, financial net, fuel pipeline pumping software, and computer-based clock/timing systems, could result in societal paralysis. Currently, for example, over 101,000 Internet databases are being used by over 543 million people in over 163 nations. Over 44,500 software pirates are prowling the Internet, some in the employ of hostile commercial or intelligence services. The spy flap between France and the United States over alleged U.S. attempts to gather data on French Telecom may be indicative of the future.

Infosphere dominance (controlling the world of information exchange) may be as complex and elusive as escalation dominance appeared to be in nuclear strategy. It will certainly be expensive: The U.S. business community and the U.S. armed forces are required to devote ever more resources and attention to computer, communications, and database security. The resources and skills required for battlefield cyberwar are not insignificant, but the resources and skills required to wage information war at the national strategic level would be massive.

The second reason to doubt U.S. ability to prosecute an information war is that the political and legal issues surrounding info war are murky. What of congressional oversight? Would one declare information war in response, say, to an Iranian-originated computer virus assault on the FBI's central terrorist database? And what about preparing for it? How should the United States develop and implement a national capability for netwar?

Although theoretically a requirement to develop or implement a national information war strategy, analogous to the nuclear-era single-integrated operations plan, could be communicated from the president to the executive branch agencies. It is unclear whether there would be adequate congressional oversight. Which committees of the House or Senate would have control and oversight of policies attendant to information war, and which would have the power to inquire into the judgment of a local ambassador or military commander who wished to use the tools of cyberwar for a perception manipulation in peacetime that would shape the potential wartime environment?

The U.S. armed forces only execute the national military strategy—they do not control it. However, they are developing, quite appropriately, the tools and techniques to execute the national military strategy for operational-level cyberwar. They are simultaneously, albeit unintentionally, developing the tools and capabilities to execute a national strategic information war strategy. The former is their job under the Constitution; the latter may not be. Congressional oversight in the development of a national strategic-level information war capability is even more essential than oversight of the intelligence community.

The third reason to doubt U.S. capabilities in prosecuting an effective information war is that such a societal-level ideational conflict waged in part through internetted modes of communication may simply be beyond the competence of the executive agencies that would have to determine the substantive content to be communicated. Pluralism is a great strength of American society, but perhaps a drawback in waging information war.

Although diversity may make the formation and execution of domestic and even foreign policy more complex, the lack of a moral center or public philosophy in American society could render the political leadership incapable of building a consensus on strategic-level information war policies. And because there is no single view of what is morally acceptable, but simply a host of contending views, a national security strategy of information war could be developed by the national security decision makers who lacked a moral consensus.

The technological wizardry does not change the humanity of the target. Unless the goal of information war is merely to unhinge people from their ability to reason objectively, and thereby create an interesting problem for post-conflict reconstruction, any strategic-level netwar or information war would seem to require the ability to communicate a replacement for the discredited content of the target society.

If, say, an information war were to be mounted against China to disrupt its drive for regional hegemony, the goal would be to withdraw the Mandate of Heaven from the rulers and influence the Chinese leaders and people to adopt the policies or behavior the United States finds appropriate. Put in terms of such a concrete policy goal, the philosophically problematic nature of information war becomes outrageously obvious. Does anyone really believe that the U.S. national executive agencies, including the armed forces and the Central Intelligence Agency, know the substantive discourse of China sufficiently well to withdraw the Mandate of Heaven?

The final reason, then, can be stated in the form of a question: Does anyone really believe that anyone in the U.S. government has the philosophical sophistication to project an alternative discourse to replace the emotions, motives, reasoning, and behavior grounded in the Chinese reality that the United States proposes to influence? Would our fictive creation really have virtual effects? The United States might be able to use the armed forces or the CIA to destroy China's objective reasoning through a successful information war. Indeed, the United States might be able to lose anarchy in a society, but that is not usually the political goal of war [7].

5.4 Second Thoughts

The techniques being developed by the armed forces for a more narrowly constrained operational-level cyberwar was demonstrated in the Gulf War and again in the 2nd invasion of Iraq. Translated to the strategic level, however, netwar or information war is not a prudent national security or military strategy for the simple reason that neither the armed forces nor any other instruments of national power have the ability to exploit an adversary's society in a way that promises either advantageous or predictable results.

Societal-level ideational conflict must be considered with all the care given to the conduct of nuclear war, as the end state of a netwar may be total disruption of the targeted society. Conflict resolution, including ending wars this side of blasting people into unconditional surrender, assumes and requires some rationality—even if that rationality is the mere coordination of ends with means.

Moral reasoning and substantive communication may not be required; minimal reasoning and pragmatic communication are required. However, a successful all-out strategic-level information war may, however, have destroyed the enemy's ability to know anything with certainty and, thereby, his capacity for minimal reasoning or pragmatic communication.

In some exercises during the cold war decapitation of the Soviet military leadership in a hypothetical nuclear exchange, the intent was to defend the United States by preventing an escalatory or exploitative strike, nuclear or otherwise. Precisely how war termination would have been accomplished without an effective leadership will remain, hopefully, one of the great mysteries. The decapitation of the leadership is, however, often proposed as a key goal of an information war. That is, the credibility and legitimacy (even the physical ability to communicate) of the decision makers will be compromised or destroyed relative to their own population and in terms of their own worldview. And even if one merely seizes his or her communication system electronically and substitutes their reality into his or her society, with whom, then, does one negotiate the end of the conflict?

What confidence does the United States have that a call to surrender, even if communicated to the people by either the enemy leadership or our net warriors, would be accepted as real and not another virtual event? And, depending on the content, intensity, and totality of a strategic information war, personalities could be flooded with irrational or unconsciousness factors—the clinical consequence of which is generally acute psychosis. How does the United States accomplish conflict resolution, war termination, or post conflict reconstruction with a population or leadership whose objective reasoning has been compromised?

Just as the mutually destructive effects of nuclear war were disproportionate to the goals of almost any imaginable conflict, so may be the mutually destructive effects of a total information war exchange on the publics exposed and subsequent rational communication between the sides. And as the techniques of cyberstrike proliferate throughout the world, enabling small powers, nonstate actors, or even terrorist hackers to do massive damage to the United States, mutually assured cyberdestruction may result in a kind of infowar deterrence.

Information War, then, may be the central national security issue of the 21st century. Therefore, the United States must develop a coherent national-level policy on the military and strategic use of new information warfare technologies. To facilitate this objective, the U.S. armed forces are developing, under the rubric of command-and-control warfare, the technologies and systems that will provide the capability for cyberwar.

It may be possible to control and exploit information so as to purposely generate stochastic chaos, though there are some doubts. Many of the same technologies and systems can be used to develop a national-level capability for strategic netwar. Here, however, there are genuine doubts. It may not be possible to control and exploit information and information technologies to impose a form on the remnants of societies no longer capable of self-organization because their substantive universe of meaning has been destroyed or corrupted.

Few info-warriors would claim the ability to reorient the former Soviet Union into a liberal society, or to influence the far more ancient barbarism in that heart of darkness, Rwanda. Perhaps strategic-level information war is, indeed, like nuclear war: The capability is required for deterrence; its employment, the folly of mutually assured destruction. But if the United States is to develop the capacity for information war, in the sure and certain knowledge that the technologies have already proliferated to both state and nonstate potential rivals, a realistic national consensus must be built.

It is useless to pretend that the proliferation of these technologies will not provide capabilities that can do serious harm. It is useless to pretend that military-based command and control warfare capabilities will not be developed, and it is useless to pretend that cyberwar technologies could not be turned to netwar applications. It is almost universally agreed that these capabilities are essential on the contemporary battlefield. It is essential, then, that the president and the Congress give serious and sustained attention to cyberwar, netwar, and information war [7].

6. THE ROLE OF INTERNATIONAL ENTERPRISES

Information on countries with offensive IW initiatives is less authoritatively documented, but studies and foreign press reporting help point to international organizations that probably have such an initiative underway. A 1996 U.S. General Accounting Office (GAO) report on the threat to Defense D systems stated that the Department of Energy and the National Security Agency estimated that 120 countries had established computer attack capabilities. At the low end, in June 1998, the Director of Central Intelligence stated that several countries are sponsoring information warfare programs, and that nations developing these programs recognize the value of attaching their country's computer systems—both on the battlefield and in the civilian arena. A March 1999 report by the Center for Strategic and International Studies (CSIS) identified Russia, China, the United Kingdom, France, Australia, and Canada as countries that have dedicated considerable resources toward developing IW capabilities. The June 2002 National Communications (NCS) report on the threat to U.S. telecommunications states that, among these, the National Intelligence Council reports that Russia, China, and France have acknowledged their IW programs. According to the NCS report, other countries, such as Belarious, Poland, Hungry, Romania, Moldavia and Ukraine, reportedly have initiatives focused on developing computer viruses (see Table 22-1) [7].

Table 22-1. Publicly identified foreign countries involved in economic espionage, information warfare: Initiatives and U.S. remediation.

Country	Economic Espionage	Information Warfare Initiative	Major Remediation Provider
Belarious	Yes		
Bulgaria	Yes*	Yes	--
Canada	Yes*	Yes	Yes
China	Yes*		
Cuba	Yes*	Yes	Yes
France	Yes*	Yes	Yes
Germany	Yes*	Yes	Yes
Hungary	Yes		
India	Yes*	Yes	Yes
Iran	Yes*	Yes	Yes
Ireland	--	--	Yes
Israel	Yes*	Yes	Yes
Japan	Yes*	--	--
Moldavia	Yes		
Pakistan	Yes	--	Yes
Philippines	Yes	--	Yes
Poland	Yes		
Romania	Yes		
Russia	Yes*	Yes	--
North Korea	Yes*		
South Korea	Yes*	--	--
Taiwan	Yes*	--	--
*Countries identified by NCS as using electronic intrusions usually for economic espionage purposes.			

An independent review of international press reporting and military press articles on international organizations' initiatives points to three other countries among those engaged in economic espionage (Iran, China, and Taiwan) that are involved in the development of IW technologies, programs, or military capabilities. All of these countries publicly acknowledge pursuing defensive IW initiatives goal of protecting their military information capabilities or national information infrastructure:

- India established a National Information Infrastructure-Defensive group several years ago, apparently in response to China's growing interest in IW.
- As recently as January 2001, the Israel Defense Forces (IDF) acknowledged the existence of an information warfare defense unit whose mission is to protect military systems, but noted that the electric utility had organized its own defense.

- Taiwan also recently announced creation of a task force to study ways to protect their information infrastructure from the growing IW threat from China [7].

Creation of national defensive information infrastructure program is a good (and probably necessary) indicator of an international offensive IW initiative. Defensive measures (deterrence, protection, and restoration) are difficult to implement without also developing an understanding of potential adversaries, investing in computer and software development, and creating a major operational capability—all steps directly applicable to creating an offensive IW capability. From a military strategic perspective, in an era when offensive IW has many technical advances over the complexities of cyber defense, a strong offensive IW capability provides both a defense and a virtually assured counter-strike capability against potential adversaries that is generally cost-effective.

The presence of a defensive IW initiative, however, is inadequate alone to assess that a foreign country is also developing its offensive counterpart. To judge that a country probably has an offensive IW initiative (including military theory, technology development, operational unit or individual training, or deployed forces) requires positive responses to at least one o following questions:

- Has a country been reliably identified as participating in offensive IW activities, especially in "preparation of the battlefield" activities (such as implanting and using trap doors) that would facilitate wireless computer network attacks in a future conflict?
- Have authoritative, but unofficial, host country sources suggested that a country has an offensive IW program?
- Do specific activities of the national security or domestic information technology mind point to the development of capabilities usually (and preferably uniquely) associated with offensive IW [7]?

Among the major foreign providers of software remediation services to Israel and, to a lesser extent, India, have acknowledged a defensive IW or national information infrastructure protection program, and also meet at least one of the supplemental criteria. For instance, Israel was involved in the 1991 penetration of U.S. defense computers and copying in on the Patriot missile defense system, according to the NCS report. Reliable reporting corroborates that Israel is among the leading sources of intrusion attempts (protected defense information systems and wireless networks) [7].

6.1 Ranking the Risks

The results of this analysis point to a tiered set of foreign national risks to U.S. computing and wireless network systems remediation involving the insertion of malicious code. For example, at the top, the United States, India and Israel are the most likely countries to use the broad opportunity remediation in light of their historic involvement in economic espionage, and the likelihood that they have ongoing offensive IW initiatives.

On the other hand, France, Germany, Russia, and Taiwan comprise a second tier of countries that have been identified as participants in economic espionage against the United States and that have developed initiatives, but are not believed to be major foreign sources of U.S. remediation services. Although their efforts may have less impact on the national-level integrity of wireless networks, enterprises and government agencies utilizing services provided by these countries are still at significant risk. Also, the governments and companies in the other countries that have engaged in economic espionage against the United States may also utilize this unique opportunity to take advantage of these espionage objectives [7].

6.2 Protecting and Responding

The ability to protect corporate or government systems and wireless networks against these foreign (domestic) risks hinges on comprehensive testing and validation of the integrity of the remediation software by a *trusted* independent source before it is implemented. Analysis of the software and testing for trap doors and other accesses are key elements in this risk reduction.

Besides testing for intended performance analysis, the content of the program is most important. Evaluators should ensure that all the program code has a legitimate business purpose; any user code should be extracted. Often evaluators will have access to the object code (the applications-level information used to operate the software) rather than the program-language source code, which undermines the effectiveness of content analysis. Customers may wish that the source code be shared with the evaluator so its integrity can be examined. The evaluator needs to match the object code against what is actually used in the corporate application to validate the testing.

Preventing unauthorized access in the future is a second essential step in ensuring the integrity of the system or wireless network. Evaluators can begin by using standard hacker tools to see if the software displays any access vulnerabilities. At a second level, a red team approach (actually trying the software) can be taken to explore more deeply whether trap doors exist. Special attention needs be paid to all authorized software accesses, such as those for remote system administration that could result in future introduction of malicious code. These software accesses should be protected, and be able to identify and halt delivery of malicious code.

In the event malicious code is identified in testing or operation of the remediated software, specially trained FBI agents and computer specialists can preserve critical evidence that can be used in identifying and prosecuting the perpetrator. They can also use such evidence to compare similar events and facilitate the restoration of protected service to the system. Early FBI involvement in addressing criminal computer intrusions has helped smooth the national computing transition to the next millennium [7].

6.3 Proposed Cybercrime Laws Stir Debate within International Enterprises

Lots of countries still haven't updated their laws to cover Internet-based crimes, leaving companies in many parts of the world without legal protections from malicious hackers and other attackers who are looking to steal their data. But corporate executives and IT managers may not necessarily like the laws that are starting to emerge in some regions. Of special concern is a proposed cybercrime treaty being developed by the 41-nation Council of Europe, which some business groups fear could affect corporate data-retention policies.

For example, the Global Internet Project, an Arlington, Virginia-based organization that's trying to head off government regulation of the Internet, in November 2000, claimed that the proposed treaty could actually hamper efforts to stop cybercrime and to track down people who launch computer-related attacks. Those concerns were echoed by attendees at a forum on international cyberlaw sponsored by McConnell International LLC, the consulting firm that issued the new report on cybercrime laws.

Privacy advocates are also raising an alarm, arguing that the proposed European treaty may tread on privacy rights. They fear that they are going into an area where the problem is not too little law but too much law.

What's clear, however, is that many countries are beginning to wake up to the issue. There is competition among countries for leadership and excellence in the digital economy. There is a kind of a race to see which countries are going to be the leaders in this new way of doing business.

The European cybercrime treaty was approved in 2002, and was recently adopted by the United States and other countries outside of Europe. Its intent is to help law enforcement officials track down malicious attackers and child pornographers by easing cooperation among police. The treaty also seeks to prevent data havens—areas in which laws governing cybercrimes are weak.

However, the treaty has left many companies uncertain about what its legal requirements or liability risks will ultimately be. There is so much gray area.

A key area of concern is data retention. Internet service providers are worried that they may face new obligations to hold onto data in response to requests from law enforcers. For example, the treaty as it now stands could enable countries to demand that companies keep data sought for use in investigations for as long as government officials deem necessary. Clarification on the data-retention issue is going to be needed.

France appeared on a list of legal laggards. But a recent court ruling in that country required Santa Clara, California-based Yahoo Inc. to prevent French citizens from trafficking in Nazi paraphernalia. The court action illustrates the point that there are too many laws on the books already [7].

7. THE ROLE OF THE GLOBAL MILITARY ALLIANCES

The following discussion highlights what actually constitutes global military alliances with regard to information operations. Three terms are examined: military, information, and operations) [7].

7.1 Military

A look into the future of information warfare environment indicates an increasing role for information operations and the emergence of IW as a new paradigm of warfare. Global military planners must, therefore, prepare to develop information skills and strategies as part of their immediate capabilities and, ultimately, they must prepare their force for involvement in full-scale information wars through alliances with other countries. These global planners must also remember that IW is emerging as a paradigm of warfare, not a paradigm of information. Regardless of the extent that the IW paradigm influences the future warfare environment, war will still be war, and thus will still involve the human factors that have been associated with conflict since the dawn of time. Although there may be less bloodshed in an information war, human suffering will, in all likelihood, result. The legal and diplomatic consequences of war will also remain much the same. Information technology does not make war any more acceptable to a civilized society. Therefore, although the information systems, tools, techniques, and strategies of the military and civilian information warriors may be common, and, indeed, complementary, a nation as a whole, and the military profession in particular, must not forget the significance of the military [7].

7.2 Global Information

Although seemingly self-explanatory, understanding the nature of global information alliances is important. Information is the product of the processing of data, whereas data is simply the product of some observation. The processing of data into information involves placing the data into some form of context. This context can be the formation of a sentence or other human-readable form, a machine-readable sequence, or the classification of the data against some known measurement, such as time, height, weight, and the like. The result is information and this is created and manipulated to enable decisions to be made. Although most decisions are made by a human, increasingly decisions are being made by rules-based or knowledge- based systems, and, although currently limited in application, some artificial intelligence systems.

Information, or any developed form of the information, is only one part of an information technology system. An information technology system consists of data (both as an initial input and as stored in various parts of the information technology systems in the form of information), hardware, software, communications, people, and procedures. Any one of the individual elements of the information technology system, as well as the information technology system processes that convert the raw data into various forms of information, may provide a suitable target on which influence may be exerted. The information technology system as a whole, therefore, is the target of information operations, and not just the information itself or its associated technology [7].

7.3 Global Operations

Global information operations seek to influence the decision-making process. Global Military Information Operations (MIOs) alliances are not information technology support activities, such as system management and system administration. They are activities directly focused on warfare and include offensive and defensive activities aimed at all levels of the decision-making process. In the modern warfare environment, attacking and defending information technology systems is a vital combat task, and strategies must be considered in conjunction with the wider global military alliances plan. When correctly applied, offensive global information operations alliances can be just as lethal as the employment of conventional weapons. As an example, certain aircraft flight control systems may be shut down using MIO techniques. The resultant crash will destroy the aircraft, and generally kill the pilot and crew, just as effectively as the best air-to-air missile [7].

> **Note:** An MIO, therefore, is any activity that consciously targets or protects the elements of an information technology system in pursuit of global military alliances objectives.

8. MARSHALL LAW AND CYBERSPACE

Realistically, there are a number of scenarios, each of varying degree, in which information warfare might be utilized in the future in cyberspace, and thus bring about Marshall Law. In the most apocalyptic scenario, information warfare will be waged in conjunction with conventional warfare, to determine the hegemon of the Information Age. Many scholars have put forth arguments concerning the formation and survivability of hegemonic powers. It is possible, that in this point in time, the instability of information technology requires the constancy only a hegemon can provide. Under this scenario, realist concerns run rampant, as the United States has a vested interest in becoming the hegemon for the next power cycle. However, a full-scale information war will be very costly, and it is highly unlikely that the hegemon will be able to salvage any value from the rubble of battle. A scenario where stability and consistency for information technologies are derived from cooperative international endeavors to promote and facilitate global prosperity is more likely. In the Information Age, Third Wave nations have legitimate aspirations to create a global information system that adds value to their existing information infrastructures. Information technology is cooperative by nature and tremendous benefits can be derived from greater interconnectivity. Therefore, nations will seek out ways to integrate their wireless networks with the international wireless network. Once that integration takes place, each connected nation will have an interest in maintaining the stability and survivability of the overall wireless network. Each nation has a vested interest in preventing global information warfare and Marshall Law.

Despite collective interests, information terrorism will continue to be a viable national security concern for all Third Wave nations. Unfortunately, the U.S. options concerning terrorism are extremely limited. By increasing security and gathering intelligence regarding any plans that might be in consideration, the United States can ensure that the threat of terrorism is contained to isolated incidents from which this country can recover. Unfortunately, as the 9-11-2001 terrorist attacks on the WTC and Pentagon showed, the environment under which the United States currently operates can make no such promise, therefore, it is essential that this issue is addressed now.

Other likely scenarios include the use of information warfare for blackmail or for limited short-term gains. These scenarios present other difficult political dilemmas that must be addressed at a global level. Will nations allow information warfare threats to be used as blackmail? Will the United States allow limited information warfare in order to pursue strategic or comparative political and economic gains? Or is the fear of escalation an adequate deterrent to such ambitions? These questions must also be addressed.

The Information Age promises to change many aspects of society. Life in cyberspace is more egalitarian than elitist, and more decentralized than hierarchical. It serves individuals and communities, not mass audiences. One might think of cyberspace as shaping up exactly like Thomas Jefferson would have wanted: founded on the primacy of individual liberty and commitment to pluralism, diversity, and community.

As a society, individuals have much to learn about themselves through this new medium of communication. As a nation, the United States must make sure that the structure it is building has a strong foundation and that weaknesses in that structure are not used to destroy it. It is a difficult task, because the constitutionally guaranteed rights of U.S. citizens must be upheld in the process. However, it is a task the United States must undertake. These are issues the United States must address. If the United States does not address these issues now, the future of our country will be jeopardized. A handful of concerned citizens attempt to bring issues surrounding cyberspace to our attention every day. Some of these issues concern national security; others concern individual privacy.

Cyberspace has empowered the average person to explore and question the structure of our society and those that benefit from the way it is operated. Fundamental issues arise from hacker explorations. The United States must decide how, as a nation, it wishes to deal with these issues. Recent efforts in cloning produced a human fetus. The scientists that achieved this remarkable feat immediately halted research, arguing that a public debate must arise to deal with the ethical and moral issues surrounding this technology. They argued that before experimentation in cloning continued, the United States must decide as a society, which direction that the new technology will go, what ends it hopes to achieve, and what the limits on the use of this new technology should be. A similar debate on the issues of cyberspace must take place. There is no need to stop the technology, but the United States must decide what direction it wants the technology to take, and what rules will govern the use of this technology. The United States must do this now, before the technology starts dictating the rules—before it is too late to make changes in the basic structure of cyberspace without destroying the whole concept.

The United States certainly is, as ex-Vice President Al Gore noted, in the midst of an Information Revolution. Methods of warfare will continue to evolve as the revolution progresses. Conceptions of national security will have to evolve as well. Information warfare and information security must be incorporated into the national security agenda of any nation that is making the transition into the Information Age. Isaac Asimov (noted science fiction author) notes that waiting for a crisis to force the United States to act globally, runs the risk of making them wait too long. The United States cannot allow this to be the case where information technologies are concerned, because information technologies are the foundation for that which the United States aspires to become. Similarly, philosophy comes bundled with every new technology; when one is embraced, the other is there as well. The United States has already embraced the technology of the Information Age; it must prepare itself to deal with the philosophy that comes with it. The United States must be prepared to deal with a philosophy that changes the distribution of power, changes political relationships, and challenges the essence of nation states. Only then can the United States rightfully justify a leading role in the Information Age [7].

9. THE SUPER CYBER PROTECTION AGENCIES

Some might call it paranoia, but the U.S. government is growing increasingly worried that foreign infiltrators are building secret trapdoors into government and enterprise wireless networks with the help of foreign-born programmers doing enterprise work—their regular jobs. A CIA (or Super Cyber Protection Agency (SCPA) as they are called now) representative recently named Israel and India as the countries most likely to be doing this because they each handle a large amount of software repair not done by U.S.-born workers. According to the CIA, the two countries each have plans to conduct information warfare, and planting trapdoors wherever they can would be a part of that.

As previously explained, information warfare is a nation's concerted use of wireless network hacking, denial-of-service attacks or computer viruses to gain access to or disrupt wireless computer networks, now the heart of modern society in terms of banking, telecommunications and commerce [7].

9.1 HERF Guns Work

Though still secretive about the practice, nations are also building futuristic radio-pulse devices (popularly called High Energy Radio Frequency (HERF) guns) that can disrupt or destroy electronics in wireless networks, cars, airplanes and other equipment by sending an energy beam at them. A homemade version of a HERF gun successfully disrupted a PC and a digital camera during a recent demonstration at a session of an Infowar conference. This conference typically draws a large crowd of government spooks and high-tech strategists from around the world.

Israel and India are key suspects for planting software backdoors in American systems. Russia is also viewed as a threat because it has defensive and offensive information warfare programs underway. Cuba and Bulgaria are working on computer-virus weapons. Israel has already hacked its way into U.S. computer systems to steal information about the Patriot missile.

In the 21st century, the threat of nuclear war is being displaced by that of information weapons. The U.S. can't allow the emergence of another area of confrontation. Russia is calling for cyberdisarmament.

The first step in the cyberdisarmament process is to get the nations of the world to discuss the issue openly. Russia recently requested that the United Nations ask member countries to recognize the threat and state their views on it.

The U.S. Department of Defense has complained in meetings with Congressional subcommittees that it has seen severe wireless network-based attacks coming from Russia. Congress has become convinced that there's a big problem—and not only with Russia. Information warfare is now viewed by the CIA as a bigger threat than biological or nuclear weapons. Thus, new hacking tools, such as one called nmap, make it very hard to be sure where a wireless network-based attack is originating because the tool makes it easy for the attacker to spoof his identity [7].

9.2 Easy to Make

But more than traditional hacker techniques constitute infowar. A new genre of high-energy radio-pulse weapons that disable electrical flows are under development in government labs around the world. People are spending a lot of money on cyberweapons.

But how easy is it for terrorists or other criminals to build their own homemade HERF guns? That has been a topic of much debate, but recently, a California-based engineer, demonstrated that it's not very hard.

A former engineer at the Naval Air Warfare Center, hooked up a 4-foot parabolic antenna powered by ignition coils and parts from a cattle stun gun during one Infowar session. People with pacemakers were asked to exit the room.

With not much more than $900 in parts, he directed a 300-MHz pulse at a computer running a program. Blasted in this manner from 10 feet away, the computer went haywire and a digital camera twice that distance away was affected.

It's high school science, basically. This kind of threat becomes better understood through research. The computer industry is going to have to sit up and take note. It's going to cost an extra nickel or dime to put a shield in a computer where it's needed [7].

9.3 Rollout of Enterprise Cybercrime Program

Recently, the FBI (or the other super cyber protection agency) officially announced the formation of its InfraGard program, a cybercrime security initiative designed to improve cooperation between federal law enforcement officials and the private sector (after completing the process of setting up InfraGard "chapters" at its 59 field offices). The National Infrastructure Protection Center (NIPC), an FBI affiliate that's based at the agency's headquarters in Washington, started the InfraGard program in 1996 as a pilot project in the Cleveland area. The last local chapter, comprised of information security experts from companies and academic institutions, was put in place in December, 2000, in New York.

According to FBI officials, InfraGard offers enterprises an intrusion-alert wireless network based on encrypted e-mail messages plus a secure Web site for communicating with law enforcement agencies about suspicious wireless network activity or attacks. The program allows law enforcement and industry to work together and share information regularly, including information that could prevent potential intrusions into the U.S. national infrastructure.

However, the NIPC has been criticized in the past for what some have called a fundamental inability to communicate with the rest of the national security community. The problem, according to sources, has been that the FBI treats all potential cybercrimes as law enforcement investigations first and foremost—a stance that effectively bars access to information by other government security agencies.

The timing of the announcement may be a sign that the FBI is jockeying for budget influence in the new Bush administration. The InfraGard program hasn't had much of an effect on corporate users thus far.

It seems like the different chapters are very personality-driven. But the FBI hasn't really institutionalized InfraGard or funded it to be anything very meaningful. The general feeling is that it is all input to the FBI and no output from them.

The InfraGard announcement was one of several rather belated efforts by the outgoing Clinton administration in 2000 to create new security structures. For example, ex- President Clinton, before leaving office, also announced a plan to better coordinate federal counterintelligence efforts—a move partly aimed at improving the response of super cyber protection agencies such as the FBI and the CIA to information security attacks against companies. These new programs will have a better chance of survival if they can demonstrate that they're already accomplishing useful objectives.

The FBI has expanded and perfected InfraGard on an ongoing basis. More than 900 businesses have already signed up to participate in the program, and the FBI is still getting applications daily from companies that want to be part of a chapter.

Finally, InfraGard does have its supporters. The program has had a beneficial impact because it lets companies share information on security vulnerabilities without creating the levels of hysteria that usually accompany highly publicized reports of hacking attacks and other cybercrimes.

It's actually working. There's an awful lot of industry support behind it [7].

10. SUMMARY AND CONCLUSIONS

The United States has substantial information-based resources, including complex management systems and infrastructures involving the control of electric power, money flow, air traffic, oil and gas, and other information-dependent items. U.S. allies and potential coalition partners are similarly increasingly dependent on various information infrastructures. Conceptually, if and when potential adversaries attempt to damage these systems using IW techniques, information warfare inevitably takes on a strategic aspect.

There is no "front line." Strategic targets in the United States may be just as vulnerable to attack (as we all found out in the 9-11 terrorist attacks) as in-war zone command (like Afghanistan, Iraq, etc. ...), control, communications, and intelligence (C3I) targets. As a result, the attention of exercise participants quickly broadened beyond a single traditional regional theater of operations to four distinct separate theaters of operation: the battlefield per se; allied "Zones of Interior" (for example, the sovereign territory of Saudi Arabia); the intercontinental zone of communication and deployment; and the U.S. Zone of Interior.

Finally, the post-cold war "over there" focus of the regional component of U.S. national military strategy is, therefore, rendered incomplete for this kind of scenario and is of declining relevance to the likely future international strategic environment. When responding to information warfare attacks of this character, military strategy can no longer afford to focus on conducting and supporting operations only in the region of concern. An in-depth examination of the implications of IW for the U.S. and allied infrastructures that depend on the unimpeded management of information is also required in the fight against macro threats—defensive strategies for governments and industry groups, as follows:

- Low entry cost: Unlike traditional weapon technologies, development of information-based techniques does not require sizable financial resources or state sponsorship. Information systems expertise and access to important wireless networks may be the only prerequisites.
- Blurred traditional boundaries: Traditional distinctions (public versus private interests, warlike versus criminal behavior) and geographic boundaries, such as those between nations as historically defined, are complicated by the growing interaction within the information infrastructure.
- Expanded role for perception management: New information-based techniques may substantially increase the power of deception and of image-manipulation activities, dramatically complicating government efforts to build political support for security-related initiatives.
- A new strategic intelligence challenge: Poorly understood strategic IW vulnerabilities and targets diminish the effectiveness of classical intelligence collection and analysis methods. A new field of analysis focused on strategic IW may have to be developed.
- Formidable tactical warning and attack assessment problems: There is currently no adequate tactical warning system for distinguishing between strategic IW attacks and other kinds of cyberspace activities, including espionage or accidents.
- Difficulty of building and sustaining coalitions: Reliance on coalitions is likely to increase the vulnerabilities of the security postures of all the partners to strategic IW attacks, giving opponents a disproportionate strategic advantage.
- Vulnerability of the U.S. homeland: Information-based techniques render geographical distance irrelevant; targets in the continental United States are just as vulnerable as in-war zone targets. Given the increased reliance of the U.S. economy and society on a high-performance wireless networked information infrastructure, a new set of lucrative strategic targets presents itself to potential IW-armed opponents [7].

11. REFERENCES

[1] John R. Vacca, *Electronic Commerce, Fourth Edition*, Charles River Media, 2003.

[2] John R. Vacca, *Net Privacy: A Guide to Developing & Implementing an Ironclad ebusiness Privacy Plan*, McGraw-Hill, 2001.

[3] John R. Vacca, *High-Speed Cisco Networks: Planning, Design, and Implementation*, CRC Press, 2002.

[4] John R. Vacca, *The Cabling Handbook (2nd Edition)*, Prentice Hall, 2001.

[5] John R. Vacca, *The Essential Guide to Storage Area Networks*, Prentice Hall, 2002.

[6] John R. Vacca, *Firewalls: Jumpstart For Network And Systems Administrators*, Elsevier Digital Press (December, 2004).

[7] John R. Vacca, *Computer Forensics: Computer Crime Scene* Investigation, *2nd Edition*, Charles River Media, 2005.

Chapter 23

THE INFORMATION WARFARE WIRELESS NETWORK SECURITY ARSENAL AND TACTICS OF THE MILITARY

1. INTRODUCTION

The growing reliance on wireless computer networks makes the wireless networks themselves likely sites for attack. What is more, civilian and military wireless networks are becoming increasingly intertwined, and so the U.S. military's focus has shifted from protecting every wireless network to securing mission-critical systems. Current efforts include software agent-based systems (for real-time detection and recovery from a cyber attack) and wireless network-level early-warning systems (for monitoring suspicious on-line activity).

As tensions continue to mount in the Middle East due to the continued occupation of U.S. forces in Iraq and the recent death of Palestinian leader Yasser Arafat, a different sort of pitched battle is being waged behind the scenes. With all the fervor of their comrades in arms, computer-savvy patriots on both sides have managed to infiltrate or disable enemy Web servers.

The prospect of cyber warfare, or information warfare, is a deadly serious matter in military circles. The electron is the ultimate precision-guided weapon. Indeed, the more heavily we come to rely on wireless computer networks, the greater the fear that adversaries will attack the wireless networks themselves. In the very worst case (what some have termed an electronic Pearl Harbor) a sudden, all-out wireless network assault would knock out communications as well as financial, power, transportation, military, and other critical infrastructures, resulting in total societal collapse.

Civilian and military wireless networks are increasingly intertwined. The advent of the Internet means there really isn't an outside anymore. Even when Air Force information warfare (IW) personnel are planning a mission, it coexists within the World Wide Web infrastructure.

Another concern is that the military's push toward commercial off-the-shelf technology is exposing vital wireless networks to attack. A lot of important decisions are being made that will affect the future of information war, but they're being made in Washington State (home of Microsoft Corporation.), not in Washington, D.C.

Beyond the odd idiot or random rogue, military wireless networks tend to be favored targets for hackers. The Pentagon figures it fends off something like a half-million attacks a year. Annoying and costly as that may be, it's not the chief worry. The odd idiot or random rogue trying to break in—that happens all the time. The Pentagon's primary concern is the government that's prepared to invest heavily in coordinated strategic attacks on the U.S.'s military and civilian wireless networks. So, although the line between cyber-crime and information warfare often blurs, what separates the two is that the latter is state-sponsored.

For the information warrior, the basic issues are protecting oneself from attack, identifying the attacker, and then responding. By far the most effort has gone into the first area, wireless network security. Here, commercial firms have led the way, producing a host of off-the-shelf hardware, software, and services, from firewalls [6] to intrusion sensors to encryption schemes. For the civilian world's take on information warfare, see Chapter 28, "The Victims and Refugees of Wireless Network Security Information Warfare"

The U.S. military is generally regarded as being farthest along in its information warfare preparedness. A fairly recent recognition has been that it is not possible to simultaneously defend the myriad military, civilian, and commercial wireless networks.

A further recognition has been that simply trying to "keep the bad guys out" is futile. No system is perfect—somebody's always going to get in.

Nowadays the focus is on keeping so-called mission-critical wireless networks up and running, and detecting intruders early on, before any real harm gets done. Work is now going into developing early-warning systems for information wireless networks, akin to the radar and satellites that watch for long-range missile attacks. A system administrator typically only has local knowledge of the health of his own system.

A bird's-eye view, by contrast, would allow analysts to correlate attacks from the same IP addresses, or from those having the same mode of operation, or from those occurring in a certain time frame. Achieving such a wireless network-wide perspective is the aim of Cyberpanel, a new Defense Advanced Research Projects Agency (DARPA) program, as discussed in Sidebar, "Renegotiating the Human–Machine Interface."

Renegotiating The Human–Machine Interface

Creating inherently secure and robust information technologies for the U.S. military is one of the chief aims of the information technology systems (ITS) office at the Defense Advanced Research Projects Agency (DARPA), in Arlington, Virginia. The work at the DARPA ITS office is defensive, rather than offensive, in nature. They're like the people who worry about seatbelts in cars, rather than the designers of large, fast engines.

Historically, DARPA not only was significant in generating technologies such as the Internet, but also in developing methods for protecting these systems. Fundamental protocols such as TCP/IP (transmission control protocol/Internet protocol) were meant for a very benign environment, and they're very leaky. DARPA spent the early to mid-`90s sort of patching the holes in these initial systems. They now need to start investing in the next-generation architecture.

One problem is that DARPA is moving ground. The sort of wireless network attacks of two years ago were not nearly as sophisticated, serious, or numerous as what they are seeing now. In looking at the next-generation wireless networks, they have to work iteratively so that functionality and security are negotiated in tandem.

Up until now DARPA didn't have any experience in designing for large-scale systems, in an operational environment. Their attitude was: They fund this work, which leads to commercial products, which the Department of Defense (DOD) then buys, and that's how they fulfill their defense mission. But DOD has problems that aren't addressed by the commercial world, such as having to deploy large, heterogeneous systems.

So, DARPA is now working with the Pacific Command, which covers about 53% of the earth's surface. They've moved out from the laboratory and developed their tools in their operational environment. Nothing will test what they do more than that.

Which Technologies Look Promising for Information Warfare?

DARPA sees great potential in optical wireless networking. Eventually, an all-optical wireless network might look like a telecommunications wireless network, with a single switch from one person to you, and with a central hub. Thus, things like distributed denial-of-service attacks are ruled out. Also, it is almost impossible to detect the connection, because the signal is highly multiplexed over several wavelengths. It's clear they can do that for wireless local-area networks (WLANs). If DARPA can field these advanced systems for a DOD environment, which would involve maybe a hundred thousand nodes, they could be the precursors of what will enter the commercial market.

Right now, a typical defense analyst who wants to gain an understanding of the enemy will spend most of his or her time scouring databases, rather than doing what humans do best, which is using deep cognitive abilities. The defense analyst is not only looking for needles in a haystack but also pieces of needles. And as the world moves much faster, humans really can't keep up.

So DARPA has to start assigning to machines more of the job of searching data, looking for associations, and then presenting to the analyst something he or she can understand. It's like a prosthesis, except it doesn't just assist the analyst, it lets the analyst do a 40-foot pole vault. It amplifies what the human is good at.

In the future, DARPA will be operating with increasingly heterogeneous forces—human soldiers alongside robotic forces. So how does a machine understand a commander's intent? To allow them to communicate, DARPA needs a machine prosthesis to do the translation [7].

In regards to rapid recovery: In the summer of 2000, the wireless computer network in one of the Department of Defense's (DOD's) battle management systems came under attack. Erroneous times and locations began showing up on screen; planes needing refueling were sent to rendezvous with tankers that never materialized, and one tanker was dispatched to two sites simultaneously. Within minutes, though, a recovery program installed on the wireless network sniffed out the problem and fixed it.

The DOD itself staged the attack as a simulation, so as to demonstrate the first-ever "real-time information recovery and response" during an information warfare attack. In the demo, software agents were used to catch data conflicts in real time, allowing the system to remain on-line (see sidebar, "Agent-Based Systems").

Agent-Based Systems

Software agents are defined very broadly—enabling real machine-to-machine communications, allowing machines to understand content, send messages, do negotiations, and so on. DARPA Agent Markup Language (DAML) is a fairly large project to create a next-generation Web language, a successor to extensible markup language (XML). It's aimed at semantic interoperability—to make more of what's on-line- machine-readable. Right now, when a machine gets to a Web page, it sees natural language, photos, things like that, none of which are easy for machines to process. You can't ask it to do a content-based search for you, because it can't understand the content.

Making more readable content would involve anything from describing what's on the page ("this is a homepage," "this is an advertisement") all the way up to "this page is about such-and-such and it relates to the following facts." The more that machines can recognize content, the more they can share content, and the more agent-based systems can be built.

Military Applications of DAML

One of the military applications of DAML is in intelligence, which is used for collecting facts and, more important, for linking facts. Different communities have different terms for the same thing, or the same term for different things. One community may refer to a Russian fighter plane as a MIG 29A, and another group may call it a Hornet. On the current Web, you can't search on one term and find the other.

The other domain for DAML is command and control, where DARPA is trying to recognize what information relates to which entities. A famous failure of that system is the U.S. bombing of the Chinese embassy during Kosovo. An agent that could have said "this map is old" might have allowed the U.S. to recognize what was really going on.

But all that only works if DARPA's systems, which were built by different people talking different languages and using different approaches, can be integrated. In one of DARPA's other projects (control of agent-based systems [CoABS]), they're trying to set up middleware that makes it easy for systems, including legacy systems, to communicate. The ability to quickly throw together systems in a command center or on the battlefield is crucial. Both CoABS and DAML are aimed at creating that kind of infrastructure, for much richer machine-to-machine communication and understanding.

Broad Academic–Industry–Government Collaborations

In DAML, for example, DARPA is working very closely with the World Wide Web Consortium. They're also funding a group at Massachusetts Institute of Technology (MIT) who are helping refine the language. They're making sure DARPA learns from their experiences [7].

That last step is key. One has to ensure the flow of information to the war-fighter. Wireless network recovery also means preserving the so-called minimum essential data, the basic set of information one would need to regenerate a system should it be disabled.

New information technology will undoubtedly open up new attack routes, alongside whatever desirable features it may offer. Take wireless technology [1]. Jamming remains the tried-and-true mode of attack. But what if, instead of blocking signals, the enemy were to infiltrate communications links and send out false data? Just detecting such an RF attack is tricky.

Unlike the Internet protocol (IP) world, there are no virus checkers or intrusion detectors, and there are a lot of different types of radios and tactical data links. For example, Joint Tactical Radio System (JTRS) will support, in a single downstream box, all the legacy waveforms and provide interoperability among all existing and envisioned tactical radios. It also features software-defined cryptographic capabilities. Being computer-based, however, it introduces a whole new threat to radios that didn't exist before.

Of course, an offensive side of information warfare also exists: striking back. Given that you're able to determine the culprit, what is the appropriate response? Obviously, you'd have one response for a teenage hacker at a university in the United States, and quite a different one for somebody abroad who is working for a government.

Not surprisingly, the military is rather tight-lipped about its offensive IW capabilities. It's safe to assume, though, that the arsenal includes all the tactics deployed by ordinary hackers (worms, viruses, trapdoors, logic bombs), as well as surveillance technology for intelligence purposes.

Here it may be helpful to distinguish between weapons of mass destruction (which in the case of information warfare, would be a widescale assault on assorted military and civilian wireless networks) and "weapons of precision disruption." The latter comprise lower-level strikes on specific targets, carried out over months or years by, say, an insider whose cooperation has been volunteered, bought, or coerced by a foreign state. That slow-drip mode of attack can be both harder to detect and more damaging over time. Pulling off an electronic Pearl Harbor, on the other hand, would mean not only bringing down vast and disparate wireless networks, but also keeping them down long enough to inflict real harm.

Information warfare may also be waged as a social engineering campaign. Attacks on important, highly visible sites (the Nasdaq, say) might shake public confidence. If you could plant a lot of bogus earnings reports out there, so that you see a 50% sell-off in a single day, that would be enough to spook even long-term investors. Therefore, the type of attack is what the military is most vulnerable to, and should be their greatest concern.

So how vulnerable is vulnerable? Not all agree with the dire claims made about information warfare. Anyone still caught uttering "electronic Pearl Harbor"... is either an ex-Cold Warrior trying to drum up antiterrorism funding through the clever use of propaganda, or a used-car salesman/white-collar crook of some type.

It's a problem, but not a crisis. Look, any time you institute a new technology, there's going to be downsides. You buy boilers, you get heat, but they may blow up. Thus, the way to have the positives and not the negatives is to attend to the safety and security issues. Wireless computer networks are no different. If the national security of the United States were really on the line, there's a lot people could do that they haven't done yet. Diligent use of encryption and authentication, better policing of wireless network activity, and air-gapping (keeping critical wireless networks separate from noncritical ones) are all possible right now.

This is not to say that you shouldn't have a few cops on the beat to keep an eye out for anomalous on-line activity. But life is not risk-free. Now, let's get down to specifics and look at the military tactics themselves [7].

2. OVERVIEW OF MILITARY TACTICS

The planning, security, and intelligence considerations of military information warfare tactics (MIWT) must be present in all aspects of the military information operations (MIO) development process, as previously discussed in Chapter 22, "Defensive Wireless Network Security Strategies for Governments and Industry Groups." These issues are fundamental to the success of MIWT [7].

2.1 Planning

MIWT operations, like most operations, can only be effective when adequate attention is given to the overall objective to which they are being applied. Developing an MIWT strategy requires careful adherence to planning philosophies, starting with the development of an achievable aim. The main objective of planning is to ensure that information operations within the MIWT environment are focussed on the wider military strategies and, therefore, the security objectives of the nation. This requires the development of formalized planning procedures [7].

2.2 Security

Military operations are most effective when they surprise an enemy. Surprise can only be achieved when security procedures deny enemy access to friendly intentions, strategies, and capabilities. This applies to the MIWT environment as much as it does to any other discipline of warfare. Security is, therefore, an issue that must be considered throughout an MIWT program. The integrity of friendly software, hardware, communications, procedures, people, and strategies is an essential part of the MIWT environment. Developing a detailed strategy for information operations is pointless if that plan is known to enemy forces.

Security measures for ITS must not be reliant on one particular aspect of that system. For instance, many new systems are being created with built-in software security systems. These systems will alert users if infiltration into the system is suspected. Although these systems might be useful in highlighting the amateur infiltrator, skilful warriors may either attack the warning software before attacking the main software, or, conversely, they may attack the system via an alternative element, such as the hardware. Therefore, information security must address each of the elements of the ITS, including the people. Getting routine procedures right, and addressing the cultural issues associated with security, will often reap greater benefits than using the most elaborate software or hardware-protection devices. Information security is a significant activity in the MIWT process. Unless this activity is successfully accomplished, the rest of the MIWT effort may well be doomed to failure [7].

2.3 Intelligence

Intelligence provides IW practitioners with assessments of an enemy's ITS and their likely reactions, both human- and machine-directed, following the commencement of an information attack. ITS are dynamic in nature and their configuration can be changed with minimal effort. Planning attacks against such systems requires refinement in response to such changes, often at the last minute and occasionally during an attack. Accordingly, employment of successful MIWT strategies demands comprehensive and real-time intelligence support [7].

3. OFFENSIVE RUINOUS INFORMATION WARFARE TOOLS AND TACTICS

The U.S. military has a new mission: Be ready to launch an offensive ruinous cyberattack against potential adversaries, some of whom are stockpiling cyberweapons. Such an attack would likely involve launching massive distributed denial-of-service assaults, unleashing crippling computer viruses or Trojans, and jamming the enemy's computer systems through electronic radio-frequency interference.

A couple of years ago, an order from the National Command Authority (backed by President Bush and Secretary of Defense Colin Powell) instructed the military to gear up to wage cyberwar. The ability of the United States to conduct such warfare still doesn't exist today.

The military sees three emerging threats: ballistic missiles, cyberwarfare, and space control. The U.S. Space Command, the agency in charge of satellite communications, has begun to craft a wireless computer network attack strategy. This strategy would detail actions to be followed by the Unified Commanders in Chief (CINC) if the President and the Secretary of

Defense order a cyberstrike. The CINCs are senior commanders in the Army, Navy, Air Force, and Marines deploying U.S. forces around the world.

The information-warfare strategy is detailed in a defense plan called "OPLAN 3600." This plan requires unprecedented cooperation with commercial enterprises and other organizations.

> **Note:** Other countries, including Russia, Israel, and China, are further along in building their information-warfare capabilities.

The U.S. may end up with a new type of weaponry for launching massive distributed denial-of-service attacks and computer viruses. The Chinese are already moving along with this.

In addition to the possibility of cybercombat between nations, the military acknowledges that terrorists without the backing of any country can potentially use cyberweapons to disrupt U.S. telecommunications or banking systems that are largely electronic. That's one reason the U.S. Space Command is joining with the FBI to build an information-warfare strategy.

This requires a close relationship between military and law enforcement. The FBI will have to help determine if any cyberattack (see sidebar, "Cyberoffense Mired in Cold War") suffered by U.S. military or business entities calls for a military or law enforcement response.

Cyberoffense Mired In Cold War

The absence of a catastrophic cyberattack against the United States has created a false sense of cybersecurity and has allowed costly Cold War-era Pentagon programs to siphon money from critically needed information technology and security programs. The United States is still mired in a Cold War-era defense-spending mentality.

The rapid advance of IT has created real and potentially catastrophic vulnerabilities. The consequences of a cyberterrorist attack "could be devastating."

Eye of the Beholder

However, senior security officials are battling a perception problem, according to IW experts. Without a clear-cut example of an "electronic Pearl Harbor," where a surprise cyberattack cripples financial markets and other critical systems, it's difficult to convince top military and political leaders that IT research and development should be a bigger priority in the budget process.

Cyberterrorism is not an abstract concept. Although attacks historically have been labeled as "nuisances," that may not be the correct way to look at the problem. The government is dealing with an "enormous educational deficit" when it comes to IT security.

Part of the problem is the fact that the Defense Department remains committed to lobbying Congress for money to pay for programs such as the F-22 Joint Strike Fighter instead of increasing funding for IT programs. That is not affordable even in this age of surpluses. DOD's assumptions about future budget gains are "wrong."

More money should be spent on advanced sensors, precision-guided weapons, and other IT programs. That type of investment would preclude the need to buy costly systems such as the F-22.

But even events such as the outbreak of the "love bug," which reportedly cost the U.S. economy billions of dollars, have not convinced people in and out of government that the problem is real. Usually, when a major crisis costs people a lot of money, it leads to many visits to Capitol Hill and requests for help. But, that never happened after the love bug outbreak.

Some experts have questioned the government's liberal use of the term *terrorism* to describe acts of mass disruption on the Internet. However, when asked about the seeming lack of interest in cyberattacks by well-known terrorists such as Osama bin Laden, a senior White House official said the focus should not be on what bin Laden does or does not do, but on being proactive and understanding that a major attack may be coming.

The U.S. is attempting to be proactive. But, many believe that the U.S. is going to get seriously nailed.

The National Security Agency is one of the federal entities that has taken a proactive approach toward security cooperation between government and industry. But one of the biggest challenges facing the nation, highlighted during the love bug incident, remains to be convincing industry that security is as important as making money.

Vendors and users have to treat information assurance as a fundamental precept of doing business. It has to become part of the business case [7].

The Internet is ubiquitous. It allows attacks from anywhere in the world. Attackers can loop in from many different Internet providers.

Note: A cyberattack can include espionage using wireless computer networks.

It could start across the street but appear to be coming from China. And something that might look like a hacker attack could be the beginning of cyberwarfare.

The growing bullets-and-guns conflict in the Middle East between Israel and the Palestinians, with Islamic supporters elsewhere, is being accompanied by cyberattacks from each side against the other. It's serious enough that the FBI issued an alert about it to the U.S. Space Command, giving U.S. forces warning that the action on the cyber front could affect them as well [7].

4. OFFENSIVE CONTAINMENT INFORMATION WARFARE TOOLS AND TACTICS

Of all the activities that have emerged with the evolution of IW and information operations, Command and Control Warfare (C2W) has attracted the most attention. The U.S.'s approach to C2W is comprehensive. This country has committed substantial resources to the development of technologies, doctrine, strategies, and organizations that will equip it to meet an information threat in any future conventional war. Countries like Australia, however, like most non-superpower nations of the world, will not be able to commit the substantial resources needed to follow the American model. Therefore, the general approach discussed in this chapter is tempered by the economic realities that will dictate the degree to which mid-level powers can invest in their own strategies.

Note: Command and Control Warfare (C2W) is the approach to military operations that employs all measures (including but not

> limited to Electronic Warfare (EW), military deception, psychological operations [PSYOPS], operations security, and targeting), in a deliberate and integrated manner, mutually supported by intelligence and ITS, to disrupt or inhibit an adversary's ability to command and control his or her forces while protecting and enhancing our own.

C2W is the war-fighting or tactical application of MIWT and is usually aimed at a specific and defined battlespace, although it may be conducted in conjunction with other MIWT that may be focused on strategic information targets. There are five individual elements of C2W, covering both offensive and defensive applications.

- Operations security
- Military deception
- Psychological operations
- Electronic warfare
- Targeting [7]

4.1 Operations Security

Operations Security (OPSEC) is a term that appears in many military documents in almost as many contexts, with several apparently different meanings. OPSEC is a process used for denying adversaries information about friendly disposition, intentions, capabilities, or limitations. It requires the employment of specialist equipment, including software, the adoption of suitable procedures, and most important, the development of a pro-security organizational culture. OPSEC is equally important as a defensive posture as it is in developing offensive strategies. By denying a potential enemy an understanding of the capabilities of friendly systems, possible hostile C2W will be more likely to miscalculate the friendly information capabilities and be ineffective [7].

4.2 Military Deception

Military deception is used to inject ambiguity and create false assessments in the decision- making process of the enemy. The objectives of employing military deception are to create a false deduction of friendly intentions, capabilities, and/or dispositions by the enemy. The target of deception is the enemy decision maker, that is, the individual who has the necessary authority to make a decision. There is no point influencing a decision if, in the event of ambiguity, the decision maker passes the decision to a higher authority. In this case, the higher authority must also be the target of deception [7].

4.3 Psychological Operations

Psychological Operations (PSYOPS) are operations that are planned activities in peace and war directed to enemy, friendly, and neutral audiences to influence attitudes and behavior affecting the achievement of political and military objectives. The objective of PSYOPS is to cause enemy, friendly, and neutral personnel to act favorably toward friendly organizations. PSYOPS have been used throughout history to influence adversary leaders and groups. The expansion and development of information technology (IT), and associated global media coverage, has enhanced modern PSYOPS opportunities [7].

4.4 Electronic Warfare

Electronic Warfare (EW) is the military action involving the use of electromagnetic energy to determine, exploit, reduce, or prevent hostile use of the electromagnetic spectrum. This action retains friendly use of the electromagnetic spectrum [7].

4.5 Targeting

Targeting is not just a process, nor is it just focused on destructive ends. Targeting is a capability that emphasizes the requirement to collect, process, and interpret information regarding decisive points in an enemy's command and control system; and, then selects the most effective option of incapacitating them. There are many hard- and soft-kill options available to a commander. Soft-kill options include the use of EW, strategic computer operations and information weapons, whereas hard-kill options refer to the various means of physically destroying targets.

Hard or soft destruction requires the capability to remove selected targets from an enemy's order of battle. These targets include vital communication nodes, national infrastructure, vital personnel, and specific military equipment. Destruction may be achieved by any arm of the military. Physical destruction has the highest risk associated with its application, and, unlike the other elements of C2W, physical destruction tends to be permanent, that is, buildings are destroyed and people are killed. This can be either a desirable or undesirable outcome, and so must be considered when strategies are being developed. The diplomatic recovery time for physical destruction is usually considerably longer than that of the other elements. Accordingly, even though it is often the most effective method of demonstrating resolve, physical destruction is generally used as a last resort.

However, a commander must have the option of employing hard- and soft-kill options to accomplish a desired C2W effect [7].

4.6 The Objective of C2W

Until the 1991 Gulf War, the C2W elements had rarely been used in conjunction with each other to specifically target an enemy's ability to command and control its forces. In the Gulf War post-mortem, the advantages of combining the five elements in pursuit of a single objective were realized and true C2W was born.

The ultimate objective of C2W is to decapitate the enemy's command structure from its body of combat forces while ensuring the integrity of friendly command and control systems. C2W aims to reduce the uncertainty of combat by creating a battlespace that becomes more predictable for friendly forces as the C2W effort increases, while becoming exponentially less predictable for the enemy. C2W activities seek to lift the fog of war for friendly forces while thickening the fog for the enemy. C2W strategies focus the five elements specifically on the decision cycles of both friendly and enemy forces. Therefore, the aim of C2W is to gain, maintain, or widen a gap in the effectiveness of C2 in favor of friendly forces throughout a campaign and particularly at decisive points in a battle [7].

4.7 C2W And The OODA Loop

The often-quoted Observation, Orientation, Decision, Action (OODA) loop has been adopted as the focal point of C 2W. The concept of the OODA loop had its origins in the Korean war where an American pilot identified the advantages of having good visibility and sensitive controls on board the US Sabre jet fighters. Although the Russian MiG 15s were faster, more powerful, more maneuverable, and could sustain greater bank angles, the American jets were consistently victorious in air to air engagements. The US pilots simply had a shorter total period between observing an event, orientating themselves to the possible ramifications of the event, making a decision, and acting. The value of a relatively short decision cycle was realized. Since the inception of air to air combat, staying inside the enemy's decision loop has been a consistent objective. This has been a recognized objective of many forms of warfare.

The OODA loop concept is now applied to most aspects of modern warfare, from land maneuvers to strategic missile developments. The OODA loop can also be seen to operate in the business world. Those who are quick to observe an opportunity, recognize the opportunity, and exploit the opportunity, are more frequently the successful or victorious business persons. The OODA loop theories can be found in the heart of modern C 2 systems and, consequently, in modern C 2W strategies. Successful C 2W operations will, therefore, increase the enemy's decision cycle (his or her OODA loop) to such a point that he or she will become increasingly vulnerable to attack [7].

4.8 C2W in the Gulf War

In the 1991 Gulf War, the Coalition forces attacked the Iraqi C 2 system from the outset. Even before the war had commenced, EW, PSYOPS, and deception were employed for influencing the Iraqi people and hierarchy. During the first hours of the air attacks in Iraq, ITS and communication devices were targeted, and, in many cases physically destroyed, leaving the huge force that had occupied Kuwait completely cut off from the commanders in Baghdad. The Iraqi air defense system was virtually shut down by coalition activity within hours of the commencement of Operation Desert Storm. The extant Iraqi air defense system was amongst the most extensive in the world. Shutting such an extensive system down with apparent ease was a significant achievement and the result of a calculated offensive involving all of the C 2W elements. This early success gave the coalition forces air supremacy. In turn, this supremacy significantly reduced the potential for coalition air fatalities and allowed the coalition air forces to strike Iraqi ground targets almost at will. The coalition forces effectively destroyed the ability of the Iraqi military commanders to observe, and the Iraqi OODA loop was significantly increased.

In the 2003 Operation Iraq Freedom, defense of friendly C 2 systems and attacks on enemy systems was of paramount importance. The U.S. had the smallest decision cycle, and thus had a decisive advantage. A confused army leads to another's victory [7].

5. DEFENSIVE PREVENTIVE INFORMATION WARFARE TOOLS AND TACTICS

Some eight years after the military pioneered intrusion-detection systems, the Defense Department now requires its massive wireless networked systems to be protected by round-the-clock intrusion-detection monitoring to defend against hacker or denial-of-service attacks. The Defense Department is developed a policy that mandates the use of intrusion-detection systems in all military wireless networks.

The Defense Department has more than 69,000 wireless computer networks that handle everything from weapons systems command-and-control to inventory to payroll. Roughly 15% of Defense Department wireless networks, such as satellite links, are considered mission-critical.

Under this draft policy, every Defense Department entity needs to have a wireless computer network-detection service provider, which could be a Defense Department entity or a commercial entity. Thus, the Defense Information Systems Agency (DISA) is responsible for defining the intrusion-detection plan. Whether the Navy, Army, or Air Force should buy commercial intrusion-detection software or entrust wireless network protection to an outside service provider should be decided on a case-by-case basis.

The military helped pioneer intrusion-detection systems by building its own software from scratch back in 1996. But since then, various parts of the military have deployed products from vendors that include Internet Security Systems, Symantec, Cisco [2], and Network Ice. Today, still only a small percentage of the military's overall wireless networked systems are guarded by any form of intrusion detection.

When the final decision on the mandatory intrusion-detection systems will arrive is still unclear. But deliberations taking place among the military's Joint Chiefs of Staff underscore their determination to do whatever it takes to prevent hackers and denial-of-service attacks from disrupting its wireless networks.

Some defense-related agencies, such as the secretive National Security Agency (NSA) in Fort Meade, Maryland, already require round-the-clock monitoring of computer hosts and wireless networks. Every system within NSA is monitored. In the Defense Intelligence Agency, it's the same sort of situation.

One difficulty in deploying intrusion-detection software is that it must be regularly updated to include new attack signatures, because new hacker exploits are discovered all the time. In addition, intrusion-detection software can record "false positives," a false alarm about trouble, and software occasionally needs to be fine-tuned to work correctly. These types of challenges, along with the difficulty in hiring security experts to manage intrusion detection, is spurring security services in which intrusion detection is done remotely in the service provider's data centers or with hired help on-site.

Not all attempts by the federal government to put large-scale intrusion-detection systems in place have succeeded. Back in 2000, President Clinton unveiled his goal of creating the Federal Intrusion Detection Network as part of what was called the National Plan for Information Systems Protection. FIDNet, as it was called, was envisioned by the White House as a government-wide intrusion-detection wireless network to monitor activities across civilian and defense wireless networks.

The idea, though, generated a firestorm of criticism from civil liberties groups that argued FIDNet's monitoring of citizens would constitute an invasion of privacy [3]. Although the General Services Administration issued a draft RFP for FIDNet, GSA indicates the idea has been shelved.

Others are just not sold on the idea of outsourcing security to services providers. They've opted not to go with managed security. With managed security services, you're giving away the keys to the castle in some respects. Therefore, any organization that wants to take advantage of managed security services has to share detailed knowledge about its operations so that intrusion-detection systems can be properly used [7].

6. DEFENSIVE RUINOUS INFORMATION WARFARE TOOLS AND TACTICS

In 2002, the Pentagon formed five technology centers, that are staffed by reservists who work in the private sector by day and spend one weekend per month defending the Defense Department against cyberattacks, through the use of defensive ruinous information warfare tools and tactics. The Deputy Secretary of Defense approved a plan that would establish five Joint Reserve Virtual Information Operations and information assurance organizations (JRVIO). The centers' mission is to ensure that American war fighters dominate the computer information domain in future conflicts, according to the Pentagon.

Information operations has emerged as an area that is extremely well suited to integration of reserve capabilities. Members of the reserves and National Guard are often way ahead by the very nature of their civilian employment, trained in their workplaces to exploit technology.

The Defense Department has long been battling a high-tech brain drain spurred by a booming economy and the lure of higher-paying jobs in the private sector. The change has made the National Guard and reserves a repository of high-tech skills. At the same time, the Pentagon is facing an increase in cyberattacks and intrusions and has increased its focus on using cybertactics to fight future conflicts. The teams could be involved in a wide range of efforts, including enemy wireless computer network attacks, defense of U.S. critical infrastructures, psychological operations, intelligence support, vulnerability assessments, and reviews of Pentagon Web sites for sensitive information.

The Pentagon expects 526 reserve officers and enlisted personnel to staff the five JRVIOs during fiscal 2005 and 2006 in Maryland, Virginia, and Texas. However, from 2007 to 2011, that number is expected to expand to more than 1,000.

The initiative is a result of a two-year Pentagon study called "Reserve Component Employment 2006." That study recommended the formation of a cyberdefense unit that would consist of people with IT skills who could work in different locations instead of at a single center. The study also urged the department to recruit high-tech savvy people from the private sector [7].

7. DEFENSIVE RESPONSIVE CONTAINMENT INFORMATION WARFARE TOOLS AND TACTICS

One of the more recent additions to the military commander's toolbox are defensive responsive containment information warfare tools. Computers and associated technology have helped change the face of modern information warfare tactics by providing the capabilities to generate and process massive amounts of data, and disseminate the resultant information throughout the battlespace. However, computers provide more than just an information-processing capability. They may also be used as weapons in their own right. The most common examples of computer operations include hacking, virus planting, and chipping. These techniques are primarily aimed at targeting the enemy's broad information environment. However, they may also be used to attack the enemy's computer-based weapon systems and computer-based platforms, such as "fly-by-wire" aircraft. Although generally strategic in nature, computer operations may be applied to the tactical and operational components of the conventional warfare environment, either in support of C2W operations or in direct support of air, land or sea operations [7].

7.1 Hacking

The term computer hacker is now synonymous with computer criminal although, arguably, this merging of terms is not justified. Someone who uses a computer to rob a bank is a criminal, not a hacker. The genuine computer hackers are still doing what the original computer hackers were doing 43 years ago—simply exploring the bounds of computer science.

Unfortunately, exploring today's computer science often means entering other people's systems. There are many computer hackers around the world who enter other people's systems on a daily basis. Most simply gain access to the systems, "snoop" around for a while, and leave. Some hackers like to explore the logic flow in systems. A few like to exploit these systems for either their own gain or simply to make life difficult for the users of that system. The genuine hackers, while invading system privacy, rarely damage the systems into which they have hacked. However, most users of systems understandably find it an unacceptable invasion of their privacy to have people intruding into their systems.

Hackers present a genuine problem to most organizations today, and a specific threat to military security. Hackers have historically found the challenge of breaking into so called "secure" military systems one of the more satisfying aspects of their hobby. Accordingly, the first and foremost aim of any information strategy for military forces must be to defend their own system integrity.

Once access is gained into a system, hackers can generally manipulate whatever files they wish. They will often set up personal accounts for themselves in case they wish to return again in the future. A hacker can, of course, collect very important information. In the business domain, intelligence can be gained about a competitor's product. In the government service domain, sensitive personal information can be obtained (or altered), which can later be used against individuals. In the military domain, classified information such as capabilities, vulnerability, strategies, and dispositions may be extracted or manipulated. A hacker can also change the file structure, amend the logic flow, and even destroy parts of the system.

Hacking is no longer simply a pursuit of misfits and computer scientists; it is now a genuine method of obtaining information by government agencies, criminals, or subversive organizations. There have been several reports about government sponsorship of such activity. Many of the world's secret security organizations are now passing industrial secrets to their nation's domestic businesses. The basic tool kit of today's industrial spy contains a PC and a modem. The industrial spy is simply a hacker who intrudes into someone else's computer system and then exploits the information obtained. Neither domestic nor international laws adequately address all of the issues surrounding hacking. Therefore, in the unlikely event that hackers are caught, in many situations prosecution is impossible.

The impact on those involved in developing MIWT is that hacking presents a genuine threat to the security and integrity of both military and civilian information systems. Defense against hacking can be successful to varying degrees. Most defensive strategies are system-dependent; therefore, listing them in this chapter would be pointless. However, defense against hacking needs to be considered by anyone who manages or operates an information technology system.

The other reason that national security forces should become involved in hacking is the potential benefits that can be derived by employing hacking techniques as an offensive tactic. Intelligence collection against information stored in an enemy's databases as well as the specific system capabilities, vulnerability, and architecture, can be accomplished successfully using hacking techniques. In future wars, information derived from hacking will form a large part of intelligence databases and, thus, manipulation of the enemy's decision-making support systems will become routine [7].

7.2 Viruses

A virus is a code fragment that copies itself into a larger program, modifying that program'. A virus executes only when its host program begins to run. The virus then replicates itself, infecting other programs as it reproduces. Protecting against computer viruses has become a part of using modern ITS. Viruses are passed from computer to computer via disks and reportedly via the more recent practice of electronic file transfer, such as e-mail. Although statistics concerning viruses are often difficult to substantiate, some specialists estimate that there are as many as 11,233 viruses currently existing on the Internet, with cures being available for only 4,583. Although virus screening software should prevent known viruses being brought into a system, they will not prevent all virus attacks. The most effective method of minimizing the risk of virus attack, and minimizing the damage caused by viruses in the event of an attack, is by employing sound and rigorous information-management procedures.

Isolating Internet systems from operating systems where practical is vital, and minimizing computer-to-disk-to-computer transfers, particularly if the origin of that data is the Internet, will reduce the chances of picking up a virus. The use of the most recent antivirus software and the screening of disks every time that they are placed in a computer will reduce the risk of disk infections being passed onto systems. Careful selection and management of passwords may deter a potential intruder from accessing a system and planting a virus, while the maintenance of comprehensive system back-ups can minimize the impact of viruses, should one find its way onto a system. Viruses, however, can also be backed-up and a dormant virus can infest any back-up files and can be reintroduced when a system is recovered. Accordingly, a layered back-up strategy is imperative. Antivirus strategies are aimed at minimizing the chances of getting a virus and minimizing the damage that viruses can cause if they are introduced. Users of today's ITS must be aware of the virus threat. Simple procedures will often be enough to avoid viruses, but a single failure to comply with antivirus procedures can result in systems becoming inoperable.

Virus planting is clearly a suitable and attractive weapon for military forces and is a valuable addition to the offensive information operations inventory. If a simple virus can be injected into the systems of a potential enemy, the need to expend effort in physically attacking that system may be eliminated [7].

7.3 Chipping

Most people are aware of the vulnerability of software to hostile invasions, such as a virus attack. Few, however, are aware of the risk to the essential hardware components of an ITS. Chipping is a term that refers to unexpected events that can be engineered into computer chips. Today's chips contain billions of integrated circuits that can easily be configured by the manufacturer so that they can initiate unexpected events at a specific time, or at the occurrence of specific circumstances. This may explain why some electronic goods fail a short time after the warranty has expired. There is almost no way of detecting whether a chip contained within a piece of equipment has been corrupted.

One way to minimize the risk of chipping is to self-manufacture all important chips, such as those that are used as part of an aircraft's flight control system. Economically, this is often not feasible. Most chips used within today's high-technology equipment are manufactured in countries where labor costs are low. Establishing an indigenous manufacturing capability would increase the cost of acquiring the equipment. A risk assessment must be made when purchasing vital equipment from overseas, by comparing the risk of vital equipment failing once hostilities commence to the cost of producing chips internally or developing rigorous quality control of imported chips.

Chipping represents a simple way to develop a conventional military advantage by those countries that regularly export military equipment. In the event of any hostilities with recipients of their "chipped" equipment, that equipment may be incapacitated without having to use conventional force. This makes economic as well as military sense. The legal and ethical aspects are a separate issue.

There are many other computer weapons that can be used in conjunction with or instead of chipping, viruses, and hacking. These weapons have many different descriptive names such as "worms," "trojan horses," and "logic bombs," and are commonplace in today's information society. They are all examples of computer operations that may be adapted to suit the information-warfare environment. A detailed description of all of these techniques is beyond the scope of this chapter. Suffice to say that computer weapons should be an integral part of any information-warfare operations strategy. They should be considered as valid alternatives to conventional weapons both in offense and defense [7].

8. COUNTERING SUSTAINED TERRORIST INFORMATION WARFARE TACTICS

Terrorism is, among other things, a weapon used by the weak against the strong. The United States has moved into the 21st century as a preeminent, global power in a period of tremendous flux within societies, among nations, and across states and regions. Terrorism will accompany changes at each of these levels, as it has in other periods of flux in the international environment. To the extent that the United States continues to be engaged as a global power, terrorism will have the potential to affect American interests directly and indirectly, from attacks on U.S. territory (including low-probability but high-consequence "superterrorism" with weapons of mass destruction) to attacks affecting the U.S.'s diplomatic and economic ties abroad, or the U.S.'s ability to maintain a forward military presence or project power in times of crisis. The United States will also have a unique, systemic interest in terrorism as a global problem (including acts of "domestic" terrorism confined within state borders that make up the bulk of terrorism worldwide) even where the United States is not directly or even indirectly targeted. In one way or another, terrorism can affect the U.S.'s freedom of action, not just with regard to national security strategy narrowly defined, but across a range of compelling issues, from drugs and money laundering to information and energy policy.

Many of the U.S.'s high-priority national objectives have been shaken by the recent experience of terrorism. The Oklahoma bombing and World Trade Center and Pentagon 9-11 terrorist attacks, struck at the U.S.'s sense of security within its borders. Attacks against U.S. forces in Saudi Arabia raise questions about the U.S.'s strategy for presence and stability in an area of critical importance for world energy supply. The U.S. embassy bombings in Kenya and Tanzania, and the U.S.S. Cole in Yemen raise questions about the exposure that comes with active engagement in world affairs, and point to the risks of privately sponsored terrorism. The assassination of Prime Minister Rabin, the increased campaign of suicide bombings in Israel, the death of Yasser Arafat, and the refusal of the Bush administration to deal seriously with the Palestinians, continues put the Middle East peace process in serious jeopardy, threatening a critical and long-standing U.S. diplomatic objective. Elsewhere, terrorism has destabilized allies (in Saudi Arabia, Egypt and Turkey), and has rendered counternarcotics relationships difficult (in Colombia and Mexico). Where societies and regions are fundamentally unstable, and where political outcomes are delicately poised, terrorism will have a particular ability to affect strategic futures [7].

8.1 Overall Observations

Most contemporary analyses of terrorism focus on terrorist political violence as a stand-alone phenomenon, without reference to its geopolitical and strategic context. Similarly, counterterrorism policy is rarely discussed in terms of its place in broader national security planning. Prior to the specter of "superterrorism," using weapons of mass destruction (WMD), terrorism, however horrible, never posed an existential threat to U.S. security. With the important exception of WMD, terrorism still does not pose a grave threat to America's future as it does to many other societies around the world. But many types of terrorism do pose a threat to U.S. interests, from homeland defense to regional security and the stability of the international system. As a global power, the U.S. perspective on terrorism is bound to differ in substantial ways from that of others, including allies such as Britain, France, and Israel, whose experiences provide lessons, but not necessarily direction for U.S. counterterrorism policy. In light of the preceding IW arsenal and tactics analysis of the military, certain overall sustained terrorist information-warfare tactics observations stand out:

- Terrorism
- Geopolitics of terrorism
- Counterterrorism versus new terrorism
- U.S. Exposure
- Comprehensive counterterrorism strategy [7]

8.1.1 Terrorism

Terrorism is becoming a more diverse and lethal problem. Contemporary terrorism occupies an expanded place on the conflict spectrum, from connections to drug trafficking and crime to its use as an "asymmetric strategy" by state and non-state adversaries in a war paradigm. For a variety of reasons, primarily the rise of religious cults with transcendent agendas, but also the hardening of established political groups, terrorism has become more lethal. With the potential for catastrophic terrorism using weapons of mass destruction, lethality could increase dramatically [7].

8.1.2 Geopolitics Of Terrorism

The geopolitics of terrorism are changing. Over the next decades, the prevailing image of terrorism affecting U.S. interests as a problem emanating largely from the Middle East is likely to be overtaken by a more diverse set of risks. The Balkans, the former Soviet Union, and Latin America are set to emerge as significant sources of terrorism aimed at or affecting U.S. civilian and military activities. Moreover, the vast bulk of global terrorism will continue to be confined within the borders of affected states. More anarchic futures in the Third World could fuel this type of terrorism, threatening America's systemic interests as a global power and placing constraints on the U.S.'s international engagement [7].

8.1.3 Counterterrorism Versus New Terrorism

Much counterterrorism experience is losing its relevance in light of the "new" terrorism. Many established images of counterterrorism policy,especially the use of force against state sponsors, are losing their relevance as traditional forms of terrorist behavior and organization (largely a product of the ideological and national liberation movements of the 1960s–1980s) give way to new patterns. The new terrorism often lacks a detailed political agenda against which the use of violence can be calibrated. It is, therefore, more lethal. It is less hierarchical in organization, more highly wireless networked, more diffuse in membership and sponsorship, and may aim at disruption as well as destruction. The absence of clear-cut sponsorship, above all, will complicate the task of deterrence and response. It will also compel a reorientation of policy to target nonstate sponsors and individual suspects [7].

8.1.4 U.S. Exposure

Foreign experts see U.S. exposure increasing, but view the problem in narrower terms. A survey of expert British, French, and Israeli perspectives yields a gloomy outlook with regard to U.S. exposure to terrorist risks, which are widely seen as deepening, particularly with regard to U.S. forces in the Gulf. Policy makers and observers in these allied countries are not surprisingly focused on specific national risks, few of which are analogous to risks facing the United States at home and abroad. With the limited exception of France, which shares a global and expeditionary outlook in strategic terms, terrorist challenges are generally viewed in narrower, but starker, terms. Notably, experts in all three countries share a degree of skepticism about technology as a "solution" in counterterrorism [7].

8.1.5 Comprehensive Counterterrorism Strategy

A comprehensive counterterrorism strategy should have core, environment shaping, and hedging components. Treating terrorism as one of many national security challenges suggests a multidimensional approach. Core, longer-term strategy must address the political, economic, and social roots of international terrorism; make deterrence relevant to nonstate actors as well as state sponsors; and reduce the risk of truly catastrophic terrorism using weapons of mass destruction. The environment- shaping aspect aims to create conditions for successfully managing terrorist risks, making terrorism more transparent, shrinking "zones of chaos," harnessing key alliances to the counterterrorism effort, reducing U.S. exposure, and cutting off terrorism's resources. Finally, the United States can hedge against inevitable terrorism by hardening policies as well as targets, and preparing to mitigate the effects of increasingly lethal terrorist acts [7].

8.2 Implications for Military Strategy and the U.S. Air Force

In many instances, air and space power will not be the best instruments in the U.S. counterterrorism arsenal, and air power will rarely be used independently against terrorism. However, air and space power can play a role in intelligence and covert action. There will also be instances, as in the past, where air and space power will be instruments of choice in the fight against terrorism. Moreover, terrorism and counterterrorism policy are changing in ways that will significantly affect the future contribution of air- and space-based instruments.

Events in Sigonella and Afghanistan as well as Operation El Dorado Canyon may be key models for the future. Air power in the service of counterterrorism will include, but will also go beyond, the surveillance and punishment of state sponsors. Deterrence and response will likely evolve in the direction of a more "personalized" approach, emphasizing the monitoring and attack of key nodes in terrorist wireless networks and the forcible apprehension of terrorist suspects—with or without the cooperation of local states. Future demands on air power may be driven as much by requirements for intercepting and extracting suspects as by the need to attack terrorist training camps and strike regimes supporting the export of terrorism.

Air and space power will help make terrorism (an inherently amorphous phenomenon) more transparent. The ability to identify and to target terrorist-related activity and to help expose terrorism and its sponsors for policy makers and international opinion will be key contributions of air- and space-based assets. As terrorism becomes more diffuse and its sponsorship increasingly hazy, finding the "smoking gun" will become more difficult, but essential to determine strategies and build a consensus for action. Space-based sensors, surveillance by UAVs, and signals intelligence (SIGINT) will facilitate the application of air power and other instruments in the service of counterterrorism.

Gaining leverage in addressing the new terrorism will be a key strategic and technical challenge. Future requirements for counterterrorism will be part of a broader need to tailor air and space power to challenges posed by nonstate actors, including networks of individuals. At the same time, policy instruments, including air and space power, will need to concentrate on detecting and preventing the use of weapons of mass destruction by terrorists—whether as a stand-alone apocalyptic act or as a low-tech delivery system in the hands of adversaries.

Much terrorism (and counterterrorism action) will focus on urban areas, with strong political and operational constraints. Terrorism is increasingly an urban phenomenon, worldwide. One explanation for this is that the political fate of most modern societies is determined by what happens in cities. Terrorists seeking to influence political conditions have many incentives to attack urban targets. Terrorists with transcendental objectives will, similarly, find symbolic and vulnerable targets in urban settings. The use of air power in a counterterrorist mode faces the more general problem of operating in an urban environment (the difficult Israeli experience in Beirut and South Lebanon is instructive). Terrorists and their facilities will be difficult to locate and target. Operations against them or to rescue hostages will pose severe challenges for the use of air power, not least the risk of placing uninvolved civilians in harm's way. The viability of air power as an instrument in such settings may depend on the capacity for discriminate targeting and the use of less-than-lethal technologies.

Air power's pervasiveness and speed are advantages in the face of transnational and transregional terrorism. In an era in which terrorist acts may take place across the globe and where sponsors cross national and regional lines, counterterrorism strategies will become "horizontal" in character. Where terrorists and their sponsors can be identified and attacked with purpose, the global sight and reach of air- and space-based assets will be valuable to national decision makers.

Air and space power will have a synergistic effect with other counterterrorism instruments. Air and space power can be used in concert with covert action, diplomacy, economic instruments, and joint military operations. The notion of "parallel warfare," developed in relation to attacks on infrastructure in war, will also be relevant to counterterrorism operations.

Operations using a range of instruments can be designed to act, in parallel, on terrorist supporters, terrorist infrastructure and networks, and the terrorists themselves [7].

9. DEALING WITH RANDOM TERRORIST INFORMATION WARFARE

During the 1970s and 1980s, political extremism and terrorism frequently focused on 'national liberation' and economic issues. The collapse of the Soviet bloc, and the ending of its covert funding and encouragement of terrorism led to a decline in the militant and violent left-wing terrorist groups that were a feature of the age.

The 1990s through the present time have seen the development of a new terrorism: Random Terrorist Information Warfare. This is not to say that state-backed terrorism has ceased, but rather that the spectrum of terrorism has widened. This new extremism is frequently driven by religious fervor, is transnational, sanctions extreme violence, and may often be millenialist. The new terrorism may seek out military or government targets, but it also seeks out symbolic civilian targets, and the victims have mostly been innocent civilians (Alfred P. Murrah Building, Oklahoma City; World Trade Center, New York; AMIA Headquarters, Buenos Aires, etc.).

Growing concern about this new terrorism has been paralleled by concern about the employment of the new information and communication technologies (ICTs). ICTs offer a new dimension for political extremists and terrorists. They allow the diffusion of command and control; they allow boundless new opportunities for communication; and they allow the players to target the information stores, processes, and communications of their opponents. The sophistication of the modern nation-state, and its dependency on computer-based ICTs, make the state ever more vulnerable.

The use of ICTs to influence, modify, disrupt, or damage a nation-state, its institutions, or population by influencing the media, or by subversion, has been called "netwar." The full range of weapons in the cyberspace armory can be employed in netwar—from propaganda campaigns at one level to interference with databases and wireless networks at the other. What particularly distinguishes netwar from other forms of war is that it targets information and communications, and may be used to alter thinking or disrupt planned actions. In this sense, it can be distinguished from earlier forms of warfare—economic wars that target the means of production, and political wars that target leadership and government.

Netwar is, therefore, of particular interest to those engaged in non-military war, or those operating at sub-state level. Clearly, nation-states might also consider it as an adjunct to military war or as an option prior to moving on to military war. So far, however, it appears to be of greater interest to extremist advocacy groups and terrorists. Because there are no physical limits or boundaries, netwar has been adopted by groups who operate across great distances or transnationally. The growth of such groups, and their growing powers in relation to those of nation-states, suggests an evolving power-based relationship for both. War in the future is more likely to be waged between such groups and states rather than between states.

Most modern adversaries of nation-states, in the realm of low-intensity conflict—such as international terrorists, single-issue extremists, and ethnic and religious extremists—are organized in wireless networks, although their leadership may sometimes be hierarchical. Law enforcement and security agencies, therefore, often have difficulty in engaging in low-intensity conflict against such wireless networks because they are ill-suited to do so. Their doctrine, training, and modus operandi has, all too often, been predicated on combating a hierarchy of command, like their own.

Only now are low-intensity conflict and terrorism recognized as "strategic" threats to nation-states, and countries that, until very recently, thought that terrorism was something that happened elsewhere, have become victims themselves. The Tokyo subway attack by the Aum Shinriko, the Oklahoma City bombing and the 9-11 terrorist attacks, would have been unthinkable a generation ago. Not only was the civil population unprepared but also the law enforcement population. And this was true despite clear warning signs that such attacks were in the offing.

Cyberspace is becoming a new arena for political extremists: The potential for physical conflict to be replaced by attacks on information infrastructures has caused states to rethink their concepts of warfare, threats, and national assets at a time when information is recognized as a national asset. The adoption of new information technologies and the use of new communication media, such as the Internet, creates vulnerabilities that can be exploited by individuals, organizations, and states.

Also, the arrival of the Internet has provided the first forum in history for all the disaffected to gather in one place to exchange views and reinforce prejudices. It is hardly surprising, for example, that the right-wing militias favorite method of communication is e-mail and that forums on the Internet are the source of many wild conspiracy theories that drive the media.

Preeminent amongst the extremists and terrorist groupings who have entered cyberspace faster and more enthusiastically than others, are the Far Right, that is white supremacists and neo-Nazis, and radical Islamists.

Others, such as eco-extremists and the Far Left appear to be slower in seizing the opportunities available.

What characterizes these two groupings are their transnational natures. The Far Right is increasingly active in the USA and Europe, but, in contrast to its ideological roots in the 1920s and 1930s, it seeks now to unite a white Anglo-Saxon, or European-originating, entity in a rear-guard action to oppose centralized democratic government and return to some imagined past world in which an armed, racially pure, white man can live untroubled by the police, the Inland Revenue, and the world banking system. The Islamist diaspora, now spread worldwide, seeks a return to divine-ruled states (or even one transnational state) in which all Muslims will live under the norms and laws of the Saudi Arabian peninsula in the first and second centuries of the Common Era. These types of organizations make them ideal users of wireless networks and proponents of netwar. Their ideas and their use of cyberspace will be further discussed in Chapter 24, "The Information Warfare Wireless Network Security Arsenal and Tactics of Terrorist Rogues."

Although the use of ICTs to enhance command and control and enhance communication is apparent among Islamist extremists and among the militia movement and Far Right in America, it is less so amongst Far Right and other extremists in other parts of the world. This clearly reflects the higher ICT access in North America. Fears by western governments that their national infrastructures may be a target for information warfare or cyberterrorism may be well-founded, but the evidence so far is that sub-state groups at least, use ICTs mainly for propaganda, secure communications, intelligence gathering, and funds management.

It has been noted by one observer that the Internet has not replaced other communications media for the Far Right, and that its largest use in this regard has been to advertise the sale of non-Internet-related propaganda, such as books, audiotapes, and videos. Nor has the Internet led to an increase in mobilization. The Seattle-based Coalition For Human Dignity observed that Far Right events in the United States, which were heavily promoted on the Internet only, were in fact failures.

For some on the American Far Right, the Internet has become an end in itself. Surfing the Net has replaced real action. It is a measure of how degenerate and weak the U.S.'s movement has become, that some people actually think this is a good thing. Not only do individuals want risk-free revolution, they now want people-free revolution. Here lies the great danger of the computer for everyone who uses it. It allows individuals to live and work interacting with a machine rather than with people.

However, it does not pay to be complacent; extremists and terrorists are increasingly information-technology literate. Unless law enforcement and national security agencies can move quickly, they will leave national infrastructures defenseless. For example, these terror networks understand the Internet, and know that law enforcement agencies lag far behind in both skills and available technologies.

Therefore, what is significant for the Far Right and its use of the Internet is that it possesses the potential to offer the relatively small numbers of people involved a means to communicate, develop a sense of common purpose, and create a virtual home symbolically—the Internet combines both intimacy and remoteness. These properties make it uniquely suitable for maintaining relationships among groups that are prone to attrition, because forms of association can be established at a social and geographical distance.

Although some futurologists warn of an electronic Pearl Harbor, the reality is that terrorists have not yet resorted to strategic information warfare. What is apparent, however, is that warfare is shifting toward attacking civilian targets and that sub-state terrorists and other extremists are increasingly targeting civilian infrastructures.

Increasingly, the perpetrators and the victims of netwar will be from the civilian sphere. It is, therefore, the civilian infrastructure that is the most vulnerable; the military can protect its own infrastructure, despite media reports that it is vulnerable and a constant victim of hacking.

Governments are becoming increasingly concerned about protecting their own national infrastructures, but global connectivity has grown to such an extent that it is now possible to talk only of a global informational infrastructure. There is only a global information infrastructure. There is no way to draw a line around the Continental United States and say that the information infrastructure belongs to the United States. This is true because there is no way to sever the United States from the information infrastructure that connects the rest of the world. What that means is that the U.S.'s infrastructure is accessible not only to their friends around the world but also to their potential foes. It is just as easy now to engage in a cyberattack from Tehran as it is from Pomeroy, Ohio [7].

9.1 Countering Sustained Rogue Information Warfare

Countering sustained rogue information warfare (IW) is envisioned as a new dimension of information warfare, bringing rogue conflict into the Information Age. Rogue IW offers combatants the ability to execute asymmetrical attacks that have nonlinear effects against an adversary. By targeting or exploiting information and information processes, an attacker can use limited resources to reap disproportionate gains. Furthermore, rogue

IW offers weaker enemies (even at the sub-state level) strategies alternative to attrition, an attractive feature especially when facing an opponent with significantly stronger conventional forces. Such potential adversaries could perpetrate a rogue IW attack against the United States, using relatively limited resources, exploiting the U.S. reliance on information systems. Targets of such attacks might include Command and Control (C2) networks, satellite systems [4], and even the power grids of the continental United States. Such an attack could potentially have a strategic impact on the national security of the United States.

In contrast, terrorism has been used by states and sub-state groups for millennia. As an instrument to pursue political or social objectives where the user lacks the strength or the political wherewithal to use conventional military means, terrorism has been especially attractive. The intended target of a terrorist act goes beyond the immediate victims. Terrorists create a climate of fear by exploiting the information dissemination channels of its target population, reaching many by physically affecting only a few. The United States experienced a tragic example of this effect in the 1983 bombing of the U.S. Marine barracks in Beirut; the USS Cole in Yemen in 2000; and, the 9-11-2001 terrorist attacks--where a small terrorist group, clearly weaker than the U.S. military, nevertheless executed an effective strategic attack against the United States.

The problem of rogue Information Warfare was not lack of capabilities, but of management and organization: The capabilities are out there already, they just are not being tapped. This "problem" has only recently emerged as a potentially new warfare area for most defense planners. The problem of terrorism, on the other hand, has been in the headlines and in the social consciousness for decades, especially since the technological advance of intercontinental flight. This part of the chapter, therefore, briefly examines these two phenomena conceptually, operationally and organizationally, seeking commonalties. If comparisons are substantiated as more than circumstantial, then the lessons that might be applied to rogue IW defense from successes and failures of 33 years of countering terrorism should be examined closely. Within the context of these comparisons, this part of the chapter will also attempt to ascertain whether there is an emergent structure or organization that suggests a "correct" approach to countering sustained rogue Information Warfare.

The bombing of the Murrah Building in Oklahoma City and the 9-11 terrorist attacks were one of many major events to remind the military that the continental United States no longer offers sanctuary from terrorism. Yet geographical borders probably will never offer sanctuary from rogue information warfare attacks. The military should organize and prepare for potential rogue IW attacks against them without necessarily having a formal definition and without having to experience a massive information attack. Establishing a rogue IW focal point involves a partial framing of the problem inasmuch as identifying key contributors to its solution. A wide-scale information attack could involve systems under the responsibility of agencies across the government, and even the commercial sector. A solution will draw on contributions from areas broader than simply military or law enforcement. In the case of the Oklahoma City bombing, organizations such as ATF and FBI investigated the incident, and FEMA responded with crisis mitigation using both federal and local resources. In a "digital OKC," who would take FEMA's place for crisis mitigation? Will local support be available? At present, no framework coordinates a response to rogue IW attacks, and establishing an ex post facto framework in response to an attack is unwise.

Clearly, rogue IW defense will demand many resources throughout the federal government. This does not, however, justify creation of an all-encompassing body tasked with jurisdiction and execution over all aspects of rogue IW.

For example, in the past, terrorism policies under President Reagan, suggested that such an organized U.S. counterterrorism agency (whether newly created or placed within an existing agency) would not have been feasible: This solution fails to take into account the nature of terrorism and the influence of bureaucratic politics. Terrorism is a complex phenomenon requiring a comprehensive response. No agency within the U.S. government possesses the vast array of capabilities needed to combat terrorism effectively. It would be difficult, if not impossible, to create a single department with the needed jurisdiction to control the U.S. response to terrorism, and would lead to even greater policy and process problems.

These problems are also inherent in organizing for rogue IW defense (IW-D). Furthermore, the distributed nature of the problem implies a distributed response from the respective agencies owning the appropriate capabilities. This distributed response, however, should be overseen by a higher office so that the left hand knows what the right hand is doing and that these complex activities are coordinated. An IW-D Oversight Office should be endowed with an independent budget and tasking authority to coordinate the decision-making process, identify capabilities needed to respond, and inform those agencies owning the capabilities as to their defensive rogue IW roles. Staffing this office would be "point members" of the represented agencies, who would then coordinate requirements within their respective agencies. This type of organization resembles, at a much broader range, the Joint Staff of the DoD, but with a budget as well as tasking authority for IW-D. Furthermore, the office could solicit and coordinate intelligence requirements from the various members of the intelligence community.

DoD has also articulated a similar concept for an office within the Executive Office of the President, organized for countering terrorism, as a potential focal point for the oversight of the U.S. antiterrorist program. This office would be a permanent body with a White House perspective; such a staff could monitor and coordinate activities of the line agency and departments; identify needed capabilities; identify special resources that might be mobilized if an international incident occurs; pull together current intelligence and ongoing analysis and research efforts; identify terrorist incidents; develop scenarios; and formulate plans. It would see to it that the necessary resources and capabilities are there when they are needed. In an actual crisis, it could function as a small battle staff for decision makers.

An Executive IW-D Oversight Office, as outlined in this chapter, would be in a prime position to identify and coordinate the investigative agencies, defense organizations, and all elements of the intelligence community that would be in positions to recognize and respond to attack. An IW-D Oversight Office might be led by a director having cabinet rank and a seat on the National Security Council (NSC). Such an office should also interact with the commercial sector, reflecting the extent to which commercial interests would be affected in IW, and the contribution industry can make toward solutions. Such interaction with the private sector might not be possible with existing agencies, due to the baggage that extant agencies might bring to the table.

In addition to reorganizing the bureaucracy, an IW-D Oversight Office might also reorganize priorities. Response strategies should not focus on protection as the only priority. One-hundred percent (100%) protection of an infrastructure is virtually impossible. Detection capabilities must drastically improve, along with crisis response and mitigation. These capabilities are fundamental to any indications and warnings (I and W) system, and are especially crucial in IW because protection is so fluid. Finally, not all crisis response and mitigation is technical. A policy for public awareness and education in the event of an information crisis (regionally coordinated in an organization similar to FEMA) might stave off panic, alert the public to measures they could take to assist, and lessen immediate public pressure on government officials to "do something." Such pressure in the history of countering terrorism has resulted in hasty responses of overbearing lawmaking and bloody reprisals.

The past 37 years have shown the United States the paradox that "low-intensity conflict" has posed to the world's mightiest military power. However, it is as yet unclear exactly where rogue IW falls in the spectrum of violence. As stated in the beginning, analogies can be useful, but at a certain point, relying on them for analysis becomes harmful. Although the organizational issues of rogue IW defense and counter-terrorism might be similar; this similarity might fail for solutions to other common issues. The unfortunate lesson of terrorism is that, as long as the United States is unwilling to cede their liberty to extortionate violence, there are no total solutions.

What the United States has achieved from the lessons of terrorism is improved crisis control, and policies that demonstrate an awareness of the complex nature of terrorism: its ability to affect any sector or jurisdiction of a free society, and the implications that come with those sobering realities. Information warfare has yet to emerge from its dogmatic stage, and still offers more slogans than lessons. Yet in retrospect of 34 years of fighting terrorism in a concentrated national and international effort, it is unclear whether an "electronic Pearl Harbor" would elicit a federal response other than the ad hoc overreactions and short-term task forces that have characterized U.S. counterterrorism policy. Such knee-jerk reactions have the potential to do much greater harm in IW than they have in countering terrorism: Heavy-handed, short-sighted, and hasty government measures in the information space might have unintended consequences ranging from stymied economic development to unconstitutional regulation to disastrous technical failures. Preempting an rogue IW attack with a multiagency policy of coordination could save the United Statesw from their adversaries, and it might even save them from themselves [7].

9.2 Fighting against Random Rogue Information Warfare

History shows that terrorism more often than not has little political impact, and that when it has an effect it is often the opposite of the one desired. Terrorism in the 1990s and the present time is no exception. The 1991 assassination of Rajiv Gandhi as he campaigned to retake the prime ministership neither hastened nor inhibited the decline of India's Congress Party. Hamas' and Hezbollah's stepped-up terrorism in Israel undoubtedly influenced the outcome of Israeli elections, but although it achieved its immediate objective of setting back the peace process on which Palestine Authority President Yasir Arafat has gambled his future, is a hard-line Likud government really in these groups' best interests? On the other side, Yigal Amir, the right-wing orthodox Jewish student who assassinated Prime Minister Yitzhak Rabin in 1996, because he disapproved of the peace agreement with the Palestinians, might well have helped elect Rabin's dovish second-in-command, Shimon Peres, to a full term had the Muslim terrorists not made Israeli security an issue again.

Terrorists caused disruption and destabilization in other parts of the world, such as Sri Lanka, where economic decline has accompanied the war between the government and the Tamil Tigers. But in Israel and in Spain, where Basque extremists have been staging attacks for decades, terrorism has had no effect on the economy. Even in Algeria, where terrorism has exacted the highest toll in human lives, Muslim extremists have made little headway since 1993, when many predicted the demise of the unpopular military regime.

Some argue that terrorism must be effective because certain terrorist leaders have become president or prime minister of their country. In those cases, however, the terrorists had first forsworn violence and adjusted to the political process. Finally, the common wisdom holds that terrorism can spark a war or, at least, prevent peace. That is true, but only where there is much inflammable material: as in Sarajevo in 1914, so in the Middle East and elsewhere today. Nor can one ever say with certainty that the conflagration would not have occurred sooner or later in any case.

Nevertheless, terrorism's prospects, often overrated by the media, the public, and some politicians, are improving as its destructive potential increases. This has to do both with the rise of groups and individuals that practice or might take up terrorism and with the weapons available to them. The past few decades have witnessed the birth of dozens of aggressive movements espousing varieties of nationalism, religious fundamentalism, fascism, and apocalyptic millenarianism, from Hindu nationalists in India to neofascists in Europe and the developing world to the Branch Davidian cult of Waco, Texas. The earlier fascists believed in military aggression and engaged in a huge military buildup, but such a strategy has become too expensive even for superpowers. Now, mail-order catalogs tempt militants with readily available, far cheaper, unconventional as well as conventional weapons—the poor man's nuclear bomb, Iranian President Ali Akbar Hashemi Rafsanjani called them.

In addition to nuclear arms, the weapons of mass destruction include biological agents and man-made chemical compounds that attack the nervous system, skin, or blood. Governments have engaged in the production of chemical weapons for almost a century and in the production of nuclear and biological weapons for many decades, during which time proliferation has been continuous and access ever easier. The means of delivery (ballistic missiles, cruise missiles, and aerosols) have also become far more effective. While in the past missiles were deployed only in wars between states, recently they have played a role in civil wars in Afghanistan and Yemen. Use by terrorist groups would be but one step further.

Until the 1970s, most observers believed that stolen nuclear material constituted the greatest threat in the escalation of terrorist weapons, but many now think the danger could lie elsewhere. An April 2000 Defense Department report says that "most terrorist groups do not have the financial and technical resources to acquire nuclear weapons but could gather materials to make radiological dispersion devices and some biological and chemical agents." Some groups have state sponsors that possess or can obtain weapons of the latter three types. Terrorist groups themselves have investigated the use of poisons since the 19th century. The Aum Shinrikyo cult staged a poison gas attack in March 1995 in the Tokyo subway; exposure to the nerve gas sarin killed ten people and injured 5,000. Other, more amateurish attempts in the United States and abroad to experiment with chemical substances and biological agents for use in terrorism have involved the toxin that causes botulism, the poisonous protein rycin (twice), sarin (twice), bubonic plague bacteria, typhoid bacteria, hydrogen cyanide, vx (another nerve gas), and possibly the Ebola virus [7].

9.3 To Use or Not to Use?

If terrorists have used chemical weapons only once and nuclear material never, to some extent the reasons are technical. The scientific literature is replete with the technical problems inherent in the production, manufacture, storage, and delivery of each of the three classes of unconventional weapons.

The manufacture of nuclear weapons is not that simple, nor is delivery to their target. Nuclear material, of which a limited supply exists, is monitored by the U.N.-affiliated International Atomic Energy Agency. Only governments can legally procure it, so that even in this age of proliferation, investigators could trace those abetting nuclear terrorists without great difficulty. Monitoring can overlook a more primitive nuclear weapon: nonfissile but radioactive nuclear material. Iranian agents in Turkey, Kazakhstan, and elsewhere are known to have tried to buy such material originating in the former Soviet Union.

Chemical agents are much easier to produce or obtain, but not so easy to keep safely in stable condition; their dispersal depends largely on climatic factors. The terrorists behind the 1995 attack in Tokyo chose a convenient target where crowds of people gather, but their sarin was apparently dilute. The biological agents are far and away the most dangerous (see sidebar, "Is Our Government Experimenting On Human Research Subjects With Biological Agents To Help Thwart The Effects Of A Biological Terror Attack?"): They could kill hundreds of thousands of people whereas chemicals might kill only thousands. They are relatively easy to procure, but storage and dispersal are even trickier than for nerve gases. The risk of contamination for the people handling them is high, and many of the most lethal bacteria and spores do not survive well outside the laboratory. Aum Shinrikyo reportedly released anthrax bacteria (among the most toxic agents known) on two occasions from a building in Tokyo without harming anyone.

Is Our Government Experimenting On Human Research Subjects With Biological Agents To Help Thwart The Effects Of A Biological Terror Attack?

Is our government experimenting on human research subjects with biological agents to help thwart the effects of a biological terror attack? Inquiring minds want to know--especially if you are a U.S. Armed Services Veteran. This author received the following letter on August 19, 2005, as shown in Fig. 23-1 [8].

Upon receiving the letter, I contacted my AMVETS (American Veterans) representative, and asked him if the letter was some kind of a scam. He instructed me to send him a copy of the letter. One week later, I received a pink note back from him in the shape of an odd looking "Happy Face." The note said:

John;

Not a scam! It's for real!

Well, that sent a chill through me; as well as, what was in the enclosed brochure, as shown in Fig. 23-2 [8]. Why does the VA want only veterans "who do not currently utilize any VA hospitals for medical care, to participate in the study?" Also, why would the VA want you to sign a consent form, if all you were doing was evaluating a survey on the effectiveness of bio-terrorism. Remember, consent forms are usually signed if you are voluntarily participating as a human test subject; in this case, possibly being subjected to biological agents.

All of this sounds very suspicious--like if something catastrophic happens to you during your participation, there is no official paper trail linking you to the VA, because you were never in the VA system. Anyway, I'll leave the rest to your imagination, with one simple warning:

Warning: If you are a veteran, and want to participate anyway, do not; and, I emphasize, do not sign any consent form until you have shown this to an attorney. And, this is very important, make sure that you have copies of everything you sign, and put those copies in a safe place. Also, do not sign anything without legal council present while participating. Remember, there is no compensation high enough to compensate you or your relatives if you voluntarily forfeit your health or your life [8].

DEPARTMENT OF VETERANS AFFAIRS
Bronx Veterans Affairs Medical Center
130 West Kingsbridge Road
Bronx, New York 10468

Dear Honorable Veteran:

The Veterans' Health Association (VHA) is interested in recruiting veterans in your geographic region who do **not** currently utilize any VA hospitals for medical care, to participate in a research study.

This very important study aims to evaluate the effectiveness of bio-terrorism educational material produced for veterans and based on suggestions provided by other veterans. Your voluntary participation would be of great value to the VA in its ability to produce the best educational material possible in preparing all veterans in the unlikely event of a future bio-terrorism incident.

Please review the attached information about the study and feel free to contact the individuals listed in the brochure for further information.

Sincerely,

Mary Sano, Ph.D.
ACOS/R&D
Principal Investigator
"Evaluating a Bio-Terrorism Preparedness Campaign
For Veterans"

WEST VIRGINIA UNIVERSITY
Institution Review Board for the
Protection of Human Research Subjects

JUN 2 2 2005

APPROVED
X ______
EXPIRES ______
H.S. # ______

Figure 23-1. Bio-Terrorism Letter.

In early Spring of 2004, veterans from all across the United States helped VA researchers develop new educational material on Bio-terrorism specifically for you...

...FOR VETERANS.

Now, you can help the VA test this material by enrolling in an important research study. Your participation and feedback are vital to ensuring this material is of the highest quality possible for you...

...FOR VETERANS.

If you choose to enroll*, there are 3 parts of the study in which you will participate

INITIAL SURVEY ON BIO-TERRORISM

Submit a survey on your knowledge, attitudes, and beliefs about bio-terrorism.

EDUCATIONAL MATERIAL

Review educational material in a brochure or on the internet. A certain number of veterans may only receive material on veteran's benefits.

FOLLOW-UP SURVEY ON BIO-TERRORISM

After 6 months, submit a follow-up survey similar to the first.

***ENROLLMENT**

Contact your local VA medical center. A Research Assistant is waiting to enroll you and answer any questions you may have. In your area, please contact.

Amy Carozza
Research Assistant
Clarksburg VAMC
1 Medical Center Drive
Clarksburg, WV 26301
1-800-733-0512 ext. 3986
304-623-3461 ext. 3986
amy.carozza@med.va.gov

If you agree to participate, you will be mailed a consent form to read and sign. Upon returning the consent form, you will be eligible to participate.
All eligible participants will receive compensation

Figure 23-2. Bio-Terrorism Brochure.

Given the technical difficulties, terrorists are probably less likely to use nuclear devices than chemical weapons, and least likely to attempt to use biological weapons. But difficulties could be overcome, and the choice of unconventional weapons will in the end come down to the specialties of the terrorists and their access to deadly substances.

The political arguments for shunning unconventional weapons are equally weighty. The risk of detection and subsequent severe retaliation or punishment is great, and although this may not deter terrorists, it may put off their sponsors and suppliers. Terrorists eager to use weapons of mass destruction may alienate at least some supporters, not so much because the dissenters hate the enemy less or have greater moral qualms, but because they think the use of such violence counter productive. Unconventional weapon strikes could render whole regions uninhabitable for long periods. Use of biological arms poses the additional risk of an uncontrollable epidemic. And although terrorism seems to be tending toward more indiscriminate killing and mayhem, terrorists may draw the line at weapons of superviolence likely to harm both foes and large numbers of relatives and friends — say, Kurds in Turkey, Tamils in Sri Lanka, or Arabs in Israel.

Furthermore, traditional terrorism rests on the heroic gesture, on the willingness to sacrifice one's own life as proof of one's idealism. Obviously there is not much heroism in spreading botulism or anthrax. Because most terrorist groups are as interested in publicity as in violence, and because publicity for a mass poisoning or nuclear bombing would be far more unfavorable than for a focused conventional attack, only terrorists who do not care about publicity will even consider the applications of unconventional weapons.

Broadly speaking, terrorists will not engage in overkill if their traditional weapons (the submachine gun and the conventional bomb) are sufficient to continue the struggle and achieve their aims. But the decision to use terrorist violence is not always a rational one; if it were, there would be much less terrorism, because terrorist activity seldom achieves its aims. What if, after years of armed struggle and the loss of many of their militants, terrorist groups see no progress? Despair could lead to giving up the armed struggle, or to suicide. But it might also lead to a last desperate attempt to defeat the hated enemy by arms not tried before. Their only hope lies in their despair [7].

9.4 Post Apocalypse

Terrorist groups traditionally contain strong quasi-religious, fanatical elements, for only total certainty of belief (or total moral relativism) provides justification for taking lives. That element was strong among the prerevolutionary Russian terrorists and the Romanian fascists of the Iron Guard in the 1930s, as it is among today's Tamil Tigers. Fanatical Muslims consider the killing of the enemies of God a religious commandment, and believe that the secularists at home as well as the State of Israel will be annihilated because it is Allah's will. Aum Shinrikyo doctrine held that murder could help both victim and murderer to salvation. Sectarian fanaticism has surged during the past decade, and, in general, the smaller the group, the more fanatical the group.

As humankind survived the end of the second millennium of the Christian era, apocalyptic movements failed to rise to the occasion. Nevertheless, the belief in the impending end of the world is probably as old as history, but for reasons not entirely clear, sects and movements preaching the end of the world gain influence toward the end of a century, and all the more at the close of a millennium. Most of the preachers of doom do not advocate violence, and some even herald a renaissance, the birth of a new kind of man and woman. Others, however, believe that the sooner the reign of the Antichrist is established, the sooner this corrupt world will be destroyed and the new heaven and earth foreseen by St. John in the Book of Revelation, Nostradamus, and a host of other prophets will be realized.

Extremist millenarians would like to give history a push, helping create world-ending havoc replete with universal war, famine, pestilence, and other scourges. Those who subscribe to such beliefs number in the millions. They have their own subcultures, produce books, and CDs by the thousands, and have built temples and communities of whose existence most of their contemporaries are unaware. They have substantial financial means at their disposal. Although the more extreme apocalyptic groups are potentially terrorist, intelligence services have generally overlooked their activities; hence, the shock over the subway attack in Tokyo and Rabin's assassination, to name but two recent events.

Apocalyptic elements crop up in contemporary intellectual fashions and extremist politics as well. For instance, extreme environmentalists, particularly the so-called restoration ecologists, believe that environmental disasters will destroy civilization as they know it (no loss, in their view) and regard the vast majority of human beings as expendable. From such beliefs and values, it is not a large step to engaging in acts of terrorism to expedite the process. If the eradication of smallpox upset ecosystems, why not restore the balance by bringing back the virus? The motto of "Chaos International," one of many journals in this field, is a quotation from Hassan I. Sabbah, the master of the Assassins, a medieval sect whose members killed Crusaders and others in a "religious" ecstasy; everything is permitted, the master says. The premodern world and postmodernism meet at this point [7].

9.5 Future Shock

Scanning the contemporary scene, one encounters a bewildering multiplicity of terrorist and potentially terrorist groups and sects. The practitioners of terrorism, up to the present time, were nationalists and anarchists, extremists of the left and the right. But the new age has brought new inspiration for the users of violence.

In the past, terrorism was almost always the province of groups of militants that had the backing of political forces such as the Irish and Russian social revolutionary movements of 1900. In the future, terrorists will be individuals or like-minded people working in very small groups (like the 9-11 terrorists), on the pattern of the technology-hating Unabomber, who apparently worked alone sending out parcel bombs over two decades, or the perpetrators of the 1995 bombing of the federal building in Oklahoma City. An individual may possess the technical competence to steal, buy, or manufacture the weapons he or she needs for a terrorist purpose; he or she may or may not require help from one or two others in delivering these weapons to the designated target. The ideologies such individuals and minigroups espouse are likely to be even more aberrant than those of larger groups. And terrorists working alone or in very small groups will be more difficult to detect (like the 9-11 terrorists) unless they make a major mistake or are discovered by accident.

Thus, at one end of the scale, the lone rogue terrorist has appeared, and at the other, state-sponsored terrorism is quietly flourishing in these days when wars of aggression have become too expensive and too risky. As the century draws to a close, terrorism is becoming the substitute for the great wars of the 1800s and early 1900s.

Proliferation of the weapons of mass destruction does not mean that most terrorist groups are likely to use them in the foreseeable future, but some almost certainly will, in spite of all the reasons militating against it. Governments, however ruthless, ambitious, and ideologically extreme, will be reluctant to pass on unconventional weapons to terrorist groups over which they cannot have full control; the governments may be tempted to use such arms themselves in a first strike, but it is more probable that they would employ them in blackmail than in actual warfare. Individuals and small groups, however, will not be bound by the constraints that hold back even the most reckless government.

Society has also become vulnerable to a new kind of terrorism in which the destructive power of both the individual terrorist and terrorism as a tactic are infinitely greater. Earlier terrorists could kill kings or high officials, but others only too eager to inherit their mantle quickly stepped in. The advanced societies of today are more dependent every day on the electronic storage [5], retrieval, analysis, and transmission of information. Defense, the police, banking, trade, transportation, scientific work, and a large percentage of the government's and the private sector's transactions are on-line. That exposes enormous vital areas of national life to mischief or sabotage by any computer hacker, and concerted sabotage could render a country unable to function. Hence, the growing speculation about infoterrorism and cyberwarfare.

An unnamed U.S. intelligence official has boasted that with $6 billion and 70 capable hackers, he could shut down America. What he could achieve, a terrorist could too. There is little secrecy in the wired society, and protective measures have proved of limited value: teenage hackers have penetrated highly secret systems in every field. The possibilities for creating chaos are almost unlimited even now, and vulnerability will almost certainly increase. Terrorists' targets will change: Why assassinate a politician or indiscriminately kill people when an attack on electronic switching will produce far more dramatic and lasting results? The switch at the Culpeper, Virginia, headquarters of the Federal Reserve's electronic network, which handles all federal funds and transactions, would be an obvious place to hit. If the new terrorism directs its energies toward information warfare, its destructive power will be exponentially greater than any it wielded in the past — greater even than it would be with biological and chemical weapons.

Still, the vulnerability of states and societies will be of less interest to terrorists than to ordinary criminals and organized crime, disgruntled employees of big corporations, and, of course, spies and hostile governments. Electronic thieves, whether engaged in credit card fraud or industrial espionage, are part of the system, using it rather than destroying it; its destruction would cost them their livelihood. Politically motivated terrorist groups, above all separatists bent on establishing states of their own, have limited aims. The Kurdish Workers Party, the IRA, the Basque ETA, and the Tamil Tigers want to weaken their enemies and compel them to make far-reaching concessions, but they cannot realistically hope to destroy them. It is also possible, however, that terrorist groups on the verge of defeat or acting on apocalyptic visions may not hesitate to apply all destructive means at their disposal.

All that leads well beyond terrorism as has the military has known it. New definitions and new terms may have to be developed for new realities, and intelligence services and policy makers must learn to discern the significant differences among terrorists' motivations, approaches, and aims. The Bible says that when the Old Testament hero Samson brought down the temple, burying himself along with the Philistines in the ruins, "the dead which he slew at his death were more than he slew in his life." The Samsons of a society have been relatively few in all ages. But with the new technologies and the changed nature of the world in which they operate, a handful of angry Samsons and disciples of apocalypse would suffice to cause havoc. Chances are that of 100 attempts at terrorist superviolence, 99 would fail. But the single successful one could claim many more victims like it did on 9-11, do more material damage, and unleash far greater panic than anything the world has yet experienced [7].

9.6 The Menace of Amateur Rogue Information Warfare

With a member base of 79,000, the amateur rogue CyberArmy may have the biggest armament the Net has ever seen, rallying to take down Web sites that "abuse" the World Wide Web—and removing power from governments. Some missions include hunting for, and taking down, child pornography Web sites.

The CyberArmy wants to self-regulate the Internet so that the government doesn't come in and regulate it. CyberArmy started off as a small group of advocates promoting free speech and Internet deregulation. Growing to a full size army of "Netizens," the group has since shifted its views—due to privacy issues and government intervention.

Now they believe in Internet self-regulation. If you deregulate, you end up with anarchy. In other words, the CyberArmy is set up just like a game. Members have to solve puzzles (which is usually breaking codes and encryption) to move on to the next commanding level.

Missions Possible

Commanding ranks give a member more power and involvement in the organization's missions. As previously mentioned, some missions include hunting for, and taking down, child pornography Web sites. They're really trying to get rid of child pornography on the Internet.

The commanding structure begins at the bottom with troopers, rising through the ranks of 2nd Lieutenant, Lieutenant, Captain, Major, Lt. Kernal, Kernal, General, and Marshal. Each division within CyberArmy has its own job to complete, with one of the divisions devoted solely to child pornography Web sites.

The division has taken down around four dozen child porn sites in the last year, and was also instrumental in bringing down the Wonderland Club child porn ring recently. The group is an advocate of ordinary citizens policing the Internet. Due to the Internet being global, governments aren't the right authority to police it [7].

9.7 Hacktivists

In defending the "hacktivist" title that the group has been branded with, the group doesn't believe in defacing a Web site just for the fun of it. If a site is defaced it's usually in the form of protest.

The group was a bit more "hackerish" in 2000, however—they were considered an amateur menace for a time. However, they're moving away from that. There are more social minded people on the Net now, which is good. The reason for being in CyberArmy is because many members are sick and tired of child pornography and Net censorship.

Finally, CyberArmy's mission is to prove that there are good hackers, not just Script Kiddies out there. The CyberArmy site also posts discussion boards and Internet tools for users, and has a section dedicated to teaching wireless network security [7].

10. SUMMARY AND CONCLUSIONS

The problem of defending against an information warfare arsenal and tactics is real. U.S. citizens and the organizations that provide them with the vital services they need can find no sanctuary from these attacks. The low cost of mounting these attacks has enlarged the field of potential adversaries and complicated efforts to collect intelligence and array U.S. military defenses. The consequences of a well-planned and coordinated attack by a relatively sophisticated foe could be serious. Even the threat of such an attack or digital blackmail is a distinct possibility. How the public will respond to the threat of IW infrastructure attacks or to actual attacks is unclear, but is a major determinant of future policy and actions.

This situation is getting worse with the rapid proliferation of information technology and know-how. U.S. citizens are becoming increasingly dependent on automation in every aspect of their lives. As information technology becomes an essential part of the way organizations and individuals create products and provide services, the need for interconnectivity and interoperability increases. With this increased need for exchanges of information (and products), vulnerabilities increase. Finally, the increased reliance on commercial-off-the-shelf products or commercial services makes it more and more difficult for organizations and individuals to control their own security environment.

Finally, given this situation, you need to focus on two goals. First, you need to find a way to protect yourself against catastrophic events. Second, you need to build a firm foundation upon which you can make steady progress by continually raising the cost of mounting an attack and mitigating the expected damage of the information warfare arsenal and tactics of the military. The conclusions are as follows:

- Information warfare (IW) has become virtually synonymous with the revolution in information technologies and its potential to transform military strategies and capabilities.
- There is a growing consensus that national prosperity, if not survival, depends on one's ability to effectively leverage information technology. Without being able to defend vital information, information processes, and information systems, such a strategy is doomed to failure.
- Information warfare is often thought of as being defined by a particular target set of decision makers, information, information processes, and information systems.
- The "battlespace" associated with IW has been a constantly expanding one, moving far beyond traditional military situations.
- In some quarters, IW has even been associated with the leveraging of information technologies to achieve greater effectiveness and efficiency. This has stretched the meaning of IW to the breaking point and has sowed more confusion than enlightenment. For this reason, this treatment of the subject uses the term "information strategies" to refer to the recognition and utilization of information and information technologies as an instrument of national power that can be independent of, or complementary to, military presence and operations.
- The scope, or battlespace, of information warfare and strategy (IWS) can be defined by the players and three dimensions of the nature of their

interactions, the level of their interactions, and the arena of their interactions.

- Nation states or combinations of nation states are not the only players. Nonstate actors (including political, ethnic, and religious groups; organized crime; international and transnational organizations; and even individuals empowered by information technology) are able to engage in information attacks and to develop information strategies to achieve their desired ends.
- The term "war" has been used so loosely in recent times (War on Poverty, War on Drugs, War on Crime) that it should be no surprise that IW has evolved over the past several years to become a "catch-all" term that encompasses many disparate activities, some of which have long been associated with competition, conflict, and warfare, and others that are of more recent origin. These include activities that range from propaganda campaigns (including Media War), to attacks (both physical and nonphysical) against commanders, their information sources, and the means of communicating with their forces.
- Under this rather large umbrella that has become known as IW, one can find activities long associated with military concepts and operations, including deception, command and control warfare (C2W), and psychological operations (Psyops).
- Technological advances have added new forms such as electronic warfare (EW) and "hacker warfare."
- The term "defensive information warfare" (IW-D) is used here to refer to all actions taken to defend against information attacks, that is, attacks on decision makers, the information and information-based processes they rely on, and their means of communicating their decisions.
- Strictly speaking, because these attacks can be launched during peace time at nonmilitary targets by nonmilitary groups, both foreign and domestic, the term IW-D should be IWS-D. However, IW-D is currently in wide use.
- This overview of IW-D does not attempt to deal with the problems of defending against all of the different kinds of information attacks, but rather focuses its attention on the subset of IW that involves attacks against information infrastructure, including what has become known as "hacker warfare" and in its more serious form, "digital warfare [7]."

11. REFERENCES

[1] John R. Vacca, *Wireless Broadband Networks Handbook*, McGraw-Hill, 2001.

[2] John R. Vacca, *High-Speed Cisco Networks: Planning, Design, and Implementation*, CRC Press, 2002.

[3] John R. Vacca, *Net Privacy: A Guide to Developing & Implementing an Ironclad ebusiness Privacy Plan*, McGraw-Hill, 2001.

[4] John R. Vacca, "*Satellite Encryption*," Academic Press, 1999.

[5] John R. Vacca, *The Essential Guide to Storage Area Networks*, Prentice Hall, 2002.

[6] John R. Vacca, *Firewalls: Jumpstart For Network And Systems Administrators,* Elsevier Digital Press (December, 2004).

[7] John R. Vacca, *Computer Forensics: Computer Crime Scene Investigation, 2nd Edition*, Charles River Media, 2005.

[8] Department Of Veterans Affairs, Bronx Veterans Affairs Medical Center, 130 West Kingsbridge Road, Bronx, New York 10468, 2005

Chapter 24

THE INFORMATION WARFARE WIRELESS NETWORK SECURITY ARSENAL AND TACTICS OF TERRORISTS AND ROGUES

1. INTRODUCTION

The information warfare arsenal and tactics of terrorist and rogues have become increasingly transnational as the networked organizational form has expanded. When terrorism's mentors were the Soviet Union and the Eastern Bloc, they imposed their own rigid hierarchical structure on terrorist groups. Now that terrorism is increasingly substate, or semidetached, networking and interconnectivity are necessary to find allies and influence others, as well as to effect command and control.

As discussed in Chapter 23, information and communication technologies (ICTs) have facilitated this, and have also enabled multiple leaders to operate parallel to one another in different countries. It, therefore, might be said that a shift is taking place from absolute hierarchies to hydra-headed networks, which are less easy to decapitate. An analogy, using the Palestinian example, may be that the more networked form of Hamas now that Arafat is dead, is replacing the hierarchical structure of the PLO. In many ways the Afghan War was a seminal event in promoting the networked form in that it showed that fluidly organized groups, driven in this case by a religious imperative, could defeat an experienced hierarchically structured army.

Geographical dispersion, both physical and in cyberspace, provides extra security. A rigid hierarchical structure is more easily penetrated and neutralized. Israel's admission that it had not yet found a way to deal with

Hamas's decentralized and internationalized command and control structure, which uses encrypted Internet messages, suggests it has had difficulty in this matter. An investigation by the Federal Bureau of Investigation into terrorist activity in the United States indicated that part of Palestinian Islamic Jihad's command and control system was located in Tampa, Florida. Likewise, Hamas allegedly has some of its fundraising infrastructure in London and the USA, and publishes its main Arabic journal, *Filistin al Muslima,* in London.

Islamist terrorists may be said to fit the network ideal. Many supportive expatriate communities are based in sympathetic or neutral states enabling political activists and terrorists to operate within the safe haven that modern democracies provide.

> **Note:** It is not the intention here that the term "Islamists" should refer only to terrorist organizations, but rather to those Muslim militants who believe that Islam is incomplete without its own state, one in which Shariah provides the system of governance, and who campaign for its imposition.

Among Islamists, it is the Jihadists (religious warriors) who are of particular interest in this chapter. The followers of Hasan al Banna, Sayyid Qutb, and Abdul Ala Maududi, the organizations they founded, Ikhwan al Muslimoon and Jamaat Islami, and the ideological off-shoots these have spawned, give rise to the "Jihadist" ideology. And, although the concept of Jihad may be interpreted on different levels, it often incorporates violence when applied to Islamists.

> **Note:** The ultimate experience is, of course, Jihad, which for Islamists means armed battles against communists (Afghanistan) or Zionists (Palestine) or, for the radicals, against renegades and the impious.

Jihad in the modern Islamist sense knows no political space or state; its space is that of the Umma, the community of Muslims, wherever they may be. An example of the networked form amongst such Islamist organizations is that of the Algerian Armed Islamic Group, the GIA. Allegedly responsible for a bombing campaign in France, it appears to have had a command and control center in Britain for some years prior to the expulsion of some members by the British authorities. At the same time, sympathizers were also safe-housing some of its weapons and explosives in Belgium.

Algerian terrorists have been able to communicate with their sympathizers and members by use of the Internet and have used the services of Muslim news agencies, which republish their postings. Foremost amongst them is MSANEWS. On their site were published communiqués from the GIA, Front Islamique de Salut (FIS), and many other Islamists.

Note: The MSANEWS also posts articles and communiqués from non-Islamist Muslim and non-Muslim sources, claiming that it has condemned terrorism, and that it no longer reposts communiqués of organizations that advocate terrorism.

The site of the Campaign for the Defense of Legitimate Rights (CDLR), the Saudi opposition group, also contains postings from groups not directly connected with it, as do London-net@Muslimsonline and the pro-Iranian Muslimedia International, which, like other sites, reposts interviews with Osama bin Laden, the exiled and wanted *dead or alive* Saudi terrorist leader (see sidebar,"Bin Laden Uses Web to Plan"). As with some other Islamists groups, Muslimedia International also promotes antisemitism and Holocaust denial and provides links with the American Holocaust denier, Michael Hoffman II, and his Campaign for Radical Truth in History, thereby highlighting the interconnectivity possibilities between totally different ideologies sharing a perceived common enemy.

Bin Laden Uses Web To Plan

Osama bin Laden and other Muslim extremists are using the Internet to plan more terrorist activities against the United States and its allies. Recently, U.S. law enforcement officials and other experts disclosed details of how extremists hide maps and photographs of terrorist targets in sports chat rooms, on pornographic bulletin boards and other popular Web sites. Instructions for terrorist activities also are posted on the sites, which the officials declined to name. To a greater and greater degree, terrorist groups, including Hezbollah, Hamas, and bin Laden's al Qaeda group, are using computerized files, e-mail, and encryption to support their operations--like the train bombing in Madrid in the summer of 2004. According to various unnamed officials and investigators, the messages are scrambled using free encryption programs set up by groups that advocate privacy on the Internet. It's something the intelligence, law-enforcement, and military communities are really struggling to deal with. The operational details and future targets, in many cases, are hidden in plain view on the Internet. Only the members of the terrorist organizations, knowing the hidden signals, are able to extract the information [5].

An Islamist site that particularly aims its message to the outside world is that of Hizb-ut-Tahrir, the Islamic Liberation Party. Their first UK-based site was hosted by Imperial College, London, but following complaints to the college authorities, the site was closed down. They now post in their own name as Hizb-ut-Tahrir, and as Khilafah, providing Internet-based access to their hard copy material, literature, and their regional activities. Al-Muhajiroun (the Emigrants) whose UK leader, Omar Bakri Mohammed, was the founding leader of Hizb-ut-Tahrir in Britain, and from which he split claiming differences with the Middle-East-based leadership, also provides details of its activities, as well as lists of its hardcopy publications and contacts. In 1998, Mohammed reported the communiqués of Osama bin Laden, for whom he claims to act as a spokesman. As a consequence of his endorsement of the bombings of the U.S. embassies in Dar-es-Salaam and Nairobi, his postings are no longer carried by MSANEWS.

Hamas and its supporters and sympathizers have been among the most prolific users of the Internet. MSANEWS provides a list of Internet resources about Hamas including copies of its covenant, its official communiqués (at Assabeel On-line), and communiqués of its military wing, the Izz al-Din Al-Kassam Brigades. Information about Hamas, in fact, may also be accessed in various different ways: via MSANEWS, the Palestine site, and the Islamic Association for Palestine. Hamas' own site, which posts in Arabic, is the Palestine Information Centre.

Religious luminaries from one country sometimes act as the higher legal and moral authority in another country. Sheikh Yusuf al-Qaradawri of the Egyptian Ikhwan al-Muslimoon (Muslim Brotherhood) lives in Qatar and serves as the Imam (religious leader) for the Palestinian Hamas. Sheikh Ibn Qatada, a Jordanian Palestinian living in London, serves as the Imam for the Algerian GIA. Sheikh Abu Hamza, an Egyptian national and former Afghan Jihad volunteer, serves as a propagandist for the Algerian GIA and Imam for the Yemeni Jihad group, but lives in London. Now, their messages of guidance and support find an outlet most frequently via ICTs.

Although some commentators have argued that modern cultural forces, such as ICTs, serve to undermine Islamisation in Muslim society, it is equally easy to argue that they provide a new and growing medium by which Islamism is disseminated. Even if they do not reach the poorer sections of Muslim society, they certainly reach many educated expatriate communities, among whom they find support. The growing number of advertisements, on the Internet and in Muslim papers and journals, for conferences to discuss the use of the Internet to promote Islam, or Islamism, supports the thesis that many activists and religious teachers see these developments as positive ones to be recommended and encouraged.

Combining religious injunctions with strategic commands is a noticeable feature of such Islamist leaders and their groups. Calls to carry out Jihad are frequently cloaked in religious and pseudo-religious language, but the implication is clear for the target audience. Thus, for example, Osama bin Laden's *Ladenese Epistle,* which was originally faxed to his London contact, Khalid al Fawaz, and then posted to MSANEWS in August 1996 by the London-based Saudi dissident groups CDLR and MIRA, is recognized as providing general guidance for anti-American terrorism.

For example, bin Laden's *Ladenese Epistle* reads, "*The sons of the land of the two Holy Places had come out to fight against the Russian in Afghanistan, the Serb in Bosnia-Herzegovina, and today they are fighting in Chechenia and—by the Permission of Allah—they have been made victorious over your partner, the Russians. By the command of Allah, they are also fighting in Tajakistan.*

I say: Since the sons of the land of the two Holy Places feel and strongly believe that fighting (Jihad) against the Kuffar in every part of the world, is absolutely essential; then they would be even more enthusiastic, more powerful and larger in number upon fighting on their own land."

The Nida'ul Islam site, based in Australia, promotes an uncompromising message of both Jihad and of suicide terrorism. A recent posting, *The Islamic Legitimacy of the Martyrdom Operations,* states that martyrdom is forbidden in Islam, but cites approvingly those martyrs who willingly gave their lives for Muslim causes and then transposes these causes to contemporary issues. It attempts to demonstrate with quotes from the Quran and the Sunnah that Islamic bombing assaults and martyrdom attacks are legitimate and fall within the framework of Islam.

Azzam Publications, named after Abdullah Azzam, a Palestinian who became a military leader in Afghanistan and who was assassinated in Pakistan in 1989, has also published calls for Jihad volunteers:

"The Saudi Government does not intend to help the Muslims in Kosova and it has prevented its nationals from going there to fight. This means that the Jihad in Kosova is now a greater responsibility on Muslims with western nationalities ... Redistribute this e-mail message all over the world ... telephone the nearest Saudi Embassy or Consulate to protest against this crack-down and tell everyone to do so until it jams the lines of the Saudi Consulates around the world ... e-mail the Saudi Embassy in Washington with messages of protest ... begin to prepare yourselves to go and fight in Kosova to make up for the lack of manpower that was heading there from Saudi Arabia. Wait for the Kosova bulletin from Azzam Publications."

Among the Far Right, the UK-based national revolutionary group, The International Third Position, illustrates graphically the adoption of ICTs to enhance a position. The group is tiny, but its foreign contacts are numerous, widespread and growing. In the space of just over one year its *Final Conflict* e-mail newsletter has grown in size and scope to reflect the news of, and messages from, its worldwide contacts.

Final Conflict also acts as a news agency for Holocaust deniers (in much the same way as MSANEWS does for Islamists), many of whom are also Far Right extremists. For example, the e-mail newsletter reposts communiqués from David Irving and Fredrick Toben's Australian Adelaide Institute, which like the California-based Institute for Historical Review, attempts to provide a scholarly veneer for denial. Some invitees to a conference held by the Adelaide Institute were refused permission to visit Australia by its Department of Immigration, but the easy access to the Internet and video links facilitated conference presentations that otherwise might not have taken place.

The Far Right has also used the Internet to post bomb-making manuals that are not otherwise available in Europe. The British neo-Nazi, David Myatt, of the National Socialist Movement posted his *Practical Guide to Aryan Revolution* at the end of November 1997 at the Web site of Canadian Bernard Klatt in order to evade police scrutiny. The chapter headings included: *Methods of Covert Direct Action, Escape and Evasion, Assassination, Terror Bombing, Sabotage, Racial War, How to Create a Revolutionary Situation, Direct Action Groups,* and so on. The contents provided a detailed step-by-step guide for terrorist insurrection with advice on assassination targets, rationale for bombing and sabotage campaigns, and rules of engagement. Although he may have committed no indictable offence in Canada, Klatt was forced to close down his site in April 1998. Myatt is currently the subject of a British criminal investigation for incitement to murder and promotion of race hatred.

Police forces in Britain and France also recently investigated an international neo-Nazi network that issued death threats against French celebrities and politicians from their British-based Internet site. Herve Guttuso, the French leader of the Charlemagne Hammer Skins, was arrested in Essex at the same time as eight members were arrested in the South of France. The French members of the network were charged with making death threats, and Guttuso was the subject of a French extradition request to the British courts. According to the French Interior Ministry, police in Toulon traced the London address of the Internet site, which was being accessed about 7,000 times a month. The investigation enabled the police to identify 3,500 people sympathetic to the neo-Nazi group in various countries including Britain, Greece, Canada, America, and Poland. The investigators found that the Charlemagne group appeared to be one of the largest and best organized neo-Nazi groups yet uncovered, with a coordinated international structure and logistical centers for disseminating violent racist propaganda, based principally in Britain and America. Although the group gave a postal address in London as their center, their material was disseminated via Klatt's

FTC Net (as have been the postings of Marc Lemire, Paul Fromm, Doug Christie, The Heritage Front, and other neo-Nazi and white supremacist groups).

The British Far Right may have been slower to realize the command and control possibilities of ICTs than their U.S. or German co-ideologies, but they appear to be catching up. Although in recent years it is the violent skinhead music scene that has provided the main medium through which they promote liaison, it is clear that for some the future lies with ICTs.

In 1999, the Pentagon had to admit that there had been a major assault on its computer systems. Zionist Occupational Government (ZOG) observers have increasingly warned that the frequency and sophistication of the hack attacks will only increase as dissident groups realize that they can strike at the very heart of ZOG at the touch of a few buttons. It doesn't matter what government specialists invent to counter the techno-terrorist, there is always a way around their antihacker programs and the more ZOG relies on computers, the more damage can be done by attacking their systems. So all you techno-terrorists out there, get working on your "hack-attacks [5]."

2. THE TERRORIST PROFILE

Sid-Ra, a 6-foot-4-inch, 350-pound giant of a man, paces between his "subjects" in the smoke-filled Goth club Click + Drag, located in the old meat-packing district of Manhattan. Inside the club are leather-clad, black-lipped females and young men dressed in women's underwear.

Sid is a hacker-terrorist and an acknowledged "social engineer" with curious nocturnal habits. There are thousands of people like him, who by day are system and wireless network administrators, security analysts, and start-up cofounders. When night comes, they transform into something quite different.

But, is this the profile of a "wanna-be terrorist? " Perhaps!

These are the self-proclaimed freedom fighters of cyberspace. They've even got a name for it: hactivism. And political parties and human rights groups are circling around to recruit hactivists into their many causes.

Recently, for example, the Libertarian Party set up a table at the HOPE (Hackers on Planet Earth) conference. The San Francisco-based Electronic Frontier Foundation (EFF) collected donations. And members of civil-rights groups, including the Zapatistas, a Mexican rebel group, spoke up at one of two sessions on hactivism.

But even without such civil-liberties groups trying to organize them, hactivists have been busy on their own. They have formed privacy-related software companies such as ZeroKnowledge Systems USA Inc. in Montreal.

They're developing anonymous, inexpensive e-mail and Web-hosting services through the DataHaven Project Inc. (*http://www.dhp.com*). And they're trying to get the Internet out to Third World human rights organizations through groups such as Cult of the Dead Cow Communications (cDc; *http://www.cultdeadcow.com/*).

Caution: URLs are subject to change without notice.

In fact, Sid feels hactivism's pull so strongly that he makes a dramatic claim: "The Internet is the next Kent State, and we're the ones who are probably going to get shot [5]."

2.1 From Vietnam Marches to Cyberdisobedience

Like any social engineer, Sid exaggerates. Except for the four-year jail terms handed down to Kevin Mitnick and Kevin Poulsen, sentencing for even criminal hacking in the past two years has been relatively light (mostly probation and fines) because of the suspects' young ages.

But the comparison to the psychedelic hippies of the 1960s who spoke out against the Vietnam War may not be so far off the mark. Only this time, the hackers are Goths and hedonists. And they're using the Internet to rid the world of tyranny.

The government tries to put electronic activism into the peg of cyberterrorism and crime with its Infowar eulogies. But E-Hippies, cDc, and others aren't criminals. The Internet just multiplies their voice.

Another group reaching out to hackers and technologists is the EFF. In 1999, the EFF successfully argued in the infamous Bernstein ruling, which stated that software code is protected as a form of speech.

Hackers question conventional models. They don't just look at technology and say, *"This is how it works." They say, "How can I make it better?" They look at society that way too—their government, their schools, or their social situation. They say, "I know how to make this better,"* and they go for it.

In the MPAA case, staffers at 2600 Enterprises Inc., based in Middle Island, New York, were threatened with imprisonment if they didn't remove a link on the 2600 Web site to the code used to crack DVD encryption. Because the link was editorial content, it sets Sid off on another diatribe.

The Libertarian Party also recruits hackers and technologists. At HOPE, the party's New York State committee (*http://www.cownow.com*) handed out fliers, signed up recruits, and took a "sticker" poll of party affiliations. The poll got hacked, but about half the stickers were yellow — for libertarian, anarchist or independent.

Many party members are programmers. They're trying to rally hackers around encryption, privacy and freedom-of-communication planks. Hackers can offer them freedom, because the Internet routes around tyranny.

But hackers have ways beyond the Internet to electronically spread their message. Take a young dude named Alpha Underflow, for instance, who late one night broke the lock to a lit-up roadside-construction sign and reprogrammed it to read, "Hack Planet Earth" in support of the 2600 Magazine staff. But then, he also likes to use his reprogrammed garage-door opener to pop open his neighbor's garage doors [5].

2.2 Grow Up

This moral confusion is typical of the younger hacking crowd. But most of the older hackers (30 years old and up) have grown up.

In the mid-1990s, there was more disillusionment as more bleeding-edge hackers ended up going to jail for cracking. That bummed out their whole theme. But now they've learned some limits, and they can still operate within them.

That means the older hackers do develop some scruples. For example, the EFF Web site (*http://www.eff.org*) was a popular target of punk hackers back in the mid-1990s, with hacks and defacements occurring weekly. Now, it's rarely being hacked. When the site did get hacked, a message was posted about it on 2600's bulletin board, and the hackers who responded called that hacker a lamer.

The process that turned the hippie of 1968 into the employed investor of 1985 is similarly going on here today. Hopefully, that the hippie-to-yuppie disillusionment that took place historically doesn't happen to hackers, too.

So, who are the real cyberterrorists? Are they for real [5]?

2.3 Will The Real Cyberterrorists Stand-up

The debate over whether the United States faces imminent danger from cyberterrorist attacks took a new turn recently when the National Security Council declared that the term "terrorism" may be too strong a word when describing potential cyberthreats.

Although it would be a tough call to tell the difference between an attack by hackers and one launched by terrorists intent on disrupting national security, the administration's cyberdefense programs are battling a perception problem that stems from the misuse of the word "terrorism."

Maybe you shouldn't be saying "cyberterrorism." Maybe you should be saying "information warfare." In the end, you're going to know it when you see it—the difference between joy-riding hackers and state-sponsored cyberattacks.

Experts agree that, to date, most of the major cybersecurity incidents are best described as nuisance attacks, although many fear that a devastating surprise attack, sometimes referred to as an "electronic Pearl Harbor," is inevitable. Although the government tries to be proactive, the United States is going to get nailed seriously—sooner rather than later.

By not preparing for the worst-case scenario, the United States may be endangering the public's civil liberties. A lot of people are going to be willing to throw civil liberties out the window in an effort to recover from an attack that cripples large portions of the nation's critical infrastructure.

Preparation is crucial, and, in the current legal system, defensive measures are more "workable" than offensive ones. Overall, however, cyberdefense is not well understood and is not talked about sufficiently.

Pretending the threats are not there is not a solution. There are numerous efforts by rogue groups to acquire encryption algorithms and sophisticated tools. One presidential administration after another has lulled the American people into a false sense of security.

Anyway, the Internet has become a new form of the "dead drop" (a Cold War-era term for where spies left information) for terrorists. And, bin Laden, the dissident and wanted Saudi businessman who has been indicted for the 1998 bombing of two U.S. embassies in East Africa; the 9-11 attacks in 2001; the bombing of the USS Cole destroyer in Yemen; and, the 2004 train bombing in Madrid, Spain, has taken advantage of that Internet "dead drop" zone.

Four alleged bin Laden associates went on trial recently in federal court in New York for the embassy bombings. Officials say bin Laden began using encryption in 1996, but recently increased its use after U.S. officials revealed they were tapping his satellite [1] phone calls in Afghanistan and tracking his activities.

Thus, bin Laden meets the requirements for the new terrorist profile: He will use whatever tools he can (e-mails, the Internet, etc.) to facilitate jihad against the Israeli occupiers and their supporters, according to Ahmed Yassin, the founder of the militant Muslim group Hamas. Bin Laden (dead or alive) has the best minds working for him [5].

3. WHY TERRORIST AND ROGUES HAVE AN ADVANTAGE IN IW

Governments have neither the financial resources nor the technical know-how to stay on top of hackers and computer terrorists. Therefore, this is why terrorist and rogues have an advantage in IW.

The private sector must (provide for) itself much of the action that is necessary to prevent attacks being made on the Internet. It's no longer possible for governments to provide the kind of resources and investment necessary to deal with these kinds of issues.

There are no cookie-cutter solutions; every wireless network is different. At the top of CIOs' concerns here was denial of service (DoS) attacks, which recently brought Yahoo, Amazon.com, eBay, and other high-profile Web sites to their knees. DoS attacks are a key concern because the only way that is currently available to prevent them is to catch the perpetrators.

Second on the list of concerns was attacks that reach into wireless networks to steal valuable corporate data. Firewalls [4] are the best way to prevent data theft that originates outside of a wireless network, whereas cryptography can help to protect data from internal theft.

There is a "real danger" of terrorists and hostile rogue nations using wireless computer networks to wage international warfare. In other words, most of the major terrorist organizations have their own Web sites, and, therefore, have the facility to carry out the same sort of action that was carried out with the release of the I Love You virus.

Cyberterrorism can be more effective and more costly to governments than the classic methods of bomb attacks and assassination. It is really a serious threat to everyone in all societies.

Solutions seemed harder to come by today than the problems just discussed. Governments, businesses, and research institutions must band together to find the best technologies and courses of action to defeat cybercrimes. And companies must be more willing to invest in security systems to protect their wireless networks.

A few participants called on software companies and service providers to make their products more secure. Default settings for software products sold to consumers should be at the highest level of security.

You wouldn't build a swimming pool in the center of town and not put a fence around it. Basically, that's just what the software companies are doing.

Although security firms have financial incentives for promoting security issues, for the average corporation, the benefits of spending millions of dollars to bolster security in wireless networks aren't immediately obvious, thus making them slow to act. If you have a choice of spending five million dollars on getting 693,000 new customers, or five million dollars on serving the ones you already have, that's a difficult value proposition. Most companies would take the additional customers.

But the severity of attacks could get worse, and enterprises would be wise to make precautionary investments now. Most enterprises have been lucky so far [5].

3.1 Cyberattack Risks if You're a Superpower

Information warfare and other security threats simply come with the territory when your country is the world's only remaining superpower. This is what is called a "superpower paradox." There is no other country that can challenge the United States directly. Instead, some countries look for indirect ways to challenge the United States. This challenge could come in the form of nuclear (see sidebar, "Stopping Nuclear Blackmail"), chemical or biological (see sidebar, "Chemical And Biological Terrorism"), or even cyberwarfare (see sidebar, "Hacker-Controlled Tanks, Planes, and Warships").

Stopping Nuclear Blackmail

Bill Clinton used to say that no Russian missiles are targeted at the United States. But there is every reason to believe that there are, or soon will be, North Korean missiles targeted at this country—missiles capable of delivering nuclear or chemical and biological warheads. In a few years, and without much warning, Iranian and Iraqi missiles could also be targeted at us and our allies. What can the U.S. do to stop such missiles once they are launched? Not a thing.

None of this was clear in 1998; it is undeniable now. The question is whether the U.S. government will build a missile defense system to protect their cities, military bases, and oil fields—and to block the kind of nuclear blackmail suggested by China's threat, during the Taiwan Strait crisis of 1996, to bomb Los Angeles.

A full warning came from a report in 1998 of the commission on missile threats headed by Defense Secretary Donald Rumsfeld. This was a bipartisan commission, with members who have often disagreed on weapons issues. The panel had access to all U.S. intelligence sources, and its conclusion was unanimous: Rogue states could "inflict major destruction on the United States." within five years of deciding to do so, and with little or no notice to us.

This contradicted the Clinton administration line that the United States. would have plenty of notice of a missile attack. That conclusion was based on a 1995 national intelligence estimate that said there would be no threat to the 48 contiguous states for the next 15 years.

Note: Evidently, the administration didn't think that the constitutional obligation to "provide for the common defense" applied to Alaska and Hawaii.

The Rumsfeld report at first seemed to do little to change the views of President Clinton's top defense advisers. Five weeks after the report was released, Gen. Henry Shelton, the chairman of the Joint Chiefs of Staff, wrote that "the intelligence community can provide the necessary warning" of hostile missile development and added, "We view this as an unlikely development." A week after that, North Korea launched a 3,000-kilometer range, two-stage Taepo Dong 1 missile over Japan. The launch indicates that North Korea has made progress in building the Taepo Dong 2, whose 10,000-kilometer range includes not only Alaska and Hawaii but also much of the continental United States. No matter: All but four Senate Democrats blocked action on a bill sponsored by Thad Cochran, a Republican from Mississippi, and Daniel Inouye, a Democrat from Hawaii, that would have forced the administration to deploy a missile defense system as soon as technologically feasible.

A New World

The case against rapid deployment rests on three arguments: (1) the threat isn't real, (2) the technology is impossible, and (3) it is more important to maintain the antiballistic missile treaty signed with the Soviet Union in 1972, which bars most missile defense systems. The Rumsfeld report demolished Argument 1. Argument 2 is still raised by some who note that the United States has spent large sums on missile defense since Ronald Reagan proposed it in 1983, with disappointing results. But stopping a few rogue-state missiles with the computers of 2005 is much easier than stopping hundreds of Soviet missiles with the computers of 1983. As for Argument 3, the strategic environment in which the ABM treaty was adopted no longer exists. The argument for the treaty was that a missile defense system might provoke a Soviet or American first strike. However, the proximate missile threats now come from states that might risk such a strike [5].

Chemical And Biological Terrorism

In "For Your Eyes Only," James Bond's irrepressible quartermaster, Major Boothroyd (a.k.a. Q) demonstrates his latest toy: a rather lethal umbrella. Using a faceless mannequin, one of Q's assistants illustrates how the umbrella looks and acts like it should until struck by water (as umbrellas are wont to do from time to time). Suddenly, sharp metal hooks extend all along the edge of the umbrella as it swiftly closes upon the victim's neck. The motion is quick and precise, but one can't help but imagine the far messier spectacle if a human being were caught under it in a rainstorm.

Unfortunately, the fictional version of MI6 portrayed in the James Bond films is not the only place one can find a deadly device masquerading as protection against the elements. In September 1978, the Bulgarian secret service shot a Bulgarian exile, Georgi Markov, with just such a device. Disguised as an umbrella, the surreptitious gun inserted a small pellet into Markov's thigh. The pellet contained only a few hundred millionths of a gram of the deadly poison ricin (supplied by the KGB), but it was enough. Markov died four days later in a London hospital. Another Bulgarian defector, Vladimir Kostov, was similarly attacked in Paris the month before. Kostov was shot in the back and suffered a high fever, but survived. He sought medical treatment after hearing of Markov's death and doctors removed from his back a small pellet identical to the one used to kill Markov.

Not satisfied with leaving such methods solely in the hands of the secret agent-types, the *Aum Shinrikyo* cult tried a simpler version during their chemical and biological escapades. In their infamous sarin gas attack on the Tokyo subway, *Aum* operatives chose the decidedly low-tech dissemination method of dropping bags of liquid sarin on the floor, puncturing them with the sharpened ends of their umbrellas, and then beating a hasty retreat as the nasty stuff spilled out onto the ground. Despite their primitive dissemination methods, *Aum* managed to murder 12 people, injure over a thousand, terrorize several thousand more, and spark a national weapons of mass destruction (WMD) counterterrorism industry in the United States.

Analysts have long commented on the copycat nature of terrorists and terrorist groups. Once a new method of attack (from car bombings to airplane hijackings to planes being used as bombs to hostage-taking for ransom money) has met with success, other terrorist groups are bound to emulate it. Given such a phenomenon among terrorists, is the United States witnessing any evidence of an increase in the use of umbrellas in terrorist operations—especially those involving chemical and biological weapons? Should the United States be calling for an international embargo on umbrella sales to Afghanistan to prevent Osama bin Laden and his al-Qaida organization from acquiring such dangerous, dual-use technology? Probably yes. For one thing, *Aum* has now inspired many follow-up attacks than many analysts had predicted shortly after their March 1995 attack. Although the jury is still out, *Aum* may have been unique. Even the Minnesota Patriots Council, which developed ricin because they believed it to be used by the CIA and the KGB, never conceived of using it in the same manner as the Bulgarian secret service. Rather than use an umbrella, the MPC experimented with using hand lotion as a means of dissemination.

For another, an umbrella (even one involving a chemical or biological weapon) simply does not offer the same level of destruction, the same "bang for the buck" as other terrorist methods. Not even the Weather Underground, whose name would seem to imply an interest in such methods, showed any evidence of ever considering using umbrellas in any of their attacks. Instead, they chose the symbolic bombing of the imperialist power structure. So the answer is yes, the standard terrorist arsenal is now the gun, the bomb, the plane bomb, box cutters and even the umbrella or anything else they can get their hands on. As all of us witnessed on 9-11, Osama bin Laden did try such a method of attack, and it brought a whole new meaning to his "umbrella terrorist network [5]."

Hacker-Controlled Tanks, Planes, And Warships

Army officials are worried that sophisticated hackers and other cybercriminals, including military adversaries, may soon have the ability to hack their way into and take control of major military weapon systems such as tanks and ships. The potential exists for hackers to infiltrate the computer systems used in tanks and other armored vehicles. Unlike in the past, today's modern tanks and ships are almost entirely dependent on computers, software, and data communications links for functions such as navigation, targeting, and command and control.

Although the Pentagon has always had computer security issues to deal with, they've never had computers in tanks and armored personnel carriers before. In fact, the Defense Department has already tested and proven that hackers have the ability to infiltrate the command and control systems of major weapons, including Navy warships. According to a training CD-ROM on information assurance, published by the Defense Information Systems Agency, an Air Force officer sitting in a hotel room in Boston used a laptop computer to hack into a Navy ship at sea and implant false navigation data into the ship's steering system.

Yes, this actually happened. The CD-ROM instructs military personnel taking the course. Fortunately, this was only a controlled test to see what could be done. In reality, the type of crime and its objective is limited only by people's imagination and ability.

> Although there are well-known security gaps in the commercial systems that the Army plans to use on the battlefield, hacking into tanks and other weapons may prove to be too difficult for an enemy engaged in battle. The problem for the enemy is that computer security vulnerabilities will almost certainly prove fleeting and unpredictable. Such tactics would be nearly impossible to employ beyond the random harassment level [5].

It is imperative for the United States to study what it means to be a superpower in the Information Age. In addition to the two dozen countries known to be pursuing technologies that would enable them to produce weapons of mass destruction, threats to the nation's critical infrastructure from cyberattacks are also high on the present administration's list of things to prepare for.

Other countries are forming cells of professionals dedicated to finding ways to interrupt the U.S.'s information infrastructure. If you can shut down the U.S.'s financial system; if you could shut down the transportation system; if you could cause the collapse of energy production and distribution system just by typing on a computer and causing those links to this globalization to break down, then you're able to wage successful warfare, and the United States has to be able to defend against that. The United States is presently taking on those defense measures [5].

3.2 U.S. Government Agencies Shape Cyberwarning Strategy Against Terrorists and Rogues

Under pressure from Congress to better coordinate the government's response to computer viruses and other cyberattacks by terrorists and rogue states, the National Security Council (NSC) has developed a plan outlining roles and responsibilities for federal cybersecurity organizations. Under the plan, the National Infrastructure Protection Center (NIPC), working with the General Services Administration's Federal Computer Incident Response Capability Office, will take the lead in alerting agencies to cyberattacks and will coordinate any immediate response.

The memo identifies the organizations and agencies to be involved in various kinds of attacks and defines the criteria for NIPC to call a meeting of the full cybersecurity community. The NSC will step in whenever a security response requires a broad policy decision, according to the plan.

This institutionalizes how the United States will share information both at an operations level and a policy level when cyberincidents occur. Many observers have called for coordination among organizations such as NIPC, the Critical Infrastructure Assurance Office (CIAO), and NSC itself.

NIPC, based at the FBI, was established in 1998 to serve as the government's central organization to assess cyberthreats, issue warnings, and coordinate responses. The CIAO was set up to help agencies develop and coordinate security policies and plans.

The proliferation of organizations with overlapping oversight and assistance responsibilities is a source of potential confusion among agency personnel and may be an inefficient use of scarce technical resources. The calls for coordination became louder after the "I Love You" virus affected almost every federal e-mail server and taxed many agencies' resources. The lack of formal coordination and communication led to many more agencies being affected by the incident than necessary, according to GAO.

Although the many warning and response organizations work together, the NSC memo lays out a standard process for coordination. In the past, that type of coordination happened on an ad hoc basis. Now, as laid out in the memo, the process is set so that it can last into the next administration in 2005 and possibly to 2009.

Some of the formal mechanisms that existed were frankly ineffective in the tasks they were meant to do. For circumstances that are extraordinary, the U.S. now has a process where the NIPC will coordinate the operational response, and the National Security Council will head the policy response [5].

4. THE DARK WORLD OF THE CYBER UNDERGROUND

It was nearly Christmas (1998) when Dionne Smith received an alarming letter that dampened her holiday spirit—to say the least. The anonymous note warned Smith, 31, an employee of a Los Angeles parking company, that by opening the envelope she had just exposed herself to the biotoxin anthrax, livestock bacteria that can be fatal if inhaled. The 1998 Christmas incident was a horrible and frightening experience--which was one of approximately 220 nuclear, biological (see sidebar, "Bioterrorists On The Green"), and chemical scares (including some 140 anthrax false alarms) in this country alone.

Bioterrorists On The Green

Will the next terrorist attack be against plants, not people? At the urging of the White House, the U.S. Department of Agriculture and the FBI are looking at the threat of agricultural bioterrorism—an assault on the country's efficient but fragile system of giant single-crop farms.

The fear is that if some party wanted to, they could damage a major crop—and the economy—by introducing a plant pathogen that doesn't normally exist here. Likely bioweapons include plant-killing fungi, such as soybean rust, or infectious microbes that induce plants to produce toxins. If the group were sophisticated enough, they could genetically engineer a highly pathogenic strain, produce it in large quantities, and sneak a lot of it in.

In wild plants, natural genetic diversity helps limit the spread of disease. Ninety-nine percent or more of the genes in crops are the same across the United States, and that uniformity makes an epidemic much more likely. Once unleashed here, a superbug could spread like wildfire before researchers identified it and figured out how to keep it in check. Even then, spores could survive and infect the next year's crop. They could also be spread by the wind, from field to field, or even state to state. It would be a continuing, recurring problem, like a permanent bomb going off [5].

Even though so far they've all been fakes, the feds are on edge. Their major worry is that terrorists are adding chemical and biological weapons to their arsenal of arms, and that, one day, they'll make good on their threats—thus, enter the dark world of the cyber underground. So, the government has begun taking precautions, and pours billions of dollars into creating a network of programs designed to respond to such attacks. The ambitious plans include amassing antidotes to potential bioagents such as anthrax and other bacteria and viruses, and to chemical weapons such as the nerve agent sarin. The government is training medical response, fire, police, and rescue teams; beefing up local health departments to handle civilians in case of a major attack; and gathering intelligence on terrorist groups believed to be interested in acquiring such weapons. These new programs have helped make counterterrorism one of the fastest-growing parts of the federal budget, even as terrorist acts plunged to a 33-year low prior to the 9-11 attacks, say congressional budget analysts. Total U.S. antiterrorism spending could exceed $90 billion in 2005, up from $60 billion in 2002. The question is whether it's money well spent.

A recent report by the General Accounting Office, Congress's watchdog, says no, claiming that lawmakers have dumped money into fighting a threat yet to be fully assessed. and probably less dangerous than widely believed, considering how tough it is to acquire, process, and use the deadly toxins. A growing number of government and private counterterrorism experts agree. They say that federal officials are so spooked by the possibility of a chemical or biological attack that they are deliberately hyping the threat to get Congress to cough up coveted cash for prevention programs. And most lawmakers are buying it wholesale. It's Mom, apple pie, and terrorism. In 1997, a jittery Congress ordered the Department of Defense to conduct multiagency training exercises in the nation's 120 largest cities against so-called weapons of mass destruction. Today, there are some 400 training courses run by myriad agencies, including the Energy and Justice Departments, the Environmental Protection Agency, and the Federal Emergency Management Agency. In just how many different ways is the United States going to set out to accomplish the same thing, because many of the programs are redundant?

The most eye-popping example of out-of-control spending, detractors say, is the Department of Health and Human Services (HHS). In 1996, HHS spent $7 million on its "bioterrorism" initiative. In 2004, it requested $785 million. Most notably, the department intends to create a national stockpile of millions of doses of vaccines and antibiotics, a potential boon for pharmaceutical companies that are among those eagerly lobbying for more antiterrorism measures. GAO investigators have repeatedly questioned the department's emphasis on vaccines for smallpox, pneumonic plague (airborne bacteria that cause respiratory failure), and tularemia (bacteria that cause a disabling fever in humans). None of these potential killers appear on the CIA's list of biggest germ threats from terrorist groups. Still, HHS is doing the right thing by focusing on them. Tularemia and pneumonic plague are very easy to develop. The easiest to develop is anthrax.

Other agencies are clamoring for a piece of the pie, leading to tremendous internecine fighting. FEMA wants a chunk of the training and equipment money, as does the Justice Department's Office of Justice Programs, whose mission is to dole out federal anticrime money to states and localities. The Department of Veterans Affairs wants to wrest stockpiling duties away from the Centers for Disease Control and Prevention. And the National Guard, a powerful lobby on Capitol Hill, is creating its own hazardous materials response teams, even though there are already more than 800 state and local hazardous material (HAZMAT) units, plus additional crews in the Army, Marine Corps, EPA, and Coast Guard. Then there's the Energy Department, which is pushing for $80 million to research palm-size bug and poison detectors and other antiterrorism products. Not to be left out, the United States Holocaust Memorial Museum and the Office of Personnel Management want $6 million apiece, and the Smithsonian Institution is asking for $7 million to bolster security against potential terrorist attacks.

When Congress first began considering this issue in 1995, the debate was driven by the belief that terrorists, although more likely to use guns and bombs, would eventually turn to lethal chemical and biological agents. The 1995 Tokyo subway gas attack by the cult Aum Shinrikyo was a shot across the bow. So were reports that Osama bin Laden—accused of masterminding the bombings of two U.S. embassies in East Africa—has tried to get his hands on unconventional weapons.

The only major case of bioterrorism in the United States was in 1984 by followers of the Indian guru Bhagwan Shree Rajneesh, who had set up a commune in Oregon. Hoping to sway a local election, they unleashed salmonella poisoning in 10 restaurants in a nearby town, sickening 751 people but killing none. Still, law enforcement officials are convinced that the risk merits whatever preventive measures the government can afford. This is not on the top 100 list of things you're going to die from. But if you're a national security expert, this is on the top 3 list of things to worry about.

One reason there have been no attacks is that it's so tough to effectively use biological weapons. But a dozen hostile nations now either possess or are actively pursuing bioweapons. Most counterterrorism and intelligence experts agree other countries would think hard before striking, because they know the United States would retaliate with stunning force. They also agree that terrorists cannot carry out large-scale lethal attacks without the backing of a foreign government. However, they can do damage. The question is: how much? Nobody knows, because few have bothered to assess how real the threats are. No one, though, wants to be caught asleep at the switch—just in case. It's one of those things it's hard to say no to. It's like fallout shelters in the 1950s. Was that wrong to do? You have to look at the world you're operating in [5].

5. THE CRIMINAL CAFÉ IN CYBERSPACE

Not long ago, if a terrorist wanted to cause a blackout in, say, New York, it would have taken some work. He or she might have packed a truck with explosives and sent it careening into a power plant. Or he or she might have sought a job as a utility worker so he or she could sabotage the electrical system. But now, intelligence experts say, it's possible for a trained computer hacker to darken Gotham from the comfort of home or a cybercafé (at a coffee house). Worse, his other home might be as far away as Tehran, Iran. Worse yet, he or she may enjoy the full backing and technical support of a foreign government.

In a closed briefing to Congress, the CIA reported that at least a dozen countries, some hostile to America, are developing programs to attack other nations' information and computer systems. China, Cuba, Russia, Korea and Iran are among those deemed a threat, sources later declared. Reflecting official thinking, no doubt, the People's Liberation Daily in China notes that a foe of the United States only has to mess up the computer systems of its banks by hi-tech means. This would disrupt and destroy the U.S. economy. Although the specifics are classified, a new National Intelligence Estimate reports at least one instance to date of active cybertargeting of the United States by a foreign nation.

Officials are worried because so much of America's infrastructure is either driven or connected by computers. Computers run financial wireless networks, regulate the flow of oil and gas through pipelines, control water reservoirs and sewage treatment plants, power air traffic control systems, and sustain telecommunications networks, emergency services, and power grids. All are vulnerable. An adversary capable of implanting the right virus or accessing the right terminal can cause massive damage.

In 1996, a Swedish hacker wormed his way through cyberspace from London to Atlanta to Florida, where he rerouted and tied up telephone lines to 11 counties, put 911 emergency service systems out of commission, and impeded the emergency responses of police, fire, and ambulance services. There have been many domestic cyberattacks as well. The number of pending FBI cases involving computer crimes (a category that includes computer infrastructure attacks and financial crimes) increased from 451 in 1999 to about 1,100 in 2004.

In 1997, intelligence officials got a glimpse of what's possible during an information-warfare exercise named "Eligible Receiver." The secret war game began with a set of written scenarios in which energy and telecommunications utilities were disrupted by computer attacks. In one scenario, the attackers targeted the 911 emergency phone system by telling Internet users there was a problem with the system. The scenario posited that people, driven by curiosity, would phone 911 and overwhelm the system.

"Eligible Receiver" culminated when three two-person "red teams" from the National Security Agency actually used hacker techniques that can be learned on the Internet to penetrate Department of Defense computers. After gaining access to the military's electronic message systems, the teams were poised to intercept, delete, and modify all messages on the wireless networks. Ultimately, the hackers achieved access to the DoD's classified wireless network (see sidebar, "Espionage By Keystroke?") and, if they had wished, could have denied the Pentagon the ability to deploy forces. In another exercise, the DoD found that 74% of test attacks on its own systems went undetected.

Espionage By Keystroke?

Forget about signal sites and dead drops (like the recent FBI Russian mole case of suspected spy Robert Phillip Hanssen). The classic tropes of the spy game have gone the way of the Model T. When an FBI computer jock finally hacked his way into Aldrich Ames's personal computer a few years ago, investigators were dumbfounded by the number of secrets he'd purloined from the CIA—hundreds of stolen documents and classified reports. FBI brass called Ames the worst case of treason in U.S. history.

But the preceding could be peanuts compared to the Wen Ho Lee case. Government officials confirmed that scientist Wen Ho Lee suspected of stealing classified data from a secret weapons laboratory downloaded reams of classified nuclear weapons information from a high-security computer system to one that could be accessed with relative ease.

Actually, reams doesn't begin to describe the dimensions of it. The FBI is talking about millions of lines of computer code here, data bits gathered during the course of 53 years of research and more than 5,000 nuclear tests—information that shows how the nation's most sophisticated nuclear weapons work. With a few simple computer strokes, in other words, someone made America's national-security crown jewels available to any reasonably sophisticated person in possession of a computer, a modem, and the file names under which the information was stored. It is flabbergasting. There's just no other word for it.

The someone in question was Wen Ho Lee, a Taiwan-born scientist employed, until recently, at the Department of Energy's weapons laboratory in Los Alamos, New Mexico. Lee was dismissed from his job in 1999 for security breaches after it was disclosed that he was the subject of an FBI espionage investigation. Prosecutors have not charged Lee with spying, and he has asserted his innocence. But when FBI agents searched Lee's computer after his dismissal, officials say, they discovered that he had transferred an incredibly large amount of nuclear data from the Energy Department's high-security computers to the more accessible wireless network, dumped the information under bogus file names, then tried to erase the evidence from his hard drive. The transfers occurred between 1983 and 1995, but accelerated in 1994 and 1995, when Los Alamos began installing a new system designed to impede such transfers. He was really racing right there at the end.

The evidence gathered to date does not show that the security breach resulted in damage on a massive scale. But it is huge nonetheless. The FBI is still investigating whether anyone accessed the data from the low-security wireless network to which Lee transferred the information. Some officials say that may never be known for sure.

Like every espionage investigation, the Lee case is rife with peculiarities. Lee first came under suspicion in 1996, after the CIA obtained a document showing that China's military had obtained classified information about the size and shape of America's newest miniaturized nuclear warhead, the W-88. The FBI was slow to investigate Lee, in part because Lee's wife was working as a confidential informant for the bureau. But there were other problems. When agents in the FBI's Albuquerque field office pressed for a search warrant in Washington, lawyers at the Justice Department rebuffed the request. The Foreign Intelligence Surveillance Court has almost never rejected a search warrant request, and bureau officials indicate the rejection here was unwarranted. In any case, by that time the damage was done [5].

In 1998, the FBI raided the homes of two California high school sophomores. Their hacker assaults on the Pentagon, NASA (which was very easy), and a U.S. nuclear weapons research lab were described as "the most organized and systematic attack" on U.S. computers ever discovered. To make the Pentagon attack hard to trace, the hackers routed it through the United Arab Emirates. They were directed by a teenage hacker in Israel.

To help industries fend off hacker attacks, both foreign and domestic, the government has created the National Infrastructure Protection Center, to be staffed by 458 people from the FBI, other agencies, and industry. Recent events make clear that tighter defenses are needed. In 1997, a 13-year-old boy with a home computer disabled control-tower communications at a Worcester, Massachusetts, airport for nine hours. The loopholes the teenager exploited have been closed. But no computer environment is totally secure. Preventing hacker attacks is like a never-ending journey. You will never get to your destination [5].

5.1 Chinese Cyber Criminal Cafe Hacktivists Spin a Web of Trouble at Home

In the university district of Beijing, a bunch of 20-year-olds calling themselves the "Web Worms" slouch around in an apartment stacked with old issues of *PC* magazine. Chinese wireless computer networks are so easy to break into nowadays. Ninety (90%) percent of them are not secure.

From the moment in 1995 that a commercial Internet provider first gave Chinese citizens access to the Web, the government has tried to maintain what some cybersurfers derisively call "the Great Firewall of China." This elaborate control system is supposed to block sites that the Communist Party considers morally or politically degenerate, from Penthouse to Amnesty International and CNN. But with a few simple tricks, ordinary Internet users are now making a mockery of the Great Firewall, tapping easily into forbidden foreign sites [5].

5.2 Sabotage

Sophisticated hackers, meanwhile, are breaking into sensitive Chinese computers (see sidebar, "Cyberspace Incidents on the Rise in China"). Members of the Hong Kong Blondes, a covert group, claim to have gotten into Chinese military computers and to have temporarily shut down a communications satellite last year in a "hacktivist" protest. The ultimate aim is to use hacktivism to ameliorate human rights conditions.

Cyberspace Incidents On The Rise In China

Intelligence and security experts are warning foreign firms in China of a growing threat of Internet-related crimes, government surveillance, and loss of proprietary data. But some U.S. companies said they view those threats as exaggerated.

The latest warning comes from a report published in 2004 by a network security firm founded by two former U.S. Navy intelligence officers. The report cautions companies that the government-controlled Internet environment in China could put the integrity of their wireless networks at risk.

The most important consideration is that, in one way or another, the government is involved in the operation, regulation, and monitoring of the country's (China) wireless networks. As a result of this and other factors, such as tensions with Taiwan, U.S. enterprises could see an increase in scans, probes, and attacks that could be aimed at gaining technical information.

But representatives from companies with major operations in China indicate that they have never had problems and don't plan to run scared now. The companies discount most of the alarmist reports.

The real focus of their control efforts is what the Chinese call "black and yellow," or political and pornographic material. How serious an issue economic espionage is depends on who you are and what business you're in. And economic espionage isn't unique to China.

Nevertheless, there are other companies who are not convinced that the Chinese government is overtly (or, for that matter, covertly) engaging in corporate espionage via the Internet. Yet, U.S. intelligence experts warn that China's vast intelligence-collecting apparatus has a voracious appetite for any U.S. technology that could help speed the People's Republic's military modernization and boost the country's economy. That puts high-tech vendor companies particularly at risk.

Businesses operating in China are up against a national government that has essentially unlimited resources and a long track record of industrial and economic espionage. Every business in China is run by the government; any effort to develop intelligence and promote those industries is a national effort. Scans, probes, and attacks against U.S. firms in China are statistically confirmed and growing, and could be Chinese tests of offensive information warfare tactics or the work of Chinese virus writers.

The U.S. firms that may be at the greatest risk of losing proprietary data include companies that have set up development laboratories in China. But those companies, eager to gain a foothold in China's burgeoning IT market, don't necessarily share the fears of intelligence experts.

Nonetheless, there are more controls in place in China than in some other countries, but they have not been put in place to foster espionage. Although the Chinese view controls and regulations as necessary to facilitate an orderly Internet market and to protect the country from subversion and other Internet crimes, the controls are partially the result of political rigidity and bureaucratic inertia.

Human nature is the same everywhere in the world. The thought that there are lots of people with time on their hands to explore what the 50 million Internet users in China are doing is totally impractical [5].

Free speech also is proliferating. A political journal called *Tunnel* (*http://www.geocities.com/SiliconValley/Bay/5598/*) is said to be edited secretly in China and sent by e-mail each week to an address in the United States, where it is then e-mailed anonymously back to thousands of Chinese readers. *Big Reference* (*http://www.ifcss.org/home/*) is another on-line challenge to the authorities. One recent issue extolled individualism and paid tribute to the mother of a student killed when troops crushed the pro-democracy protest in Tiananmen Square in 1989.

The Internet provides not only speed and efficiency but also cover. If you tried to do a traditional newsletter promoting democracy in China, you'd surely get arrested. If only the authorities were smart enough to realize what's going on, all the political activities on the Internet would really have them scared.

Perhaps they are smart enough. New regulations introduced in 1997 imposed stiff penalties (including jail sentences) for using the Internet to damage state interests, spread rumors, or publicly insult others. Nonetheless, China's wired population has grown to 7.731 million, according to government figures. Although that is a tiny portion of an overall population approaching 4 billion, China's Internet users are virtually by definition the country's most educated and modern elite. To watch over them, a new force of more than 500 "Internet security guards" has been assigned to patrol wireless computer networks at state companies and ministries. What the Chinese government is really afraid of is political infiltration. The government's goal is to have a security guard in every work unit.

Perhaps most worrisome to the authorities, young Chinese are using the Net to coordinate political campaigns. On August 17, 1998, Indonesia's independence day, hackers in China broke into Indonesian government Web sites and posted messages protesting violence against ethnic Chinese there. Chinese security officials ignored the demonstration until it reached the streets. That day, about 200 students rallied outside the Indonesian Embassy, carrying photographs of rape and murder victims that they had downloaded from the Web. The incidents weren't written up in the Chinese news, but were posted on the Web.

Recently, the government has taken even more drastic action. In Shanghai, a computer engineer named Lin Hai faces charges of inciting the overthrow of state power by providing 60,000 e-mail addresses to *Big Reference*. And at the end of 2000, the publishers of *Tunnel* went into hiding [5].

6. THE SUPER COMPUTER LITERATE TERRORIST

During the next 20 years, the United States will face a new breed of Internet-enabled terrorists, super computer literate criminals, and nation/state adversaries who will launch attacks not with planes and tanks, but with computer viruses and logic bombs. Americans adversaries around the world are hard at work developing tools to bring down the United States' private sector infrastructure. The United States faces an increasingly wired but dangerous world, as evidenced by the following:

- Many countries have programs to develop cyberattack technologies and could develop such capabilities over the next decade and beyond.
- The U.S., Russia, China, France, and Israel are developing cyberarsenals and the means to wage all-out cyberwarfare.
- Terrorist groups are developing weapons of mass destruction.
- Russia has become a breeding ground for computer hackers. The Russian equivalent of the U.S. National Security Agency and organized crime groups recruit the best talent.
- Electronic stock scams, robberies, and extortions are proliferating [5].

A report by the Washington-based Center for Strategic and International Studies (CSIS) went even further, warning of a future cyberarms race and the rise of terrorist groups supported by super computer-literate youngsters bent on disrupting the Internet. China is of particular concern here, because it's devising strategies for unrestricted electronic warfare. Officials said critical infrastructures in the United States could be targeted in the future as revenge for incidents like the 1999 accidental bombing of the Chinese embassy in Serbia. The Chinese government has even suggested having every person in China send one e-mail to an address of interest in the United States or use hacker tools easily available on the Internet to support a mass denial-of-service attack.

On-line extortion and falsification of shipping manifests by criminals, and attempts by countries to use hacking techniques to evade trade sanctions are a rising concern. DoD officials are also becoming increasingly concerned with the proliferation of "always-on" Internet appliances such as modems and wireless network printers. Hackers are finding ways to penetrate these devices and possibly use them as launching pads for more devastating distributed denial-of-service attacks.

In 2000, a hacker cracked into a printer at the Navy's Space and Naval Warfare Center and rerouted a potentially sensitive document to a server in Russia. Therefore, the real threat comes from the design of the U.S. infrastructure and the people who run it.

Companies build these systems and their business models on the assumption that things will always work. If a major attack is made on the infrastructure, it's going to happen from the inside.

The aforementioned reports hold a "powerful message" for the national cybersecurity effort. However, that future preparedness will be determined by how much emphasis companies and the government place on fixing known vulnerabilities, training and education, and enforcing good security policies.

Although the threat of terrorists groups attacking the infrastructure is real, a word of caution is needed. It's scary, but it's really hard to bring down the Internet [5].

7. THE BRILLIANT AND NASTY ROGUE

At a table equipped with two computers, Mark Coletta (not his real name to protect his privacy) plies his trade. The intense, lanky 27-year-old is hunting for holes in an enterprise wireless network.

Mark seeks clues that will reveal operating systems, firewalls, or user names. Any one of these could become a key for breaking into the system. He thinks purely as a brilliant but nasty rogue hacker.

But Mark is no malicious rogue hacker. He's a security engineer at an information security company where he's paid to tinker with clients' wireless networks and uncover their vulnerabilities.

With cybercrime increasingly making news headlines, services such as vulnerability assessments, integration of firewalls and other security components, and subscription-based managed security are in high demand. That demand is spawning a lucrative market. IT security services will generate up to $13.1 billion worldwide in 2006 and are growing at a per-year compound rate of 54%, predicts research firm GartnerGroup.

IT security services are definitely growing, both in demand and in supply. There's a lot of user demand and an enormous number of companies starting in this space.

Most other security people are vulnerability experts who have switched their black rogue hacker hats for white hats. Many ex-hackers have become security consultants. Their vision now is to simplify the security process using the Internet by documenting the best security practices and then providing them to clients over the Web. Tying security into a company's e-commerce [2] strategy also is key. Most people in this business are just out of school.

Meetings with prospective clients begin with a knowledge test of what and whom you know. Once some sort of connection is established, the client will pose technical questions about any number of areas, such as databases, Unix, or the Novell platform.

The thing that makes the security professional different from other IT professionals is that you have to know something about everything. Some clients, influenced by media reports of computer crimes or by upper management pushing a security plan, are ready to "pull the trigger" immediately. Others, though, are hit with sticker shock.

They don't understand how much security costs. Mark has to provide a clear return on investment statement. It's the same problem an insurance agent has. Mark has to identify the probability that something will happen—it's the downstream effect.

To devise a security plan for a client, Mark makes a technical assessment of a wireless network and combines it with his own interviews and observations. He assesses a company's "pain threshold," or how much security risk it can endure before the business would shut down. Once completed, a security plan can be 700 pages long. Mark then either implements the plan or recommends how the client can enact it [5].

8. HOW THEY WATCH AND WHAT THEY KNOW

Palestinian supporters are using a combination of hacking tools and viruses to gain what appears to be the upper hand in the Middle East's ongoing cyber war. How Palestinians hackers watch and what they know will determine the success of this cyber war for them.

They are distributing the tools and viruses for destroying Israeli sites using a recently created attack site. Visitors to the site are greeted with the message, "I swear that I will not use these programs on anyone but Jews and Israelis." The site comes complete with a list of directions on how to use the attack tools.

LoveLetter, CIH, and the Melissa Virus (along with 12 *Word* macro viruses) form the arsenal for attacking Israeli sites. Apparently, it's an effective system.

According to sources at iDefense, an international security firm monitoring the situation, pro-Palestinian hackers are using a variety of tools to orchestrate a well-organized attack against the 400 or more Israeli Web sites that have been hit during the conflict. It is hard to say for sure who is winning. But it appears that the pro-Palestinian hackers have successfully affected more sites.

The pro-Palestinians have been much more aggressive in scope. Instead of just targeting specific sites, they've been methodically working through all the sites, thus broadening their agenda.

Over 559 Web sites have been targeted by both sides for denial-of-service attacks, attempts to gain root access, system penetrations, defacements, and a variety of other attacks. Many sites have been indirectly affected, due to the strain that the attacks have placed on the Net infrastructure in the Middle East.

The conflict began on October 6, 2000, when pro-Israeli hackers created a Web site to host FloodNet attacks. Since then, both sides have sustained blows to vital-information and financial-resource sites such as the Palestinian National Authority site and the Tel Aviv Stock Exchange.

Sixteen tools have been identified as those actively distributed among attackers, with many others being discussed or suspected of already being deployed. One such tool is called the *EvilPing,* believed to have been created especially for this war. The tool launches a "ping of death attack" that, when utilized by several users against the same target, crashes the system.

Then there is *QuickFire,* an attack tool that sends 32,000 e-mails to the victim from what appears as the same address. Used simultaneously by multiple attackers, the tool crashes an e-mail server.

QuickFire strength is that it does not relent, continually firing off thousands of e-mails until the server is shut down and the address blocked. It is believed to be the tool used for hack attacks on the Israeli Foreign Ministry site and its Webmaster's e-mail address.

A group called "Hackers of Israel Unite" originally used another popular tool called *WinSmurf,* which also uses mass pinging to bring down a site. Borrowing amplifying power from broadcast sites, the hackers send out pings that are boosted 10,000-fold, or more. According to the group, they were able to shut down Almanar.org using one computer with a 56K modem and an ADSL line.

According to Netscan.org, a site that provides a list of broadcast sites with an average amplification of times five, a dial-up user with 28.8 Kbps of bandwidth, using a combination of broadcast sites with an amplification of 40, could generate 1152.0 Kbps of traffic, about two-thirds of a T1 link. With tools like these, a 56K can become a powerful weapon and your bandwidth is irrelevant.

Netscan.org creators call themselves a small group of concerned wireless network administrators who got fed up with being smurfed all day. But they recognize the fact that their site has also become a hacking tool.

Pro-Palestinians recently turned the tables by using broadcast-site attack tools against Israeli sites. Although the leaders in the war (groups such as UNITY, DoDi, and G-Force Pakistan) remain in the limelight, many previously unknown hackers are taking the cyberwar to another level.

Hackers are making moves to gain root access to Israeli computers and servers. Root access is the ultimate possession, it means doing whatever you want with a system. In essence, a hacker who gains root access control of a computer can scan, delete, and add files, use it as an attack tool against others, and even view and hear users whose computers are equipped with cameras and microphones.

With no end in sight to the Middle East cyberwar, talk of targeting U.S. interests on the Web has been popping up in chat rooms and IRC channels frequented by pro-Palestinian hackers. Hackers such as DoDi have come out and said that the current war isn't just against Israeli, but the U.S. as well. But Arab activists such as Mustapha Merza believe the American media continues to portray Arabs as terrorist aggressors, even in cyberspace.

The irony of the matter is that the number of times that U.S. government sites have been targeted by Israelis are more numerous than those times they were targeted by pro-Palestinians. Yet, the American media fails to identify its real perpetrators and victimizes the Arabs as usual. For its part, the National Infrastructure Protection Center (a division of the FBI concerned with cyber warfare, threat assessment, warning, and investigation) lists both Israeli and Arab sites that promote the cyberwar [5].

8.1 How Israelis Watch and What They Know Too

A group of self-described ethical hackers are taking the reins of the Israelis' Web wireless networks into their own hands in the Middle East's cyberwar. Known as the Israeli Internet Underground, the coalition of anonymous on-line activists from various Israeli technology companies has set up a Web site to disseminate information concerning the ongoing battle in cyberspace.

According to the IIU mantra, they are dedicated to the Israeli spirit and united to protect Israel on the Internet against any kind of attacks from malicious hacking groups. The site claims to provide a comprehensive list of sites that were hacked by Arab attackers since the cyberwar went into full swing in October 2000.

Listed are over 80 Israeli sites that have been defaced and vandalized by various hacking groups. The number coincides with estimates provided by officials at iDefense, an international private intelligence outfit in Washington that is monitoring the ongoing war. IIU also provides a list of Israeli sites that they believe run services with commonly known security holes such as BIND NXT overflow, IIS 4 holes, and FTP format string bugs.

Examples of defacements by Arab hackers such as the one perpetrated on the homepage of Jerusalembooks.com, one of the largest Jewish booksellers on the Web, serve as a warning to those Israeli sites with suspect security. The Jerusalembooks.com text and graphics were recently replaced with the word "Palestine" in flaming letters and with text asking Israelis if the torah teaches them to kill innocent kids and rape women. The site is currently under construction due to the attack.

Taking credit for the attack is the group "GForce Pakistan," a well-known activist group that has joined forces with Palestinians and other Arab hackers in fighting the cyberwar against Israeli interests.

Working alongside the group is the highly skilled Arab hacker named DoDi. On November 3, 2000, DoDi defaced an Israeli site and stated he could shut down the Israeli ISP NetVision, host of almost 80% of the country's Internet traffic.

Though petty defacements and racial slurs have been the norm on both sides of the battle, Arab hackers like DoDi have promised to kick the war into high gear in the coming years, implementing what they refer to as phases three and four of their "cyber-jihad."

The Muslim extremist group "UNITY," with ties to "Hezbollah," laid out a four-part plan for destroying the Israeli Internet infrastructure at the onset of the cyberwar. Phase four culminates in blitzing attacks on e-commerce sites, causing millions of dollars of losses in transactions. IIU said there is already evidence of phase-four attacks, such as the destruction of business sites with e-commerce capabilities, which they believe caused a recent 12% dip in the Israeli stock exchange.

The current onslaught of cyber attacks against Israel's key Web sites is perhaps the most extensive, coordinated, and malicious hacking effort in history. ISPs and e-businesses must recognize the need to install protection that goes beyond firewalls to provide real security against application-level assaults.

In order to thwart future attacks, IIU has created what they call the "SODA project" (sod is Hebrew for secret). The stated goal of the project is to inform and provide solutions wherever the IIU can and, therefore, protect their sites against political cyber-vandalism. It lists those Web sites with security vulnerabilities, making them susceptible to future attacks by Islamic groups.

The SODA project formed an alliance with the Internet security firm 2XS Ltd., which is linked to the site and agreed to provide security advice for casualties of the cyberwar. 2XS Ltd., however, does not accept responsibility for IIU actions. On November 3, 2000, IIU contacted 2XS Ltd. to share their idea of creating a site for publishing vulnerability alerts.

Another link on the SODA project is the Internet security information forum SecurityFocus.com, a resource guide to on-line security links and services based in San Mateo, California. The site is not taking any sides in the Middle Eastern war.

Typically, the odds are heavily in the attackers' favor — the attacker can launch attacks against any number of sites for little to no cost. They only need to find one vulnerable victim to succeed, perhaps after checking thousands of potential victims.

Because both Arabs and Israelis are launching volley after volley against the others' sites, neither faction gets to play the victim in this war. The victims end up being citizens and businesses in the affected area. Unfortunately, that's not uncommon in that part of the world [5].

9. HOW AND WHERE THEY GET THEIR TOOLS

Despite increasing concern about cyberterrorism, the tactics and goals of the world's terrorist organizations remain low-tech. Although the terrorist's toolbox has changed with the advent of the Information Age, the objectives of the world's terrorist organizations have not.

A growing percentage of terrorist attacks are designed to kill as many people as possible. Guns and conventional explosives have so far remained the weapons of choice for most terrorists.

However, terrorists are adopting information technology as an indispensable command-and-control tool. Raids on terrorist hideouts, for example, are increasingly likely to result in the seizure of computers and other IT equipment. Instead of just finding a few handwritten notebooks and address books, counterterrorism authorities are faced with dozens of CD-ROMs and hard drives. Likewise, terrorists' increasing use of advanced encryption tools often delays the process of finding key files and information.

Terrorists groups, such as the Osama bin Laden organization, have yet to demonstrate that they value the relatively bloodless outcome of a cyberattack on the nation's critical infrastructure, but the threat remains real. There are warning signals out there. If the United States fails to recognize this, then the United States will pay another high price like they did on 9-11 [5].

9.1 Information Weapons

There are several weapons or tools currently available that can negate, destroy, or incapacitate information systems, with many more being rapidly developed. Within this part of the chapter, these are broadly grouped into three main types: High Energy Radio Frequency (HERF) guns, Electro-Magnetic Pulse (EMP), and other information weapons [5].

9.2 HERF Guns

A HERF gun (as discussed briefly in Chapter 22) is a device that directs high-power radio energy at an electronic target. Electronic circuits are vulnerable to overload; a HERF gun simply overloads particular circuits to disable specific pieces of equipment that are dependent on that circuit. A HERF gun can be designed to cause varying degrees of damage from simply shutting a system down to physically destroying equipment. Pointed at a computer, a HERF gun may either permanently or temporarily terminate its operations. A HERF gun pointed at a "fly-by-wire" aircraft may trigger a catastrophic failure.

Although currently limited in range and destructive capacity, in the near future, HERF guns are likely to be substantially more capable and freely available and, therefore, must be taken seriously. HERF guns represent an excellent addition to the offensive military inventory of a nation, and also a significant threat if possessed by an enemy. The defensive measures that can be employed to reduce the risks of HERF attacks are not well developed at this stage, but include using Gaussian shielding, gaseous discharge devices, and the maintenance of physical separation [5].

9.3 Electro-Magnetic Pulse

Electro-Magnetic Pulse (EMP) has been described as "the next great weapon to evolve in modern warfare." Initially discovered as a side effect of nuclear tests, the phenomenon has now been extended to non-nuclear generators. Such generators can create an EMP that will disable unshielded electronic systems. A development beam generator with a one gigawatt capacity could be used to develop a line of sight EMP that would knock-out most unshielded electronic devices within a radius measurable in tens to hundreds of meters, depending on the employment method. High power microwaves, communications, computers, navigation, and data processing systems would be most affected by such weapons. The current limitations of these weapons are power generation and capacitor storage capability, but these can be expected to be overcome in the future.

Research is well advanced with EMP warheads recently being fitted on USAF air- launched cruise missiles. EMP weapons are less discriminatory than HERF guns and could be used to shut down a general area rather than a specific system. Again, with the exception of screening techniques such as Gaussian shielding, defensive measures are not common [5].

9.4 Other Information Weapons

There are several weapons that are currently being developed that do not fit in the HERF or EMP categories. Some already are in service with various military forces, others remain on the drawing board. The following weapons are described in a variety of freely available publications and give an indication of the technologies being developed and the possible capabilities that may result [5].

9.4.1 Low-Energy Lasers

These lasers can be used to damage the optical systems of sensors (including data collection devices), thus attacking the information systems at the data collection level. Low-energy lasers have already been fitted on rifles and armored vehicles and were deployed during the Gulf War. A number of systems are reported to be under further development in the United States and United Kingdom [5].

9.4.2 Electrical Power Disruption Technologies

An electric power disruption munition was first used during the Gulf War in 1991. The technology originated after an accident on the U.S. West Coast when chaff cut power supplies to the city of San Diego in 1985. The weapon uses light conductive carbon fibers that wrap around transmission lines and distribution points to cause a massive short circuit. Even when power is restored, the fibers must be removed because any breeze can result in another short circuit. This weapon can be delivered by cruise missiles, as was the case in the Gulf War, or from manned aircraft.

Individually, each of the military information operations (MIO) tools (previously discussed in Chapter 22) and techniques just described, will present a military commander, whether operating in the conventional or IW environment, with a substantial force multiplier. Collectively, they offer a decisive addition to military power. As more MIO capabilities are developed, the effectiveness of the MIO strategy will increase exponentially, reflecting the synergistic relationship that exists between individual elements of the MIO environment. Accordingly, nations developing information strategies should consider investment, both intellectually and financially, across the gamut of information operations [5].

9.5 New Arsenals, Old Rivalries

Could a small country develop the capability to hit the United States with a long-range unconventional weapon? Most certainly one could. But whether such a state would be inclined to try is an entirely different matter.

The risk is real. Congress learned in 1998, according to The Report of The Commission to Assess the Ballistic Missile Threat to the United States, that the United States has entered "a new nonproliferation environment" in which there is a far greater availability of ballistic missiles and weapons of mass destruction. As previously mentioned, the report was the work of a private commission headed by Donald Rumsfeld, secretary of defense under President Reagan and now Bush.

To begin with, the report refers to a club of renegade nations that appear to work with one another to DoDge the strictures of nonproliferation agreements. The states trade with one another, and build on the progress of other members of the club to advance their own systems. Indeed, it is arguable that the recent North Korean firing of a Taepo Dong missile was meant to further its own missile development and to serve, in effect, as a marketing demonstration to attract buyers from other countries outside the international nonproliferation framework.

Another factor, the report points out, is that access to information on a global scale keeps growing exponentially, as the bounds of the Internet in particular remain uncharted. What's more, there has been an easing of access to what the Rumsfeld Commission terms the rudimentary technologies that were employed in early generations of U.S. and Soviet missile systems.

There is yet a fourth factor: the flexibility with which technical personnel from the West, and especially from the former Soviet Union, can move to a potential proliferator. Because so many third-world countries now have ballistic missiles of their own and, therefore, are interested in upgrades, whether of guidance or range, they are less pressed to acquire whole systems.

They are also well aware that it is the acquisition of such whole systems that garners the most international attention and is most easily policed by the Web of agreements, such as the Missile Technology Control Regime, that the United States and its allies have spun to guard against proliferation.

Instead, what many nations are focusing on is brainpower, people who are intimately familiar with technical data packages, who can advise on both long-term improvements and quick fixes, and who can offer recommendations on everything from guidance systems to materials technology to quality control to integration.

Many such scientists and technicians, particularly in the former Soviet states, are willing and eager to improve their material lot by helping renegade nations enhance systems that were often acquired from the Soviet Union or that are derivatives of such systems.

Although Western nations recognize the destabilizing impact of peripatetic unemployed scientists working in countries that "show them the money," they can produce few options. Other than propose alternative employment, the United States and its allies have little to offer, particularly to those motivated by ideological or religious ideals. A significant number of missile owners are potential adversaries of the United State. And, many of the third-world powers have mutual supranational interests. Should a Muslim nation, for example, be taken over by extremists, it could seek support in other Muslim nations from like-minded elements that might not necessarily have seized power, but would be in a position to offer the new regime intellectual assistance and perhaps financial aid.

Certainly, even the availability of resources, and of willing foreign technologists, combined with nefarious intentions, does not in itself suffice for the successful pursuit of a program for intercontinental ballistic missiles. Otherwise, Libya would long ago have been in a position to threaten the United States. Nevertheless, the ability of lesser powers to mount such a threat over the next two decades cannot be ruled out.

Furthermore, as the recent North Korean and Iranian missile tests demonstrated yet again, a third-world country whose leadership is determined to advance its capabilities will not be deterred by nonproliferation regimes. It will find ways to draw upon outside resources in support of its program; and, no less important, it will do so well in advance of the timetables set forth by Western intelligence [5].

9.6 Stolen Thunder Tools

Like a neutron bomb (whose design Chinese agents allegedly stole), the Cox report demolished any doubt that China engages in espionage against the United States (see sidebar, "China Grabs U.S. Technology to Modernize Its Military"). But it left standing a whole array of big questions and small mysteries.

China Grabs U.S. Technology To Modernize Its Military

The request, to a Massachusetts defense contractor, seemed innocent enough: China needed fiber-optic gyroscopes, the latest in navigation equipment, for a new high-speed rail system, the buyers allegedly said. The deal might have gone through if not for a small hitch: The manufacturer recalled that the men, using a different company name, had tried earlier to get a U.S. license to export the gyroscopes to China—and had been turned down.

In 1999, U.S. Customs agents in San Diego arrested a Chinese national named Yao Yi for criminal export violations. Yi has pleaded not guilty; as well as his co-conspirator, Collin Shu, a Canadian, who was also arrested and pleaded not guilty. The two are accused of conspiracy to illegally export items designed for military purposes. The gyroscopes are generally used for guiding missiles or maneuvering fighter jets. To put these in a train, is like putting an F-14 engine in a Cessna.

Intense Debate

The gyroscope case is one of the latest incidents illuminating Beijing's voracious appetite for high-end U.S. technology that has military capabilities. Also in 1999, a man was arrested in Detroit for allegedly trying to illegally ship to China a riot-control vehicle. A report by a panel chaired by Rep. Christopher Cox (R-Calif.), in 1999, suggested that China may have married U.S. computer technology with nuclear weapons designs it stole in the 1980s from U.S. labs. The report presented no hard evidence of this. But it will almost certainly add fuel to an already intense debate over exports of high-speed computers to Beijing.

Proponents of the sale of high-tech goods to China say they help open the country to influences like American television shows beamed off U.S.-manufactured satellites. And, they add, U.S. electronics firms need foreign markets like China if they are to stay healthy in the face of stiff foreign competition.

The present and past administrations have generally supported this view. But in 1999, in a surprising turnaround, past Clinton advisers blocked California satellite maker Hughes Electronics Corp. from sending two $670 million satellites to be launched in China. Various officials offered different explanations for the decision, but the government told Hughes the launches could transfer too much militarily significant know-how.

Critics of high-tech exports to China say they have other concerns as well: The same technology that is already turning China into a land of ATM machines and cell phones could help the People's Liberation Army begin to master information warfare. Pentagon officials counter that nobody is assessing the impact of the fiber-optic lines, electronic-switching gear, computers, and satellites pouring into China. Some examples:

In March 1996, as Beijing was threatening Taiwan with missiles, the State and Defense departments approved the export of two satellite receiving stations worth $7.3 million. The recipient, documents show, was China Electronic Systems Engineering Company, part of China's military. The stations came equipped with ports to plug in Chinese-made encryption devices. The National Security Agency signed off on the deal, but congressional critics say the sale deserves a second look.

China buys nearly half of the supercomputers exported to high-risk countries. Experts point out that the Chinese can evade U.S. export controls by harnessing together less powerful machines—or buying high-capacity machines on a Russian Internet site. Industry groups plan to lobby Congress to allow more powerful machines to be exported, arguing that 1995 limits are already outdated. The Cox report calls for greater scrutiny, including spot checks in China to ensure the best machines are used only for civilian purposes. There are two trains rushing down the track directly at each other on this.

Experts say China's rapidly modernizing military is still years from catching up with the United States, at best. But some worry that China will put high-tech imports to their best military uses and turn into a surprising adversary [5].

The most sensational charge in the 872-page report—that China has obtained secret data on every warhead in the U.S. nuclear arsenal—is based on a single document that a Chinese agent deliberately fed the CIA in 1995. Why would China's spy masters tip their hand? Maybe they bungled, giving away too much in an effort to plant a double agent. Maybe they were warning Washington to butt out of China's touchy relationship with Taiwan. Or, maybe, they were just following the 1,500-year-old advice of the military philosopher Sun Tzu to "sow confusion in the enemy's camp."

The release of the bipartisan Cox report in 1999 certainly did that. Its overall conclusions are chilling. For two decades, it says, China has used spies, front companies, and scientific exchanges to filch some of America's most precious secrets. But on closer reading, it is still unclear how much damage has been done to U.S. national security. In most cases, it seems, Beijing got helpful hints, not blueprints.

Democrats on the congressional panel, which was led by Republican Rep. Christopher Cox of California, unanimously approved the report. But they also questioned its alarmist tone. There are, unfortunately, a number of places where the report reaches to make a point and, frankly, exaggerates.

On the other hand, concrete advances from spying sometimes don't show up in weapon systems until years later. It's possible, as Cox and some Pentagon officials argue, that the sum of China's technological thievery is even larger than its parts. So how worried should Americans be? Here's what the report does and does not say [5]:

9.7 Nuclear Warheads

China stole classified design data on the W-88, a miniaturized nuclear warhead that is the most advanced in the U.S. arsenal. The CIA discovered this in 1995 when a Chinese "walk-in" (an agent who came forward voluntarily) handed over a Chinese document stamped "secret." The unclassified version of the report does not reveal the contents of the document, but an administration official at the time said it contained two "quite specific and detailed" bits of data on the W-88. One was the size of the "package" containing the nuclear device, whose yield (explosive power) was already available from open sources. Although useful, that knowledge is a far cry from a detailed plan for a nuclear weapon. It's more like looking at a car's engine compartment and knowing how much horsepower the block can produce.

Because the CIA later determined that the "walk-in" was a double agent acting on the orders of China's intelligence service, it is unclear whether the Chinese had already milked the information or never considered it all that important. The Chinese document, dated 1988, also described the size and yield of four other U.S. warheads. But that may have come from publicly available sources.

Why does China covet America's nuclear secrets? The Cox committee concluded that U.S. technology would help China build smaller warheads to sit atop a new generation of lighter, mobile missiles. But the upgrade has been in the works for 23 years. And most experts think that its goal is to ensure China's "second-strike capability"–the ability to retaliate for a nuclear attack, not to launch a first strike. Beijing's leaders have good reason to worry about the reliability of their current strategic-missile force: fewer than 20 aging, 1950s-era rockets.

The first of the new missiles, the DF-31, won't be able to reach most of the territory of the United States. But could it intimidate China's neighbors or make the United States hesitate to defend Taiwan in a crisis? Definitely [5]!

9.8 Rocket Technology

The Cox panel was established partly to look into allegations that two U.S. aerospace firms, Hughes and Loral, helped China to improve the reliability of its Long March booster rockets. The report says that the two companies ignored restrictions on technology transfers and gave away sensitive information while helping China investigate a series of failed attempts to launch the firms' satellites into space.

What, exactly, did Chinese scientists learn? How to build better "fairings," the nose cone that protects the satellite during launch. How to compensate for the violent winds that buffet rockets in flight. How to fix the Long March 3B's guidance system. How to better investigate failed launches. This information has improved the reliability of Chinese rockets useful for civilian and military purposes.

Still, it is unclear how quickly China will be able to make those improvements. In the past, China has sometimes had difficulty absorbing Western technology. The spying and technology transfer is of enormous concern. But, having it in your hand doesn't mean you know how to use it or effectively deploy it [5].

9.9 Computers

There is no mystery about how China got advanced computers. The question is what it does with them. Under relaxed export rules, China has legally bought 1,236 high-speed, American-made computers since 1996. The Cox report says that they have been used in nuclear weapons applications, such as modeling hypothetical explosions rather than conducting real ones after Beijing signed the Comprehensive Test Ban Treaty in 1996. But the congressional panel recommended spot checks to monitor the use of U.S. computers in the future, rather than cutting off sales [5].

9.10 Radar

The Cox report also asserts that classified U.S. radar research stolen by the People's Republic of China could be used to threaten U.S. submarines. But the White House produced a letter from the Navy to the Justice Department stating, "It is difficult to make a case that significant damage has occurred" from the alleged disclosure.

China has never aspired to a large nuclear arsenal. One possible explanation for Beijing's disclosure of its own espionage is that Chinese leaders wanted the world to know they could build a large, modern arsenal—if they wanted to. It's deterrence on the cheap. If that was the plan, it just might have worked [5].

10. WHY TOOLS ARE EASY TO GET AND USE

Why are hacking and information warfare tools and weapons of mass destruction easy to get and use? Easy answer: They can be stolen!

For example, investigators at Los Alamos National Laboratory, in 2000, discovered that computer hard drives containing nuclear weapons data and other highly sensitive material stored in a vault at the laboratory had disappeared, according to several United States Government officials. The hard drives were stored in locked containers inside a vault in the nuclear weapons division of the national laboratory. Officials reported that the hard drives were missing on June 1, 2000 after officials went to search for them following the forest fires in the area. The containers remained in the vault, but the hard drives were gone.

The material, stored in the vault of the laboratory's X Division, where nuclear weapons are designed, contained what officials described as nuclear weapons data used by the government's Nuclear Emergency Search Team, or NEST, which responds to nuclear accidents and nuclear-related threats from terrorists. The material includes all the data on American nuclear weapons that the team needs to render nuclear devices safe in emergencies. In addition, the missing material included intelligence information concerning the Russian nuclear weapons program.

The Energy Department's security czar at the time, Eugene E. Habiger, conducted an intensive search and investigation at Los Alamos but did not find the data. He has written a secret report on the matter, and the F.B.I. has been brought in to assist with the investigation. Officials said they remained uncertain whether the data has been misplaced or stolen.

The disappearance of the nuclear weapons data represents a major embarrassment for a laboratory that had already spent the past year under scrutiny for lax security in connection with the Wen Ho Lee case. As previously mentioned, Dr. Lee was a scientist at Los Alamos who was fired in March 1999 for security violations after being the subject of a counter-intelligence investigation that looked into evidence that China had stolen American nuclear secrets.

Dr. Lee was never charged with espionage, but after he was dismissed, investigators accused him of downloading and copying vast amounts of secret nuclear weapons data from a secure computer at Los Alamos into an unclassified wireless computer network and onto portable tapes. Dr. Lee was arrested in December 1999, on charges of mishandling classified material.

The discovery of Dr. Lee's unauthorized downloads in April 1999, prompted then Energy Secretary Bill Richardson to order a shutdown of the lab's computer systems, while mandating security training sessions for Los Alamos employees. Congress later passed legislation creating a new nuclear weapons agency within the Energy Department to oversee Los Alamos and the nation's other nuclear weapons laboratories. The new security breach is believed to have occurred long after Dr. Lee was dismissed.

Energy Department officials indicated that they notified the F.B.I. as soon as they discovered that the material was missing. But some law enforcement officials say that officials at the lab downplayed the fact that the data was missing from the vault, and assumed that the hard drives would turn up somewhere else in the lab. The officials are said to have assumed that the material was in use somewhere in the lab by Los Alamos scientists. The fact that the missing data included intelligence reports, has led law enforcement officials to become skeptical that the material was simply misplaced. The hard drives were eventually found near a trash-can. Questions remain: Who took them? And how much classified material was downloaded before they were eventually found [5]?

11. WHY NASTY PEOPLE ARE SO HARD TO TRACK DOWN AND CAPTURE

When a pair of suicide bombers crippled the destroyer USS Cole in a Yemeni port in mid-October 2000, killing 17 American sailors, top U.S. counterterrorism officials had a fearful intuition: This is revenge for Albania. In mid-1998, the CIA, working with Albania's intelligence service, had rolled up a terrorist cell guided by wanted dead or alive Saudi exile Osama bin Laden. A deadly bombing of the U.S. Embassy in Albania's capital, Tirana, was barely averted. The Middle Eastern plotters were sent home to face prosecution. They have not forgotten that. And they're still looking for payback.

The United States has yet to nail all of those responsible for the Cole attack (even though a few suspects have been detained), but that first guess made a macabre sort of sense for those waging the interminable war against terrorism. Few Americans realize the full extent and intensity of what has become an around-the-clock, across-the-globe campaign against fundamentalist Islamic terrorists, a confusing web of groups and names fused only by their hatred for the United States and, often, their shared experience fighting the Soviet Union's occupation of Afghanistan in the 1980s.

This war is waged largely in the shadows, a cat-and-mouse contest between terrorists and intelligence agencies that only rarely comes into public view. But much of the vast U.S. national security machinery (and a ballooning budget) is trained on the threat. Fragmentary bits of data gleaned from eavesdropping satellites, human informers, friendly governments, and old-fashioned police work are pieced together to deter and disrupt terrorist attacks on a regular basis [5].

11.1 High Alert

Recently, U.S. intelligence agencies warned that U.S. naval forces in Italy and the airbase in Incirlik, Turkey, were being targeted. The posts went on high alert, and the aircraft carrier USS Truman was diverted from Naples to Crete. Meanwhile, the State Department sent a global alert to all its posts, ordering them to review security procedures.

The war comes out of the shadows when U.S. intelligence and law enforcement agents lose a battle, such as in the Cole attack or the 1998 bombings of two U.S. embassies in East Africa (see sidebar, "Putting Terror Inc. on Trial"). But as traumatic as they are, for each such loss there is many an unheralded success—dashing terrorists' hopes of more bloodied bodies and battered buildings on the world's TV screens.

Putting Terror Inc. On Trial

Ali Mohamed is a man of many faces: Egyptian intelligence agent, U.S. Army paratrooper, FBI informant, and aide to wanted dead or alive terrorist mastermind Osama bin Laden. Before bombs shattered U.S. embassies in Kenya and Tanzania, Mohamed says, he scouted possible targets and personally brought bin Laden photos of Nairobi sites. Bin Laden looked at the picture of the American Embassy, and pointed to where a truck could go as a suicide bomber.

Mohamed, 51, is now poised to play a new role—as the Justice Department's star witness in the long-awaited trial of bin Laden's alleged followers which started in January, 2001 in New York City. A sweeping 319-count indictment charges bin Laden and 20 others with a terrorist crime spree dating back to 1991. Among the charges: bombings, perjury, and conspiracy to murder Americans around the globe. The attacks include not only those on the U.S. embassies in 1998 (which left over 220 dead and 5,000 injured) but also on U.S. troops in Somalia and Saudi Arabia. Although bin Laden remains at large (with a $100 million U.S. reward on his head), five of those indicted are now in U.S. custody—as is Mohamed, who pleaded guilty in October 2000.

Infidels

October's (2000) suicide bombing of the USS Cole and the 9-11 attacks (tied by investigators to bin Laden's network) has added fresh urgency to the government's efforts to thwart the wanted dead or alive Saudi exile, now hiding in the badlands of Afghanistan or nearby in Pakistan. Led by Mohamed's likely testimony, the trial promised an unprecedented look at America's most wanted terrorist and at al-Qaeda, the fanatic organization that he guides. The indictment imparts an image of a paranoid, virulently anti-American network determined to purge Muslim lands of "infidels." To achieve this, bin Laden's men strove to obtain chemical and even nuclear weapons, according to prosecutors.

Proving a grand conspiracy may be difficult. Prosecuting international terrorists is often a delicate balance between law enforcement's need for evidence and the intelligence world's need to protect sources and methods. Through electronic eavesdropping, for example, U.S. officials quickly learned of bin Laden's involvement in the embassy blasts, but they are loath to introduce such sensitive records into court.

Holy War

Such concerns may explain the indictment's at times tenuous links among the alleged terrorists. Prosecutors tie bin Laden to the conspiracy largely through his funding of al-Qaeda and his calls for holy war against the West. For some defendants, their work with al-Qaeda appears to be enough. For others, it is their work in his businesses in Sudan, from construction and agriculture to an investment house, which prosecutors call fronts for terror. Still others are tied to al-Qaeda's ruling council, where terrorist plots (like the 9-11 attacks) are said to be hatched.

With his guilty plea, Mohamed has now made the prosecution's job far easier. Under oath, Mohamed already has done more than tie bin Laden directly to the embassy bombings. He strongly hinted he could connect the dots to the five others in custody, who have all pleaded not guilty.

Two of the defendants, Mohamed Rashed Daoud al-Owali and Khalfan Khamis Mohamed, face the death penalty if convicted, as prosecutors have the strongest evidence tying them to the embassy attacks. Al-Owali, a Saudi Arabian, allegedly filmed a statement before the bombing celebrating his "martyrdom," and rode in the pickup carrying the Nairobi bomb; he was found later in a hospital with keys to the truck's padlock nearby. Prosecutors say Khalfan Mohamed, a Tanzanian, helped grind up TNT and load the truck used in the Dar es Salaam bombing. A third defendant, Saddiq Odeh, a Jordanian, is allegedly tied to TNT and detonators used in Tanzania.

A fourth man, Mamdouh Mahmud Salim, allegedly purchased the 1998 Toyota Dyna truck that carried the bomb in Nairobi. His case was recently severed from the others after he stabbed a prison guard in the eye. Investigators are hoping a fifth defendant, Wadih el-Hage, will follow Ali Mohamed's lead and cooperate. A tire store manager in Arlington, Texas, he acted, prosecutors contend, as a bag man and passport fixer while working as bin Laden's personal secretary.

Targets

Ali Mohamed's testimony, which will likely earn him a reduced sentence, may prove particularly damning to el-Hage. The former U.S. Army sergeant, a naturalized American citizen born in Egypt, claims he worked with el-Hage in Nairobi and that during a visit to the man's house, bin Laden's security chief told him to conduct surveillance on American, British, French, and Israeli "targets" in Senegal.

Defense attorneys on the case know they're facing tough odds. Mohamed's guilty plea has thrown "a wrench" into their strategies. For defendants facing the death penalty, their lawyers' primary focus is to stop them from getting killed. If Ali Mohamed does indeed take the stand, his credibility will likely come under fire. The talkative terrorist has a record of shifting loyalties and admits to lying to investigators in the past.

El-Hage, a naturalized U.S. citizen, certainly seems to be feeling the pressure. Five days after Mohamed's testimony, he suddenly also attempted to plead guilty. The plea, offered without consulting with prosecutors, was thrown out because el-

Hage told the judge he was acting not out of guilt, but because he wanted to escape the humiliation of a trial. Should el-Hage decide to flip with prosecutorial blessing, his testimony could offer a trove of information. Court documents place the 43-year-old el-Hage within a rogues' gallery of terrorists. The Lebanese native is allegedly tied not only to the embassy bombs but also to a string of criminal acts, including attempted arms sales to those later convicted in the 1990 murder of radical Rabbi Meir Kahane and the 1993 World Trade Center bombing.

Further revelations may come from Ali Mohamed, who is cooperating with the FBI. Terrorism experts already are pondering his assertion that through the mid-1990s, bin Laden's al-Qaeda maintained close ties to Hezbollah, the Iranian-backed militia, and to Iranian security forces. Al-Qaeda and its allies received explosives training at Hezbollah camps in Lebanon, Mohamed claimed, and received bombs disguised to look like rocks from the Iranians. The implications are troubling. Iran is an untold story in this. How many elements have they kept out of this indictment?

Perhaps several. Ties to the USS Cole bombing may well emerge from trial testimony. And a further indictment in New York (this one under seal) names even more alleged bin Laden conspirators. Clearly, the trial will be but one act in an ongoing and altogether grim play [5].

In November, 2000, for instance, authorities in Kuwait, who thought they had radical Islamists under control, got a nasty shock. They uncovered a tiny terrorist cell plotting to bomb U.S. and Kuwaiti facilities, and quietly called in the CIA, which helped trace the plot beyond Kuwait's borders. A suspect, Mohammed al-Dosary, led investigators to a desert weapons cache that held 293 pounds of high explosives, 1,450 detonators, and, for good measure, five hand grenades. They were in the final stages of casing targets, claims a U.S. official. Even more worrisome, the plotters had helpers in Kuwait's government, one of the closest U.S. allies in the Persian Gulf.

Publicly, the face of the adversary is bin Laden's. But focusing on one man misses the full picture. Bin Laden has tapped into what U.S. officials sardonically call the "Afghan Veterans" Association, Arabs who answered the call to holy war against the Soviets two decades ago—at the time, with backing from the CIA. The threat posed by the Jihadist network didn't become clear until five years ago, with the U.S. arrests of those behind the

1993 World Trade Center bombing. That case sounded the first broad alarms that thousands of Arab veterans from the Afghan war had now trained their sights on the West [5].

11.2 Terror Inc.

Bin Laden finances and motivates a "network of networks," co-opting homegrown terrorist groups, from Egyptian Islamic Jihad to the Abu Sayyaf group in the Philippines to the Islamic Movement of Uzbekistan. It's like you're winding up little dolls and sending them back to their own countries and letting them create their own movements. The United States was slow as a government to recognize what bin Laden was doing. Bin Laden was doing something much more than spreading the money around.

Without fanfare, Washington in 1999 opened a new front in the war. The strategy: Go on the offensive, and hound and disrupt terrorist cells wherever they can be found. U.S. intelligence agencies routinely tip off local security services to a problem they didn't even know they had. Cells are placed under surveillance or, using a legal process called "rendition," suspects are forcibly returned to their home country. In Albania, Algeria, Pakistan, Syria, and elsewhere, bin Laden devotees have been booted out, often on immigration charges and with little publicity. More than two dozen suspects have been brought to justice. The war on terror is fought down in the weeds. It's guys talking to their sources, pulling people in for questioning, and digging for telephone records. It's the slow, dirty, grunt police work that goes on every day.

In 2000, the CIA launched the largest counterterror operation in U.S. history, working with counterparts in Jordan and other countries to thwart a multicontinent "terrorist spectacular" during the millennium celebrations. CIA operatives in more than 60 countries pressured, pleaded, and paid local authorities to crack down on Islamic radicals. The message was: It's crunch time. This cost the agency a great deal of money and resources.

But like battling the mythical Hydra, an eliminated terrorist cell only seems to regenerate, sometimes in the same place. Bin Laden and his organization, al-Qaeda, are still itching to pull off an attack in pro-Western Jordan. And though U.S. and Kenyan authorities busted up an al-Qaeda operation in Nairobi in early 1998, the victory was only temporary. They immediately came back in. They were able to use the infrastructure that was in place, spin up a new cell, and go after the target. The August 7, 1998 blasts at the U.S. embassies in Nairobi and Dar es Salaam, Tanzania, killed more than 220 people.

Just as sobering is what officials call the "mujahideen underground railroad," a vast effort to move young recruits to terrorist training camps in Afghanistan. An estimated 40,000 recruits have gone to Afghanistan since 1996. Using professionally forged documents and Hotmail Internet accounts to keep in touch, network members move people through Italy, the Balkans, Turkey, and Dubai [5].

11.3 Recruitment

The Finsbury Park Mosque sits in a gritty part of London, not far from the Arsenal soccer stadium. Inside, the only sign of Islamic activism is a hand-made sign protesting Russia's war in Muslim Chechnya. But U.S., British, and Yemeni officials say the mosque is a recruitment station for terror camps and that its fundamentalist imam, Abu Hamza al-Masri, has ties with terror groups abroad.

Often, the process begins when a potential recruit visits a local mosque and makes a small donation. While some of the funds may go to legitimate Islamic charity work, bin Laden receives a steady stream from mosques, charities, and schools. The way to get rich is to come up with a scheme to get everyone to pay you 5 cents a month. That's what he's done. The United States has tried to block the money flow but has made little progress.

Despite the nature of the quarry, progress is being made in the war on terrorism. The U.S. is certainly holding their own. There's a chance that the U.S. is gaining ground, but not very much. Bin Laden's very success (due to the 9-11 attacks) has spurred unprecedented international law enforcement cooperation. Jordanian officials alerted U.S. agencies to the millennium threat. Even Russia, which fears radical Islam in Chechnya and Central Asia, now works regularly, if quietly, with U.S. counterparts. This is a dramatic turnaround. Nevertheless, some U.S. agencies still *dropped the ball* in not being able to prevent the 9-11 terrorist attacks.

Through eavesdropping and, increasingly, informants, Western spy agencies are gaining a clearer picture of the structure of bin Laden's network and his inner circle. But penetrating the distinct cultures in which the terrorists operate is difficult. The CIA has begun a special program to recruit Muslims, in hopes of worming its way inside. One advantage: the terrorist networks' decentralized structure. They are vast, but they're not real tight. The CIA is more aggressive but is hampered by 1995 regulations restricting recruitment of sources with unsavory backgrounds. The CIA claims the rules don't block operations.

Nor have the terror fighters been able to get bin Laden himself, who still moves between homes, residences, and underground bunkers in Afghanistan or in neighboring Pakistan. He is protected by what a Pentagon official calls "double walls of security." One is provided by his dwindling Taliban hosts, and his own personal security detail includes an elder son who is said to rarely leave his side, but now is presumed dead.

> **Note:** Recently, terror fighters were further embarrassed by the elusiveness of bin Laden, when he showed up at his elder son's wedding in full view of international media cameras.

The terrorists do their own spying. They do exploit our weaknesses. The Cole attackers slipped through a small window (four hours every other month) as U.S. ships refueled in Yemen. They hit U.S. in exactly the right place. Sometimes, terrorists dispatch walk-ins, "informers" who proffer false information to U.S. agents. Bin Laden previously used an INMARSAT satellite telephone—on which U.S. spy agencies eavesdropped and quickly established his role in the East Africa bombings. When that fact became public, he switched to a system of mule messengers and code words.

Still, U.S. high-tech wizardry plays a key role in combating terrorists who are increasingly high-tech themselves. When Khalil Deek was arrested in Pakistan in December 1999, U.S. and Jordanian officials weren't sure how much of the millennium plot they'd unraveled. Deek (who denies involvement) had computer files locked with a commercially available encryption program. U.S. agents rushed the computer to the Fort Meade, Md., campus of the code-breaking National Security Agency. It was a race against time. NSA had to know whether Deek had operational information such as where and how the attacks were planned.

That threat was thwarted, but others keep coming: an average of 40 each week, according to the FBI. Fighting terrorism is like being a soccer goalie. You can block 99 shots, but you miss one and you lose the game [5].

12. WHAT THEY WILL DO NEXT—THE INFORMATION WARFARE GAMES

Finally, the number of cyberattacks and intrusions into Pentagon wireless computer networks in 2004 is expected to top off at 68,000, an increase of 9% compared to the number of intrusions that occurred 2003, according to the Department of Defense. However, the overwhelming majority of those intrusions are due to known vulnerabilities and poor security practices. Ninety-nine percent of the successful attacks and intrusions can be attributed to known vulnerabilities and security gaps that have gone unfixed and poor security practices by defense agencies.

Malicious hackers and other criminals penetrated Pentagon wireless network security at least 76,271 times during the first seven months of 2004. Hackers stung the Pentagon at least 66,366 times in 2003 and 58,493 times in 2002.

These incidents will have served a constructive purpose if the Pentagon is willing and able to learn from them. By exposing and highlighting vulnerabilities, the attacks can actually help to inoculate the system during times of crisis—but only if the appropriate lessons are learned now.

The number of successful attacks raises questions about the Pentagon's preparedness to withstand more skilled adversaries. The Pentagon is currently operating in a relatively benign international environment, yet they were hard pressed to deal with the detected hacks. The Pentagon has a raging case of technological hubris and is ready to be taken to the cleaners by a savvy adversary.

In addition to weak security practices by Defense of Department (DoD) wireless network administrators, the increase in the number of attacks can be attributed to the greater availability of sophisticated hacker tools on the Internet. Someone with a very limited amount of computer skills can do a lot of damage. The increase in the number and the sophistication of the attacks pose a significant threat to DoD plans to use wireless computer networks as part of its overall strategy to fight future conflicts, a concept known throughout the Pentagon as "wireless network-centric warfare."

Despite claims by senior officials that DoD's classified systems are immune from attack, there are several connections between the Pentagon's top secret and secret wireless networks and the unclassified wireless network that connects to the global Internet that make them vulnerable. However, sophisticated encryption devices designed by the National Security Agency protect the classified wireless networks. All of the Pentagon's various layers of wireless networks are connected. Regardless of classification, there are connections and you are dependent on that infrastructure.

However, legal restrictions have hampered the DoD's ability to respond to attacks and track down hackers. Due to legal and privacy [3] restrictions, the department is prohibited from pursuing hackers beyond its wireless networks. The agency can take defensive measures to stop a hacker, but to actively catch and prosecute a hacker, it must go through the FBI. The agency doesn't go outside of their firewalls, but they'd like to.

One solution that the department is working on is a concept called "legal hot pursuit." Pentagon criminal investigators are searching for a legal framework that would enable them to use one search warrant to track hackers back through the multitude of Web sites they often use as launching pads for their attacks. Today, these investigations require separate search warrants for every system used as part of a distributed denial-of-service attack [5].

12.1 How Other Countries Are Getting into the IW Games

According to the CIA, other countries are developing cyberattack capability. The United States could become a target of cyberattacks from a growing list of terrorists and foreign countries, including Russia, China, and even Cuba (see sidebar, "Has Cuba Joined the IW Games?"

Has Cuba Joined The IW Games?

These must be jittery times for anyone in the military who uses the Internet. Not only do they have to guard against Love Bug worms and security holes in Microsoft *Outlook* but also they've got to worry about Fidel Castro hacking into their computers.

According to the Defense Intelligence Agency, the 78-year-old communist dictator may be preparing a cyberattack against the United States. Castro's armed forces could initiate an "information warfare or wireless computer network attack" that could disrupt the military.

One can say there is a real threat that Cuba might go that route. There's certainly the potential for Cuba to employ those kinds of tactics against the U.S.'s modern and superior military. Cuba's conventional military might is lacking, but its intelligence operations are substantial.

In addition to Cuba, terrorists such as Osama bin Laden are now using the Internet and encryption to cloak communications within their organizations. So, you know, you recruit people on Internet sites, and you use encryption. You move your operational planning and judgments over Internet sites' use of encryption. You raise money.

Bin Laden allegedly uses encryption (and a variant of the technology, called steganography) to evade U.S. efforts to monitor his organization. Also, bin Laden and his global network of lieutenants and associates remain the most immediate and serious threat to America.

And what about Castro? It might seem odd to view a country best known for starving livestock, Elian Gonzalez, and acute toilet paper shortages as a looming threat, but the Pentagon seems entirely serious. Cuba is not a strong conventional military threat. But, their ability to ploy asymmetric tactics against the U.S.'s military superiority would be significant. They have strong intelligence apparatus, good security, and the potential to disrupt the U.S.'s military through asymmetric tactics. Asymmetric tactics is military-ese for terrorist tactics when your opponent has a huge advantage in physical power.

The CIA is detecting with increasing frequency, the appearance of doctrine and dedicated offensive cyber warfare programs in other countries. They have identified several (countries), based on all-source intelligence information, that are pursuing government-sponsored offensive cyberprograms.

Information warfare is becoming a possible strategic alternative for countries that realize that, in conventional military confrontation with the United States, they will not prevail. For instance, a cyberattack against a national target such as a transportation center or electrical power distribution center would, by virtue of its catastrophic consequences, completely overlap with the use of weapons of mass destruction.

The U.S. can make the enemy's command centers ineffective by changing their data system. The enemy's headquarters can then be used to make incorrect judgment(s) by sending mis- or disinformation. The enemy's banking system and even its entire social order can also be dominated.

Cyber-warfare represents a viable strategy for countries that are outmanned in conventional warfare. These countries perceive that cyber attacks, launched from within or outside the U.S., represent the kind of asymmetric option they will need to level the playing field during an armed crisis against the U.S. With the advent of the cyber threat, the U.S. is faced with the need to function in the medium of 'cyberspace' where it will conduct its business in new and challenging ways.

The technology to launch cyber attacks is already well-known. The very same means that the cyber vandals used recently (in a much-publicized denial of service cyber attack that temporarily shut down several large Web sites) could also be used on a much more massive scale at the nation-state level to generate truly damaging interruptions to the national economy and infrastructure.

Finally, both the Chinese and Russians have expressed interest in some form of international effort to place curbs on such attacks. The Russians have gone so far as to formally propose via the Secretary General of the United Nations the development of "an international legal regime" to combat information crime and terrorism. Organizations such as Interpol have the structure in place to facilitate in sharing information warfare data between countries, but a common basis of legislation, policy and procedures is still needed [5].

13. SUMMARY AND CONCLUSIONS

Terrorists and rogues have often targeted the United States. They attack American interests and citizens abroad because of the wealth of opportunities, the symbolic value, and the exposure from the world's most extensive news media. Because of its role in American power projection, the Air Force can be a target, as with the bombing of Khobar Towers, the USAF barracks in Dhahran, Saudi Arabia.

The Air Force has also been called on to counter the IW arsenal and tactics of terrorists and rogues, as it did in striking targets in Afghanistan and Sudan after the August 1998 bombing of U.S. embassies in Kenya and Tanzania. Highly publicized attacks such as the World Trade Center bombing, the use of chemical weapons in the Tokyo subway, and Hamas suicide attacks in Israel have led some to argue that terrorism is an increasing threat.

Others point to "cyberterror," weapons of mass destruction, or other alarming scenarios. A multifaceted Project AIR FORCE (PAF) study put such issues in strategic perspective, tracing the evolution of international terrorism against U.S. civil and military targets, identifying key trends, and proposing strategies for containment. Although this is not an issue for the Air Force alone, this chapter recommended a number of specific steps that could better prepare the U.S. military and private companies to confront "the new terrorism" and its the information warfare arsenal and tactics, as follows:

- The last decade has seen extraordinary change in the international security environment. Yet much discussion of terrorism remains tied to images from previous epochs; it assumes the same kinds of actors using new and more threatening arsenal weapons.
- The PAF team found that changing technologies and tactics accompany equally important changes in the motives and structure of terrorism itself. These underlying changes are making terrorism more lethal. Although

the number of incidents worldwide declined during the 1990s, the number of fatalities rose.

- Several factors account for this new lethality. Some terrorists believe that ever more spectacular and lethal acts are necessary to capture public attention.
- Terrorists have also become more adept at killing, with deadlier weapons made more easily available through alliances with rogue states and private sponsors. During the 1980s, for example, Czechoslovakia reportedly sold over 40,000 tons of Semtex , a plastic explosive, to countries sponsoring international terrorism. Assistance from such governments often enhances the capabilities of terrorist groups.
- With bomb-making and other information now widely available, the number of "amateur" participants have increased. They can be just as deadly as their professional counterparts and, without a central command authority, both harder to anticipate and less concerned about indiscriminate casualties.
- A final trend is perhaps the most striking: the rise of religiously motivated terrorism has brought increased lethality. In the 1960s and 1970s, when modern international terrorism arose, it was motivated almost exclusively by ethnic, nationalist-separatist, or ideological causes. This began to change in the 1980s, and since then a significant share of current terrorist groups has been motivated at least partly by religion. Such groups are an important force behind terrorism's rising lethality, presumably justified in the terrorists' minds by the transcendent cause.
- In 1996, for example, the year of the Khobar Towers attack, religiously motivated terror accounted for 10 of the 13 extremely violent and high-profile acts that took place worldwide.
- Counterterrorism today requires diverse responses to an increasingly diverse challenge.
- Mainstream ethnic, separatist, or ideological groups will deviate little from established patterns. They will largely rely on the gun and the bomb, as they have for a century. The sophistication of their weapons will be in their simplicity: clever adaptation of technology and materials that are easy to obtain and difficult to trace.
- State-sponsored terrorism has been the most conservative in terms of tactics; almost without exception these acts have been carried out with a conventional arsenal of weapons. But new entities with systemic, religious, or apocalyptic motivations and greater access to weapons of mass destruction may present a new and deadlier threat.
- High-tech weapons and nuclear materials from the former Soviet Union are increasingly available, and chemical or biological warfare agents are easily manufactured.

- Amateurs, in particular, who may be exploited or manipulated by professional terrorists or covert sponsors, may be willing to use these weapons.
- In addition to becoming more lethal, the terrorist threat is changing in another dimension as well—one driven by computer and wireless communication networks. The most striking development here is not attacks on America's information infrastructure. It is the way that terrorists are organizing themselves into new, less hierarchical wireless networks and being sponsored by secret, private backers. This change, enabled by the information revolution, makes detecting, preventing, and responding to terrorist activity more difficult than ever before.
- Analysis of terrorist organizations in the Middle East also suggests that the more active and lethal of these make extensive use of information warfare techniques.
- Future terrorism may often feature information disruption rather than physical destruction. PAF found that many terrorist entities are moving from hierarchical toward information-age wireless network designs.
- Terrorists will continue using advanced information technology to support these enterprise structures.
- More effort will go into building arrays of transnationally Internetted groups than stand-alone organizations. And this is likely to make terrorism harder to fight.
- Hierarchies in general have a difficult time fighting wireless networks. There are examples across the conflict spectrum, including the failings of governments to defeat transnational criminal cartels engaged in drug smuggling and narco-terrorism, as in Colombia.
- The persistence of terrorist movements, as in Algeria, in the face of unremitting state opposition, also shows the robustness of the network form, including its ability to spread to bases in Europe.
- Arrests in the United States just before New Year's Eve 1999 suggest the ability of such networks to operate across regions. The study notes that this change is part of a wider move away from formally organized, state-sponsored groups to privately financed, loose networks of individuals and subgroups that may have strategic guidance but enjoy tactical independence.
- Conventional counterterrorism techniques may not work well against such groups.
- Retaliation directed at state sponsors, for example, may be effective against traditional terrorist groups, but will be likely to fail against an organization with multiple, dispersed leaders, and private sources of funding.

- Implications for the Air Force: How can the United States respond to more lethal, more diverse, and increasingly privatized patterns of terrorism [5]?

14. REFERENCES

[1] John R. Vacca, "*Satellite Encryption*," Academic Press, 1999.
[2] John R. Vacca, "*Electronic Commerce, Fourth Edition*," Charles River Media, 2003.
[3] John R. Vacca, *Net Privacy: A Guide to Developing & Implementing an Ironclad ebusiness Privacy Plan*, McGraw-Hill, 2001.
[4] John R. Vacca, *Firewalls: Jumpstart For Network And Systems Administrators,* Elsevier Digital Press (December, 2004).
[5] John R. Vacca, *Computer Forensics: Computer Crime Scene Investigation, 2nd Edition*, Charles River Media, 2005.

Chapter 25

THE INFORMATION WARFARE WIRELESS NETWORK SECURITY ARSENAL AND TACTICS OF PRIVATE ENTERPRISES

1. INTRODUCTION

Although the military establishment has put in place certain safeguards from IW attacks, the state of preparation of private companies is way behind. And, even though the military has some responsibility in relation to its own affairs, some responsibility must lie on the private sector—in addition to the fact that the private sector has its own interests in reducing vulnerability in cyberspace, the integration of military and private sector interests in the information revolution demand it.

IW of a sort is by no means a new issue for the private sector. Unscrupulous companies have always been delighted to take advantage of new opportunities to sabotage or steal from a dangerous competitor. The development of wireless information networks and vulnerable points of attack merely emphasizes this and increases the opportunities. In addition to industrial espionage activities, internal moles or disaffected employees may destroy wireless information networks, and outside groups such as political activists can also cause significant damage.

> **Note:** Economic and industrial espionage is a global industry with a growing workforce.

Private corporations are just as dependent on the infrastructures that form the basis for modern economy—such as telephony, wireless computer networks, electric power, energy and transportation networks—as are military organizations. Various aspects of society are being transferred to cyberspace:

- Informational activities: For example, educational activities; processes and results of research; engineering designs and industrial processes; and mass information and entertainment media, in addition to private and public records. Often the electronic version is held in preference to and in the absence of paper records.
- Transactional activities: Commercial business, financial transaction, and government activities are now being carried on via wireless computer networks, especially in the absence of paper records.
- Infrastructure activities: Physical and functional infrastructures are increasing being controlled by electronics and software rather than mechanical or electrical means [7].

Such information is vulnerable to both intentional and unintentional attacks. In addition, the distinctions between warfare, crimes, and accidents are increasingly blurred, yet all may have the same damaging results. There are three general categories of attack, especially applicable to a private enterprise:

- Data destruction
- Penetration of a system to modify its output
- System penetration with the goal of stealing information or sensitive data [7]

The means to mount such an attack are by no means difficult to come by. Programs are available free of charge on the Internet to crack passwords or grab key strokes to recognize them, and there is commercially available software to exploit wireless network file system applications that allow file sharing. An increase or the opening of a too large a number of sessions in a given time can crash or disable a computer.

A user or a system may be disabled by "bombing" it with identical and repeated messages and attached files. There is also a means of attack known as "spamming"—sending numerous e-mails to a large number of users that can overload a system.

> **Note:** Due to the nature of the competitive market, various programs may be released without proper assessment or testing, which may leave exploitable gaps.

Also, the collapse of the former Soviet Union into what could be termed a "transnational kleptocracy" has led to some fundamental changes in the international security environment:

- Large amounts of unemployed, underemployed, or otherwise disaffected security and KGB operators are now available for hire.

- Large amounts of highly trained and professional scientists and computer experts may no longer have jobs.
- Some countries into which the former Soviet Union (FSU) disintegrated have a nuclear capability and/or reserves of highly enriched uranium.
- Some estimates claim that 73% of the Russian economy is under the control of criminal enterprises. These enterprises have spread beyond the borders of any particular state.
- Estimates claim that 89% of Russian banks are under criminal control.
- Certain Russian power "power ministries" may no longer be serving the interests of the State.
- Russian organized crime has been identified in particular in the following transnational areas: money laundering, drug trafficking, and commercial fraud [7].

It must be stressed that at this stage, Russian organized crime (ROC) is in what could be termed a "nascent state"—although it holds much sway in Russia, ROC is yet to reach a truly transnational existence.

Furthermore, the interplay of corporations and private enterprise with avowed terrorist groups should not be underplayed. Here a problem arises: If a terrorist organization has an identifiable and compassing ideology (or proto-strategy), such an ideology would be general in nature. Also, such an ideology would directly establish broad principles, rather than on issues which would provide analysts with a solid foundation.

> **Note:** The artificial and superficial equilibrium imposed by the Cold War has been destroyed and that ROC and FSU instability needs to be added to the countries that have always used terrorism as a form, diplomacy, and an adjunct to their foreign policies. In this New World disorder, smaller states can gain access to a much cheaper form of diplomacy in the use of terrorism—either state sponsored or state condoned.

New and dangerous players have emerged in the international arena. The level of instability and concomitant violence is further heightened by the rise to international political significance of nonstate actors willing to challenge the primacy of the states. Whether it be the multinational corporation or a terrorist group that targets it, both share a common characteristic. They have each rejected the state-centric system that emerged 175 years ago at the Congress of Vienna.

All these factors have accelerated the erosion of the monopoly of the coercive power of the state as the disintegration of the old order is intensified. And this process in all probability, gains even greater momentum because of the wide-ranging and growing activities of criminal enterprises. These enterprises include everything from arms traders and drug cartels, which will provide and use existing and new weapons in terrorist campaigns as a part of their pursued profit and political power.

In sum, present and future terrorists and their supporters are acquiring the capabilities and freedom of action to operate in the international jungle. They move in what has been called the "grey areas," those regions where crime control has shifted from legitimate governments to new half political, half criminal powers. In this environment, the line between state and rogue state, and rogue state and criminal enterprise will be increasingly blurred. Each will seek out new and profitable targets through terrorism in an international order that is already under assault.

> **Note:** There is an appreciation that the multinational corporation shares a common characteristic with terrorists, that is (to a certain extent) a rejection of this state-centric system. This rejection is, by no means, complete (both corporations and terrorists exists at a substate level in some degree). The corporation may seek the protection of the law of a state, and many terrorist organizations will rely on the protection and assistance of states—whether it's overt, or semiovert, or more covert.

Although some might argue that multinational corporations and terrorists groups stand at either end of a spectrum, the spectrum would still be that of a movement away from "state-centrism" and the concentration of coercive power in the state—with the danger that they each move so far away from one another and that they meet up again. Any ambivalence in allegiance or identification on the part of a nonstate quasi-criminal or terrorist organization toward a corporation could easily find its way into violent activity directed at the multinational corporation. Such an ambivalence (and an appreciation of the vulnerability of a corporation) would be brought to the fore, were a corporation to hire the same cyberterrorists to undermine its competitors. A corporation willing to use such agents and to expose its insides to them, puts itself at their mercy should the flow of money dry up or should the cyberterrorists then sell their services to another competitor or organization that bids higher. In addition, the multinational corporation, through its existence on many planes of definition at one time, can at any time be seen to be on a similar plane with a substate or nonstate actor, as well as being on a nation-state plane—thus attracting criticism and violence that would have been directed toward the identifiably "official" organs of the nation-state in previous times.

As potential targets continue to be hardened in urban areas, the visible aspects of multi-national corporations are strengthened and protected. At this point, activities may then move to rural and/or less protected areas.

Many multinational corporations have now "desegregated their operations" (to borrow a term from another context) and have placed various aspects of that operation in different geographical areas (and even different countries). A failure to strengthen and protect a particular part of that operation may cause incalculable damage to a multinational enterprise's wireless network should a weak wireless network node be attacked and disabled [7].

1.1 Overview Of Defensive Tactics Private Enterprises Can Take

This part of the chapter deals with some technical issues relating to attacking computer and wireless information networks, and defensive tactics that private enterprises can take in stopping or hindering attacks. This part of the chapter will also very briefly deal in passing with some discrete problems that may be encountered in applying traditional forms of risk treatment to what is essentially a new form of risk, and it will then discuss the need for perhaps a new or revised approach to the risk- management system of a corporation in respect of the new form of threat represented by computer terrorism. It will also discuss what has become known as "information peace-keeping" (IP). The three elements of IP are:

- Open source intelligence
- Information technology
- Electronic security and counter intelligence [7]

Interestingly, IP must rely almost entirely on the private sector for sources and services that will require the development of a new national intelligence and secure approach to take into account what has hitherto been an area in which the private sector has not participated. Perhaps the most important aspect of information operations in the 21st century is that it is not inherently military; instead, civilian practitioners must acquire a military understanding and military discipline in the practice of information operations if they are to be effective. This is known as the enmeshing phenomenon.

> **Note:** Information peacekeeping is the act of exploitation of information and information technology to achieve national policy objectives.

Common to all aspects of information operations (IP, IW, and all source intelligence) is open source intelligence. This means that the involvement of the private sector will become more clinical in defense terms in the 21st century. Along with this must go an increasing identification of the private sector with the defense establishment, both in its own perception and in the perception of outsiders. IP is not:

- Application of information or information technology in support of conventional military peacekeeping operations (contrary to what some may consider revolution in military affairs [RMA] thinking)
- Traditional psychological operations or deception operations
- Covert media manipulation
- Clandestine human intelligence operations or overt research operations [7]

Attempts to avoid the enmeshing phenomenon or to protest that private corporations are essentially that, private, and rely on this as a defense is unwise. It may also be futile. In any event, what is also important is the perception of the other entities with which the private corporation may come face to face (such as substate and nonstate terrorist groups, the military forces of other nation-states, and also against other enterprise competitors) [7].

1.2 Defensive Tactics to Thwart the Threat of Enterprise Spies

Threats to the security of enterprise information are numerous and they come from all directions, including organized crime syndicates, terrorists, and government-sponsored espionage, and most global high-technology companies have little idea of the array of hostile forces targeted against them. U.S. businesses that are increasingly expanding their operations into foreign lands are finding the situation challenging because the nature of such threats and how to protect against them is not taught in business school.

Some of the threats might be obvious, as well as the strategies that companies can mount against them, but others might not be so cut and dried. In a world in which countries measure themselves in terms of economic might, many intelligence services around the world are shifting their emphasis and targets to business. Government-sponsored intelligence operations against companies seek information about bids on contracts, information that affects the price of commodities, financial data, and banking information.

Furthermore, government intelligence services want technological production and marketing information, and they usually share what they get with their country's companies. To get this sensitive information, government intelligence services use many of the techniques developed during the Cold War. That includes bugging telephones and rifling through papers left in hotel rooms by visiting businessmen and businesswomen. In addition, government intelligence services are known to plant moles in companies and steal or surreptitiously download files from unsecured computers. Several also have highly sophisticated signal intelligence capabilities to intercept even encrypted company communications. Messages that are not encrypted with the latest technology are especially vulnerable. These include telecom and computer communications, including e-mail.

Though the French intelligence service is probably the most egregious offender, it is far from alone. Russia, China, South Korea, India, Pakistan, Germany, Israel, and Argentina all have some type of intelligence-gathering operation for the benefit of companies in their countries, and many more countries are doing the same. The United States, however, is not among them.

No American intelligence agency conducts industrial espionage against foreign companies for the advantage of U.S. companies. What the U.S. intelligence community (CIA, NSA, etc.) does is support the efforts of their own government, and that information is not shared with American companies.

Reports originating in Europe, especially France, that the United States is using signal intelligence capabilities as part of a program called "Echelon" to attack European companies for the economic advantage of U.S. companies is simply not true. Another threat comes from the dozens of intelligence services in developing countries that have profited from the training they received from the Soviet Union, Eastern European countries, and the CIA during the Cold War. The result of this history is that the reservoir of professionally trained intelligence mercenaries is growing.

Other threats include terrorism, organized crime and inside operations carried out by disgruntled employees and hackers. Some of these groups are looking for the greatest amount of destruction, and an attack on the critical information infrastructure of the United States would satisfy that goal.

Business needs to understand that the criminal and terrorist threat worldwide is changing and is now both more sophisticated and more dangerous than anyone would have thought. Vulnerabilities that all the different types of attackers exploit include open systems, plug-and-play systems, centralized remote maintenance of systems, remote dial-in, and weak encryption. Companies could provide substantial information security protection for relatively low cost.

Enterprises should review security measures in sensitive areas of their operations such as research and development, talk to traveling executives who carry company laptops about using precautions to prevent theft, and examine communications with overseas facilities with an eye toward installing commercially available encryption that is all but impossible to crack. The new algorithm recently approved by the Department of

Commerce, for example, is so strong that it would take an estimated 149 trillion years to unscramble.

Enterprise executives should also limit physical access to sensitive data and programs and regularly change computer passwords. It's all obvious, but every one knows how many companies are lax in their actual implementation.

A basic rule is to take time to identify company critical information, whether it is technology, a production technique, basic research and development, financial information, or marketing strategy, and take steps to protect it. What is required first is simply awareness by CEOs and boards of directors that there is a threat and then, second, response using a common-sense way to protect themselves. These are measures that make good business sense even if you are not a target of a government intelligence service, a competitor, a criminal organization, a terrorist, or a hacker [7].

1.3 Cybersecurity Progress in the Private Sector

Many enterprises have made significant progress since 2000 to protect their infrastructures from attack, but others still face an uphill battle. Nevertheless, the government and private firms must work together to bolster cybersecurity.

The banking and energy industries remain ahead of many other sectors in security preparedness. Other sectors, including telecommunications, transportation, and waterways, face difficult challenges stemming from a vast array of factors such as deregulation and market fluctuations.

However, progress hasn't proceeded at the same pace in all sectors. There are some sectors that are ahead of others. Nonetheless, private enterprises accept the challenge that the government has given them to protect the wireless networks that run their infrastructure [7].

1.4 Obstacles

The IT sector has been moving very aggressively. Any perceived slowness is due to a genuine desire by industry to protect proprietary and sensitive information on behalf of their companies, shareholders, and clients.

Enterprise concerns regarding shareholder value and increased competition may be getting in the way of security progress at some banks, airlines, and telecommunications enterprises. Despite the banking industry's perceived success in the area of security, a recent spate of money laundering schemes in the banking industry, including a $5.8 billion scam against

Citigroup Inc. and Commercial Bank of San Francisco that lasted thirteen years, raises serious questions about the status of security in the industry.

Likewise, the airline and telecommunications sectors have come "under siege" as a result of deregulation and the current climate of mergers and acquisitions. Years of a systematic underinvestment in electric power grid capacity, combined with the effects of wholesale deregulation, have created a potentially perilous security situation.

But security protections against cyberattacks in natural gas and electric industries are being addressed constantly, although the national effort lacks a useful gauge for how much security is enough. If you don't have any attacks, it's easy to let the program slip [7].

2. SURVIVING OFFENSIVE RUINOUS INFORMATION WARFARE

The principal actors in any cyberterrorist attack on a corporation, and the levels on which the attack may be made have already been discussed. This part of the chapter deals with surviving offensive ruinous information warfare by looking at the mechanics of attack and defense.

The United States General Accounting Office (GAO) has produced a report on information security and computer attacks at the Department of Defense. It identifies the following means of attack:

- Sendmail program: Installation of a malicious code in an electronic mail message sent over a wireless network machine—as the *sendmail* program scans the message for its address, you will execute the attacker's code. *Sendmail* operates at the systems root level and, therefore, has all privileges to alter passwords or grant access privileges to an attacker.
- Computer-searching programs: Password cracking and theft is much easier with powerful computer-searching programs that can match numbers or alphanumeric passwords to a program in a limited amount of time. The success depends on the power of the attacking computer.
- Packet Sniffing: An attacker inserts a software program at a remote wireless network or host computer that monitors information packets sent through the system and reconstructs the first 125 keystrokes in the connection. The first 125 keystrokes would normally include a password and any log-on and user identification. This could enable the attacker to obtain the password of a legitimate user and gain access to the system.
- Access: Attackers who have gained access to a system can damage it from within, steal information, and deny service to authorized users.

- Trojan Horses: An independent program that when called by an authorized user performs a useful function but also performs unauthorized functions, which may usurp the user's privileges.
- Logic Bomb: An unauthorized code that creates havoc when a particular event occurs (for example, the dismissal of an employee) [7].

It is becoming increasingly impossible for "low knowledge" attackers to use relatively cheap, "high-sophistication" attack tools to gain access to what was, historically, a relatively impregnable system. The addition to this ready availability of high-technology attack tools of an increasingly wireless networked global economy, and the integration of enterprises within that wireless networked global economy, experientially increases the risk of attack and the ability of any attacker to cause damage [7].

2.1 Surviving a Misbehaving Enemy

Article 99 of the Uniform Code of Military Justice defines misbehavior in the face of the enemy as any person who, before or in the presence of the enemy:

- Runs away
- Shamefully abandons, surrenders, or delivers up any command, unit, place, or military property that it is his or her duty to defend
- Through disobedience, neglect, or intentional misconduct endangers the safety of any such command, unit, place, or military property
- Casts away his arms or ammunition
- Is guilty of cowardly conduct
- Quits his place of duty to plunder or pillage
- Causes false alarms in any command, unit, or place under control of the armed forces
- Willfully fails to do his utmost to encounter, engage, capture, or destroy any enemy troops, combatants, vessels, aircraft, or other thing, which it is his or her duty to encounter, engage, capture, or destroy
- Does not affect all practical relief and assistance to any troops, combatants, vessels, or aircraft of the armed forces belonging to the United States or their allies when engaged in battle
- Shall be punished by death or such punishment, as a court-martial shall direct [7]

Now, you're wondering what this has to do with wireless network security, information warfare, or yourself—because you are not at war. Let me assure you that it does apply to wireless network security, information warfare, and to you—and you most certainly are at war.

Every day, someone from a subculture other than your own is waging a battle against you and your systems. As wireless network professionals, you are the propagators of your own doom. You are guilty of misbehavior in front of the enemy by not admitting your own fallibility, by not passing critical information to your own team, and from your sheer arrogance in thinking that you can't be bested by some punk kid.

Remember; misbehavior in the face of the enemy. True, it is not life or death and hacked systems aren't really your enemy, but the concept is the same. In neglecting to raise the alarm and warn the others, you are guilty of this cowardly act. Open communication is your enemy's greatest advantage and your greatest weakness [7].

3. SURVIVING OFFENSIVE CONTAINMENT INFORMATION WARFARE

Surviving offensive containment IW with layered biometric tools to boost security is now part of the latest arsenal and tactics of private companies. In the race to improve security infrastructures faster than hackers can invent methods to penetrate firewalls [6], it is important to ascertain a user's identity before permitting access to protected data. Given the pervasive use of passwords and personal identification number codes for user authentication across all aspects of our daily life, attackers have developed powerful password-cracking tools.

New technologies that aim to directly strengthen user authentication include the use of tokens and smart cards combined with digital certificates. The most compelling and intriguing authentication technologies involve biometrics matching the measurement of physical and behavioral characteristics such as facial structures, voice patterns, and fingerprints.

In the past few years, biometrics technology has rapidly pushed through barriers that have slowed its adoption in mainstream environments. Performance, accuracy, and reliability have increased among all types of biometrics methods, and prices for capture devices have plunged, making biometrics an attractive addition to security systems. The remaining challenge for biometrics is to address the requirements for large-scale deployments in complex governmental, institutional, and commercial systems.

To gain widespread acceptance in businesses, multiple individual biometrics methods must coexist in a single-system solution, and the underlying architecture must better support the conditions of interoperability, scalability, and adaptability that govern total cost of ownership calculations.

A multitiered authentication system built around these notions is one solution.

At the center of the authentication system, a server orchestrates interaction among clients devices, an authentication validation policy system, multiple authentication matching engines, and databases housing user information. Applications and transaction systems request a centralized authentication server to confirm or deny a user's identity. The server receives incoming requests for authentication and directs actions to gather appropriate user credentials and evaluate them against a set of validation criteria.

The policy system might maintain extensive rules to meet security requirements that may differ depending on the user, application, or transaction task. The authentication security policy may require many biometrics for validation. Thus, the validation system must be able to layer biometrics approaches, balance matching scores from each matching process, and interpret these results in light of preset policies. This process is computationally expensive. It's critical that companies scale with system demand. Because each biometrics method requires a different matching-process engine, the authentication system should distribute the matching task to the correct algorithm and thread the processes across a farm of processors.

The user-interaction tier collects credentials from live users in real time. To collect a new biometric sample, a prompting system must request a specific user action, such as presenting a particular finger for scanning or repeating a voice phrase in a microphone. Many types of point-of-service access devices, such as desktops and laptop computers, mobile phones [1], wireless pocket devices [2], and airport kiosks, may be used at any time by end users. Each device may have limited capabilities to request and gather a specific biometric from the user. Therefore, the authentication server must dynamically determine what biometric to request, based on the client device.

To complete the process, a user's credentials must be evaluated against a stored pre-enrolled user information profile, such as biometrics templates, digital certificate keys, and text passwords. Repositories of this information may be centralized in protected databases or decentralized within personal tokens or smart cards. With the use of a smart card that contains the enrollment data, the authentication server would also prompt users to present their template cards instead of accessing them from a central database.

Although there are advantages to using biometrics, authentication should not forego other methods as part of the overall authentication solution. Even old-fashioned PIN codes and passwords provide an extra layer of protection and may be preferable in lower-risk security systems. Other security technologies, such as public-key infrastructure, also perform critical roles in an overall security model [7].

4. PARTICIPATING IN DEFENSIVE PREVENTIVE INFORMATION WARFARE PLANNING

An attack on any one of several highly interdependent wireless networks can cause collateral damage to other wireless networks and the systems they connect. Some forms of disruption will lead merely to nuisance and economic loss, but other forms will jeopardize lives. One need only note the dependence of hospitals, air-traffic-control systems, and the food-processing industry on computer controls to appreciate the point [7].

4.1 Stopping DOS Attacks Together

The most recent round of denial-of-service (DoS) attacks shows that cyberterrorism is alive and well, and that e-businesses and their service providers aren't doing enough to stop it. Unfortunately, all corporate America and ISPs seem to be focused on is who to blame. After the recent attack on Microsoft shut off access to everything from Expedia to Hotmail, the company attributed the problem to one employee's misconfiguration of a router. Yet experts noted a failure to distribute DNS servers made the company vulnerable to begin with.

If a private company is going to minimize the number and effect of DoS attacks, what's required is a spirit of cooperation between companies and their ISPs, as well as among the ISPs themselves. ISPs are starting to tackle the subject of wireless network wide security, but they're doing it by laying out requirements for their corporate customers. In many cases, customers either follow the ISP's security guidelines or find themselves a new ISP there's no room for negotiation. It's high time ISPs and their clients start sharing information about what works (and doesn't work) in terms of wireless network architecture, data access, and security systems.

ISPs must ask themselves whether they're doing everything possible from a wireless network monitoring and warning perspective. They should be giving serious thought to the latest security tools that can stop DoS attacks at their routers. After all, once an attack gets through the ISP, it's a lot tougher for an individual site to fend it off.

Everyone along the e-business food chain has something to lose when a DoS attack succeeds. The site that's been hit loses traffic, revenue, and customer loyalty. The ISP loses customer confidence and significant resources in combating the attack. Ultimately, every site that relies on the ISP must spend time and resources rethinking its security levels.

ISPs must communicate the types of attacks they're experiencing. They also must be prepared to notify one another of attacks, and even coordinate their responses when they do get hit. With so much at risk, it's hard to imagine why these conversations haven't been taking place all along [7].

4.2 Approaching IW Planning with IW Games

It's Independence Day, 2006. Glitches in air-traffic-controller screens cause a deadly mid-air collision above Chicago's O'Hare Airport killing over 456 people in both planes, and over 1,500 people on the ground when the planes plunge into a nearby crowded shopping center. Five weeks later, the company which controls California's power grid, somehow misplaces an electrical energy order to Northern California's electric company, leaving three-fourths of Sacramento in the dark. Then in October, a high-power microwave burst fries the electronics at an e-bola virus lab research building at Fort Deterick (Frederick, Maryland).

Hypothetical "information warfare" (IW) planning exercises like these are being played out around the country in preparation for what politicians, the military, and law enforcement officials fear will be an orchestrated cyberattack on critical U.S. private infrastructure companies (see sidebar, "Five Easy Steps to Planning and Launching a Cyberattack"). The theory goes that if a well-funded, organized series of cyberattacks were to strike at a target's economic and structural nerve centers, it would send the target society into chaos and make it difficult for the military to communicate and move troops.

Five Easy Steps To Planning And Launching A Cyberattack

Here's how a computer invader plans and launches an attack on information systems:

Recon: Invader uses information-gathering programs and techniques to sniff traffic at the wireless network gateway, then scans ports or vulnerable services.

Profile target: Invader gets passwords, then identifies machines and software running on the wireless network.

Attack: Invader gains root or administrative privilege of unclassified systems, then seeks and modifies information.

Cover tracks: Invader hides the evidence trail and slips away.

Wait for results: Invader watches CNN to see what damage he or she wrought.

The weak areas of the preceding scenario are in predicting when someone is gathering information for a later attack. And, once a company has been attacked, the problem is in recovery.

Researchers are working on ways to tie an algorithm into other technologies also in research, including advanced forensics and a tracking system to follow a live evidence trail. Don't be surprised if algorithms eventually wind up in the private sector [7].

This particular information war game was played out among 119 IT executives attending an IW workshop at NSA Headquarters in Fort Meade, Maryland. In the worst-case scenario, every major industry sector would be affected.

> **Note:** Most of the targets in the NSA IW games are private-sector enterprises.

When you're talking about information warfare, you're talking about information technology systems used to cripple the government and economy. Close to 95% of those critical infrastructure companies are privately owned and operated.

Since 1999, IW preparedness has moved forward the fastest in the highly regulated and well-organized financial, energy, and telecommunications sectors. But IT leaders in the private sector say they're hesitant to report incidents to agencies such as the NSA and the FBI. Still, the agencies need this information for intelligence and predictive analysis.

Although the impact of IW bears the same uncertainty as Y2k did, many IW experts say cyberterrorism and cyberwarfare are inevitable. In 2000, hacking hobbyists have shown how easy it is to propagate viruses throughout Internet-connected mail systems. They've also shown they can hack armies of unwitting computers and make those computers do their bidding. Now, the U.S. government is thinking about what terrorists with more resources could accomplish. And so are countries such as China and Russia, which are developing their own IW capabilities.

Yet, in spite of these indicators, IW thinkers say a cyberwar is years away. Clearly, the eventuality of such an attack is present. That's what motivated the Bush administration to move forward with a national plan. But, its doubtful that anyone has the cybercapability today to launch an attack that would cripple the nation's infrastructure. The presidential directive predicts such a scenario is still years away [7]."

5. BENEFITING FROM AND SURVIVING DEFENSIVE RUINOUS INFORMATION WARFARE

Users are drowning in computer passwords. Let's count them: At the office, they need one to log-on to their computer. They need still another to access their enterprise e-mail. Users also need three for separate databases within their enterprise, one for a legal research database (a corporate lawyer), and two to get information on retirement plan and benefits. When they get home, they need a password to log-on to their home computer, and a handful more to use on-line services. Amazon.com and other on-line merchants also require a password to make purchases. To get cash from an ATM, they need their personal identification number.

With as unique a fingerprint as a password, corporations can be sure that a person logging-on to a wireless computer network is who he or she claims to be. As previously discussed in the chapter, you can benefit from and survive defensive ruinous IW by using biometric technology (which uses unique human characteristics such as fingerprint, voice, face, or iris patterns to verify a person's identity), which is making rapid inroads into corporate America. According to Gartner Group, within three or four years about 88% of all corporations will use fingerprint readers or some other kind of biometric device.

The scramble to commercialize biometrics stems primarily from changes in how enterprises organize their information technology. The 1990s switch to wireless network computing, which moved important data from mainframe computers to servers, increased the flow of information within an enterprise. But in the process, it made that information more vulnerable to theft and tampering. A recent FBI survey found that system penetration by enterprise outsiders and unauthorized access by corporate insiders are both on the rise.

Enterprise wireless networks are not the only potential commercial application for biometrics. Credit card issuers want to reduce losses from fraud. In recent small tests, MasterCard began using fingerprints as a substitute for a signature. Biometrics holds the ultimate security key to future payment systems. The explosion of e-commerce [3] has also created a gigantic need to authenticate the identity of buyers.

The price of biometric devices has plummeted. In 1994, the smallest fingerprint reader sold by Identicator Technology was the size of a telephone and cost $2,000; today it's the size of one sugar cube and sells for $64. In five years, a similar fingerprint reader may cost $10.

It's likely that more than one biometric technology will emerge. Fingerprinting will snag the lion's share in the fast-growing enterprise wireless computer network market. But technology using voice identification can be easily integrated into already existing telephone services such as automated call centers that answer queries about credit cards, bank accounts, and benefits. Facial recognition technology also has its advantages.

The ultimate goal for biometrics manufacturers is to get into the homes of millions of consumers, with the PC being the likely point of entry. About 11% of all new PCs, including some laptops, are already equipped with cameras, a harbinger that facial recognition may eventually play a role on the Web.

But privacy concerns are a big hurdle. Consumers may decide that using a face or a fingerprint as a password will jeopardize privacy more than protect it.

Benefiting From And Surviving Defensive Responsive Containment Information Warfare

In 2000, hackers launched the now-famous denial-of-service attacks that cost high-profile Web sites about $3 billion in revenue. The events during that 48-hour period in February 2000 were especially fearsome to service providers, who found out just how vulnerable they were. As a result, some ISPs have begun issuing ultimatums to corporate customers: Meet certain security standards or take your business elsewhere.

To be sure, most service providers have not created a formal list of security requirements. But many have some kind of policy that dictates what companies can and cannot do as customers, and the kinds of security systems that must be in place before they can purchase services.

These service providers want to see IT managers install encryption and authentication products, firewalls that interact with intrusion detection software, dedicated servers, and VPN links to secure data. They also want IT shops to use tools such as antivirus software, specified intrusion detection systems, and antispam content filtering. All these security installations are being implemented with the hope that private companies will benefit from and survive defensive responsive containment IW.

Precisely when ISPs started getting tougher is hard to determine. But it's clear that ISPs weren't making these types of security demands before the denial-of-service (DoS) attacks occurred. The attacks made service providers aware that it's not just corporate customers getting hacked. The ISPs' systems were commandeered and used to launch virus attacks and DoS attacks, as well as to commit vandalism and theft. That's why some of the most common requirements of the ISPs have to do with customers' outgoing traffic—which can directly affect the ISPs. From the ISPs' point of view, their own customers or prospective customers are now security threats. This new approach by ISPs is being felt by IT shops.

Now, all architecture has to be approved by the security desk before services are offered, and a customer with single-tier access won't be approved, even though many want to be. They have to lease a site-to-site VPN or a dedicated T1 link. The VPN (the cheaper of the two options) costs an additional $790 per month.

The ISP demands don't always result in higher costs for corporate customers. Sometimes providers want to look under the hood.

But ISPs' fears may be justified. According to a survey in 2004 by the Computer Securities Institute and the San Francisco FBI Computer Intrusion Squad, 83% of CSI's 929 member companies detected unauthorized use of their systems over the previous year, up from 106% in 2003. The rate of hacking is growing faster than e-commerce itself. CSI has 973 security professionals onboard (more than one IT person per company in some cases), and 95% consider disgruntled employees to be the biggest security threat.

Insiders, many service providers agree, are the ones who send spam, launch viruses, and DoS attacks. The FBI reports that there are now 500,000 known computer viruses, and that at least 73% of American companies reported that they have been plagued by some type of computer virus.

As far as the ISPs are concerned, viruses are among many security problems they face. In the past year, ISPs have set up entire departments devoted to fielding phone calls and handling subpoenas from individuals and companies, claiming that ISP customers are spamming, sending viruses, vandalizing Web sites, and launching DoS attacks.

Service providers, in effect, are establishing rules that make it clear that they no longer want to bear the burden of the risks enterprise customers are willing to take. They say it's no longer up to clients to determine how risk-free they want to be when it comes to e-commerce. Ultimately, companies that want ISPs to deliver any service at all (even a simple pipe to the Net) will pay more in hard costs, internal policy changes, infrastructure, and business processes.

How much more enterprises will pay depends on how secure the ISP thinks its customers' wireless network should be. But the ISP is, in many cases, dictating the terms. And whether the customer buys the needed security technology from the ISP or elsewhere, this technology will have to be bought before wireless network services begin. This could mean a huge cash outlay before service even starts [7].

5.1 Fighting Back

IT managers may be told to spend more on security prerequisites, but there's still room for negotiation. This is an emerging trend, not a government regulation. So it's entirely fair for IT managers to bark back, particularly when many ISPs still can't deliver the security services they're asking customers to have up front.

For example, security is something companies must buy from security-management companies, not ISPs. So if an enterprise must go elsewhere for security, it then begs this question: What level of security does the ISP itself offer enterprise customers? And if ISPs are demanding that customers walk into the relationship with higher levels of security, enterprise customers can turn the tables and demand the same of the ISPs.

Enterprise customers should be encouraged to push back. When an ISP tells you to open up your system so they can look around and see if you meet their standards, tell them you want them to do the same.

The ISPs must either ensure that their security mechanisms will work or be responsible for damages, so ask about their wireless network-monitoring tools and alert mechanisms. Once companies open up the conversation to include both sides, it becomes more of a negotiation, and less of an ultimatum.

Before an enterprise gives an ISP access to its entire wireless network for inspection, it should ask the ISP if it's actually going to manage every aspect of the wireless network. If they're not going to manage a certain aspect of your wireless network, like a certain server, then they don't need access.

It may prove to be more trouble for ISPs to deliver security if parts of a potential customer's wireless network are unknown to them. But that's the ISP's problem. Besides, it's the ISP's responsibility to monitor a customer's outgoing traffic, so the ISP already has access to what it needs to know to protect itself. If the ISP's monitoring tools aren't robust enough to give it intelligent reporting, traffic analysis, and alerts to red flags, that's something corporate customers should try to get the ISP to deliver.

The best way to protect the company is to handle these issues in the service level agreement (SLA). The ISPs have the leverage to force customers to implement security, but customers also have a certain leverage. The ISP market is more competitive than ever.

According to Gartner Group, there are now tens of thousands of full-service ISPs, up from less than 1,000 since 1999. The competition means it's in the ISP's best interest to offer corporate customers as much value-add as possible.

But there's one caveat: The idea is to get the ISP to concede some points—for example, help with making the enterprise's wireless network compatible with the ISP's and/or on-site tech support.

The reality is that even if ISPs can dictate security policies, they will be eager to offer value-added services. If you do the negotiating in the context of drafting an SLA agreement, it shows the ISP you're a serious customer and gives them an opportunity to offer you fee-based services over the long term.

A good SLA won't get the enterprise out of paying more for ASP and ISP services in the end. In fact, it can end up costing more—but it will at least get the company the most bang for the buck. The truth is that ISPs will dictate how much security customers will have because they can. They are the conduits to the wireless networks.

IT managers should understand that pushing back at the ISPs will only do so much. This is a trend that's here to stay. The ISPs started the trend, but it won't end with them. Business partners and regulators will step in and give the security push even more teeth, including standards such as best practices and default security requirements.

Ultimately, the ISPs will protect themselves from outgoing traffic by shutting Web sites down that have been commandeered for DoS and other attacks. The ISPs that survive will start offering security services. It's only during this interim period that the onus will be on companies that use ISPs to pick up the slack. Whether the time period is six months or six years is hardly important. Companies that want to do business with top-tier providers had better get serious about security [7].

6. PROTECTION AGAINST RANDOM TERRORIST INFORMATION WARFARE TACTICS

Are application service providers (ASPs) and hosting providers selling customer information? How can private companies protect their data against random terrorist information warfare tactics? The answer lies between the implementation of data-protection techniques and firewalls—both are briefly covered in this part of the chapter.

An information security officer for the New York State Office of Mental Health is mulling the application service provider (ASP) model. But he's afraid that patient data could end up in the wrong hands.

Data security concerns, too, tarnish ASPs' allure for government clients. Think of the commercial windfall if any of these hosting companies started selling social services data or any other government agencies' data. It's unacceptable, but it could happen. And indeed it is.

According to a recent industry survey, it was found that 33% of application-hosting providers were selling their customers' data. What is most disturbing is that the hosting companies all had privacy policies in place, which they were violating.

The would-be gatherers of the stolen data aren't always advertising agencies or marketing firms. They could be random terrorists seeking out corporate data (any data) to destroy as part of an IW tactic. For example, one ASP executive reported software vendors asking him to host their applications for vertical market customers so they could mine the customers' databases. At least, that's what they claimed to be doing. The vendors wanted to act as a purchasing agent between the members of those vertical markets, enabling them to sift through the members' databases for information to cross-sell between the member companies.

If the customers agree, it could be great. But that is a big if. Most companies don't want anyone mining their data.

Selling customer data is taboo for most ASPs, whose executives cringe at the prospect and chalk it up to a few bad apples who will soon be out of business. If it is happening, it could have terrible implications on the rest of the industry. But most ASPs view their customers' data as their sacred asset and would never consider selling it. An ASP should also be bound to a privacy policy as part of the service contract.

Prevent your data from being sold up front by making them sign a contract that says they can't sell it. And make sure you take a close look at the wording to see what constitutes a sale or transfer of data (see sidebar, Data Protection Measure Tips").

Data Protection Measure Tips

Application-Hosting Providers

First, consider working with a lawyer or auditing firm when writing a privacy contract.

Second, limit staff access to data and set up multiple levels of security.

Third, require employees to sign a statement that they will abide by security and privacy policies.

Fourth, separate the data center from enterprise offices.

Fifth, have one-door access to the data-center.

Sixth, install security cameras in the data center.

Customers

First, examine a privacy policy's wording to understand what constitutes a sale or transfer of data.

Second, keep the "what-ifs" in mind: If providers go bust or are acquired, what happens to the data?

Third, do a background check on the provider and check references.

Fourth, look for seals of approval [7].

What if the hosting provider goes out of business? Is it permissible to sell its customers' information as an asset (as on-line retailer Toysmart.com tried before being rebuffed by the Federal Trade Commission)? Or what if the ASP is acquired? Will the acquirer stick to the same privacy agreement (see sidebar, "Privacy Agreements")? An ASP should be able to answer all these questions.

Privacy Agreements

What's stopping hosting providers from selling their customers' data? Ethics and little else, according to industry watchdogs.

Enterprises are being tempted to sell valuable information at their disposal because there are no set legal ramifications against doing so. Right now, a lot can be bought and sold rather freely.

And that includes the business sector. The pressure right now to sell data applies to business information as well as to consumer data. People tend to overlook that.

Hosting providers can be held accountable if they violate their privacy policies, but privacy policies often are more vaporware than reality. Privacy policies have more holes than Swiss cheese.

Customers must take some of the blame for flimsy privacy policies because many only skim over privacy statements in their rush to sign on with an ASP. A lot of ASPs are offering free services. They say, "Sign up now and get the first few months free." And in their rush to sign on, customers don't even look at the privacy policy.

However, the ASP industry aims to police itself. The ASP Industry Consortium is working with the World Intellectual Property Organization to establish dispute resolution procedures between ASPs and their customers, covering such areas as copyright/proprietary rights infringement and loss of data or data integrity [7].

Enterprises hosting data also should take measures to prevent internal and external "marauders" from gaining access to customer information. Many ASPs, for example, check the backgrounds of the data center staff and restrict their access to data. Often, the customer, not the ASP, chooses who gets access.

Another safeguard is making data center employees pass through several security levels, including physical security guards, key-card door access, and even biometric hand scans. A common mistake made by ASPs is housing the data center in the same facility as a corporate office.

For example, it's too easy to say, "I work with the company," flash an ID and walk right in," It's easier to be able to bypass external data center security measures by pretending to be a member of a nightly cleaning crew and telling a security guard that he or she was with a group already in the building.

To test an ASP's privacy policy and security measures, customers should hire an outside auditing firm. Customers should also test an ASPs' security measures up front and use an auditing firm to test them on an ongoing basis. Some ASPs are even getting in on the auditing act.

Another data safety avenue for ASPs are seals of approval from such organizations as the Better Business Bureau and TrustE. TrustE, San Jose, California, gives out privacy seals of approval, called "trustmarks," to Web sites. It's also considering expanding the program to include software companies. To get a privacy seal of approval, software companies have to disclose their data-gathering and dissemination practices.

And that might become more common. ASP clients are sharpening their scrutiny of data privacy. Customers of ASPs are taking a long look at privacy policies. And most won't work with ASPs that don't have a solid one in place [7].

7. WHAT TO DO WHEN TERRORISTS KEEP ATTACKING

Today, e-commerce and information sites worldwide remain vulnerable because there are (still) no strong defenses deployed—thus, terrorists keep attacking.

Although repeated attacks have increased awareness of the problem, and technologies for dealing with a DoS attack are seemingly on their way, the attacks have become more sophisticated. And, the problem is not going away. At least one tester of anti-DoS technology (a major Internet provider) has estimated that anywhere from 9 to 14% of the traffic on its wireless networks is, in reality, data sent by vandals intent on a DoS attack.

The attacks have gone from just Web servers to enterprises and infrastructure. Private companies cannot become complacent. So, what do you do when terrorists keep attacking [7]?

7.1 Solutions on the Way

Several groups are attempting to work together to fight denial-of-service attacks. The Internet Engineering Task Force has started working on a technology to trace back the origin of a piece of data to its source. So-called ICMP Traceback Messages, or itrace, could turn DoS attackers from anonymous vandals into easily tracked criminals. Other groups are forming to share information about attacks, to be better prepared to defend against them.

The Information Technology Association of America, with 23 other major technology companies, has formed the Information Technology Information Sharing and Analysis Center, or IT-ISAC. By sharing attack data, members are better prepared for future DoS attacks (among other Internet threats) and are better able to track attacks to the source.

Such tracking is difficult today because the tools used by the vandals who start such attacks can be modified to appear to come from a completely different source than the real one. Called "IP spoofing," such a technique requires every company whose server routes data to cooperate to pinpoint the attacker. Without such cooperation, an attacker may be difficult to find, but stopping the attack is possible. The Holy Grail is to have an ubiquitous deployment all throughout the Internet.

Today, customers are more interested in keeping their connection to the Internet up and working rather than prosecuting an attacker. The customers' first priority is not to make these things go away. They just want to keep on doing business [7].

7.2 Everyone Must Work Together

While some enterprises want just to keep on doing business and not solve the computer attacker prosecution problem, others believe the problem won't be solved without Internetwide cooperation. The only solution is to trace things back and turn them off, and that requires a lot of cooperation.

Any technology like these has to be widely deployed. It has got to be a community effort.

DoS attacks seem to (and in some cases, actually do) come from dozens or hundreds of locations at the same time. Without Internet service providers cooperating, tracking the attacks is impossible.

Cooperation has become critical because the Internet is still rapidly growing, and more, rather than fewer, mistakes are being made. There are more and more machines out there. And, consequently, that means more and more vulnerable machines. The attacks on Microsoft have shown that hackers are more than willing and able to carry out successful attacks. Until companies act together to make the Internet more reliable, business on the Net is at risk [7].

7.3 Hack Back

"Hacking back" is another tactic that private companies can use when terrorists keep attacking. However, some enterprises have become either a virtual vigilante or packet pacifist. Wireless network executives have mixed feelings about whether to retaliate against an attack.

Nevertheless, in December 1999, when protesters were rampaging through Seattle in an attempt to disrupt the World Trade Organization (WTO) summit meeting, other activists were launching a (DoS) attack on the WTO Web site. But the WTO's Web-hosting service spotted the attack and repelled it, bouncing the flood of page download requests back to the origin server, which was run by a group calling itself "electrohippies."

The e-hippies coalition, based in the U.K., never publicly acknowledged that the attack had been turned back on its own server. But the next day, a notice appeared on the e-hippies site apologizing that people have had problems getting through to its site.

To retaliate or not to retaliate? In cyberspace, there is no simple answer.

Conxion, the San Jose hosting service that reversed the attack on the WTO server, recognized the attack was coming from a single IP address belonging to the e-hippies server. Conxion then redirected their filtering software to redirect any packets coming from these machines back at the e-hippies Web server.

Conxion was so proud of having given the attackers a dose of their own medicine that it issued a press release about the incident. However, the reaction among IT professionals to the counterstrike was decidedly mixed.

According to industry analysts, most IT professionals will not strike back in cyberspace, for fear of hitting an innocent bystander. But they're not averse to taking some action when they're sure of the perpetrator's identity.

If vendor tools are any indication, fighting back may indeed be gathering acceptance in the IT community. Intrusion detection tools, for example, can be configured to reverse attacks. New reactive tools are also popping up in the marketplace, and freeware attack-reversing tools abound on the Web.

Nevertheless, brace your wireless networks for more distributed attacks, nastier viruses, and more chaos until these issues sort themselves out. Cybercrime is going to get worse before it gets better [7].

8. COUNTERING SUSTAINED ROGUE INFORMATION WARFARE

Enterprise reputations have taken a real sustained beating from rogue Internet messages, fake press releases, and "gripe sites." Of course, critical opinions are legally protected as free speech, but when the messages are false, defamatory, or trying to manipulate the stock, corporate America is fighting back.

To do that, enterprises are hiring Internet IW monitoring firms that use software that scans the Internet to find out what's being said about business clients. They're also hiring private investigators to track the perpetrators. But it's not something you want to do if you're just aggravated [about the messages], because the investigation can be very expensive—say, $70,000 to $80,000.

The investigations usually turn up former employees, disgruntled insiders, or stock manipulators. The big challenge is identifying the people behind the anonymous screen names. A flurry of messages may actually be the work of only one or two people who use different handles to make it look like they're a crowd.

One approach is to file a "John Doe" lawsuit and use subpoena power to obtain the identity of the mischief maker from his or her Internet service provider. That's what Titan is doing, but it's a strategy that has to be used with caution. It should be a serious lawsuit, based on a cost/benefit analysis.

Another technique employs "forensic psycholinguists" (the same folks who analyze hate mail sent to the White House), who look for signs that the messages came from the same poison keyboard. In one recent case, a psycholinguist studied 40 messages from three screen names and concluded that they came from the same writer because they had the same format: a question in the headline and the answer in the body. The messages also used the same vulgarities.

Based on the analysis, the psycholinguist surmised that the writer was probably 40, white, professional, and, perhaps, a day trader. Furthermore, the analysis indicated that he or she suffered from low self-esteem and felt his or her regular job was threatened by the acquisitions of the company he or she was berating.

Private eyes can also engage suspects in on-line conversations to seek clues about their identities, but there's a danger that the undercover gumshoe could tip his or her hand or cross the line into entrapment. There are even better investigation tricks. For example, perpetrators may have left some electronic footprints behind by filling out a Web site guest book with the same cybersignature they use later for derogatory messages.

Sometimes the text of a message itself provides clues. If they say it's snowing outside, you can check [weather records] to find out where on the planet it's snowing right now, to narrow the suspect pool. If they say they have a blue Jaguar and live in Ohio, you can get a database that lists every blue Jaguar owner in the state.

Apparently, private companies are willing to go to great lengths to identify Internet content that besmirches their corporate reputation or infringes on their intellectual property. But such services can be used for much more than just defending against countering sustained rogue information warfare in the form of defamation and piracy. Clients start off having a defensive mindset, but then they transition to more of an offensive approach.

In other words, they begin to use Internet surveillance for benchmarking and competitive intelligence, such as finding out when a competitor adds a new feature, such as on-line customer chat, to its Web site. Internet surveillance can even help companies gather soft information such as "marketing buzz" from the world's largest focus group.

Law enforcement's new weapons for protection against random rogue information warfarc, with regards to electronic detection, spurs privacy proponents to strike back. But will these shifting tactics by law enforcement agencies really protect private companies? Let's take a look [7].

9. PROTECTION AGAINST RANDOM ROGUE INFORMATION WARFARE

The growing availability of powerful encryption has, in effect, rewritten the rule book for creating, storing, and transmitting computer data [4]. People everywhere rightly regard confidentiality as essential for conducting business and protecting against random rogue information warfare and personal privacy [5]. But governments worldwide have been sent into a spin, for fear secret encryption keys will add to the weapons of terrorists and other criminals. Some nations have even attempted to control the technology by constructing a maze of regulations and laws aimed at blocking its import, export, and/or use. Such bans have largely failed, however.

In recent years, the war over encryption has moved beyond controlling the technology itself. Now, some governments are granting law enforcement agencies new powers and funding the development of new tools to get at computerized data, encrypted or otherwise. Rising to that challenge, privacy proponents are striking back with new techniques for hiding data and preserving anonymity in electronic communications [7].

9.1 Confess Up!

One legal tactic being used by states is to require owners of encrypted files to decrypt them when asked to by authorities. So far, only Singapore and Malaysia have enacted such laws, with Britain and India about to follow suit.

In Britain, two recent bills would give law enforcement officers the authority to compel individuals to decrypt an encrypted file in their possession under pain of a two-year jail term. Further, anyone given such a command would have to keep the giving of the notice, its contents, and the things done in pursuance of it secret—on penalty of a six-year jail term. The bills broadly define encryption, even including what some consider to be mere data protocol.

Straightforward though it seems, the approach is technically flawed. After all, a suspect may truly be unable to decrypt an encrypted file. He or she may have forgotten or lost the key. Or, if public-key encryption was used, the sender of a file will have the key used to encrypt the file, but rarely, if ever, the decryption key, which remains the exclusive property of the intended recipient. If symmetric key encryption was used, and the sender's hard disk crashes, the key will likely be wiped out along with all the other stored data. This flaw in the legislation was demonstrated by a British group, which mailed an ostensibly incriminating document to a government official and then destroyed the decryption key, making it impossible for that official to decode the file, even if "compelled."

Moreover, according to the latest version of the Cyberspace Electronic Security Act (CESA), police would be at liberty to present a text in court and claim it was the decrypted version of an encrypted file, without revealing to the defendant exactly how they arrived at the plaintext. This means that the defendant can have a hard time defending himself, and makes it a lot easier for the police to fabricate evidence. The ability to receive a fair trial could be at stake [7].

9.2 Escrowed Encryption

Another controversial scheme for letting law enforcement in on encrypted data is known as escrowed encryption. Here, a third party is appointed by the state to keep a copy of the decryption keys (in escrow, as it were) for the state to use to decrypt any file sent to or by any user. In other words, encrypted files would be protected—except from the state.

Needless to say, many people abhor the mere idea. Even if a sound case could be made for revealing the decryption key to government personnel, what is to prevent them from reusing that key in the future, to look at other documents by the same user? Furthermore, drug traffickers, terrorists, and others of most concern to law enforcement are the least likely to use encryption that is openly advertised as readable by the government.

Then, too, given the transnational nature of the Internet, a global key-escrow system would need to be established. Sovereign states, with their own interests to protect, would object to such a system; this, in fact, happened with the escrow scheme known as the "Clipper Chip," which was heavily promoted by the U.S. government but largely dismissed by other states. The logistics of who keeps the escrowed keys, who has authority to demand their release, under what conditions, and so on, becomes unwieldy when vast numbers of encryption keys, states, and legal systems are involved. In view of such concerns, official support for escrowed encryption has all but died in the United States and elsewhere [7].

9.3 Global Surveillance

The ineffectiveness of legal constraints on encryption appears to have persuaded many governments to change direction. They are instead seeking to capitalize on the unencrypted nature of most digital traffic and to derive information by monitoring that traffic. Even encrypted messages tend to leave unencrypted who is communicating with whom and when.

Officially, most states deny the existence of electronic surveillance wireless networks. But extensive claims of their existence persist. Echelon and FIDNet are two such alleged intelligence-gathering efforts that have been frequently described in the mainstream press and debated in official hearings by government legislatures. Echelon is, according to the Washington, D.C.-based Federation of American Scientists, a global wireless network that searches through millions of interceptions for preprogrammed keywords on fax, telex, and e-mail messages.

The same sort of public inquiries have been made about the Federal Intrusion Detection Network (FIDNet) that the U.S. National Security Council has proposed creating. It would monitor traffic on both government and commercial wireless networks, with the stated goal of safeguarding the critical U.S. information infrastructure. Although the House Appropriations Committee did away with funding for it last summer, FIDNet supporters continue to push the program, arguing that it would not intrude on individuals' communications. Meanwhile, a number of civil rights groups, including the Electronic Privacy Information Center (EPIC), in Washington, D.C., and the American Civil Liberties Union, based in New York City, have challenged FIDNet's constitutionality. The plan demonstrates that privacy concerns are being swept under the carpet [7].

9.4 Computer Forensics

As society relies increasingly on computers, the amount of crime perpetrated with the machines has risen in kind. To law enforcement's delight, electronic records have proved to be a fertile ground for detectives. Indeed, in their present shape, computers, the Internet, and e-mail are the most surveillance-friendly media ever devised.

This development has given rise to an entirely new industry: computer forensics. Its purpose is not only to find out what files are stored in a computer, but also to recover files that were created with, stored in, sent by, received from, or merely seen by that computer in the past, even if such files were subsequently "deleted" by the user.

The ability to resurrect electronic paper trails from supposedly deleted files stems, in large part, from the features built into many computer programs. For example, the delete command in most software does not delete. It merely marks the space that such a file occupied in a disk as being available in the future to be overwritten.

> **Note:** If it was really deleted, then undelete commands would not work.

Also, many Windows applications save temporary versions of a file being worked on, just in case the computer crashes. Even if a user were to deliberately overwrite the original file, the temporary version still lurks in some part of the disk, often with an unrecognizable name and occasionally even invisible from the conventional directory.

Electronic paper trails are also left behind by the fast save function, which saves the latest version of a word-processing document as the original plus the sequence of changes made to it. A recipient of the electronic end result can see how the document evolved over time—not the kind of information most people care to share.

Internet-related applications, like many other software programs, do a lot of internal housekeeping that involves writing information onto the hard disk. For example, the popular Web browser *Netscape Navigator* creates a file called *netscape.hst,* which gives a chronological listing of almost everything its user has done with the browser since it was installed.

Simply surfing the Web pushes other data into computer memory, in the guise of "cookies" and as documents "cached" on one's disk. Web sites visited can also learn the visitor's Internet service provider, Web browser, and a lot more. A remote Web site could even gain full access to a visitor's hard disk, depending on how aggressive that remote site elects to be and how extensive the protective measures taken by the visitor are.

Software tools now make it fairly straightforward to get a computer to cough up information that its owner may not realize is there. Not to be outdone, computer programmers have developed numerous tools that can defeat most computer forensics tools. Although such counter-forensics programs will remove most traces of sensitive data from a computer, it is extremely difficult to remove all traces that may have been left behind. In the absence of a thorough schooling in the esoteric details of computers, the odds favor the competent computer forensics investigator.

Also favoring the forensics expert are new laws legalizing the accessing of computers by law enforcement agencies. In December 1999 for example, the Australian Parliament passed a bill giving the Australian Security Organization the power to obtain warrants to access computers and telecommunications services, and, if necessary, to delete or alter other data in the target computer and conceal the fact that anything had been done under the warrant. As of February 2000, the Dutch authorities are now permitted to use bugging devices in computers to retrieve text [7].

9.5 Countermeasures

The various legal roadblocks and technical wizardry contrived by governments and law enforcement to block encryption's spread have, of course, curbed neither the need for the technology nor the ingenuity of privacy-loving programmers. As a result, a number of countermeasures have been engineered to augment or replace encryption. Among them are anonymizers, which conceal the identity of the person sending or receiving information, and steganography, which hides the information itself.

The need for anonymity in a democratic society has long been recognized, to shield whistleblowers and political dissenters from retaliation, to protect the records of medical patients, and so on. Less dramatic situations also justify anonymity, such as placing a personal ad or seeking employment through the Internet without jeopardizing one's current job. To be sure, anonymity can be exploited by sociopaths seeking to avoid accountability for their actions. But, in general, it serves a useful social function.

Anonymous and pseudonymous remailers are computers that are accessible through the Internet that launder the true identity of an e-mail sender. Most are operated at no cost to the user. A pseudonymous remailer replaces the sender's e-mail address with a false one and forwards the message to the intended recipient. The recipient can reply to the sender's pseudonymous address, which, in turn, forwards the response to the sender's true address.

Anonymous remailers come in three flavors: cypherpunk, mixmaster, and Web-based. Cypherpunk remailers strip away the message header, which describes where the message came from and how it got there, before forwarding the message to the recipient. Conceivably, someone with physical access to such a remailer's phone lines could correlate the incoming and outgoing traffic and make inferences.

Mixmaster remailers avoid that problem by using stronger encryption and tricks for frustrating traffic analysis, such as padding messages to disguise their true length. But even mixmasters can be compromised. For example, through a concerted effort, it would be possible to detect a correlation between Mr. A sending an encrypted message through a remailer, and Ms. B receiving a message at some variable time afterwards.

Web-based anonymizers range from sites offering conventional anonymizer services, to others where the connection between the user's computer and the anonymizer is itself encrypted with up to 128-bit encryption. The job is done using the standard Secure Socket Layer (SSL) encryption, built into all Web browsers of recent vintage.

For extra privacy, a message may be routed through a series of remailers. For example, the Onion Router project (see the site at *http://www.onion-router.net*) of the Naval Research Laboratory in Washington, D.C., offers another way to string together remailers. What's more, it allows anonymized and multiply encrypted Web browsing in real time.

Onion routing is a two-stage process. The initiator instructs router W (in this case, a proxy server at the firewall of a secure site) to create an onion, which consists of public-key-encrypted layers of instructions. Router X peels off the first layer of the onion, which indicates the next step in the path and supplies a symmetric decrypting key for use when the actual message comes through later.

The onion then goes to Routers Y and Z, depositing keys at each stop. Once the connection is established, the encrypted message is sent through and successively decrypted, arriving at the recipient as plaintext. To respond, the recipient sends the message to Router Z, which encrypts the text, onion-style, and sends it back through the already established path [7].

9.6 Hiding Data

The microdot used by German spies during World War II to transmit strategic information is an example of steganography, used to hide data in plain view. The microdot consisted of a greatly reduced photograph of a page of text, which was pasted over a period in an otherwise innocuous document. A more modern application is the digital watermark, for identifying official copies of copyrighted images and recordings. Unlike encryption, which hides the content of a message in an obvious manner, steganography hides the mere existence of anything hidden. The commercially available computer-based steganography programs popular today rely on three techniques:

- Merging the information to be hidden into a "cover" sound file by changing the least significant bit of each digitized sample of the file. The resulting file sounds the same to the human ear and is the same length as the original file.
- Merging the information to be hidden into a cover image file by changing the least significant bit of the digitized value of the brightness of each pixel. Typical images use 256 levels of brightness, with 8 bits per pixel for black-and-white images and 8 bits for each of the three primary colors (red, green, and blue) per pixel for color images. A lot of data can lurk in a 1024-by-768-pixel image.
- Hiding data in the areas of a computer floppy disk or hard drive that are normally not accessed. A computer disk is divided into clusters, each of which holds from 512 bytes to over 32,000 bytes. When a file is saved, it uses a portion of one or more clusters; because DOS and Windows store only one file per cluster, the space left over between the end of a file and the end of the cluster (called the slack) is available to hide data in. This scheme is extremely easy to detect, however [7].

9.7 The Future of Encryption

Encryption today is as strong as it is because there is no need for it to be any stronger. Of course, the underlying mathematical assumptions might be challenged by a breakthrough, such as a solution to factoring large numbers into their prime-number components. Meanwhile, an encryption method can be strengthened by merely adding bits to the encryption key.

Nevertheless, several schemes under development may eventually find use for electronic communication and storage: elliptic curve encryption; voice encryption (already freely available and used worldwide over the Internet); quantum cryptography; and DNA cryptography.

Few microprocessors have been specially designed to run encryption software. Most personal computers can accommodate the hardware and software requirements of modern encryption, but most hand-held devices cannot. For these devices, a new class of algorithms, known as elliptic curve encryption, is claimed to provide encryption strength equal to that of the standard algorithms in use today, while using a smaller key and arithmetic that is easier on microprocessors and that needs much less memory. Being a new type of encryption, its security has yet to withstand the concerted scrutiny of experts.

Voice encryption is a response to the increasing flow of audio traffic over the World Wide Web, which has led, among other things, to the merging of strong encryption with Internet telephony. Given appropriate software, anyone today can carry on fully encrypted conversations with any other user connected to the Internet.

Perhaps the most advanced such software is *Speak Freely,* which is available worldwide free of charge (see *http://www.speakfreely.org*). Some mainstream voice-over-the-Internet services do not offer encryption, though. Instead, they route the data through the company's servers, thereby opening up a security weakness.

Quantum cryptography is not in itself an encryption algorithm. Rather, it is a means for creating and securing the distribution of private keys. Based on the Heisenberg uncertainty principle, the idea is that communicating photons cannot be diverted from the intended recipient to the unsought-for interceptor without creating an irreversible change in the quantum states of the system.

The precepts of quantum cryptography date from the early 1970s, and research has been ongoing for the last decade at universities such as Johns Hopkins University, in Baltimore, Maryland, and the University of Geneva in Switzerland; at U.S. national laboratories such as Los Alamos; and in the corporate sector, at British Telecom and elsewhere.

In DNA cryptography, each letter of the alphabet is converted into a different combination of the four bases that make up human deoxyribonucleic acid (DNA). A piece of DNA spelling out the message to be encrypted is then synthesized, and the strand is slipped into a normal fragment of human DNA of similar length. The end result is dried out on paper and cut into small dots. As only one DNA strand in about 30 billion will contain the message, the detection of even the existence of the encrypted message is most unlikely [7].

9.8 Shifting Attitudes

If, as seems likely, encryption and related products will continue to develop for personal and commercial uses, countries will have to rethink their policies toward the technology. In what may be a sign of things to come, the German government announced in May 1999, that it would fund the development and free distribution of open-source encryption software that the government itself will be unable to break (see *http://www.gnupg.org*). The Federal Ministry of Economics and Technology released a report stating that Germany considers the application of secure encryption to be a crucial requirement for citizens' privacy, the development of electronic commerce, and the protection of business secrets.

Also in 1999, French Prime Minister Lionel Jospin announced a similar shift, saying that his country would scrap any key escrow plans in favor of free use of cryptography. In both cases, the motivation seems to have been the realization that protecting data from foreign parties outweighs any law enforcement concerns, and that the use of strong encryption furthers, rather than hinders, national security.

Independently, the Canadian government announced in October 1999 that it would not seek to regulate the domestic use of encryption. The significance of such trends is clear: The global reach of the Internet has made it extremely easy for encryption software to travel between countries, with or without controls, and if one or more major countries elects not to enforce controls, the technology will spread even more easily. Society's transformation into a computer-based economy makes protecting corporate and personal information not only desirable but also necessary.

How then does one balance privacy and confidentiality with security? Governments are undoubtedly obligated to protect their citizens from terrorism and out-and-out criminality. A partial solution may be to criminalize the use of encryption only in the commission of generally recognized serious crimes and to encourage its use elsewhere.

Attempting to control encryption, however, has proved to be an ineffective means of preventing crime and may actually hurt vital national interests. Similarly, the granting of new policing powers to law enforcement agencies will do less to protect a country's critical infrastructure than building better security technology. And if greater security is truly what governments are after, then much can be done with the tools already in hand: Encrypting all important data and communications makes their illegal retrieval and interception useless to the thief [7].

10. KEEPING THE AMATEUR ROGUE OUT OF THE CYBER HOUSE

Finally, how do you keep amateur rogues out of the cyberhouse? Today, you probably can't; but, tomorrow (see Chapter 26, "The Information Warfare Wireless Network Security Arsenal of The Future")—well, that's another matter.

Today however, motivated amateur rogue "hacktivists" have grabbed headlines, announcing they've collected credit card and other personal data on some 58,800 business and political leaders. Increasingly, these amateur social activists have turned to hacking to make their point, breaking into computer systems and wreaking havoc on organizations they oppose. The Internet has turned out to be a remarkable tool for nonviolent protest on a scale activists could only dream of before.

As previously explained, the term "hacktivist" was first applied to supporters of the Zapatista rebels in Mexico's southern state of Chiapas, who have sabotaged Mexican government Web sites since 1998 and held "virtual sit-ins" designed to overload servers. More recently, the tactic has been used in Serbia, Pakistan, and India—and by both Palestinians and Israelis in the Middle East. In one case, Palestinian sympathizers broke into a Web site operated by a pro-Israel lobbying group in the United States, stealing credit card information and e-mail addresses.

However, the theft of private data is a relatively new tactic, which goes beyond defacing Web sites and electronic bombardment of servers. Antiglobalist protesters contend the WTO's trade treaties benefit big corporations and rich countries at the expense of the environment and workers.

Protesters who showed up in person were largely stymied by a heavy police presence at the recent Davos meeting. On-line, however, they effectively surmounted physical barriers [7].

10.1 Another Frontier

The Net is another frontier for people to engage in these types of activities. The attacks against forum organizers showed just how far hacktivists could reach. They obtained the travel itineraries (including flight numbers) of politicians from around the world, and published them on the Web. This poses operational security problems, and goes beyond what's been seen before.

Finally, almost every major corporation and organization has been hit at one time or another by hacking, with McDonald's, Starbucks, and the WTO being favorite targets of hacktivists. In some respect, it is really quite clever and quite funny. But it is less funny when people believe it (as has been the case) and go to a lot of trouble and then are deceived [7].

11. SUMMARY AND CONCLUSIONS

It can be seen that the development of the Internet presents serious threats to the security of private companies, in addition to the much-touted opportunities it provides. It may also be that the more extreme scenarios discussed in this chapter may never eventuate—the possibility that they may, however, must be appreciated. It is not advisable for any risk-management approach to merely disregard the threats previously discussed on the basis that they are far-fetched and fanciful. In addition to the threats being technically feasible, either now or in the next two decades or so, the ability of intruders to gain entry to computer systems and disguise the very fact of entry makes this a peculiarly difficult threat to appreciate. Undetectability of many attacks per se may lead private companies to a false sense of security, and leave the companies vulnerable to serious disruption of total disablement in the event of an attack.

The possibility of means of attack this presents to aggressors, can help realistically guide the process of moving forward in dealing with the information warfare arsenal and tactics of private companies. The conclusions drawn from this follow next:

- As competition between corporations for profit increase, and consumer expectations grow, there may soon be a time that, for some private companies, even a limited disablement may be fatal, or nearly fatal, to its continued existence, surely one of the most important post-threat outcomes of any risk-management plan.
- The growth in the number of aggressors must also be appreciated.
- Added to the traditional aggressors identified by private companies, are the additional ones that may now see the companies as a visible surrogate

of an entity that is either impregnable from attack or that it is inadvisable to attack.

- Some private companies have always been the target of aggression, and the identity and number of aggressors may stay the same.
- It must be appreciated, however, that new, and very powerful, tools of aggression may now be available to those traditional aggressors.
- Traditional forms of risk management are, it is argued, not particularly suitable to the dynamic, desegregated forms of aggression.
- The approach to determining risk and how to protect against and prevent wireless network attacks must be revised. A fundamental rethinking of the way private companies organize themselves, and the way they leave themselves at risk, will also be necessary.
- Traditional forms of risk management represent an approach positioned in a hierarchical paradigm, which may not deal adequately or at all with new forms of threat posed to a dynamic wireless network.
- Until these fundamental issues are addressed, no private company can truly say that it has identified all forms of risk that are or will be relevant to that organization. Nor will it be able to say that it has treated them. These must be imperatives in an environment where any single risk could conceivably threaten the entity's very survival [7].

12. REFERENCES

[1] Vacca, John R., *i-mode Crash Course*, McGraw-Hill, 2002.

[2] Vacca, John R., *Wireless Broadband Networks Handbook*, McGraw-Hill, 2001.

[3] Vacca, John R., *Electronic Commerce, Fourth Edition*, Charles River Media, 2003.

[4] Vacca, John R., *The Essential Guide to Storage Area Networks*, Prentice Hall, 2002.

[5] Vacca, John R., *Net Privacy: A Guide to Developing & Implementing an Ironclad ebusiness Privacy Plan*, McGraw-Hill, 2001.

[6] Vacca, John R., *Firewalls: Jumpstart For Network And Systems Administrators*, Elsevier Digital Press (December, 2004).

[7] John R. Vacca, *Computer Forensics: Computer Crime Scene Investigation, 2nd Edition*, Charles River Media, 2005.

Chapter 26

THE INFORMATION WARFARE WIRELESS NETWORK SECURITY ARSENAL OF THE FUTURE

1. INTRODUCTION

Terrorists take control of the New York Stock Exchange? Terrorism over the Internet? Computer viruses in the arsenal of Hizballah? As discussed in preceding chapters, such possibilities are currently being discussed by strategic analysts under the catch-all title of "Information Warfare." To date, the defense establishment has yet to agree on the exact definition of the term "information warfare."

Only the entertainment industry, in the form of films and novels, has popularized the notion of an electronic doomsday scenario in which covert terrorist groups manage to penetrate critical nodes of the National Information Infrastructure (NII) and Defense Information Infrastructure (DII) and are able to, variously, launch nuclear weapons, crash the telephone system, cause mayhem on the railways or in the air, or bring the financial sector to a catastrophic halt (see sidebar, "Will The Real La Femme Nikita Stand Up"). Warnings also come from more sober sources. In 1999 the U.S. Joint Chiefs of Staff concluded that the convergence of vulnerable information infrastructures with traditional critical infrastructures had created a tunnel of vulnerability previously unrealized in the history of conflict. In other words, the one thing that everyone agrees on, in the digital age, information, and its dissemination, has achieved the status of a vital strategic asset.

Will The Real La Femme Nikita Stand Up

Section One (in USA's ex series "LA Femme Nikita") was the most covert antiterrorism organization on the planet. ABC's "Alias" and Fox's "24," don't hold a candle to "LA Femme Nikita" when it comes to whole scale assassinations and torture. Section One is a skilled team of operatives responsible for protecting human life around the globe from chaos and destruction.

Sound implausible? Maybe. But the creation of such covert antiterrorism organizations are currently in the planning stages by the National Security Agency (NSA) and the CIA.

NSA and the CIA realize that conventional IW tactics will not be enough in the future to thwart the very dangerous and often suicidal world of covert terrorist organizations. Like the character Nikita (USA's LA Femme Nilkita), who "transforms into a highly trained agent dedicated to fighting global terrorism by any means necessary—legal or otherwise," today's agents will have to do the same.

In the very near future, agents trained and armed with an arsenal of futuristic high-tech weapons, and trained in the most sophisticated techniques on how to carry out successful assassinations will swoop down upon deadly terrorist operatives. Like Nikita, they will all have to be the perfect weapon. They will also have to keep their wits about them, as well as ingenuity to keep themselves alive, where a single mistake could mean death. This will be their most vital weapon—and the best hope for the future of all that is good in the world [8].

If the response of the American defense establishment is any indication, strategic analysts are taking the possibilities of infowar seriously. The first global cyberwar will be like no other war ever fought before-- where the enemy is invisible, the battles virtual, and the casualties all too real. Special committees in every branch of the U.S. armed forces are studying the potential of infowar, both as a defensive and an offensive weapon. The National Security Agency (NSA) is reportedly studying a rather imaginative arsenal of "info- weapons." Among the current possible offensive weapons are:

- Computer viruses, which could be fed into an enemy's computers either remotely or by "mercenary" technicians.
- Logic bombs, another type of virus which can lie dormant for years, until, upon receiving a particular signal, it would wake up and begin attacking the host system.
- "Chipping," a plan (originally proposed by the CIA, according to some sources) to slip booby-trapped computer chips into critical systems sold by foreign contractors to potentially hostile third parties (or recalcitrant allies?).
- Worms, whose purpose is to self-replicate ad infinitum, thus eating up a system's resources.
- Trojan horses, malevolent code inserted into legitimate programming to perform a disguised function.
- Back doors and trap doors, a mechanism built into a system by the designer, to give the manufacturer or others the ability to "sneak back into the system" at a later date by circumventing the need for access privileges [8].

A few other goodies in the arsenal of information warfare are devices for disrupting data flow or damaging entire systems, hardware and all. Among these, as explained in earlier chapters, are High Energy Radio Frequency (HERF) guns, which focus a high power radio signal on target equipment, putting it out of action; and, Electromagnetic Pulse (EMP) devices, which can be detonated in the vicinity of a target system. Such devices can destroy electronics and communications equipment over a wide area.

All of the preceding current and future offensive and defensive IW weapons arsenal will be discussed in specific detail next. Let's take a look [8].

2. YOU HAVEN'T SEEN ANYTHING LIKE WHAT IS COMING

Body count: 796. Cause: midair collision. The air traffic control system was "cybotaged." News reports indicate that FAA personnel complained that their radar screens were freezing, and were switching data tags (such as aircraft altitude data) between close-flying planes. Series of near-misses in skies throughout the country—and one head-on collision between passenger jets in a thunderstorm over New York, resulting in the deaths of all aboard. It's suspected that the automated route and altitude management program's collision-avoidance algorithm was damaged.

Body count: 1,807.Cause: midair collision with a structure. The navigation system of another passenger jet was taken over by hackers, leaving the pilots helpless as the jet nose-dived into the Sears Towers in Chicago. No reports yet on how the hackers got in. A couple of hit sites have posted theories, some of them pretty good.

A message posted to 90,000 newsgroups from a group known as "The Vulture of Jihad" claimed credit for the attack. As they're an obscure Sunni sect known for abjuring the use of any technology, their claim, made during prayers in a mosque in Aleppo, was disregarded. Other Islamic splinter groups also claimed credit, along with a white supremacist faction and an anarchist syndicate. These claims were swiftly dismissed, too: All were missing the digital signature which the Islamic Liberation Army (ILA) had in both previous site hacks. The most outrageous theory as to the identity of the people responsible for the attack came on a hit site called the "Hit Theorist." It says the whole thing is a CIA, NSA, and DOD plot to generate support in Congress for increased spending of military and Black Ops operations.

Do these scenarios sound like spin-offs from Fox's *X-Files*: "The Lone Gunmen"? Perhaps! But could it happen? You bet [8]!

2.1 The E-Bomb: A Weapon of Electrical Mass Destruction

Perhaps the most dangerous of all of defensive and offensive weapons in the IW arsenal of the future is the E-Bomb. High Power Electromagnetic Pulse generation techniques and High Power Microwave (HPM) technology have matured to the point where practical E-Bombs (Electromagnetic bombs) are becoming technically feasible, with new applications in both strategic and tactical information warfare. The development of conventional E-Bomb devices allows their use in non-nuclear confrontations. This part of the chapter discusses aspects of the technology base, weapon delivery techniques and proposes a foundation for the use of such devices in warhead and bomb applications.

The prosecution of a successful Information Warfare (IW) campaign against an industrialized or post-industrial opponent will require a suitable set of tools. The efficient execution of an IW campaign against a modern industrial or post-industrial opponent will require the use of specialized tools designed to destroy information systems. Electromagnetic bombs (also popularized by USA's ex futuristic sci-fi show "Dark Angel") built for this purpose, can provide, where delivered by suitable means, an effective tool for this purpose [8].

2.2 The EMP Effect

The ElectroMagnetic Pulse (EMP) effect was first observed during the early testing of high-altitude airburst nuclear weapons. The effect is characterized by the production of a very short (hundreds of nanoseconds) but intense electromagnetic pulse, which propagates away from its source with ever-diminishing intensity, governed by the theory of electromagnetism. The ElectroMagnetic Pulse is, in effect, an electromagnetic shock wave.

> **Note:** EMP stands for electromagnetic pulse. The source can be a nuclear or a non-nuclear detonation. It can be used by special forces teams who infiltrate the enemy's and detonate a device near their electronic devices. It destroys the electronics of all computer and communication systems in a quite large area. The EMP bomb can be smaller than a HERF gun to cause a similar amount of damage and is typically used to damage not a single target (not aiming in one direction) but to damage all equipment near the bomb.

This pulse of energy produces a powerful electromagnetic field, particularly within the vicinity of the weapon burst. The field can be sufficiently strong to produce short-lived transient voltages of thousands of Volts (kiloVolts) on exposed electrical conductors, such as wires, or conductive tracks on printed circuit boards, where exposed.

It is this aspect of the EMP effect that is of military significance, as it can result in irreversible damage to a wide range of electrical and electronic equipment, particularly computers and radio or radar receivers. Subject to the electromagnetic hardness of the electronics, a measure of the equipment's resilience to this effect, and the intensity of the field produced by the weapon, the equipment can be irreversibly damaged or, in effect, electrically destroyed. The damage inflicted is not unlike that experienced through exposure to close proximity lightning strikes, and may require complete replacement of the equipment, or at least substantial portions thereof.

Commercial computer equipment is particularly vulnerable to EMP effects, as it is largely built up of high-density Metal Oxide Semiconductor (MOS) devices, which are very sensitive to exposure to high-voltage transients. What is significant about MOS devices is that very little energy is required to permanently wound or destroy them; any voltage typically in excess of ten volts can produce an effect termed "gate breakdown," which effectively destroys the device. Even if the pulse is not powerful enough to produce thermal damage, the power supply in the equipment will readily supply enough energy to complete the destructive process. Wounded devices may still function, but their reliability will be seriously impaired. Shielding electronics by equipment chassis provides only limited protection, as any cables running in and out of the equipment will behave very much like antennae, in effect, guiding the high-voltage transients into the equipment.

Computers used in data processing systems; communications systems; displays; industrial control applications, including road and rail signaling; and those embedded in military equipment, such as signal processors, electronic flight controls, and digital engine control systems, are all potentially vulnerable to the EMP effect.

Other electronic devices and electrical equipment may also be destroyed by the EMP effect. Telecommunications equipment can be highly vulnerable, due to the presence of lengthy copper cables between devices [1]. Receivers of all varieties are particularly sensitive to EMP, as the highly sensitive miniature high-frequency transistors and diodes in such equipment are easily destroyed by exposure to high-voltage electrical transients. Therefore, radar and electronic warfare equipment, satellite, microwave, UHF, VHF, HF, and low-band communications equipment and television equipment are all potentially vulnerable to the EMP effect. It is significant that modern military platforms are densely packed with electronic equipment, and unless these platforms are well hardened, an EMP device can substantially reduce their function or render them unusable [8].

2.3 The Technology Base for Conventional Electromagnetic Bombs

The technology base which may be applied to the design of electromagnetic bombs is both diverse, and in many areas quite mature. Key technologies which are extant in the area are explosively pumped Flux Compression Generators (FCG), explosive or propellant driven Magneto-Hydrodynamic (MHD) generators and a range of HPM devices, the foremost of which is the Virtual Cathode Oscillator or Vircator. A wide range of experimental designs have been tested in these technology areas, and a considerable volume of work has been published in unclassified literature.

This part of the chapter will review the basic principles and attributes of these technologies, in relation to bomb and warhead applications. It is stressed that this treatment is not exhaustive, and is only intended to illustrate how the technology base can be adapted to an operationally deployable capability [8].

2.4 The Lethality of Electromagnetic Warheads

The issue of electromagnetic weapon lethality is complex. Unlike the technology base for weapon construction, which has been widely published in the open literature, lethality-related issues have been published much less frequently.

Although the calculation of electromagnetic field strengths achievable at a given radius for a given device design is a straightforward task, determining a kill probability for a given class of target under such conditions is not.

This is for good reasons. The first is that target types are very diverse in their electromagnetic hardness, or ability to resist damage. Equipment that has been intentionally shielded and hardened against electromagnetic attack will withstand greater orders of magnitude and field strengths than standard commercially rated equipment. Moreover, various manufacturer's implementations of like types of equipment may vary significantly in hardness due the idiosyncrasies of specific electrical designs, cabling schemes, and chassis/shielding designs used.

The second major problem area in determining lethality is that of coupling efficiency, which is a measure of how much power is transferred from the field produced by the weapon into the target. Only power coupled into the target can cause useful damage [8].

2.5 Targeting Electromagnetic Bombs

The task of identifying targets for attack with electromagnetic bombs can be complex. Certain categories of target will be very easy to identify and engage. Buildings housing government offices and, thus, computer equipment, production facilities, military bases, and known radar sites and communications nodes are all targets that can be readily identified through conventional photographic, satellite, imaging radar, electronic reconnaissance, and human operations. These targets are typically geographically fixed and, thus, may be attacked providing that the aircraft can penetrate to weapon release range. With the accuracy inherent in GPS/inertially guided weapons, the electromagnetic bomb can be programmed to detonate at the optimal position to inflict a maximum of electrical damage.

Mobile and camouflaged targets that radiate overtly can also be readily engaged. Mobile and relocatable air defense equipment, mobile communications nodes [2], and naval vessels are all good examples of this category of target. While radiating, their positions can be precisely tracked with suitable Electronic Support Measures (ESM) and Emitter Locating

Systems (ELS) carried either by the launch platform or a remote surveillance platform. In the latter instance, target coordinates can be continuously datalinked to the launch platform. As most such targets move relatively slowly, they are unlikely to escape the footprint of the electromagnetic bomb during the weapon's flight time.

Mobile or hidden targets that do not overtly radiate may present a problem, particularly should conventional means of targeting be employed. A technical solution to this problem does, however, exist for many types of target. This solution is the detection and tracking of Unintentional Emission (UE). UE has attracted most attention in the context of TEMPEST surveillance, where transient emanations leaking out from equipment due to poor shielding can be detected and, in many instances, demodulated to recover useful intelligence. Termed "Van Eck radiation," such emissions can only be suppressed by rigorous shielding and emission-control techniques, such as are employed in TEMPEST rated equipment.

Although the demodulation of UE can be a technically difficult task to perform well, in the context of targeting electromagnetic bombs this problem does not arise. To target such an emitter for attack requires only the ability to identify the type of emission and, thus, target type, and to isolate its position with sufficient accuracy to deliver the bomb. Because the emissions from computer monitors, peripherals, processor equipment, switchmode power supplies, electrical motors, internal combustion engine ignition systems, variable duty cycle electrical power controllers (thyristor or triac-based), superheterodyne receiver local oscillators, and computer networking cables are all distinct in their frequencies and modulations, a suitable Emitter Locating System can be designed to detect, identify, and track such sources of emission.

A good precedent for this targeting paradigm exists. During the SEA (Vietnam) conflict, the United States Air Force (USAF) operated a number of night interdiction gunships that used direction finding receivers to track the emissions from vehicle ignition systems. Once a truck was identified and tracked, the gunship would engage it.

Because UE occurs at relatively low power levels, the use of this detection method prior to the outbreak of hostilities can be difficult, as it may be necessary to overfly hostile territory to find signals of usable intensity. The use of stealthy reconnaissance aircraft or long range, stealthy Unmanned Aerial Vehicles (UAV) may be required. The latter also raises the possibility of autonomous electromagnetic-warhead-armed expendable UAVs, fitted with appropriate homing receivers. These would be programmed to loiter in a target area until a suitable emitter is detected, upon which the UAV would home in and expend itself against the target [8].

2.6 The Delivery of Conventional Electromagnetic Bombs

As with explosive warheads, electromagnetic warheads will occupy a volume of physical space and will also have some given mass (weight) determined by the density of the internal hardware. Like explosive warheads, electromagnetic warheads may be fitted to a range of delivery vehicles.

Known existing applications involve fitting an electromagnetic warhead to a cruise missile airframe. The choice of a cruise missile airframe will restrict the weight of the weapon to about 340 kg (750 lb), although some sacrifice in airframe fuel capacity could see this size increased. A limitation in all such applications is the need to carry an electrical energy storage device (a battery), to provide the current used to charge the capacitors used to prime the FCG prior to its discharge. Therefore, the available payload capacity will be split between the electrical storage and the weapon itself.

In wholly autonomous weapons such as cruise missiles, the size of the priming current source and its battery may well impose important limitations on weapon capability. Air-delivered bombs, which have a flight time between tens of seconds to minutes, could be built to exploit the launch aircraft's power systems. In such a bomb design, the bomb's capacitor bank can be charged by the launch aircraft enroute to target, and after release a much smaller onboard power supply could be used to maintain the charge in the priming source prior to weapon initiation.

An electromagnetic bomb delivered by a conventional aircraft can offer a much better ratio of electromagnetic device mass to total bomb mass, as most of the bomb mass can be dedicated to the electromagnetic-device installation itself. It follows, therefore, that for a given technology an electromagnetic bomb of identical mass to an electromagnetic-warhead-equipped missile can have a much greater lethality, assuming equal accuracy of delivery and technologically similar electromagnetic device design.

A missile-borne electromagnetic warhead installation will comprise the electromagnetic device, an electrical energy converter, and an onboard storage device such as a battery. As the weapon is pumped, the battery is drained. The electromagnetic device will be detonated by the missile's onboard fusing system. In a cruise missile, this will be tied to the navigation system; in an antishipping missile, the radar seeker; and in an air-to-air missile, the proximity fusing system. The warhead fraction (ratio of total payload [warhead] mass to launch mass of the weapon) will be between 15% and 30%.

An electromagnetic bomb warhead will comprise an electromagnetic device, an electrical energy converter, and an energy storage device to pump and sustain the electromagnetic device charge after separation from the delivery platform. Fusing could be provided by a radar altimeter fuse to airburst the bomb, a barometric fuse or in GPS/inertially guided bombs, the navigation system. The warhead fraction could be as high as 85%, with most of the usable mass occupied by the electromagnetic device and its supporting hardware.

Due to the potentially large lethal radius of an electromagnetic device, compared to an explosive device of similar mass, standoff delivery would be prudent. Although this is an inherent characteristic of weapons such as cruise missiles, potential applications of these devices to glidebombs, antishipping missiles and air-to-air missiles would dictate fire and forget guidance of the appropriate variety, to allow the launching aircraft to gain adequate separation of several miles before warhead detonation.

The recent advent of GPS satellite [3] navigation guidance kits for conventional bombs and glidebombs has provided the optimal means for cheaply delivering such weapons. Although GPS-guided weapons without differential GPS enhancements may lack the pinpoint accuracy of laser- or television-guided munitions, they are still quite accurate (CEP \(~~ 40 ft), cheap, and autonomous all-weather weapons.

The USAF has deployed the Northrop GPS-Aided Munition (GAM) on the B-2 bomber; as well as, the GPS/inertially guided GBU-29/30 Joint Direct Attack Munition (JDAM) and the AGM-154 Joint Stand-Off Weapon (JSOW) glidebomb. Other countries are also developing this technology. For example, the Australian BAeA Agile Glide Weapon (AGW) glidebomb is achieving a glide range of about 140 km (75 nmi) when launched from that altitude.

The importance of glidebombs as delivery means for HPM warheads is threefold. First, the glidebomb can be released from outside the effective radius of target air defenses, therefore minimizing the risk to the launch aircraft. Second, the large standoff range means that the aircraft can remain well clear of the bomb's effects. Finally the bomb's autopilot may be programmed to shape the terminal trajectory of the weapon, such that a target may be engaged from the most suitable altitude and aspect.

A major advantage of using electromagnetic bombs is that they may be delivered by any tactical aircraft with a nav-attack system capable of delivering GPS-guided munitions. As you can expect GPS-guided munitions to be become the standard weapon in use by Western air forces in the 21st century, every aircraft capable of delivering a standard guided munition also becomes a potential delivery vehicle for an electromagnetic bomb. Should weapon ballistic properties be identical to the standard weapon, no software changes to the aircraft would be required.

Because of the simplicity of electromagnetic bombs in comparison with weapons such as Anti Radiation Missiles (ARM), it is not unreasonable to expect that these should be both cheaper to manufacture and easier to support in the field, thus allowing for more substantial weapon stocks. In turn, this makes saturation attacks a much more viable proposition [8].

2.7 Defense against Electromagnetic Bombs

The most effective defense against electromagnetic bombs is to prevent their delivery by destroying the launch platform or delivery vehicle, as is the case with nuclear weapons. This however may not always be possible, and, therefore, systems that can be expected to suffer exposure to the electromagnetic weapons effects must be electromagnetically hardened.

The most effective method is to wholly contain the equipment in an electrically conductive enclosure, termed a "Faraday cage," which prevents the electromagnetic field from gaining access to the protected equipment. However, most such equipment must communicate with and be fed with power from the outside world, and this can provide entry points via which electrical transients may enter the enclosure and cause damage. Although optical fibers address this requirement for transferring data in and out, electrical power feeds remain an ongoing vulnerability.

Where an electrically conductive channel must enter the enclosure, electromagnetic-arresting devices must be fitted. A range of devices exist, however, care must be taken in determining their parameters to ensure that they can deal with the rise time and strength of electrical transients produced by electromagnetic devices. Reports from the United States indicate that hardening measures attuned to the behavior of nuclear EMP bombs do not perform well when dealing with some conventional microwave electromagnetic device designs.

It is significant that hardening of systems must be carried out at a system level, as electromagnetic damage to any single element of a complex system could inhibit the function of the whole system. Hardening new build equipment and systems will add a substantial cost burden. Older equipment and systems may be impossible to harden properly and may require complete replacement. In simple terms, hardening by design is significantly easier than attempting to harden existing equipment.

An interesting aspect of electrical damage to targets is the possibility of wounding semiconductor devices, thereby causing equipment to suffer repetitive intermittent faults rather than complete failures. Such faults would tie down considerable maintenance resources while also diminishing the confidence of the operators in the equipment's reliability. Intermittent faults may not be possible to repair economically, thereby causing equipment in this state to be removed from service permanently, with considerable loss in maintenance hours during damage diagnosis. This factor must also be considered when assessing the hardness of equipment against electromagnetic attack, as partial or incomplete hardening may, in this fashion, cause more difficulties than it would solve. Indeed, shielding that is incomplete may resonate when excited by radiation and, thus, contribute to damage inflicted on the equipment contained within it.

Other than hardening against attack, facilities that are concealed should not radiate readily detectable emissions. Where radio frequency communications must be used, low probability of intercept (spread spectrum) techniques should be employed exclusively to preclude the use of site emissions for electromagnetic-targeting purposes. Appropriate suppression of UE is also mandatory.

Wireless communications networks for voice, data, and services should employ topologies with sufficient redundancy and failover mechanisms to allow operation with multiple nodes and links inoperative. This will deny a user of electromagnetic bombs the option of disabling large portions if not the whole of the wireless network by taking down one or more key nodes or links with a single or small number of attacks [8].

2.8 Limitations of Electromagnetic Bombs

The limitations of electromagnetic weapons are determined by weapon implementation and means of delivery. Weapon implementation will determine the electromagnetic field strength achievable at a given radius, and its spectral distribution. Means of delivery will constrain the accuracy with which the weapon can be positioned in relation to the intended target. Both constrain lethality.

In the context of targeting military equipment, it must be noted that thermionic technology (vacuum tube equipment) is substantially more resilient to the electromagnetic weapons effects than solid-state (transistor) technology. Therefore, a weapon optimized to destroy solid-state computers and receivers may cause little or no damage to a thermionic technology device, for instance early-1960s Soviet military equipment. Therefore, a hard electrical kill may not be achieved against such targets unless a suitable weapon is used.

This underscores another limitation of electromagnetic weapons, which is the difficulty in kill assessment. Radiating targets such as radars or communications equipment may continue to radiate after an attack even though their receivers and data processing systems have been damaged or destroyed. This means that equipment that has been successfully attacked may still appear to operate. Conversely, an opponent may shut down an emitter if attack is imminent and the absence of emissions means that the success or failure of the attack may not be immediately apparent.

Assessing whether an attack on a nonradiating emitter has been successful is more problematic. A good case can be made for developing tools specifically for the purpose of analyzing unintended emissions, not only for targeting purposes but also for kill assessment.

An important factor in assessing the lethal coverage of an electromagnetic weapon is atmospheric propagation. Although the relationship between electromagnetic field strength and distance from the weapon is one of an inverse square law in free space, the decay in lethal effect with increasing distance within the atmosphere will be greater due quantum physical absorption effects. This is particularly so at higher frequencies, and significant absorption peaks due water vapor and oxygen exist at frequencies above 20 GHz. These will, therefore, contain the effect of HPM weapons to shorter radii than are ideally achievable in the K and L frequency bands.

Means of delivery will limit the lethality of an electromagnetic bomb by introducing limits to the weapon's size and the accuracy of its delivery. Should the delivery error be of the order of the weapon's lethal radius for a given detonation altitude, lethality will be significantly diminished. This is of particular importance when assessing the lethality of unguided electromagnetic bombs, as delivery errors will be more substantial than those experienced with guided weapons such as GPS-guided bombs.

Therefore, accuracy of delivery and achievable lethal radius must be considered against the allowable collateral damage for the chosen target. Where collateral electrical damage is a consideration, accuracy of delivery and lethal radius are key parameters. An inaccurately delivered weapon of large lethal radius may be unusable against a target should the likely collateral electrical damage be beyond acceptable limits. This can be a major issue for users constrained by treaty provisions on collateral damage [8].

2.9 The Proliferation of Electromagnetic Bombs

At the time of this writing, the United States is one of several nations with the established technology base and the depth of specific experience to design weapons based upon this technology. However, the relative simplicity of the FCG and the Vircator suggests that any nation with even a 1940s technology base, once in possession of engineering drawings and specifications for such weapons, could manufacture them.

As an example, the fabrication of an effective FCG can be accomplished with basic electrical materials, common plastic explosives such as C-4 or

Semtex, and readily available machine tools such as lathes and suitable mandrels for forming coils. Disregarding the overheads of design, which do not apply in this context, a two-stage FCG could be fabricated for a cost as low as $14,000–15,000, at Western labor rates. This cost could be even lower in a third-world or newly industrialized economy.

Although the relative simplicity and, thus, low cost of such weapons can be considered of benefit to first-world nations intending to build viable war stocks or maintain production in wartime, the possibility of less developed nations mass producing such weapons is alarming. The dependence of modern economies upon first-world nations' information technology infrastructure, makes them highly vulnerable to attack with such weapons, providing that such weapons can be delivered to their targets.

Of major concern is the vulnerability resulting from increasing use of communications and data communications schemes based on copper cable media. If the copper medium were to be replaced en masse with optical fiber to achieve higher bandwidths, the communications infrastructure would become significantly more robust against electromagnetic attack as a result. However, the current trend is to exploit existing distribution media such as cable TV and telephone wiring to provide multiple Megabit/s data distribution (cable modems, ADSL/HDSL/VDSL) to premises. Moreover, the gradual replacement of coaxial Ethernet networking with 10-Base-T twisted pair equipment has further increased the vulnerability of wiring systems inside buildings. It is not unreasonable to assume that the data and services communications infrastructure in the West will remain a "soft" electromagnetic target in the foreseeable future.

At this time, no counter-proliferation regimes exist. Should treaties be agreed to limit the proliferation of electromagnetic weapons, they would be virtually impossible to enforce given the common availability of suitable materials and tools.

With the former Soviet Union suffering significant economic difficulties, the possibility of microwave and pulse power technology designs leaking out to third-world nations or terrorist organizations should not be discounted. The threat of electromagnetic bomb proliferation is very real [8].

2.10 A Doctrine for the Use of Conventional Electromagnetic Bombs

A fundamental tenet of IW is that complex organizational systems such as governments, industries, and military forces cannot function without the flow of information through their structures. Information flows within these structures in several directions, under typical conditions of function. A trivial model for this function would see commands and directives flowing outward from a central decision-making element, with information about the state of the system flowing in the opposite direction. Real systems are substantially more complex.

This is of military significance because stopping this flow of information will severely debilitate the function of any such system. Stopping the outward flow of information produces paralysis, as commands cannot reach the elements that are to execute them. Stopping the inward flow of information isolates the decision-making element from reality, and, thus, severely inhibits its capacity to make rational decisions that are sensitive to the currency of information at hand.

The recent evolution of strategic (air) warfare indicates a growing trend toward targeting strategies that exploit this most fundamental vulnerability of any large and organized system. The Desert Storm air war of 1991 is a good instance, with a substantial effort expended against such targets. Indeed, the model used for modern strategic air attack places leadership and its supporting communications in the position of highest targeting priority. No less important, modern electronic combat concentrates on the disruption and destruction of communications and information-gathering sensors used to support military operations. Again, the Desert Storm air war provides a good illustration of the application of this method.

A strategy that stresses attack on the information-processing and communications elements of the targeted systems offers a very high payoff, as it will introduce an increasing level of paralysis and disorientation within its target. Electromagnetic bombs are a powerful tool in the implementation of such a strategy [8].

2.11 Computer Viruses

A virus is a code fragment that copies itself into a larger program, modifying that program. A virus executes only when its host program begins to run. The virus then replicates itself, infecting other programs as it reproduces.

Viruses are well known in every computer-based environment, so that it is not astonishing that this type of rough program is used in information warfare. One could imagine that the CIA (or Army, Air Force, etc.) inserts computer viruses into the switching networks of the enemy's phone system. As today's telephone systems are switched by computers, you can shut them down, or at least causing massive failure, with a virus as easy as you can shut down a "normal" computer [8].

2.12 Worms

A worm is an independent program. It reproduces by copying itself in full-blown fashion from one computer to another, usually over a wireless network. Unlike a virus, it usually doesn't modify other programs.

Also, if worms don't destroy data, they can cause the loss of communication by merely eating up resources and spreading through the wireless networks. A worm can also easily be modified so that data deletion or worse occurs. With a "wildlife" like this, you could imagine breaking down a wireless networked environment such as a ATM and banking network [8].

2.13 Trojan Horses

A Trojan horse is a code fragment that hides inside a program and performs a disguised function. It's a popular mechanism for disguising a virus or a worm.

A Trojan horse could be camouflaged as a security-related tool. If someone edits this program so that it sends discovered security holes in an e-mail message back to him (Let's also include the password file? No problem.), the Cracker learns much information about vulnerable hosts and servers. A cleverly written Trojan horse does not leave traces of its presence and, because it does not cause detectable damage, it is hard to detect [8].

2.14 Logic Bombs

A logic bomb is a type of Trojan horse, used to release a virus, a worm, or some other system attack. It's either an independent program or a piece of code that's been planted by a system developer or programmer.

With the overwhelming existence of U.S.-based software (MS *Windows* or UNIX systems), the U.S. Government, or whomever you would like to imagine, could decide that no software would be allowed to be exported from that country without a Trojan horse. This hidden function could become active when a document with "war against the USA" exists on the computer. Its activation could also be triggered from the outside. An effect could be to format the computers hard disks or to mail the document to the CIA [8].

2.14.1 Trap Doors

A trap door, or a back door, is a mechanism that's built into a system by its designer. The function of a trap door is to give the designer a way to sneak back into the system, circumventing normal system protection.

As previously mentioned, all U.S. software could be equipped with a trap door that would allow IW agencies to explore systems and the stored data on foreign countries. This could be most useful in cases of military strategic simulations and plans and would provide the DoD's intelligence with vital information [8].

2.14.2 Chipping

Just as software can contain unexpected functions, it is also possible to implement similar functions in hardware. Today's chips contain millions of integrated circuits that can easily be configured by the manufacturer so that they also contain some unexpected functions. They could be built so that they fail after a certain time, blow up after they receive a signal on a specific frequency, or send radio signals that allow identification of their exact location— the number of possible scenarios exceeds, by far, the scope of this chapter. The main problem with chipping is that the specific (adapted) chip be installed in the place that is useful for the information warrior. The easiest solution is to build the additional features into all the chips manufactured in the country that is interested in this type of IW [8].

2.14.3 Nano Machines and Microbes

In the future, Nano machines [7] and microbes will provide the possibility to cause serious harm to a system. Unlike viruses, you can use these to attack not the software, but the hardware of a computer system. Nano machines are tiny robots (smaller than ants) that could be spread at an information center of the enemy. They crawl through the halls and offices until they find a computer. They are so small that they enter the computer through slots and shut down electronic circuits.

Another way to damage the hardware is a special breed of microbes. It is known that this special breed of microbes can eat oil, but what about if they were bred for eating silizium? They would destroy all integrated circuits in a computer lab, a site, a building, a town, and so on. Anyway, nano technology and microbes will be discussed in much greater detail later in the chapter [8].

2.14.4 Electronic Jamming

In the old days (and even today) electronic jamming was used to block communications channels at the enemy's equipment so that they couldn't receive any information. The next step is not to block their traffic, but, instead, overwhelm them with incorrect information—otherwise known as disinformation [8].

3. NEW TOYS FOR BIG BROTHER AND THE BOYS

GPS receivers: One of the newest and probably most important of the IW toys for Big Brother and the boys, will be everywhere soon—in cars, boats, planes, backpacks, briefcases, purses, jackets, and pants pockets. The good news is, you'll always know exactly where you are. The bad news is, so will everyone else.

Most humans who have ever lived have known roughly where they were, day-by- day, year-by-year. Not in abstract terms, of course, but in the terms of experience and familiarity—by neighborhood, not map. For eons, we've known things about ourselves that could be expressed in a statement like "I'm standing on the threshing floor in the village of my birth." or "I'm walking across the mid-morning shadow cast by Notre Dame." Or even "I'm in a part of town I've never seen before." Whether one utters it or not, this awareness of "whereness" is part of the meaning of being human. But for centuries, a dedicated band of mapmakers, navigators, astronomers, inventors, and mathematicians has tried to turn this innate sense of place into a more precise determination of position that is intelligible to anyone, not only to locals. On one level, this is like the difference between knowing you're coming to the corner where you always turn left on your way to the grocery store and knowing the names of the streets that cross at that intersection. On another level, however, the pursuit of pure position is about to lead us into a world that none of us has ever seen. The agent of change will be GPS—the Global Positioning System, which, like so many tools of the modern world, is both familiar and misunderstood at the same time.

Until recently, not a single human-made object has ever known where it was. Even a venerable tool of navigation such as a sextant knows nothing more about its location than does the *Mona Lisa* or the pigments of which she is painted. So imagine a world in which man-made objects know where they are and can communicate that information to other self-locating, communicating objects. This sounds as strange and surprising as the

Marauder's Map in the "Harry Potter" novels for children [4]. The Marauder's Map shows the position and movement of every animate creature at the school of wizardry called Hogwarts. A Marauder's Map of the world would be even stranger. It would show the position and movement (even a history of movements) of man-made objects as well. This would be an ever-changing map or a world filled with artifacts busily announcing something significant about themselves to each other and to anyone else who cared to listen.

That world is here. In August 2000, a company called SiRF Technology based in Santa Clara, California, announced that it had developed an advanced GPS chip no bigger than a postage stamp. SiRF's vision is to bring location awareness to virtually everything that moves. This is a subtle but profound change in the history of GPS technology—a change driven, like everything else these days, by increasing miniaturization and declining prices for sophisticated circuitry. In the past few years, consumers have grown used to the sight of hand-held GPS receivers, which have been marketed as individual positioning devices for anglers, hunters, hikers, and cyclists. But what SiRF and other companies have in mind, is conferring upon objects a communicable sense of place. One day soon, most GPS devices will not be stand-alone receivers used by those of us who venture off the beaten path, but integral components of everyday objects.

Some of these objects, especially the big ones, are easy enough to imagine, because they exist now. Boats and ships of every kind already incorporate GPS technology as do some automobiles made by Toyota, Honda, Lexus, and Cadillac. So do the newest farm implements, such as combines that allow farmers to map crop yields in precise detail. But some uses of GPs that are not yet widely available will soon be common in smaller devices. For instance, the Federal Communications Commission requires cellular-phone service providers to be able to identify the location of a cell-phone caller who dials 911. This means that most cell phones now include a tiny GPS chip. So do beepers and watches and handheld digital assistants and other digital devices like Game Boy Colors, Tamagotchis (virtual game animals), dog collars, and, probably, handguns as well.

The spreading of a technology such as GPS is easy enough to predict, but it's much harder to foresee what the effect of that spreading may be. There's always a limit to how far one can see into the future of the tools being used, especially into a future where those tools become interlinked. There was a time (only as long ago as Bill Gates's first book) when the value of computers was believed to lie mainly in their stand-alone power, not in the wireless networks they might form when linked together. Now there's the Internet and the World Wide Web, whose far-reaching implications are only dimly visible, but which have already transformed the way countries all over the world do business.

The development of GPS technologies may follow a similar pattern. It's already obvious how useful GPS is in discrete applications: for surveying and mapmaking, the tracking of commercial vehicles, maritime and aeronautic navigation, and for use by emergency rescue crews and archaeologists. But there is simply no telling what it will mean when, on a planet full of location-aware objects, a way is found to coordinate all the data they send out. Awareness may be a metaphor when applied to inanimate objects, but the potential of that metaphor is entirely literal.

In the meantime, for most of us, there is still a more basic question to be answered: Where did GPS technology come from and how does it work?

GPS depends on an array of 50 satellites (47 in regular use, plus spares) flying some 11,000 miles above Earth. They were put there by the Department of Defense, which began the NAVSTAR global positioning system program in 1973. A version of GPS was first tested in 1964, when the Navy deployed a five-satellite prototype, called "Transit," for submarines. It could take an hour and a half for a Transit satellite to saunter above the horizon and then another 10 or 15 minutes to fix the submarine's position. The current generation of satellites was built by The Boeing Company and Lockheed Martin, and each one orbits the planet in about 12 hours, cutting across the equatorial plane at an angle of roughly 55 degrees. The U.S. Air Force tracks the satellites from Colorado Springs, Hawaii, and three other islands: Ascension in the South Atlantic, Diego Garcia in the Indian Ocean, and Kwajalein in the South Pacific. These ground stations provide the satellites with navigational information, which the next generation of satellites will be able to supply to each other. Ordinary users can track this constellation of satellites with one of several Web sites or with an appealing public-domain software program called "Home Planet," which can map any satellite you choose, GPS or not, against a projection of the Earth's surface. Or you can track the satellites with a GPS receiver.

In the world of GPS, knowing where you are, give or take a few meters, depends on knowing precisely when you are. Just as longitude couldn't be effectively calculated until 1764, when John Harrison's chronometer was tested on a voyage to Barbados, so GPS couldn't be created until there was a way to mount highly accurate clocks in stable orbits. The problem with finding longitude in Harrison's era was making a chronometer that could keep accurate time at one location (Greenwich, England) even while the ship carrying that chronometer was halfway around the globe. The chronometer provides a constant frame of reference for the celestial events that shift as a ship moves eastward or westward.

GPS satellites effortlessly provide a constant frame of reference. Each carries three or four ultra-precise clocks synchronized to GPS time—which is, essentially, Coordinated Universal Time (UTC) without the leap seconds. The satellite clocks are accurate to within one-millionth of a second of UTC as kept by the U.S. Naval Observatory. The GPS receiver translates the time that the satellites transmit into local time. In fact, as far as most civilian users are concerned, GPS is more accurate for time than it is for position. And, in most cases, GPS is far more accurate for position than it is for altitude. In 1764, Greenwich time was available only in Greenwich (on the meridian running through England, if you knew where that was) and in the presence of a properly maintained chronometer, of which there were two. Now, GPS time is available globally to anyone with a receiver.

When you turn on a GPS receiver, it tunes itself to a radio signal called "L_1," which comes from any GPS satellite—usually one of four to eight—coasting above the horizon. The American military and other authorized users also receive two encrypted signals—one from L_1, another from a frequency called "L_2." Those extra signals are one of the reasons military users can fix their location more precisely than civilian users can. By measuring the time it takes a signal to reach it, a GPS receiver calculates what is called the pseudo-range to the transmitting satellite. With at least four satellites in view, and, hence, four pseudo-ranges (the minimum for determining accurate location plus time), a GPS receiver can compute its position using basic trigonometry. It can also calculate velocity by comparing location readings taken at different points in time.

The real value of GPS begins to emerge when you consider a GPS receiver's ability to compare where it is now with where it was moments or hours or days ago. When you begin to move, a GPS springs to life. It announces your directional bearing, average speed, approximate altitude, the estimated time to get to a named destination, the degree to which you're adhering to a planned path, and the distance to your destination—in short, it calibrates the dimensions of your dynamism or the dynamism of anything you attach it to, from a delivery truck to an outcropping of the Earth's crust. A navigator's task has always been to plot his current position, compare it with his previous day's position, and deduce from those two points some idea of tomorrow's position. These are the functions inherent, and almost instantly accessible, in a dynamic tracking system such as GPS. It's no wonder GPS has rapidly made its way into the navigation stations of recreational boats and commercial ships alike, replacing older electronic navigation systems as well as celestial navigation.

But for civilian GPS users, there is a catch. The system is purposely compromised, its accuracy intentionally degraded. GPS was designed, as the responsible federal agencies are careful to remind us, to serve "as a dual-use system with the primary purpose of enhancing the effectiveness of U.S. and allied military forces." One way to do that is to de-enhance everyone else's effectiveness—to deny nonmilitary users and foreign adversaries the kind of accuracy that military users enjoy, which in all kinds of targeting weapons is a difference of dozens of feet. This has been done by selectively and intermittently introducing error into the information GPS satellites dispense to receivers lacking access to the military's encrypted signals—in other words, to the receivers nonmilitary users and foreign adversaries can buy. One of the many ironies of GPS, however, is that a system designed mainly for military use and developed through the Department of Defense at a cost of more than $10 billion has been engulfed by the commercial market. The result is that "Selective Availability," as GPS's intentional error is called, will most likely be phased out within the next decade.

The more positional signals a GPS receives, the more accurate it is. That's one reason why in 1999, then Vice President Gore, announced a $600 million initiative that would help fund additional civilian signals on GPS satellites scheduled for launch in the next decade—a clear acknowledgment of the scientific, commercial, and economic importance of nonmilitary GPS. But even at present, there are ways around Selective Availability. Some GPS receivers have been manufactured that can also tunein to the Russian equivalent of GPS, called "GLONASS," which operates without signal compromise, but lacks the reliability of GPS. The most common solution is Differential GPS, or DGPS, in which "differential corrections" (indications of the degree of error at one station) are transmitted to GPS receivers via a radio link, greatly enhancing their accuracy regardless of Selective Availability. Even DGPS chips have shrunk to the size of postage stamps.

The U.S. Coast Guard operates a maritime DGPS service available to civilians, and the Federal Aeronautics Administration is implementing a similar system, called "Wide Area Augmentation System," which uses satellites as well as ground stations. Once a complementary system called "Local Area Augmentation Service" is in place at selected airports, the FAA will eventually be able to turn over the task of flight navigation, from takeoff to precision landings, entirely to GPS. The result of this is a bizarre irony, in which some branches of the federal government are working hard to offset error purposely created by another federal agency, the Department of Defense.

GPS, especially Differential GPS, has come as a particular godsend to geophysicists, the men and women who study the physical and dynamic parameters of planet Earth. Most of us think of the Earth as an inherently stable platform: bedrock. But to geophysicists, Earth experiences a wide range of volatility—some of it very slow, some of it occurring at the rate of days or weeks. Tectonic plates grind at each other's edges, cresting upward. Portions of the crust are still rebounding from the weight of long-vanished ice sheets. It adjusts locally to the shock of earthquakes and volcanoes. And as one geophysicist writes, "The torques from the sun, moon, and planets move the rotation axis of Earth in space." Torques from the atmosphere, ocean, and fluid core move the rotation axis relative to the crust of Earth. Both sources of torques change the rotation rate of Earth. GPS offers an extraordinary leap in the rate of data collection, with a corresponding leap in the understanding of Earth's motion.

As technology becomes more sophisticated, it seems as though freedom gets defined in more basic terms. GPS offers one version of freedom—knowing where you are. But it may ultimately threaten a more basic kind of freedom—being where you are without anyone else knowing it. Everyone would like to have a Marauder's Map, but no one wants to appear on the Marauder's Map without approving it. The value of cell- phones embedded with GPS chips is obvious when it comes to emergency services. But the fact that cell-phone service providers are able to track the location of a 911 call, means that GPS could track the location of every other kind of call as well. Already GPS is being used to monitor the movement of commercial trucks of every description. This is both a form of insight to the vehicle owners and a form of intrusion to the drivers, who find their movements visible to management in a way they never were before. GPS is also being used in experimental programs to monitor the movements of parolees. There is only a difference of emphasis between tracking a parolee with a GPS and tracking a sales representative with the same tool. All of us have learned all too well in the last few decades that even innocuous information can be assembled in ways that make it dangerous. Location, movement, and time are not innocuous forms of information.

As technology advances, it abstracts us farther and farther from the earth we live on. All of us inhabit a world of the senses, a world infinitely full of sensory clues to our location and bearing. Directionality is implicit in our being. The very factors that influence Earth's rotation (the sun, moon, planets, atmosphere, ocean) influence our sense of orientation, if only we can remember how to know them. It is far easier, after all, to navigate by pushing a single button and reading the numbers on yet another of the small gray screens that crowd our lives. GPS may mean many wonderful things, but it may also mean yet another death for the powers of human observation.

And, too, GPS may be an example of technology that reaches the market the moment it becomes unnecessary—at least where ordinary consumers are concerned. Now that the nonaqueous and nonarctic globe is mostly paved, and the population of people is as thick on the Earth as mold on month-old bread, a device has been invented at last that tells you where you are without having to ask strangers [8].

Now, let's look at the latest snoop, sniff, and snuff tools.

4. SNOOP, SNIFF, AND SNUFF TOOLS

There's a fine line in the difference between "snoop" and "sniff" tools. The meaning of "snuff" tools is obvious. Let's look at "Sniffit" First.

4.1 Snoop/Sniff Tools

Sniffit is a kind of a wireless network packet sniffer/snooper. Packet sniffers are rather intriguingly named pieces of software that monitor wireless network traffic.

Under many wireless networking protocols, data that you transmit gets split into small segments, or packets, and the Internet Protocol address of the destination computer is written into the header of each packet. These packets then get passed around by routers and eventually make their way to the wireless network segment that contains the destination computer.

As each packet travels around that destination segment, the wireless network card on each computer on the segment examines the address in the header. If the destination address on the packet is the same as the IP address of the computer, the wireless network card grabs the packet and passes it on to its host computer [8].

4.1.1 Promiscuous Wireless Network Cards

Packet sniffers work slightly differently. Instead of just picking up the packets that are addressed to them, they set their wireless network cards to what's known as "promiscuous mode" and grab a copy of every packet that goes past. This lets the packet sniffers see all data traffic on the wireless network segment to which they're attached—if they're fast enough to be able to process all that mass of data, that is. This wireless network traffic often contains very interesting information for an attacker, such as user identification numbers and passwords, confidential data—anything that isn't encrypted in some way.

This data is also useful for other purposes— wireless network engineers use packet sniffers to diagnose wireless network faults, for example, and those in security use packet sniffers for their intrusion detection software. That last one is a real case of turning the tables on the attackers: Hackers use packet sniffers to check for confidential data; companies use packet sniffers to check for hacker activity. That has a certain elegant simplicity to it.

The thing that worries most people about *Sniffit* is how easy it is to install. It takes about three commands and three minutes to get this thing installed and running on a Linux machine. It even has a GUI (not exactly pretty, but it is free).

Like *Nmap, Sniffit* is easy to use and does exactly what it says it does: It sniffs your wireless network and shows you what sort of data is getting passed around.

It is recommended that you install a packet sniffer and have a look at what sort of data you can see on your wireless local network. Better still, get one of your wireless network engineers to install it for you. They probably know of better, more professional sniffers and will be able to talk you through some of the data that you see going past. It's an interesting look into exactly what's going on within your wireless network [8].

4.2 Sniff

Security experts are still not convinced that *Carnivore* (the software created by the U.S. Federal Bureau of Investigation [FBI] to tap into Internet communications) is either ready to be used safely (without abuse) or can gather information that would be legally admissible in court. Although *Carnivore* is the best software available for the job today, it is perhaps not as good as it could be. *Carnivore*'s source code should be made available for open review.

Such a review would provide confidence in *Carnivore*'s ability to gather information accurately and fairly—confidence needed to make it a publicly accepted crime-fighting instrument. Unless it is demonstrated that *Carnivore* will enable surveillance personnel to obtain the information they are authorized to see, and not draw innocent bystanders into its net, it will remain an object of public suspicion.

The FBI publicly admitted the existence of *Carnivore* in July 2000, after it had been in use for over a year at numerous Internet service providers (ISPs) and rumors of its existence began to surface. Congress and privacy advocates then called for full disclosure of the software. Replying that such disclosure would only help criminals get around the system, the FBI offered to let it be reviewed by an outside technical group selected by the Bureau. Illinois Institute of Technology's Research Institute (IITRI) was chosen after accepting the review limits proposed by the FBI, a stipulation other institutions such as the San Diego (California) Supercomputing Center (SDSC) would not accept.

The IITRI report does not address significant technical issues. Although it looks at how *Carnivore* worked when it was used as intended, the report failed to look at "the larger issue,": its system requirements. They did not look at the interaction between *Carnivore* and its host operating system or the interaction between *Carnivore* and the ISP's setup. Thus, the vulnerability of the system to hackers is still not clearly established.

Carnivore runs on Windows and, to control it, the person who is using it must be logged on at the highest level: administrator. At that level, the operator (meant to be an FBI agent) has a great deal of freedom. For instance, he or she can access the content of all communications, and change and edit files at will. What is more, anyone logged in as administrator can hide any evidence of the activity. Thus, it would be possible for an agent or someone who hacked into the system to tamper with evidence, plant false leads, or extract confidential information for bribery, extortion, fraud, and so on.

Failure to examine the interaction between *Carnivore* and an ISP's systems may be a gap in the report. The limited nature of IITRI's review cannot support a conclusion that *Carnivore* is accurate, safe, or always consistent with legal requirements. The scope of IITRI's review was dictated by the FBI, and any additional effort would have invalidated the contract under which the work was performed [8].

5. BEWARE: E-MAIL WIRETAPS LIKE *CARNIVORE* CAN STEAL SENSITIVE CORRESPONDENCE

As part of its ongoing research, The Privacy Foundation (see sidebar, "The Privacy Foundation") found that a simple, hidden JavaScript code segment in HTML-formatted e-mail messages can effectively allow someone to monitor all succeeding messages that are forwarded with the original message included. Clearly, this can cause confidential internal communications to be compromised. Here's a look at how to identify wiretaps and protect yourself from them.

The Privacy Foundation

The Privacy Foundation at the University of Denver conducts research into communications technologies and provides the public with tools to maintain privacy in the information age. You can read the Foundation's report and commentary on e-mail wiretaps. The report cites the following possible uses for this security breach:

First, the wiretaps can provide the ability to monitor the path of a confidential e-mail message and the written comments attached.

Second, in a business negotiation conducted via e-mail, one side can learn inside information from the other side as the proposal is discussed through the recipient company's internal e-mail system.

Third, a bugged e-mail message can capture thousands of e-mail addresses as the forwarded message is sent around the world.

Fourth, commercial entities, particularly those based offshore, may seek to offer e-mail wiretapping as a service [8].

This security problem is a particularly dangerous one for organizations that conduct conversations containing sensitive internal information via e-mail. The usual scenario for such communication is that a message from an outside source is forwarded from executive to executive within a company, and it includes each person's comments. If there's an e-mail wiretap on the original external document, each time someone forwards the message to someone else, a copy of their message is automatically and invisibly e-mailed to the original sender of the external message (or someone designated by them).

This problem affects only HTML-enabled e-mail readers that have *JavaScript* turned on by default, such as Microsoft *Outlook, Outlook Express,* and *Netscape Communicator*. *Eudora* and *AOL* are not affected, nor are Web mail services such as *Yahoo* and *Hotmail [8]*.

5.1 Snuff

As hackers obtain ever more dangerous and easy-to-use tools, they are being countered by novel defense strategies. The Pentagon envisions a war in the heavens, but can it defend the ultimate high ground? You bet! Witness the experimental idea of setting up a wireless decoy network separate from your real one to fool intruders as they try to fool you [8].

5.1.1 Wireless Deception Network

This so-called wireless "deception" network is envisioned as more than just a single server set up to be a "honeypot," where hackers may break in, find a dead-end, and have their activities recorded with an eye toward prosecution. Rather, the decoy net is an entire fake wireless network, complete with host computers on a LAN with simulated traffic, to convince hackers for as long as possible that it's real.

Experts debate whether such nets will be worth the effort, but agree they can be a way to slow hackers long enough to sort the curious from the truly destructive "snuff." A group calling itself "The Honeynet Project" has quietly begun testing wireless decoy networks on the Internet.

The Honeynet Project is not intended to prosecute intruders who haplessly wander into their elaborate decoys, but to study hacker responses in depth to devise the best decoy defenses. Other wireless decoy networks do slow intruders with an eye toward collecting evidence to prosecute them. To collect evidence, you need to divert the hacker to a wireless deception network. The idea is to feed back information about what hackers do to a kind of "deception central" for wireless network administrators. The time the hackers are dealing with a deception environment, is time they're not in your wireless network.

It is possible to create a wireless deception network that has the same IP network address as your real wireless network. Deception wireless nets carry obvious administrative burdens, such as the need to generate realistic traffic to fool a hacker and maintain a wireless network no one really uses.

> **Note:** There is a risk that administrators will lose track of what's real and what's not.

These deception techniques have doubters. It's not clear yet if you can fool a lot of people with this deterrent. Meanwhile, hackers continue to learn new tricks.

It's pretty nasty stuff. For very sensitive networks, you may want to activate port-level security on your switches.

Many tools that let hackers carry out surveillance are now Web-based. Why Web-based? It's easy. No complicated downloads or zip files. They can hack from anywhere, and it's anonymous.

Although a talented few among hackers actually make attack tools, many of these tools today are freeware. And they're posted on dozens of techie sites, not the secret underground.

The tool, which involves launching an attack to determine operating system weakness, was given solely to vendors, but somehow ended up posted on the Packetstorm site in its depository for tools. In the wrong hands this tool is dangerous. But that version isn't as dangerous as other versions that will be released [8]."

5.1.2 The New IW Space Race

The war was not going well. Serbian forces were sowing terror across Kosovo. NATO pilots squinting through clouds could do little to stop them. Errant NATO bombs had killed dozens of civilians and shaken support for the alliance. Then the Pentagon saw it had another problem. A Colorado outfit, called "Space Imaging," was about to launch a picture-taking satellite with clarity nearly as good as that of U.S. spy satellites. The company could have sold photos of NATO air bases or troop encampments to, say, Serbian operatives. That had to be stopped. But how?

The brass canvassed its experts for recommendations. The U.S.-licensed firm could simply be ordered not to take pictures over a broad swath of Europe. A similar ban could be issued for a few key areas, such as northern Albania. In the end, however, no order was issued. A malfunction sent Space Imaging's satellite plunging into the Pacific Ocean 30 minutes after it lifted off.

Fortune may not be so kind next time. Space Imaging launched another satellite and started selling pictures from it. Several other companies are right behind it. Before too long, an international bazaar for high-quality satellite imagery will be open for business. And potential foes are making headway with their own satellite capabilities. There's a new proliferation of space-based capabilities. Plus, the U.S.'s Cold War-era capabilities have atrophied.

That's pushing the Pentagon into a whole new kind of warfare. In the future, the U.S. military will be responsible for countering space systems and services used for hostile purposes. That's a nice way of saying the Pentagon needs to be prepared to defend the ultimate high ground by attacking hostile satellites. The new policy also directs the Space Command to start developing tactics and doctrine for conducting warfare in the heavens. It must also come up with plans for deploying space-based lasers or other weapons that could be used against targets anywhere on Earth or above it. If the United States ultimately deploys such weaponry, not only would it break one of the great taboos of the past 53 years, but it could also transform the way America structures its military and fights wars.

But aggressive "space control," as the military calls its quest for dominance in the sky, could backfire. The military view is that it would be the neatest thing in the world to have a death ray in space. But will deploying it lead to a war with somebody?

Very possibly, some critics say. Developing space weapons would be a mistake of historic proportions that would trigger an arms race in space. Imagine scenarios in which other nations follow the U.S. example and scramble to launch their own space weapons while frantic generals, unable to tell exactly who has put what into orbit, plead for extravagant countermeasures. In Pentagon war games, just trying to defend U.S. satellites causes problems. If you defend the satellite, you often widen the war. The activity ends up being the problem and not the solution [8].

6. SPY DUST BALLS AND MECHANICAL DRAGONFLIES

Now, let's look at some really "far out" IW arsenals of the future: spy dust and tiny mechanical robots. Let's look first at spies the size of a mote of dust. This will be followed by tiny robotic insects that may soon serve as military scouts [8].

6.1 Spy Dust Balls

"If only these walls could talk" may not be an idle plea much longer. Kris Pister, expectant father of an invention he calls "smart dust," thinks that in a few years almost anything, from a wall to a mote floating in the air, may have a story to tell (see sidebar, "A Dusty Future"). Thousands of these gossipy particles, each a tiny bundle of electronic brains, laser communications system, power supply, sensors, and even a propulsion system, could lurk all around, almost undetectable. One or more of the remote sensors would fit inside the letter "O."

A Dusty Future

Scientists recently set up a wireless network of small, wireless sensors called motes that detect birds as well as measure temperature, humidity, and barometric pressure. The battery powered devices transmit their data by radio link to a solar-powered base station and then to the Internet. You can literally be anywhere in the world, and know what's going on.

Thanks to this new technology, many scientists are getting the chance to observe what was previously unobservable. Just as MRI technology revolutionized the ability to peer inside the body, the new wireless networks are expected to shed much-needed light on planetary problems like climate change and how pollutants move through the environment. Other researchers are testing the devices for modeling earthquake damage and monitoring everything from vehicle movements in war zones to water use in agricultural fields.

But, while smart dust is generating excitement, some people already are concerned about the dark side of what will undoubtedly be its expanded presence on the landscape. It's a very intrusive technology and could be abused.

Sensors and computer chips have long been embedded in consumer products, whether cars or refrigerators. What's new is that because motes are wireless and battery-powered, they can be used in previously hard-to-access places and moved around at will. Before the technology takes off, motes may have to get smaller--currently, prototypes are the size of matchboxes. The devices also will have to become cheaper, more reliable, and more energy efficient. But, there's little doubt that they could serve as ubiquitous information collectors. These wireless networks of tiny communicating computers could even function as a new kind of Internet that, by merging with the physical world, would allow us to query almost anything-buildings, roads, rivers-for information.

As previously explained, the technology was jump-started back in 1998, when Kris Pister, then an engineer at the University of California-Berkeley and now CEO of Dust Inc., got funding from the Defense Department's Defense Advanced Research Projects Agency to develop tiny, intelligent sensing devices. He had no idea what the applications would be and never in his wildest dreams expected it would lead where it has. The initial challenge was to miniaturize the components, including the sensors, radio transmitters, batteries, and computer hardware. Programming the devices also was tricky because they needed to be both smart and energy efficient. Recently, scientists solved this problem by designing software that enabled the motes to sleep most of the time, yet wake up regularly to take readings and communicate. The scientists didn't want lots of people to have to baby-sit the motes.

Indeed, an early test in March 2001 showed just how independent the devices could be. At a military base in Twentynine Palms, California, Pister and his team dropped six motes from an airplane along a road. As soon as they hit the ground, they organized themselves into a wireless network and began sensing the magnetic field around them. When that changed as a vehicle drove by, the motes cooperated to calculate its speed and direction, later transmitting the data to a laptop at a nearby base camp.

Now, several companies make prototypes with customized sensors that are showing great promise in field tests. For instance, in 2003, scientists deployed motes made by Intel to measure sunlight, temperature, and humidity in a half-dozen redwood trees in a botanical garden. In addition to eliminating miles of airing and reducing the cost of the experiments 10-fold, the motes will give scientists the first 3-D view of the redwood forest microclimate. Recently, scientists packed up the motes and moved them to a remote natural grove. Their goal is to better understand how the loss and fragmentation of redwood forests affect local climate and water resources. In an even more ambitious study of ecosystems, UCLA's CENS is setting up a wireless network with a couple hundred devices in a forested 40 acre reserve near Palm Springs.

Motes are poised to become practical tools for protecting and managing all sorts of resources. For instance, CENS will test motes to monitor an alfalfa field to see how well the plants dissipate high-nitrate wastewater. In 2003, a vineyard in British Columbia deployed a wireless network of 65 motes to closely track temperature fluctuations on its slopes. One aim is to determine when temperatures are perfect for picking grapes to make a late-harvest wine known as ice wine.

While the applications of wireless sensor networks seem endless, the first field tests have revealed shortcomings, which companies are working hard to address. Generally, the motes have needed more-robust packaging to survive rough treatment, curious animals, and frigid weather. At times the radios have been as fickle as cellphones in their signaling and reception.

Still, the biggest challenge may be dealing with the crushing load of information in smart dust. For instance, the Embedded Collaborative Computing Area at the Palo Alto Research Center in California, is trying to reduce the volume of incoming data by training the motes to pay attention only to what's important in the surrounding environment. Others are trying to ensure that the data are accurate and secure--a crucial step, if motes are ever to be used for, say, monitoring a city for signs of bioterrorism.

As motes are deployed more and more widely, the potential for the misuse of the information they collect can only grow. Scientists at Berkeley's Center for Information Technology Research in the Interest of Society are already looking at how laws might be updated to protect the privacy of individuals whose comings and goings, for instance, may one day be tracked by motes. Scientists don't think it will be difficult to draw the lines, but they do need to ask: How far do they let this go? It's a question far removed from observing seabirds on a wind-swept island [8].

Pister exemplifies the surging confidence of the leaders in a new field called "MEMS," which stands for microelectrical and mechanical systems. The idea is to build complex gadgets so small one needs a microscope to see the parts, using fabrication methods invented by the electronics industry for making silicon chips. He talks big and earnestly, even though the best his prototypes (an undustlike inch or so across) can do is exchange laser signals with a counterpart on a tennis court visible 900 yards away through his fifth-floor office window at the University of California-Berkeley. But that's enough to show that the components work. Miniaturizing them is well within current technology [8].

6.1.1 Tracking Tots

Cheap, dispersed sensors may tell farmers the exact condition of their acreage; manufacturers, the precise humidity and temperature history of their raw materials; parents, the locations and conditions of their small children all the time. Climate-control systems in buildings would know exactly where it is too cold, humid, hot, or drafty. In five years, smart dust could be linked by satellite. Eventually, you could log on to readings from smart dust almost anywhere. One dream of Pister: explore outer space with smart dust. NASA could scatter smart dust sensors into the Martian atmosphere and they'd settle all over the planet (like in the recent movie "Red Planet").

But they could also be used as tiny spies. In 1992, as a new associate professor at the University of California-Los Angeles, Pister attended a Rand Corp. workshop in Santa Monica sponsored by the Pentagon's Defense Advanced Research Projects Agency. The topic was miniaturization of novel battlefield surveillance methods. The question was whether tiny electronic sensors could be scattered in contested territory to relay vital information back to commanders. You could find out, say, if a tank had gone by, or whether there was anthrax in the air. The concept sent his imagination racing [8].

6.1.2 Poppy Seeds

He coined the label "smart dust" in 1996 and produced the first complete smart-dust particle in mid-2002, about 1 millimeter on a side, or roughly between a poppy seed and a grape seed in size. None of the unit's components seems to present major fabrication obstacles. Before building the first fully small versions, however, the team wants to be sure it can get oversize prototypes to work. One of the students is working on a variant that will sport a thin, winglike extension, like that of a maple seed, so that a modest breeze will keep it aloft. Another student is designing a solid rocket micromotor, visible with a good magnifying glass, carved out of silicon. If a smart dust particle detects a tank going by, it could hop up and hitch a ride like a little spy. Some smart dust may be equipped with solar cells for power. Others might alight on vibrating machinery to soak up energy from the motion, or charge batteries off electromagnetic pulsations leaking from power lines. Sensors, at first, would be simple (such as for temperature, humidity, a few targeted chemicals, etc.) but eventually microphones and camera systems should be possible.

Pister tells the grad students and postdocs in the engineering school's smart dust group that above all, they must have a passion for new ideas and teamwork. He warns them against giving in to the "dark side," (Big Brother) against becoming stealth researchers whose distrust of others makes them, in his words, roach motels for information. Love of freely flowing communication is appropriate from a man who expects a tomorrow suffused with tiny snoops. He knows his ideas may occasionally serve nefarious ends, invading privacy or monitoring citizens of authoritarian governments. His reply to nervous objections is simple. “Information is good, and information flow reinforces democracy and not tyranny.” Well, maybe [8]!

6.2 Mechanical Dragonflies

The military calls it “situational awareness”: the ability to detect how many hostile tanks await in the next valley, or if bombed-out buildings are filled with snipers. And it is an advantage that has proved difficult to attain: Spies, satellites, and U-2s have all failed to keep commanders from blundering into ambushes and mismatches. The worst thing is just not knowing where the enemy is. It’s having the sense that somebody’s out there trying to get you but having no idea of where the enemy might be.

Military researchers are working to free future American troops from the terror of the unknown. The researchers envision tomorrow’s soldiers coming to a hill, halting, and reaching into their packs for cigar-shaped tubes. From every tube emerges a robotic spider, or a robotic dragonfly, each no longer than 3 inches. Equipped with cameras or acoustic sensors, the mechanical insects range forward and provide data on the hazards that lie in wait on the other side: the number of machine gun nests and the position of artillery [8].

6.2.1 Robotic Conundrum

Backed by $7.7 million from the Pentagon’s Defense Advanced Research Projects Agency (DARPA), military researchers are designing such insect-inspired spies. Recently, the researchers built their first crawling bug prototypes, and they aim to perfect the design within two years. Insect-shaped “micro aerial vehicles” are next on the slate. Along with providing the military with state-of-the-art scouts, the researchers hope their project alters the way engineers approach the long-vexing problem of robotic locomotion.

In most robotic systems today, people think that if you want to move one joint, then you need to attach a motor at that joint. That makes for large, bulky, energy-hog robots. It also reduces robots to the ranks of expensive toys. Motors are only about 80% efficient in turning electrical power into movement. So, although robots may impress with their futuristic looks, most motor-driven devices have ranges limited to only a few dozen yards, rendering them useless for practical applications.

In the initial design, piezoelectric ceramics—thin, ceramic-coated metal wafers that bend when an electrical current is applied to their surfaces—were proposed. Such materials already are used commercially to make silent pagers vibrate or to make zoom lenses move strips (built from lead, zirconium, and titanium) that are sandwiched together, a structure known as a bimorph actuator. When charged, one half of the actuator expands while the other contracts, causing it to curve. When the brief energy pulse ends, the structure snaps back to its original form and then can begin the cycle anew. The researchers attached titanium legs to these vibrating strips. Vibration is translated into motion, as the crawler takes 2-millimeter-long forward strides in response to each oscillation.

Because piezoelectrics require only occasional energy boosts to keep up the vibration, the bugs promise to be up to 70% more energy efficient than traditional robots. If you're in a weight room and you lift 100 pounds up and down 10 times, that takes a lot of energy from a person (illustrating the principle behind the design). The same work could more easily be accomplished by hooking that weight to a spring on the ceiling, then displacing it a bit and letting it bounce up and down by itself. Another common analogy is a child on a swing set; once in motion, very little pumping action is required to keep moving. The bugs' energy efficiency should give them ranges of almost 600 yards, and allow them enough juice to carry such intelligence-oriented payloads as chip-size infrared detectors and quarter-size video cameras [8].

6.2.2 Natural Efficiency

The engineers' decision to model their robots after bugs was a natural choice: Biological systems are far more energy efficient than anything cooked up in the laboratory. Most things biological sort of oscillate as they walk. If you look at humans walking and the way our legs act as pendulums off our hips and swing back and forth, that's a cyclic motion. They were also impressed by the shape of daddy longlegs, whose low-slung bodies and inverted-V legs create a stable configuration—important for robots that will have to scamper across uneven, sometimes treacherous terrain. Additional hardiness comes from the solid-state legs, which are free of bearings, rods, or shafts that could get jammed by pebbles or dust. They probably won't survive being stomped on, but short of that they're pretty tough—they could actually survive four-story falls.

Before the bugs can be unleashed on the battlefield, however, a few major hurdles remain. Chief among them is a power problem; though the robots will require around 60 volts to start vibrating, the watch-size batteries being considered can provide only 3 to 6 volts. To get the bugs moving without the aid of chargers, circuitry must be developed to amplify the current, and it must be small enough to fit the 2-by-¾-inch bugs. Still, the design's voltage requirement is impressively low; rival efforts to create locomotive robots of comparable size have needed well over 1,000 volts.

Another lingering question is how a robotic swarm can be controlled. With thousands of bugs roving at once, commanding each individual unit would be close to impossible. So, a battalion leader, outfitted with a remote control, would only have to control a "mother ship," an insect at the fore that would then relay instructions to other members of the swarm. In the event of the mother ship's destruction, the leadership role could be shifted to a surviving robot. The exact details of this control, however, have yet to be worked out.

Nevertheless, DARPA officials and the researchers are optimistic that the kinks can be worked out and that assembly-line production of the bugs is nearing. Along with the crawling prototype, the researchers have already managed to construct a piezoelectrically actuated thorax for the flier. Once all design issues are resolved, the researchers believe, the insects could cost as little as $7 per unit. The required metals are readily available, and the bimorph strips and legs can be cheaply pressed from large sheets.

The low price makes the insects potential candidates for a variety of uses, including delivering lethal toxins on the battlefield or aiding police SWAT teams. Or perhaps 40,000 of the mechanical creatures could be dropped on the Martian surface to probe the nooks and crannies Pathfinder missed. But those missions are far distant; the bugs' first and foremost duty will be to give American troops an upper hand and to save them from stumbling into situations too perilous to survive [8].

Finally, let's look at how machines the size of molecules are creating the next industrial revolution in information warfare.

7. NANOTECHNOLOGY

In 2000, a group of scientists from the University of Michigan's Center for Biologic Nanotechnology traveled to the U.S. Army's Dugway Proving Ground in Utah. The purpose of their visit: to demonstrate the power of "nano-bombs." These munitions don't exactly go "Kaboom!" They're molecular-size droplets, roughly 1/5000 the head of a pin, designed to blow up various microscopic enemies of mankind, including the spores containing the deadly biological warfare agent anthrax.

The military's interest in nano-bombs is obvious. In the test, the devices achieved a remarkable 100% success rate, proving their unrivaled effectiveness as a potential defense against anthrax attacks. Yet their civilian applications are also staggering. For example, just by adjusting the bombs' ratio of soybean oil, solvents, detergents, and water, researchers can program them to kill the bugs that cause influenza and herpes. Indeed, the Michigan team is now making new, smarter nano-bombs so selective that they can attack E. coli, salmonella, or listeria before they can reach the intestine.

If you're a fan of science fiction, you've no doubt encountered the term *nanotechnology*. Over the past 23 years, scores of novels and movies have explored the implications of mankind's learning to build devices the size of molecules. In a 1999 episode of The *X-Files* titled "S. R. 819," nanotechnology even entered the banal world of Washington trade politics, with various nefarious forces conspiring to pass a Senate resolution that would permit the export of lethal "nanites" to rogue nations.

Yet since 1999, a series of breakthroughs have transformed nanotech from sci-fi fantasy into a real-world, applied science, and, in the process, inspired huge investments by business, academia, and government. In industries as diverse as health care, computers, chemicals, and aerospace, nanotech is overhauling production techniques, resulting in new and improved products—some of which may already be in your home or workplace [8].

7.1 Silicon Fingers

Meanwhile, nearly every week, corporate and academic labs report advances in nanotech with broad commercial and medical implications. In 2000, for example, IBM announced it had figured out a way to use DNA to power a primitive robot with working silicon fingers 1/50 as thick as human hair. Within a decade or so, such devices may be able to track down and destroy cancer cells. Over at Cornell University, researchers have developed a molecular-size motor, built out of a combination of organic and inorganic components, that some dub nanotech's "Model T." In tests announced in 2003, the machine's rotor spun for 120 minutes at 6 to 7 revolutions per second. When further developed, such motors will be able to pump fluids, open and close valves, and power a wide range of nanoscale devices.

These inventions and products are just the beginning of what many observers predict will be a new industrial revolution fostered by man's growing prowess at manipulating matter one atom, or molecule, at a time. Because of nanotech, all of us will see more change in our civilization in the next 30 years than we did during all of the 20th century.

> **Note:** Nanotech takes its name from the nanometer, a unit of measurement just one billionth of a meter long.

Imagine the possibilities. Materials with 10 times the strength of steel and only a small fraction of the weight. Shrinking all the information housed at the Library of Congress into a device the size of a sugar cube. Or detecting cancerous tumors when they are only a few cells in size."

To build such objects, engineers are employing a wide range of techniques, borrowed from bioengineering, chemistry, and molecular engineering. Such feats include imitating the workings of the body, where DNA not only programs cells to replicate themselves but also instructs them how to assemble individual molecules into new materials such as hair or milk. In other words, many nanotech structures build themselves [8].

7.2 Atom by Atom

The inspiration for nanotech goes back to a 1959 speech by the late physicist Richard Feynman, titled "There's Plenty of Room at the Bottom." Feynman, then a professor at the California Institute of Technology, proposed a novel concept to his colleagues. Starting in the Stone Age, all human technology, from sharpening arrowheads to etching silicon chips, has involved whittling or fusing billions of atoms at a time into useful forms. But what if we were to take another approach, Feynman asked, by starting with individual molecules or even atoms, and assembling them one by one to meet our needs? The principles of physics, as far as Feynman could see, did not speak against the possibility of maneuvering things atom by atom.

Four decades later, Chad Mirkin, a chemistry professor at Northwestern University's $45 million nanotech center, used a nanoscale device to etch most of Feynman's speech onto a surface the size of about 10 tobacco smoke particles—a feat that Feynman would no doubt have taken as vindication. But the course science took to achieve such levels of finesse has not always been straightforward. Nor has it been lacking in controversy.

Indeed, some scientists are alarmed by nanotechnology's rapid progress. In 2000, the chief scientist at Sun Microsystems, created a stir when he published an essay in *Wired* magazine warning that in the wrong hands, nanotech could be more destructive than nuclear weapons. Influenced by the work of Eric Drexler, an early and controversial nanotechnology theoretician, the scientist predicted that trillions of self-replicating nanorobots could one day spin out of control, literally reducing the earth's entire biomass to "gray goo."

Most researchers in the field don't share that type of concern. They are compelled to keep going. Researchers are knocking on the door of creating new living things, new hybrids of robotics and biology. Some may be pretty scary, but they have to keep going.

The early payoffs have already arrived. Computer makers, for example, use nanotechnology to build "read heads," a key component in the $45-billion-a-year hard disk drive market, which vastly improve the speed at which computers can scan data. Another familiar product, Dr. Scholl's brand antifungal spray, contains nano-scale zinc oxide particles—produced by a company called Nanophase Technologies—that make aerosol cans less likely to clog. Nanoparticles also help make car and floor waxes that are harder and more durable and eyeglasses that are less likely to scratch. As these examples show, one huge advantage of nanotech is its ability to create materials with novel properties not found in nature or obtainable through conventional chemistry.

What accounts for the sudden acceleration of nanotechnology? A key breakthrough came in 1990, when researchers at IBM's Almaden Research Center succeeded in rearranging individual atoms at will. Using a device known as a scanning probe microscope, the team slowly moved 35 atoms to spell the three-letter IBM logo, thus proving Feynman right. The entire logo measured less than three nanometers.

Soon, scientists were not only manipulating individual atoms but also "spray painting" with them. Using a tool known as a molecular beam epitaxy, scientists have learned to create ultrafine films of specialized crystals, built up one molecular layer at a time. This is the technology used today to build read-head components for computer hard drives.

One quality of such films, which are known as giant magnetoresistant materials, or GMRs, is that their electrical resistance changes drastically in the presence of a magnetic field. Because of this sensitivity, hard disk drives that use GMRs can read very tightly packed data and do so with extreme speed. In a few years, scientists are expected to produce memory chips built out of GMR material that can preserve 100 megabits of data without using electricity. Eventually, such chips may become so powerful that they will simply replace hard drives, thereby vastly increasing the speed at which computers can retrieve data [8].

7.3 Natural Motion

The next stage in the development of nanotechnology borrows a page from nature. Building a supercomputer no bigger than a speck of dust might seem an impossible task, until one realizes that evolution solved such problems more than a billion years ago. Living cells contain all sorts of nanoscale motors made of proteins that perform myriad mechanical and chemical functions, from muscle contraction to photosynthesis. In some instances, such motors may be re-engineered, or imitated, to produce products and processes useful to humans.

Animals such as the abalone, for example, have cellular motors that combine the crumbly substance found in schoolroom chalk with a "mortar" of proteins and carbohydrates to create elaborate, nano-structured shells so strong they can't be shattered by a hammer. Using a combination of biotechnology and molecular engineering, humans are now on the verge of being able to replicate or adapt such motors to suit their own purposes.

How are these biologically inspired machines constructed? Often, they construct themselves, manifesting a phenomenon of nature known as self-assembly. The macromolecules of such biological machines have exactly the right shape and chemical- binding preferences to ensure that, when they combine, they will snap together in predesigned ways. For example, the two strands that make up DNA's double helix match each other exactly, which means that if they are separated in a complex chemical mixture, they are still able to find each other easily.

This phenomenon is potentially very useful for fabricating nanoscale products. For instance, in 1999, a team of German scientists attached building materials such as gold spheres to individual strands of DNA and then watched as the strands found each other and bound together the components they carried, creating a wholly new material.

Similarly, the 1996 Nobel Prize in chemistry went to a team of scientists for their work with "nanotubes"—a formation of self-assembling carbon atoms about 1/50,000 the width of a human hair. Scientists expect that when they succeed in weaving nanotubes into larger strands, the resulting material will be 100 times stronger than steel, conduct electricity better than copper, and conduct heat better than diamond. Membranes of such fibers should lead to rechargeable batteries many times stronger, and smaller, than today's.

In 2000, a team of IBM scientists announced that they had used self-assembly principles to create a new class of magnetic materials that could one day allow computer hard disks and other data-storage systems [6] to store more than 100 times more data than today's products. Specifically, the researchers discovered certain chemical reactions that cause tiny magnetic particles, each uniformly containing only a few thousand atoms, to self-assemble into well-ordered arrays, with each particle separated from its neighbors by the same preset distance.

Other scientists have discovered important new self-assembling entities by accident. In 1996, Samuel Stupp, a professor at Northwestern University, was in his lab trying to develop new forms of polymer when he inadvertently came upon "nanomushrooms." He saw the potential right away. The molecules he had been experimenting with had spontaneously grouped themselves into supramolecular clusters shaped like mushrooms. Soon afterward, Stupp discovered, again accidentally, that he could easily program these supramolecules to form film that behaves like Scotch tape.

Meanwhile, researchers at UCLA and Hewlett-Packard have laid the groundwork for the world's first molecular computer. Eventually, the researchers hope to build memory chips smaller than a bacterium. Such an achievement is essential if computing power is to continue doubling every 18 to 24 months, as it has for the past four decades. This is because the more densely packed the transistors on a chip become, the faster it can process, and we are approaching the natural limit to how small transistors can be fabricated out of silicon [8].

7.4 Future Phenomena

Finally, where will it all end? Many futurists have speculated that nanotech will fundamentally change the human condition over the next generation. Swarms of programmable particles, sometimes referred to as "utility fog," will assemble themselves on command. The result could be a bottle of young wine molecularly engineered to taste as if it had aged for decades, or a faithful biomechanical dog with an on/off switch.

Meanwhile, new, superstrong, lightweight nanomaterials could make space travel cheap and easy and maybe even worth the bother, if, as some scientists predict, nanotech can be used to create an Earth-like atmosphere on Mars. And space colonization could well be necessary if the new science of "nanomedicine" extends life indefinitely, manufacturing new cells, molecule by molecule, whenever old cells wear out. It all seems hard to imagine; yet nanotech has already produced enough small wonders to make such big ideas seem plausible, if not alarming—at least to the high priests of science and the IW military strategists [8].

8. SUMMARY AND CONCLUSIONS

Information technology is being developed by strategic planners both as an offensive battlefield weapon, and as a weapon for "logistics attack," as a means to disrupt the civilian infrastructure on which an enemy's military apparatus depends. Technology has already been used effectively by American forces in the Gulf War, in Iraq and in the conflict in Haiti.

However, information warfare is a double-edged swordthose countries most capable of waging it are also the ones most vulnerable to it. The increasing dependence on sophisticated information systems brings with it an increased vulnerability to hostile elements and terrorists. The following are conclusions drawn from the information warfare arsenal of the future:

Even though the anticipated national security threats of the coming decades involve less-developed countries, the CVW threat and other methods of intrusion and disruption are not necessarily beyond their reach.

- Opportunities to deceive and confuse through an elaborate misinformation scheme along a myriad of information paths are available to anyone.
- The information warfare arsenal of the future provides a new avenue to employ deception techniques through the use of multiple paths that create the perception and validation of truth.
- There exists the prospect of an intelligence analyst manipulating an adversary's command-and-control system so that reality is distorted.
- Tomorrow's soldier will depend more than ever on the very well-known and trusted factors of mobility.
- Imagine a scenario depicting a "left hook" in the Iranian desert that fails because the systems in use were successfully attacked by CVW, or some other intrusion method, with the resulting disruption putting U.S. troops in a flailing posture—facing the unknown and losing confidence in their operation.
- One thing is sure. An Iranian "left hook" will be difficult to repeat.
- One can assume that Iran, and others, will exploit the GPS to their own advantage. The information warfare arsenal is coming of age!
- World War II set the stage, but only with today's technology can we expect action in this sphere of warfare on a grand scale.
- The necessity to prevent irresponsible groups and individuals from getting access to nanotechnological manufacturing capability is a prime concern in the near future.
- The chapter has shown how this quest for containment has shaped many aspect of society, most notably via the institution of a global surveillance wireless network [8].

9. REFERENCES

[1] John R. Vacca, The Cabling Handbook (2nd Edition), Prentice Hall, 2001.

[2] John R. Vacca, *i-mode Crash Course*, McGraw-Hill, 2002.

[3] John R. Vacca, *Satellite Encryption*, Academic Press, 1999.

[4] J.K. Rowling, , *Harry Potter and the Chamber of Secrets (Book 2)*, Scholastic Trade, 1999.

[5] John R. Vacca, *Net Privacy: A Guide to Developing & Implementing an Ironclad ebusiness Privacy Plan*, McGraw-Hill, 2001.

[6] John R. Vacca, *The Essential Guide to Storage Area Networks*, Prentice Hall, 2002.

[7] John R. Vacca, *The World's 20 Greatest Unsolved Problems*, Prentice Hall, 2004.

[8] John R. Vacca, *Computer Forensics: Computer Crime Scene Investigation, 2nd Edition*, Charles River Media, 2005.

Chapter 27

WIRELESS NETWORK SECURITY SURVEILLANCE TOOLS FOR INFORMATION WARFARE OF THE FUTURE

1. INTRODUCTION

Wireless systems [7] capable of monitoring vehicles and people all over the planet (basically everything) are leaving businesses and the military aglow with new possibilities, and some privacy advocates deeply concerned. Companies seeking to tap the commercial potential of these technologies are installing wireless location systems in vehicles, hand-held computers, cell-phones—even watchbands. Scientists have developed a chip that can be inserted beneath the skin, so that a person's location can be pinpointed anywhere.

For example, the owner of a small company in Dallas that installs automobile alarms, uses a wireless tracking service to monitor his fleet of six Dodge Dakota pickup trucks, and the equipment alerted him recently when one of his trucks turned up in the parking lot of the Million Dollar Saloon, a strip club. When he signed up for this service, he told his guys, "Big Brother's keeping an eye on you, and I'm Big Brother." After he fired that one fellow, you bet they all believed him.

These technologies have become one of the fastest-growing areas of the wireless communications industry. The market for location-based services is already estimated at nearly $1 billion and is forecast to approach $9 billion by 2007.

2. MONITORING EVERYTHING

A federal effort to make it easier to pinpoint the location of people making emergency 911 calls from mobile phones, means that cell-phones sold in the United States are now equipped with advanced wireless tracking technology. Various plans already under way include alerting cell-phone users when they approach a nearby McDonald's, telling them which items are on sale, or sending updates to travelers about hotel vacancies or nearby restaurants with available tables. One Florida company wants to provide parents with wireless watchbands that they can use to keep track of their children.

Although the commercial prospects for wireless location technology may be intriguing, and the social benefits of better mobile 911 service are undisputed, privacy-rights advocates are worried. By allowing location-based services to proliferate, you're opening the door to a new realm of privacy abuses. What if your insurer finds out you're into rock climbing or late-night carousing in the red-light district? What if your employer knows you're being treated for AIDS at a local clinic? The potential is there for inferences to be drawn about you based on knowledge of your whereabouts.

Until recently, location-based services belonged more in the realm of science fiction than to commerce. Although satellite-based Global Positioning System technology has been commercially available for some time for airplanes, boats, cars, and hikers, companies have only recently begun manufacturing GPS chips that can be embedded in wireless communications devices [1]. GPS uses satellite signals to determine geographic coordinates that indicate where the person with the receiving device is situated. GPS monitoring technology will be discussed in much greater detail later in the chapter.

Real-life improvements in the technology have come largely from research initiatives by start-up companies in the United States, Canada, and Europe as well as from large companies like IBM, which recently formed a "pervasive computing" division to focus on wireless technologies such as location-based services.

Location technology is a natural extension of e-business. It's no surprise that a whole new ecology of small companies has been formed to focus on making it all more precise.

For instance, Peter Zhou helped to create a chip called "Digital Angel" that could be implanted beneath human skin, enabling his company to track the location of a person almost anywhere using a combination of satellites and radio technology. After all, he reasoned, wouldn't the whereabouts of an Alzheimer's patient be important to relatives? Wouldn't the government want to keep track of paroled convicts? Wouldn't parents want to know where their children are at 10 p.m., 11 p.m., or any hour of the day?

A review of Digital Angel's commercial potential, though, revealed concern over the possibility of privacy abuses [2]. So Professor Zhou, the chief scientist for Applied Digital Solutions, a company in Palm Beach, Florida, which makes embedded devices for tracking livestock, altered his plans for Digital Angel, which is about the size of a dime, so that instead of being implanted it could be affixed to a watchband or a belt.

Embedding technology in people is too controversial. But that doesn't mean a system capable of tracking people wherever they go won't have great value. Digital Angel is now commercially available.

That Professor Zhou found himself in the middle of the privacy debate is no surprise, given the growing interest in location-based services. Through the use of existing cellular communications technology or the Global Positioning System, researchers' ability to track wireless devices more precisely is growing [3].

Some of the world's largest wireless carriers, such as Verizon Wireless, Vodafone of Britain, and NTT DoCoMo of Japan are promoting the technology, in addition to dozens of small companies in the United States and Europe. The SignalSoft Corporation, based in Boulder, Colorado, develops software that allows tourists or business travelers to use their mobile phones to obtain information on the closest restaurants or hotels in a given city [4]. Meanwhile Cell-Loc Inc., a Canadian company, has already installed a wireless service in Austin, Texas, and in Calgary, Alberta, that, after determining a caller's location, delivers detailed driving directions.

Some companies are even more ambitious. Webraska, a French company that secured $90 million in financing from investors in the United States and Europe, has mapped every urban area in the world and allows these maps to be retrieved in real time on wireless devices.

Yet while businesses around the world seek to improve the quality of location-based services, the biggest impetus behind the advancement of the technology has come from the federal government, through its effort to improve the precision of locating wireless 911 emergency calls. Nearly a third of the 593 million 911 calls made in 2003 came from cell phones, according to the National Emergency Number Association.

With the number of wireless users growing, carriers are now equipping either cell-phones or their wireless communications networks with technology that would allow authorities to determine the location of most callers to within 600 feet, compared with current systems that can locate them within about 900 feet. For example, Verizon Wireless and Western Wireless have developed a network-based system that pinpoints the signal on a handset using the existing cellular network to determine the location, whereas other carriers including Sprint PCS, Alltel, and Nextel favor handsets equipped with GPS chips. Supporters of the initiative, called "E-911" for "enhanced 911," expect the technology's precision to be even better than the federally mandated 600-foot radius.

If your cell-phone is on while you're driving, you can tell which intersection you're at. Although the E-911 initiative has driven wireless carriers in the United States to improve their location technology, industry groups have started to grapple with privacy issues. The Wireless Advertising Association, a group of carriers, advertising agencies and device manufacturers, encourages companies to allow consumers to choose whether they want location-based services. The association will endorse companies that adhere to the policy.

People are justifiably concerned with the rapidity with which this technology is being deployed. They need to be assured that there is no conspiracy to use this information in an underhanded way [8].

3. CYBER SURVEILLANCE

Nicodemo S. Scarfo, the son of Philadelphia's former mob boss, was almost paranoid enough. Scarfo, who has been charged with masterminding a mob-linked loan sharking operation in New Jersey, reportedly used the popular PGP encryption software to shield his computer's secrets from prying eyes or cyber surveillance.

But when the feds learned of Scarfo's security measures, they decided to do something that would bypass even the best encryption software: FBI agents sneaked into Scarfo's office in Belleville, New Jersey, on May 10, 1999, and installed a keyboard-sniffing device to record his password when he typed it in.

A seven-page court order authorized the FBI and cooperating local police to break into Scarfo's first-floor "Merchant Services of Essex County" office as many times as necessary to deploy, maintain, and then remove recovery methods that will capture the necessary key-related information and encrypted files. The case, which is still awaiting trial, appears to be the first in which the U.S. government used such aggressive surveillance techniques during an investigation; some legal observers say the FBI's breaking-and-entering procedures go too far. This case has the potential to establish some very important precedents on this issue.

Scarfo's prosecution comes at a time when the FBI's Carnivore surveillance system (previously discussed) is under increasingly heavy fire from privacy groups, and the use of data-scrambling encryption products appears to be growing. Recently, for instance, news leaked out about Yahoo's encrypted Web-based e-mail service it introduced through a deal with Zixit, a Dallas firm.

Scarfo has been charged with supervising "an illegal gambling business" in violation of state and federal law and using extortionate loan shark tactics, according to a three-count indictment filed in federal court in June 2000. He has pleaded not guilty.

The elder Scarfo, who once ran the Philadelphia mob that also dominated the Atlantic City gambling racket, was imprisoned in 1991 on racketeering charges. The spring 1999 investigation of the younger Scarfo, who is now 38 years old, may be what prompted the previous Clinton administration to recommend changing federal law to allow police to conduct electronic "black bag" jobs.

The idea first publicly surfaced in mid-1999, when the Justice Department proposed legislation that would let police obtain surreptitious warrants and "postpone" notifying the person whose property they entered for 30 days. After vocal objections from civil liberties groups, the administration backed away from the controversial bill. In the final draft of the Cyberspace Electronic Security Act submitted to Congress, the secret-search portions had disappeared.

In January 2000, the previous Clinton administration seemed to change its mind. When criminals such as drug dealers and terrorists use encryption to conceal their communications, law enforcement must be able to respond in a manner that will not thwart an investigation or tip off a suspect.

The feds didn't need a new law—and would instead rely on "general authorities" when asking judges to authorize black bag jobs. A related "secret search" proposal resurfaced in May 2000 in a Senate bankruptcy bill.

In the Scarfo case, the FBI in May 1999 asked for authority to search for and seize encryption-key-related pass phrases from his computer as well as install and leave behind software, firmware, and/or hardware equipment that will monitor the inputted data entered on Nicodemo S. Scarfo's computer by recording the key related information as they are entered. Although the government has refused to release details, this appears to indicate the FBI was using either a hardware device (inserted into the keyboard or attached to the keyboard cable) or a software program that would quietly run in the background and record keystrokes. With the PGP private key and Scarfo's secret password, the government could then view whatever documents or files he had encrypted and stored on his computer.

Ruling that normal investigative procedures to decrypt the codes and keys necessary to decipher the "factors" encrypted computer file have been tried and have failed, U.S. Magistrate Judge G. Donald Haneke granted the FBI's request. Haneke did not, under federal law, have the authority to grant such an order. The interesting issue is that they in those (court) documents specifically disclaim any reliance on the wiretap statute. If they're on record saying this isn't communications (and it isn't), then that extraordinary authority they have under the wiretap laws does not apply.

If the government is now talking about expanding (black bag jobs) to every case in which it has an interest, where the subject is using a computer and encryption, the number of break-ins is going to skyrocket. Break-ins are going to become commonplace.

However, the government could successfully argue that break-ins are constitutional. There's nothing in the Constitution that prohibits this kind of anticipatory search. In many respects, it's no different from a wiretap.

A lawyer for Scarfo told the Philadelphia Inquirer that he would file a motion challenging the legality of the FBI's black bag job. The FBI's got everything that Scarfo typed on that keyboard (a letter to his lawyer, personal or medical records, legitimate business records, etc.).

Finding a mentally impaired relative, a lost child, or a criminal in a sprawling metropolitan area would be simple if the person were equipped with a personal locator device. The next part of the chapter will take a close look at these IW tracking devices [8].

4. THE CYBER FOOT PRINT AND CRIMINAL TRACKING

At 10:00 a.m. one morning in 1999, an elderly woman in Osaka, Japan, became alarmed. Her 74-year-old husband, who suffers from dementia, had left four hours earlier and had not yet returned. She did not panic, but contacted the provider of her personal locator service, Life Service Center. Within a minute, the provider found him on the second floor of a department store, simply by paging a miniature locator device secured to the man's clothes. Forty minutes later, when the man's son arrived at the department store, his father had already left. Fortunately, the service provider continued tracking the elderly man and was able to direct the son to the fourth floor of an Osaka hotel. At 1:10 p.m., the two were reunited. Locus Corp. provided the system that made this possible.

The belief that it should be easy to find anyone, anywhere, at any time with a few pushes of a button has caught on with the advent of the global positioning system (GPS). People imagine a miniature device, attached to one's person, that reports one's whereabouts almost instantaneously. Add the highly practical need to find missing persons promptly, and the personal locator system (PLS) industry is born.

Systems of this nature, whether based on the GPS or some other technology, are being tested throughout the world. Some, in fact, are already being deployed in Japan. The service alone can be sold by cellular companies, which base it on their wireless infrastructure. But several companies looking into the technology options plan to offer a broad array of services to the public and to businesses.

In Japan, location services are now commercially available to 74% of the nation's population, including Tokyo, Osaka, Kyoto, Yokohama, Nagasaki, and Hiroshima. Initially designed to support the mentally handicapped, personal locator services have expanded to serve children, the elderly, tourist groups, and security patrols, as well. They may also be used to track valuables and recover stolen vehicles. Not surprisingly, service areas coincide with wireless infrastructure deployments, which personal locators have exploited since their beginning in 1998.

In the United States, two further factors encourage the adoption of these geolocation systems. One is the need to effectively monitor offenders on parole and probation. Tagging offenders with locator devices would tighten their supervision and enhance public safety, and could even reduce the prison population. The other is the wish to provide wireless callers with enhanced 911 (E-911) emergency services. For land-line telephony, the location of a phone from which a 911 call is made appears automatically on the 911 operator's computer screen. But callers using cellular phones could be anywhere and unlocatable, unless location technology were applied to the wireless telephone system.

Of course, wireless services for locating vehicles have been thwarting car theft and managing fleets of cars since the mid-1980s. But unlike vehicular locators, which are less constrained by size and power, locators borne on the person have to be the size of a pager, and their power output has to be less than 1 W, because they can only carry a small battery that cannot be continuously recharged. Most challenging of all, personal locators have to be able to operate in RF-shielded areas such as buildings, because people spend a lot of their time indoors [8].

4.1 One PLS Architecture

A personal locator system is likely to involve a service provider, a location center, and a wireless network. In this setting, three scenarios, each involving a different operating mode, are possible. The person bearing a locator device is either being sought by a subscriber to the service, or is seeking help from the subscriber, or, as in the case of a parolee, is having his or her whereabouts monitored continuously.

Consider again the introductory example, but from a system architecture perspective. It is representative of the first scenario, based on the paging mode, wherein the person with the locator device is sought. In this instance, the subscriber calls the service provider, giving the operator there a password and the "wanted" person's identification (user) number (ID). The operator enters the ID into a computer, which transmits it to another computer at the location center. That machine calls the locator device, in effect paging it to establish communication through the PHS wireless telephone switching office (where PHS stands for personal handy phone systems). Immediately the office forwards the call to the wireless base station nearest the locator.

Once communication is established between the center and the device, the center asks the device for the signal strength data and IDs of any base stations in its vicinity. The locator replies, and from those inputs, plus RF database information on the base stations, the center computes the locator's coordinates. Details of the geolocation technology behind this architecture follow in subheading: "Enhanced Signal Strength."

These coordinates are transmitted to the service provider's computer, which displays the missing person's position on a street map for the service operator to report to the subscriber. The user's location is continuously updated on the service provider's map as long as the location center maintains its call connection to the locator device.

In a second scenario, surrounding the emergency mode, the user of the locator is lost or in dire straits of one sort or another, and presses the device's panic button. The locator calls the location center, which computes the user's position and alerts the service provider, which in turn alerts the subscriber to the user's situation.

The system can employ either packet data or voice channel communications. If a data channel is used, the service takes about 8 seconds to obtain a geolocation fix. But if a voice channel is used, the wait could last up to 33 seconds because of processing differences between the two channel types.

Several minutes may be added by communication between the service's operator and the subscriber. Such a human interface may be necessary given the complexity of Japanese (as explained in the Japanese example earlier) city-addressing schemes. Otherwise, subscribers using personal computers may obtain the information directly from the computer of either the service provider or location center.

Both the emergency and paging operating modes of personal locator systems are characterized as intermittent. In addition, a continuous automatic mode, in which the system polls the locator nonstop, is possible. Strictly speaking, the polling is periodic rather than continuous, but the latter term is more common.

Of the three locator modes, this last requires the most RF bandwidth and battery power. If it were implemented with a continuous voice call between the system and the locator, the expense would be beyond the reach of most applications. Assuming a minimal cost of 3 cents per minute for airtime, such a connection would cost US $46.59 a day—and also drain the locator battery within a few hours.

Packet data calls between the locator and the rest of the system are far more economical. In the packet version, the locator is likely to be polled every few minutes, exchanging 100 bytes or so with the system in a fraction of a second. Given a 3-minute polling interval and a 1-cent-per-poll cost, the daily cost per locator would be only $4.80 [8].

> **Note:** For most of the time, the locator would be in standby mode, conserving valuable battery charge. Another plus, upcoming third-generation mobile wireless telephony will increase the availability of packet data communications.

4.2 Six Technologies

A personal locator system could use any of several technologies. Among the most common methods are angle and time difference of the signal's arrival, global positioning system (GPS) and the more recent assisted GPS, enhanced signal strength, and location fingerprinting [8].

4.2.1 Signal Direction

The simplest is based on measuring the direction of a signal received from an RF transmitter at a single point. This can be done by pointing a directional antenna along the line of maximum signal strength. Alternatively, signal direction can be determined from the difference in time of arrival of the incoming signals at different elements of the antenna. A two-element antenna is typically used to cover angles of ±60 degrees. To achieve 360-degree coverage, a six-element antenna can be used.

A single mobile directional antenna can give only the bearing, not the position, of a transmitting object. The single bearing can be combined with other information, such as terrain data, to provide location. Such an antenna is generally used to approach and locate objects up to several kilometers away. A common use of this technique is tracking RF-tagged wildlife. The same basic technique is used by LoJack Corp., of Dedham, Massachusetts, in its system for finding stolen vehicles.

With two directional antennas spaced well apart, however, the position of a transmitting device in a plane can be computed. In this method, also known as the angle of arrival (AOA) method, transmitter position is determined from the known (fixed) position of the receivers' antennas and the angle of arrival of the signals with respect to the antennas.

Angle measurement precision affects the accuracy of positioning calculations, as does the geometry of the transmitting device and receiving antennas. For example, if a transmitter is too near a line drawn between two receiving antennas, its measured position could be off by more than the distance between the antennas. Fortunately, multiple receiver antennas distributed throughout the area of coverage enable the cellular system to select those antennas that introduce the smallest error [8].

4.2.2 Signal Times Of Arrival

Similarly, the time difference of arrival (TDOA) between signals received at the geographically disparate antennas can be used to determine position. Given the speed of light and known transmit and receive times, the distance between the mobile locator and receiver antenna can be calculated.

Accurate clocks are of the essence here, because an error of 1 μs in time corresponds to an error of 300 meters in space. Also, all clocks used must be synchronized. But as synchronizing the mobile locator clock is usually impractical, at least three receiving antennas are required for the calculation.

Sometimes the calculations produce ambiguous results, which can be resolved by considering signals received at a fourth antenna. As with the angle of arrival method, the relative positions of receivers and transmitter affect computational errors.

In an alternative time difference scheme, the locator and the antennas reverse roles: the antennas are transmitters and the mobile locator is a receiver. This technique is known as forward link trilateration (FLT). This is relatively simple to implement in some code-division multiple access (CDMA) wireless systems, where the time difference of arrival can be determined from the phase difference between pseudo-random noise code sequences of 0s and 1s transmitted from two antennas [8].

4.2.3 Global Positioning System

As previously explained, a global positioning system (GPS) relies on a constellation of 24 satellites. It, too, employs signal timing to determine position, but the mobile locator is a receiver and the orbiting satellites are transmitters. The satellites transmit spread-spectrum signals on two frequency bands denoted L1 (1575.42 MHz) and L2 (1223.6 MHz). The signals are modulated by two pseudo-random noise codes, the precision (P) code, and coarse/acquisition (C/A) code. The GPS signal is further modulated with a data message known as the GPS navigation message.

> **Note:** Only the C/A code in the L1 band is used in civilian applications and, hence, is of interest here.

To acquire the satellites' signals, the GPS receiver generates a replica of the satellites' pseudo-random noise codes. The GPS navigation message can be demodulated only if the replica can be matched and synchronized with the pseudo-random noise codes received. If the receiver cannot match and synchronize its replica, the GPS signal appears to the receiver as noise. Matching the pseudo-random noise codes and using the satellites' navigation message also enables the receiver to calculate the signal transmit time as well as the coordinates of the satellites.

The accuracy of GPS position calculations depends partly on measurement accuracy and partly on satellite configuration. Measurement errors depend on physical parameters, such as ionospheric delays and orbital uncertainties, and on the selective availability (SA) factor, introduced by the U.S. Department of Defense to degrade satellite data for nonmilitary users. Total measurement errors are estimated at 35 meters; without selective availability, they are reduced to 8 meters.

The configuration of the GPS satellites at the time of the measurements adds further distortion. If those in sight are scattered throughout the sky, the measurement error is multiplied by about 1.5. If they are clustered together, the multiplier is 5 or more.

To estimate actual position accuracy, it is necessary to combine the measurement errors with the errors introduced by the spatial disposition of the satellites. To determine its position, a GPS receiver calculates its x, y, and z coordinates as well as the time the satellite signals arrive. Data must be acquired from at least four (and preferably more) observable GPS satellites. When fewer than four satellites are in view, in areas such as city canyons, one remedy is a hybrid approach, augmenting GPS with the land-based measurements called "forward link trilateration." To illustrate, the use of two GPS satellites and two cellular base stations would suffice to determine a locator's position.

The unobstructed line of sight to the orbiting transmitters is important. The satellite signals are weak (below 10^{-15} W) when they arrive at a receiver's antenna, and are further weakened upon entering a building. Moreover, a conventional GPS receiver could take several minutes to acquire the satellite signals and, therefore, tends to operate continuously rather than be turned on and off for each acquisition. The drain on the receiver's battery is significant [8].

4.2.4 Server-Assisted GPS

To combat the shortcomings of GPS, an innovative technique known as "server-assisted GPS" was introduced in 1998. The idea is to place stationary servers throughout the area of coverage to assist mobile receivers to acquire the GPS signals. In effect, the servers are stationary GPS receivers that enhance the mobile GPS receiver's capabilities by helping to carry their weak signals from satellites to locator. The server includes a radio interface, for communicating with the mobile GPS receiver, and its own stationary GPS receiver, whose antenna has full view of the sky and monitors signals continuously from all the satellites within view.

To ask a mobile GPS receiver for its position, the server feeds it satellite information through the radio interface. Included in this information is a list of observable GPS satellites and other data that enable the mobile receiver to synchronize and match its pseudo-random noise code replicas with those of the satellites. Within about a second, the GPS receiver collects sufficient information for geolocation computation and sends the data back to the server. The server can then combine this information with data from the satellites' navigation message to determine the position of the mobile device.

With the assisted GPS approach, the mobile receivers conserve power by not continuously tracking the satellites' signals. Moreover, they have only to track the pseudo-random noise code and not extract the satellites' navigation message from the signal, in effect, becoming sensitive enough to acquire GPS signals inside most buildings.

In addition, the assisted version of the technology attains greater accuracy. Because the actual position of the stationary GPS receiver is known, the difference between that and its measured position can be used to calculate a correction to the mobile receiver's position. In other words, assisted GPS is inherently differential GPS (DGPS), which counters some of the inaccuracy in civilian GPS service.

Note: The most accurate GPS service is reserved for military use.

In June of 2000, Lucent Technologies Inc., of Murray Hill, New Jersey, announced that its wireless assisted GPS had attained an accuracy of better than 5 meters outdoors—an achievement attributable to the differential GPS capability of assisted GPS. More good news in this field was announced by SiRF Technology Inc., of Santa Clara, California, in the form of a postage-stamp-sized chipset (Star II) with built-in DGPS. In addition to providing improved GPS capability, it also offers reduced power consumption and greater accuracy, as well as performing well at handling weak signals [8].

4.2.5 Enhanced Signal Strength

If no obstructions are present, computing the position of a mobile locator is straightforward for both the signal timing and signal strength methods. When timing is used, the speed of light is multiplied by the time a signal takes to propagate between the two points gives the distance between them.

For the signal strength method, the distance between two points can be determined from the signal attenuation between the points. However, direct line contact seldom exists inside buildings, where signal attenuation is usually unknown and many indirect paths between transmitter and receiver are likely. Although techniques exist for reducing this multipath effect, the effect cannot be eliminated, and the errors it produces are difficult to predict. Multipath effects impede signal timing methods somewhat, but affect signal strength methods even more.

In addition, signal strength is very sensitive to antenna orientation, attenuation by obstructions, and other operating conditions. In contrast, signal timing is unaffected by antenna orientation and is less sensitive to attenuation.

Nonetheless, an enhanced signal strength (ESS) method that overcomes such impediments as multipath effects, attenuation, and antenna orientation has allowed the deployment of personal locator systems in PHS service areas in Japan. Such a system takes in three-dimensional information on the lay of the land, buildings, elevated highways, railroads, and other obstructions, and uses it to simulate the RF signal propagation characteristics of every PHS wireless transmitting antenna in the area of interest. The location system center stores the results in an RF database.

The position of a mobile locator is determined by getting it to measure the signal strength of preferably three to five base stations. From this input plus information from the base stations' databases, the system can calculate the position of the locator. The mean accuracy of the ESS is 40–50 meters. Inside large public buildings, with a PHS base station on every floor, the system can indicate a specific floor level. In subway and railroad stations, the availability of base stations makes it possible to find an individual on a specific track.

The stand-alone locator used by Locus Corp.'s enhanced signal strength method weighs only 58 grams and can operate for 16 days on a single battery charge. The ESS geolocation capabilities are also available in a standard PHS phone handset, in which the firmware has been modified. Presently, researchers in Japan are investigating how to apply ESS technology to other wireless phone systems [8].

4.2.6 Location Fingerprinting

Instead of exploiting signal timing or signal strength, a new technique from U.S. Wireless Corp., of San Ramon, California, relies on signal structure characteristics. Called "location fingerprinting," it turns the multipath phenomenon to surprisingly good use: By combining the multipath pattern with other signal characteristics, it creates a signature unique to a given location.

U.S. Wireless's proprietary *RadioCamera* system includes a signal signature database of a location grid for a specific service area. To generate this database, a vehicle drives through the coverage area transmitting signals to a monitoring site. The system analyzes the incoming signals, compiles a unique signature for each square in the location grid, and stores it in the database. Neighboring grid points are spaced about 30 meters apart.

To determine the position of a mobile transmitter, the RadioCamera system matches the transmitter's signal signature to an entry in the database. Multipoint signal reception is not required, although it is highly desirable. The system can use data from only a single point to determine location. Moving traffic, including vehicles, animals, and/or people, and changes in foliage or weather do not affect the system's capabilities [8].

4.3 What's PLS Good For?

In the United States, the need to provide wireless phone users with emergency 911 services has been one of the spurs to the development of location technologies. Today, an enhanced 911 (E-911) emergency call made over a land line is routed to a public safety answering point (PSAP), which matches the caller's number to an entry in an automatic location information database. When the match is made, this database provides the PSAP with the street address plus a location in a building—maybe the floor or office of the caller handset. So quickly is the caller located that the emergency crew can respond within 5 to 7 minutes on average.

But the very mobility of wireless handsets rules out a simple database relationship between phone number and location. In fact, the response to a wireless call can be 10 times longer than for a land-line call—far from ideal in an emergency.

Accordingly, the U.S. Federal Communications Commission (FCC) in Washington, D.C., directed operators of wireless phone services to enable their E-911 services to locate callers. The directive specified two phases. The first required an accuracy of several kilometers by April 1998 and the second required an accuracy of 125 meters with 0.67 probability by 2002. Whereas the first phase needed only software changes to the system, the second required the adoption of new location technologies.

The original FCC directive for Phase II also required support for existing handsets, which implied that only wireless network upgrades would be acceptable. Yet a wireless network-only solution would preclude the use of emerging technologies, such as assisted GPS, because that would require handset modification in addition to any wireless network infrastructure and software changes. All users might not bring in their handsets for modification, severely complicating support for handsets already in service.

To ease the introduction of new technologies, in September 1999, the FCC modified its original Phase II directive to permit handset-enabled solutions and also to tighten the accuracy required. In addition to the many technical roadblocks to implementing the E-911 directive, an even greater obstacle is cost. Upgrading all the wireless networks will cost billions of dollars. Cost recovery is the central issue for cellular service providers. Although wireless subscribers are the most likely source of recouping the cost, the government has made no formal decisions yet.

Presently, only the U.S. government requires its wireless companies to add caller geolocation to their E-911 services. But as the United States is a major telecommunications market, many manufacturers of wireless telecommunications equipment elsewhere are developing approaches to meet the commission's directive.

In an international development, a working group of the European Telecommuni-cations Standards Institute (ETSI), based in Sophia Antipolis, France, is currently drafting a standard for supporting location services for the Global System for Mobile Communications (GSM). Currently, GSM is the most common mobile wireless system in the world and is available in more countries than any other wireless system [8].

4.4 Monitoring Tops Services LIST

Wireless E-911 just helps the individual. But monitoring the mentally impaired and criminals could have even greater impacts on society at large. With the changing demographics of the developed world, the percentage of individuals over age 65 will soar over the next several decades. So will the number of elderly afflicted with age-related mental impairments. Most of the eight million or so U.S. patients diagnosed with Alzheimer's disease are over age 65.

Recall how personal locator technology helped a family find a mentally impaired elderly man, fortunately within 50 minutes or so. But what if many hours passed before anyone noticed that the man was missing? What if he had run into some kind of difficulty during that time? Being mentally impaired, he would be unlikely to press the panic button. An automatic polling system could solve this problem by checking whether the man was within a defined polygonal area or not—the location service and the family would be alerted whenever the man went out of this area.

As the population ages, the need for and cost of long-term care are likely to increase, too. Today it costs over $70,000 per year in the United States to care for a patient in a nursing home. Systems that monitor the whereabouts of the mentally impaired elderly could help them live longer in their communities and spend less time in institutions.

Criminal justice is another area of social concern where personal locators could intervene. The United States leads industrial nations in the percentage of its population incarcerated. In 2003, according to U.S. Department of Justice statistics, almost 8.5 million people were serving time in U.S. jails and prisons, and a further 11 million were in parole and probation programs. In comparison, in Japan in 2003, only 468,000 were serving prison terms while 423,000 were on parole or probation.

The high human and monetary cost of corrections could be cut by new technologies, such as personal locator systems, that would reduce prison populations and improve the monitoring of parolees and persons on probation. First-generation monitoring systems, introduced in the mid-1980s, track the location of the offender in a very confined area, such as the home. They enable the corrections system to verify that a parolee stays there during specified periods, 6 p.m. to 6 a.m., say. But by day, when the offender is presumably at work, these systems can do nothing.

Second-generation monitoring systems do better. A tamperproof personal locator is fastened on the offender and tracked continuously and automatically over a wide area. The newer system compares the actual with the supposed positions of the offender, as stored in a database. If any violation or tampering with the locator occurs, the system alerts the appropriate corrections or law enforcement agencies.

The goal is to verify that parolees and probationers comply with the directives imposed by the corrections system as to where and when they should and should not be by day and night. For example, a child molester is excluded from school areas, and a stalker is excluded from areas near the home and workplace of the victim.

Storage of the offender's ongoing whereabouts in an electronic file benefits law enforcement agencies in other ways [5]. The record can be used to exclude or include a monitored offender as a suspect in a crime by comparing events at the crime scene with the file entries [8].

4.5 Privacy, Security Still Issues

Confidentiality of information about a person's whereabouts is a serious concern for location technology. Databases already store large amounts of personal information, including medical data, marketing preferences, and credit information. Lax security could lead to serious abuse of this data. Access to a database of location information could aggravate this situation by further exposing a person's movements. Moreover, it can have real-time implications. For example, someone could find and harm a victim.

The location information stored in databases needs to be secured, as does the tracking and locating process itself. Because RF communications are used, eavesdropping is a possibility. To reduce this risk, location information can be encrypted or transmitted using coded signals employing such spread-spectrum technology as CDMA.

Privacy protection can also depend on the technology used. For example, in GPS or the enhanced-signal-strength method, the location system uses information captured and transmitted by the locator. Some devices are equipped with an option to block such transmissions, preventing the system from locating the device. But in wireless network-based locator systems that measure the locator's signal characteristics without requiring its cooperation, the only safe way for users to keep their locations secret is to turn off the device [8].

4.6 More Work To Be Done

Despite the strides made in recent years in personal locator technologies, much work remains to be done on their accuracy, locator miniaturization, battery life, multipath effects, ability to penetrate buildings, and the economical use of RF bandwidth. In addition, hybrid systems may be required to provide improved coverage and open the door to new applications.

Reducing the cost of deploying location technology is essential in removing barriers to the use of location services. The concern over how to pay for E-911 services demonstrates the need for cost reduction. However, if a rich set of location services could share the expense of the additional infrastructure needed to support these services, the cost per subscriber would be reduced. The new location technologies, as well as wireless data packet services that are now emerging around the globe, offer opportunities for entrepreneurs to expand personal locator services.

In June 2002, Loc8.net (*http://www.loc8.net*), based in Seattle, Washington, provided location-based services employing the ReFLEX two-way paging wireless infrastructure. The services use wireless assisted GPS, and include personal locator services for Alzheimer patients and children, as well as commercial services such as fleet management. The two-way paging systems using ReFLEX, developed by Motorola Inc. of Schaumberg, Illinois, cover 98% of the U.S. population. The Loc8.net system operates in emergency and paging modes.

Recent advances such as assisted GPS are likely to enhance GPS-based offender-monitoring systems, reducing device size and power consumption, adding to accuracy, and offering new capabilities such as in-building tracking. In the future, personal locators could bring many other blessings. Equipping young children with personal locators may offer parents greater peace of mind. Small enough locators could even track pets.

Personal locators could also be helpful to medical patients where the locator would be combined with a detector that monitors the patient's vital signs. If the detector picked up abnormalities in the signals, it would alert the nurse or physician with both medical and location information. Such a service could offer a patient greater freedom and a shorter stay in hospital or nursing home. Its greatest contribution, however, may be peace of mind for patients, their families, and doctors.

Obviously, technical and commercial considerations will determine the success of the technology. Issues of users' privacy and confidentiality will, however, have to be addressed first [8].

5. THE IMPLICATIONS OF COOKIES AND INTEGRATED PLATFORMS

Cookies have benefits and drawbacks. Used properly, they can enhance a visitor's experience of a Web site. Used carelessly, they can poison a user's impression of a site and even prompt some users to stay away forever.

All Web site integrated platform designers will eventually face the question of whether and how to use cookies. Often, designers find themselves ill-equipped to make this decision and so they employ cookies haphazardly or without regard for user acceptance or data privacy.

This part of the chapter is for anyone involved in Web site integrated platform design, not just engineers, so it avoids addressing every low-level technical nuance of cookies. Instead, it explores technical considerations, interface design challenges, and (perhaps most importantly) ethical issues [8].

5.1 A Cookie In A Nutshell

When a visitor views a Web page, the server can assign that visitor a unique customer ID, known as a cookie. The server asks the visitor's browser program to "accept" the cookie—to save the ID number on the visitor's computer. Then the browser sends the cookie back to the Web server each time the visitor returns to that page, or in some cases, to any page on the Web site.

The ID number tells the server that the visitor has visited the site in the past. The server can use the ID number as a key to store any information the visitor has provided in past visits, or any details it has observed about the visitor's preferences or browsing behavior. The ID number can save a visitor from having to repeatedly log-in to a members-only site on each visit [8].

5.2 Why Cookies Provoke Controversy

Cookies (like any powerful data-gathering tool) can be abused. Many users fear, sometimes justifiably, that a cookie they accept may allow unscrupulous Web site operators to gather information about them and then use that data in an unauthorized manner. So, these users set their browser software to warn them of each incoming cookie—and many users reject every cookie, without exception [8].

> **Note:** Web site integrated platform designers should note this sometimes justified mistrust of cookies, and design accordingly.

5.3 Poor Support of Cookies in Browsers

Part of the climate of mistrust surrounding cookies stems from the poor cookie interface provided by current Web browser software. As noted in the preceding, both Netscape and Microsoft browsers can consult users before accepting a cookie, and many users choose to browse with this preference turned on.

Currently, a visitor who rejects a cookie on a Web site integrated platform but continues to browse experiences a relentless barrage of cookie requests. Often, even visitors who accept a cookie are still bombarded by offers of more cookies from the same site. On many sites, this badgering can include multiple cookies per page, cookies that change gratuitously even once accepted, and cookies on pages that don't even require cookies for any apparent reason. As feedback from users reaches the designers of browser software, look for browsers to add the following features to help users cope with this overuse of cookies:

- Reject all cookies option
- Better choices when asked
- Cookie management tools [8]

5.3.1 Reject All Cookies Option

Today, browsers present only the annoying false choice between "Accept all cookies without asking" and "Ask about each cookie." Users should expect to see these choices expanded to include a third "Reject all cookies without asking" option [8].

5.3.2 Better Choices When Asked

For users who choose notification, browsers should offer a more flexible set of choices regarding what happens after a cookie has been accepted or rejected. Specifically, the user should be able to say "I want to accept/reject this cookie, and then don't ask me again..."

- About this particular cookie on this Web site integrated platform
- About any cookie on this Web site integrated platform
- About any cookie on this page [8]

5.3.3 Cookie-Management Tools

Finally, expect to see browsers offer a mechanism that lets users view and manage the set of cookies they've collected. Certain browsers, such as the most recent release of Microsoft *Internet Explorer,* have begun to add these or similar cookie- management features. Until a majority of common browsers have incorporated these options, integrated platform designers should plan to minimize the number and type of cookies a visitor encounters on a site [8].

6. WINTEL INSIDE, OR HOW YOUR COMPUTER IS WATCHING YOU

A previously discussed, the Privacy Foundation has discovered that it is possible to add "Web bugs" to Microsoft *Word* documents. A "Web bug" could allow an author to track where a document is being read and how often. In addition, the author can watch how a "bugged" document is passed from one person to another or from one organization to another. Some possible uses of Web bugs in *Word* documents include:

- Detecting and tracking leaks of confidential documents from a company
- Tracking possible copyright infringement of newsletters and reports
- Monitoring the distribution of a press release
- Tracking the quoting of text when it is copied from one *Word* document to a new document [8]

Web bugs are made possible by the ability in Microsoft *Word* of a document to link to an image file that is located on a remote Web server. Because only the URL of the Web bug is stored in a document and not the actual image, Microsoft *Word* must fetch the image from a Web server each and every time the document is opened. This image-linking feature then puts a remote server in the position to monitor when and where a document file is being opened. The server knows the IP address and host name of the computer that is opening the document. A host name will typically include a company name if a computer is located at a business. The host name of a home computer usually has the name of a user's Internet Service Provider (ISP).

An additional issue, and one that could magnify the potential surveillance, is that Web bugs in *Word* documents can also read and write browser cookies belonging to *Internet Explorer.* Cookies could allow an author to match up the computer viewer of a *Word* document to their visits to the author's Web site.

Web bugs are used extensively today by Internet advertising companies on Web pages and in HTML-based e-mail messages for tracking. They are typically 1-by-1 pixel in size to make them invisible on the screen to disguise the fact that they are used for tracking.

Although the Privacy Foundation has found no evidence that Web bugs are being used in *Word* documents today, there is little to prevent their use. Short of removing the feature that allows linking to Web images in Microsoft *Word,* there does not appear to be a good preventative solution. However, the Privacy Foundation has recommended to Microsoft that cookies be disabled in Microsoft *Word* through a software patch. In addition to *Word* documents, Web bugs can also be used in *Excel* and *PowerPoint* documents [8].

6.1 Detailed Description

Microsoft *Word* has, from the beginning, supported the ability to include picture files in *Word* documents. Originally, the picture files would reside on the local hard drive and then be copied into a document as part of *Word* .DOC file. However, beginning with *Word 97,* Microsoft provided the ability to copy images from the Internet. All that is required to use this feature is to know the URL (Web address) of the image. Besides copying the Web image into the document, *Word* also allows the Web image to be linked to the document via its URL. Linking to the image results in smaller *Word* document files because only a URL needs to be stored in the file instead of the entire image. When a document contains a linked Web image, *Word* will automatically fetch the image each time the document is opened. This is necessary to display the image on the screen or to print it out as part of the document.

Because a linked Web image must be fetched from a remote Web server, the server is in a position to track when a *Word* document is opened and possibly by whom. Furthermore, it is possible to include an image in a *Word* document solely for the purpose of tracking. Such an image is called a Web Bug. Web bugs today are already used extensively by Internet marketing companies on Web pages and embedded in HTML e-mail messages.

When a Web bug is embedded in a *Word* document, the following information is sent to the remote Web server when the document containing the bug is opened:

- The full URL of the Web bug image
- The IP address and the host name of the computer requesting the Web bug
- A Web browser cookie (optional) [8]

This information is typically saved in an ordinary log file by Web server software.

Because the author of the document has control of the URL of the document, they can put whatever information they choose in this URL. For example, a URL might contain a unique document ID number or the name of the person to whom the document was originally sent.

These tracking abilities might be used in any number of ways. In most cases, the reader of a particular document will not know that the document is bugged, or that the Web bug is surreptitiously sending identifying information back through the Internet.

One example of this tracking ability is to monitor the path of a confidential document, either within or beyond an enterprise's wireless computer network. The confidential document could be "bugged" to "phone home" each time it is opened. If the company's Web server ever received a "server hit" from an IP address for the bug outside the organization, then it could learn immediately about the leak. Because the server log would include the host name of the computer where the document was opened, a company could know that the organization that received the leaked document was a competitor or media outlet, for example.

All original copies of a confidential document could also be numbered so that a company could track the source of a leak. A unique serial number could be encoded in the query string of the Web bug URL. If the document is leaked, the server hit for the Web bug will indicate which copy was leaked.

A serial number could be added to a Web bug in a document either manually (right before a copy of a document is saved) or automatically through a simple utility program. The utility program would scan a document for the Web bug URL and add a serial number in the query string. A Perl script of less than 20 lines of code could easily be written to do this sort of serialization.

Another use of Web bugs in *Word* documents is to detect copyright infringement. For example, a publishing company could "bug" all outgoing copies of its newsletter. The Web bugs in a newsletter could contain unique customer ID numbers to detect how widely an individual newsletter is copied and distributed.

A third possible use of Web bugs is for market research purposes. For example, a company could place Web bugs in a press release distributed as a *Word* document. The server log hits for the Web bugs would then tell the company what organizations have actually viewed the press release. The company could also observe how a press release is passed along within an organization or to other organizations.

In an academic setting, Web bugs might be used to detect plagiarism. A document could be bugged before it is distributed. An invisible Web bug could be placed within each paragraph in the document. If text were to be cut and pasted from the document, it is likely that a Web bug would be picked up also and copied into the new document

To place a Web bug in a *Word* document is relatively simple. These are the steps in *Word:*

1. Select the Insert | Picture | From File... menu command.
2. Type in the URL of the Web Bug in the "File ame" field of the Insert Picture dialog box.
3. Select the "Link to File" option of the "Insert" button [8].

The use of Web bugs in *Word* does point to a more general problem. Any file format that supports automatic linking to Web pages or images could lead to the same problem. Software engineers should take this privacy issue into consideration when designing new file formats.

This issue is potentially critical for music file formats such as MP3 files where piracy concerns are high. For example, it is easy to imagine an extended MP3 file format that supports embedded HTML for showing song credits, cover artwork, lyrics, and so on. The embedded HTML with embedded Web bugs could also be used to track how many times a song is played and by which computer, identified by its IP address [8].

7. DATA MINING FOR WHAT?

The use of data mining for information warfare is growing rapidly. The number of data-mining consultants, as well as the number of commercial tools available to the "nonexpert" user, are also quickly increasing. It is becoming easier than ever to collect datasets and apply data-mining tools to them. As more and more nonexperts seek to exploit this technology to help with their business, it becomes increasingly important that they understand the underlying assumptions and biases of these tools. There are a number of factors to consider before applying IW data mining to a database. In particular, there are important issues regarding the data that should be examined before proceeding with the data-mining process. Although these issues may be well-known to the data-mining expert, the nonexpert is often unaware of their importance.

Now let's focus on three specific issues. Each issue is illustrated through the use of brief examples. Also, insight is provided for each issue on how it might be problematic, and suggestions are made on which techniques can used for approaching such situations.

The purpose here is to help the nonexpert in IW data mining to better understand some of the important issues of the field. Particular concern is also established here with characteristics of the data that may affect the overall usefulness of the IW data-mining results. Some recent experiences, and the lessons learned from them, are described. These lessons, together with the accompanying discussion, will help to both guide the IW data-collection process and better understand what kinds of results to expect.

One cannot blindly "plug-and-play" in IW data mining. There are a number of factors to consider before applying data mining to any particular database. This general warning is not new. Many of these issues are well-known by both the data mining experts and a growing body of nonexpert, data "owners." For instance, the data should be "clean," with consistent values across records and containing as few errors as possible. There should not be a large number of missing or incomplete records or fields. It should be possible to represent the data in the appropriate syntax for the required data-mining tool (attribute/value pairs).

As previously mentioned, this part of the chapter will discuss three specific, but less well-known, issues. Each will be illustrated through real-world experiences. The first is the impact of *data distribution*. Many IW data-mining techniques perform class or group discrimination, and rely on the data-containing representative samples of all relevant classes. Sometimes, however, obtaining samples of all classes is surprisingly difficult. The second issue is one of *applicability and data relevance*. High-quality data, combined with good data-mining tools, does not ensure that the results can be applied to the desired goal. Finally, this part of the chapter will discuss some of the issues associated with using *text* (narrative fields in reports) in data mining. The current technology cannot fully exploit arbitrary text, but there are certain ways text can be used.

These three issues are not new to the field. Indeed, for many IW data-mining experts, these are important issues that are often well understood. For the nonexpert, however, these issues can be subtle or appear deceivingly simple or unimportant. It is tempting to collect a large amount of clean data, massage the representation into the proper format, hand the data tape to the consultant, and expect answers to the most pressing business questions. Although this part of the chapter does not describe all of the potential problems one might face, it does describe some important issues, illustrate why they might be problematic, and suggest ways to effectively deal with these situations [8].

7.1 Two Examples

The discussion of data distribution, information relevance, and use of text will be illustrated with examples from two current projects. The first involves a joint project with the Center for Advanced Aviation Systems Development (CAASD) in the domain of aviation safety. In this project, one of the primary goals is to help identify and characterize precursors to potentially dangerous situations in the aviation world. One particular way to do this is to mine accident and incident reports involving aircraft for patterns that identify common precursors to dangerous situations. For any type of flight—commercial, cargo, military, or pleasure—accidents (and often less serious incidents) are investigated. A report is filed containing a variety of information such as time of day, type of aircraft, weather, pilot age, and pilot experience. These reports often include the inspector's written summary. One task involves using collections of these reports to try to identify and characterize those situations in which accidents occur. A source of such reports is the National Transportation Safety Board (NTSB).

Another project we are currently working on involves targeting vehicles for law enforcement. In this particular instance, vehicles (mostly passenger vehicles and small trucks) arrive at an inspection stop. At this primary stop, a brief inspection is conducted to decide if further examination is necessary. There is typically a constant flow of cars to be processed, so excessive time cannot be taken. This first inspection typically takes 20 to 30 seconds. If the primary inspector feels it is warranted (and there are any number of reasons that justify this), any vehicle can be pulled out for secondary inspection. This secondary inspection and background check is more thorough. If the driver/vehicle is found to be in violation of the particular laws under consideration, then various information concerning both driver and vehicle is collected and entered into the "violators" database. The goal of this project is to find a way to better profile these violating drivers and vehicles, so that the primary inspectors can more accurately identify likely suspects and send them for secondary inspection [8].

7.2 Data Distribution

Let's first discuss the issue of data distribution. Of particular concern is the situation in which the data lacks certain types of examples. Consider the aviation safety domain. One goal of the project in this domain is to characterize situations that result in accident flights. An obvious source of information is the NTSB's database of accident reports.

> **Note:** This database does not contain records about uneventful flights (the NTSB is an accident investigation agency). That is, the data are unevenly distributed between records of accident flights and records of uneventful flights.

This lack of reports about uneventful flights has important consequences for a significant class of data-mining techniques. When given the data containing only accident flights, each of the approaches in this class concludes that all flights contain accidents. Such a hypothesis is clearly incorrect. The majority of the flights are uneventful. Also, such a hypothesis is not useful because it does not offer any new insight on how to differentiate the accident flights from the uneventful ones. Furthermore, some of the most popular IW data-mining tools, including decision tree inducers, neural networks, and nearest neighbor algorithms, fall into this class of techniques. (They assume that the absence of uneventful flights in the data implies that they do not exist in the world.)

To continue this discussion, it is necessary to first define some terms used in data mining. The "target concept" is that concept you are trying to learn. In the aviation domain, the target concept is accident flights. Consequently, each example of an accident flight (each accident report in the database) is called a member of the target concept, and each uneventful flight is a nonmember of the target concept. The NTSB data do not contain records of uneventful flights. That is, there are no descriptions of nonmembers of the target concept. The problem of learning to differentiate members from nonmembers is called a "supervised concept learning problem."

Note: It is called supervised because each example in the data contains a label indicating its membership status for the target concept.

For example, a supervised concept learner uses a training sample as input. A training sample is a list of examples, labeled as members or nonmembers, which is assumed to be representative of the whole universe. The supervised concept learner produces hypotheses that discriminate the members and nonmembers in the sample. Many IW data-mining tools use supervised concept learners to find patterns.

Let's say that a supervised concept learner makes the closed-world assumption that the absence of nonmembers in the data implies that they do not exist in the universe. Why do some of the popular learners make the closed-world assumption? The case of decision tree learners provides a good illustration. These learners partition the training sample into pure subsamples, containing either all member or all nonmembers. The partitioning of the training sample drives the rule generation. That is, the learners introduce conditions that define partitions of the training sample; each outcome of a condition represents a different subsample. Ultimately, the conditions will become part of the discrimination rules. Unfortunately, if the input sample contains only data that are members of the target class, the training sample is already pure and the decision tree learner has no need to break-up the sample further. As a consequence, the rules commit to classifying all new data as members of the target class before conducting any tests. Thus, in the aviation project, all flights would be classified as accident flights, because the learner never saw any uneventful flights. This is not to say that learners employing the closed-world assumption are inappropriate in all, or even most situations. For many problems, when representative data from all the concepts involved is available, these learners are both effective and efficient [8].

7.3 Applicability and Relevance of Data

Even when collected data is of high quality (clean, few missing values, proper form, etc.) and the IW data-mining algorithms can be successfully run, there still may be a problem of relevance. It must be possible to apply the new information to the situation at hand. For instance, if the data mining produces typical "if...then..." rules, then it must be possible to measure the values of the attributes in the condition ("if" part) of those rules. The information about those conditions must be available at the time the rules will be used. Consider a simple example where the goal is to predict if a dog is likely to bite. Assume data are collected on the internal anatomy of various dogs, and each dog is labeled by its owner as either "likely" or "unlikely" to bite. Assume further that the data-mining tools work splendidly, and it is discovered that the following (admittedly contrived) rules apply: Rule 1—If the rear molars of the dog are worn, the dog is unlikely to bite; and, Rule 2—If the mandibular muscles are over-developed, the dog is likely to bite.

These may seem like excellent rules. However, if faced with a strange, angry dog late at night, these rules would be of little help in deciding whether you are in danger. There are two reasons for this: First, there is a time constraint in applying the rules. There are only a few seconds to check if these rules apply. Second, even without such a constraint, the average person probably can't make judgments about molar wear and muscle development. The lesson here is that just because data are collected about biting (and nonbiting) dogs, it does not mean one can predict whether a dog will bite in every situation.

In the vehicle-targeting task described earlier, a similar situation occurred. The initial goal was very specific: develop a set of rules, a profile, that the primary inspectors could use to determine which vehicles to pull out for secondary inspection. As mentioned, much more information is collected concerning actual violators than for those that are just passed through the checkpoint. Thus, the initial goal was to profile likely violation suspects based on the wealth of information about that group. The problem, noticed before any analysis was done, was that the information that would make up the profiles would not be applicable to the desired task. As mentioned, the primary inspectors have only a short amount of time to decide whether a particular vehicle should be pulled out for secondary inspection. During that time, they have access to only superficial information. That is, the primary inspectors don't have quick access to much of the background knowledge concerning the driver and vehicle. Yet this is precisely the knowledge collected during seizures initially chosen to build profiles. Thus, they have no way to apply classification rules that measure features such as "number of other cars owned," "bad credit history," or "known to associate with felons" (types of data collected on violation vehicles and drivers).

The problem here is not that the data is "bad," or even that the data is all from the target concept. The problem is that the data cannot be applied to the initially specified task. How does this situation come about in general? The answer involves a fairly common situation. Often, IW data mining begins with data that has been previously collected, usually for some other purpose. The assumption is made that since the collected data is in the same general domain as the current problem, it must be usable to solve this problem. As the examples show, this is often not the case. In the vehicle- targeting task, the nature of the law enforcement system is such that a great deal of information is collected and recorded on violators. No one ever intended to use this information as a screening tool at stop points. Thus, it is important to understand the purpose for which a set of data was collected. Does it address the current situation directly? Similarly, when data is collected for the specific task at hand, careful thought must go into collecting the relevant data.

There are two primary ways to address this problem of data irrelevancy. The most obvious is to use additional data from another source. It may be that different data already exists to address the primary question. For instance, returning to the dogs example, general aggressiveness characteristics for different breeds of dogs have been determined. Using this data, rather than the original data, deciding how likely a dog is to bite is reduced to the problem of determining its breed (often done by quick visual inspection). When the necessary data does not already exist, it may be necessary to collect it. Some of this data collection will likely take place in the vehicle-targeting project. In this case, data must be collected that relates directly to the information available to the inspectors at the initial inspection. For example, the demeanor of the driver may be an important feature. Of course, collecting new data may be a very expensive process. First, the proper attributes to collect must be determined. This often involves discussions and interviews with experts in the field. Then the actual data-collection process may be quite costly. It may be that an inordinate amount of manpower is required, or that certain features are difficult to measure.

If additional data cannot be obtained, there is another, often less desirable way to address this issue. It may be possible to alter the initial goals or questions. This will clearly require problem-specific domain expertise to address a few simple questions: Is there another way to address the same issue? Is there another relevant issue that can be addressed directly with this data? In the vehicle-targeting domain, only those attributes that were directly accessible to the inspector were used. A good example would be looking at simple statistical patterns for time of day, weather, season, and holidays. This is not a deep analysis and doesn't quite "profile" likely violators, but it makes progress toward the initial goal. Another alternative is to use the violator database to profile suspects for other situations. It may be that profiles of certain types of violators bear similarities to other criminal types. Perhaps this information can be used elsewhere in law enforcement. Admittedly, this latter solution does not address the initial issue: helping the primary inspectors decide who to pull out for secondary inspection. However, it may not be possible to achieve that goal with this data and the given time constraints. It is important to understand this potential limitation early in the process, before a great deal of time, effort, and money has been invested [8].

7.4 Combining Text and Structured Data

IW data mining is most often performed on data that is highly structured. Highly structured data have a finite, well-defined set of possible values, as is most often seen in databases. An example of structured data is a database containing records describing aircraft accidents that includes fields such as the make of an airplane and the number of hours flown by the pilot. Another source of valuable yet often unused information is unstructured text. Although more difficult to immediately use than structured data, data mining should make use of these available text resources.

Text is often not used during IW data mining because it requires a preprocessing step before it can be used by available tools such as decision trees, association rule methods, or clustering. These techniques require structured fields with clearly defined sets of possible values that can be quickly counted and matched. Such techniques sometimes also assume that values are ordered and have well-defined distances between values. Text is not so well behaved. Words may have multiple meanings depending on context (polysemy), multiple words may mean the same thing (synonymy), or may be closely related (hypernymy). These are difficult issues that are not yet totally solved, but useful progress has been made and techniques have been developed so that text can be considered a resource for data mining.

One way to exploit text, borrowed from information retrieval, is to use a vector-space approach. Information retrieval is concerned with methods for efficiently retrieving documents relevant to a given request or query. The standard method for doing this is to build weight vectors describing each document and then compare the document vector to the query vector. More specifically, this method first identifies all the unique words in the document collection. Then this list of words is used to build vectors of words and associated weights for the query and each of the documents. Using the simplest weighting method, this vector has a value of 1 at position x when the xth vocabulary word is present in the document; otherwise it has a value of 0. Every document and query is now described by a vector of length equal to the size of the vocabulary. Now each document vector can be compared to every other document by comparing their word vectors. A cosine-similarity measure (which projects one vector along another in each dimension) will then provide a measure of similarity between the two corresponding documents. Surprisingly, although this approach discards the structure in the text and ignores the problems of polysemy and synonymy altogether, it has been found to be a simple, fast baseline for identifying relevant documents.

A variant of this vector-space approach was used on the airline safety data to identify similar accidents based on a textual description of the flight history. The narrative description of each accident was represented as a vector and compared to all other narratives using the approach described earlier. One group of accidents identified by this technique can be described as planes that were "veering to the left during takeoff." The following accident reports were found to be similar in this respect:

- **MIA01LA055:** During takeoff roll he or she applied normal right rudder to compensate for engine torque. The airplane did not respond to the pilot input and drifted to the left.
- **ANC00LA099:** Veered to the left during the first attempt to take off.
- **ANC00LA041:** Pilot added full power and the airplane veered to the left [8].

Identifying this kind of a group would be difficult using fixed fields alone. This technique can also be used to find all previous reports similar to a given accident, or to find records with a certain combination of words. This can be a useful tool for identifying patterns in the flight history of the accident so that the events leading up to different accidents can be more clearly identified.

The information stored in text can be extracted in other ways as well. A collection of documents and a taxonomy of terms are combined so that maximal word or category associations can be calculated. It could also be used in the airline safety domain to calculate, for example, which class of mechanical malfunctions occurred most often in winter weather.

Another approach very relevant to IW data mining from text is information extraction (IE). Information extraction is interested in techniques for extracting specific pieces of information from text and is the focus of the DARPA Message. Understanding Conference (MUC). The biggest problem with IE systems is that they are time-consuming to build and domain specific. To address this problem a number of tools have (and continue to be) developed for learning templates from examples such as CRYSTAL, RAPIER, and AutoSlog. IE tools could be used in the airline safety data to pull out information that is often more complete in the text than in the fixed fields. This work is geared toward filling templates from text alone, but often the text and structured fields overlap in content.

An example of just such an overlap can be found in the NTSB accident and incident records. The data in the NTSB accident and incident records contains structured fields which together allow the investigator to identify human factors that may be important to the accidentscene. However, it was found that these fields are rarely filled out completely enough to make a classification: 95% of the records that were identified as involving people could only be classified as "unknown." IE methods could be used to reduce this large unknown rate by pulling information out of the narrative, which described if a person in the cockpit made a mistake. Such an approach could make use of a dictionary of synonyms for "mistake" and parser for confirming if the mistake was an action made by the pilot or copilot and not in a sentence describing, for example, the maintenance methods.

Although IW data mining has primarily concerned itself with structured data, text is a valuable source of information that should not be ignored. Although automatic systems that completely understand the text are a still a long way off, one of the surprising recent results is that simple techniques, which sometimes completely ignore or only partially address the problems of polysemy, synonymy, and complex structure of text, still do provide a useful first cut for mining information from text. Useful techniques, such as the vector-space approach and learned templates from information extraction, can allow IW data miners to make use of the increasing amount of text available on-line [8].

8. THE INTERNET IS BIG BROTHER

How prepared is your business for the future? As the Internet expands its reach to the farthest corners of the globe, companies will find themselves dealing with increasingly complex challenges such as Big Brother.

In the Internet of tomorrow you can be sure of at least three things. First, the experience will not be anything like what we're familiar with. Second, despite Point One, little will change in the next 10 years. And third, the business environment of the future will be much less forgiving, so companies that do not take the new technologies seriously are putting themselves at risk [8].

8.1 A High-Fiber Diet

During the 1990s, everyone heard about fiber-optic technology's potential for increasing bandwidth and enhancing performance. Recently, better installation techniques and a wider range of fiber-compatible equipment have made fiber both easily available and less expensive than it used to be.

Although fiber will remain a good choice for backbones and carrier-class interconnections, most companies will probably continue to run copper, at least for the foreseeable future. After all, few businesses want to bear even slightly higher costs for fiber installation and wireless networking gear. Instead, expect to see more companies upgrading their existing copper technology. Expect them to get away with it, too, because for most practical purposes, copper can handle the load [8].

8.2 Broadband

High-speed Internet connections, especially the cable modem [6] and DSL technologies lumped together under the "broadband" heading, are increasing in number at an astonishing rate, at least in First World countries like the U.S. and within the European Union. The biggest problem the providers face seems to be keeping up with demand. The current waiting list of DSL and cable access orders probably won't be caught up until 2006. Even then, North America will be home to wide swaths of rural territory without high-speed access.

This scenario isn't likely to change in the next decade, either. Unless governments (Big Brother) insist that Internet carriers supply rural service at a loss (as American telephone companies were ordered to do with voice service), broadband providers will have little incentive to deploy their technologies on a wide scale.

The increasing availability of broadband wireless connections is another revolution that's already under way, although it won't make a serious difference anytime soon either. In its infant state, broadband wireless, with its ability to support certain e-business applications, is best suited to a LAN-like role. It will be a while before we see a device that combines the size of the cell-phone with the power of the laptop [8].

8.3 A Truly Global Internet

The days of Americans (Big Brother) ruling the Internet are not over by a long shot. Even so, the next decade will bring an explosion of Internet usage in places such as Asia and Africa. As the Internet becomes more pervasive, businesses will face an even greater shortage of skilled employees. American businesses that have traditionally relied on a foreign labor market may be caught short when those workers can find jobs at home.

As always, the spoils will go to those businesses that think ahead. We are going to see big changes, so start preparing yourself for a world ruled by broadband, culturally neutral, easily maintained wireless access. That means considering technical issues and different standards of civil rights, conduct, and privacy. The companies that do will have a head start on the competition for the next decade and maybe the next century [8].

9. THE WIRELESS INTERNET—FRIEND OR FOE?

The wireless networking engineer was working her way through the information warfare test range when she stopped and looked at her computer screen. Another unsecured access point, she noted. She was actually testing the roaming capabilities of 802.11b network devices; but as she moved their portable computers through the areas covered by the devices, other access points popped up.

This isn't surprising, because one of the nice things about wireless Internet is the ability to install the products quickly and easily, with a minimum amount of configuration. Clearly, some people around her test site (which is being kept nameless to protect the guilty) took advantage of the ease of installation but never got around to protecting their internetworks.

If Internetwork managers don't pay attention to the fact that the default condition of wireless access points is to let anyone into the network, then they may be doing just that. Those people who constitute "anyone" can include people across the street, your competitors parked outside, and malcontents who want to use your wireless network to shield their activities. It's like installing a wireless network port on the lamppost outside your building and asking anyone who walks by to plug in.

Fortunately, if you plan accordingly, securing your wireless Internet isn't very difficult. It just requires wireless network administrators to take a few simple steps.

First, turn off the broadcasting of your access point's extended service set identification, which lets anyone with a wireless Internet card know the address of your wireless Internet access point. Having that ID makes logging-in even easier than it already is.

Second, turn on encryption. All 802.11b access points support the wireless encryption protocol (WEP), which can handle 40- and 128-bit encryption.

Third, turn on your ability to use access control lists, available in some access points. This allows you to keep a list of acceptable users according to the MAC address of their wireless network card.

These steps will keep most wireless networks reasonably secure. It's convenient that these capabilities are built into most wireless Internet access points—the only exception being some early Apple AirPorts, which can be upgraded.

You must also deal with the fact that wireless access points are inexpensive, and that getting them running is a no-brainer. This means that pretty much anyone on your wireless network can pick up a wireless access point at Best Buy, plug it into the wireless enterprise network, and use it. You'd then have an entry point into your wireless network that's open to anyone with a wireless Internet card. Fortunately, if you already have a wireless Internet, you probably also have the management software that lets you locate all access points, including those that aren't authorized, and can either take them off the wireless network or secure them.

Another problem is that, without limits on what users are allowed to do and where they're allowed to go, you lose control. So even more security is needed.

One solution is to move to a third-party provider of wireless security products, such as WRQ, whose *NetMotion* product requires a log-in that's authenticated through Windows NT. It uses much better encryption than is available under WEP, and it offers some security management features, such as the ability to remotely disable a wireless Internet card's connection to the wireless network. Such capabilities require a bit more attention from managers, but the result can be a wireless Internet that's more secure than the wired one it's attached to.

Stories abound of employees who bought their own wireless access points, installed them, and claimed they were "just testing" when they were discovered. Meanwhile, these employees opened their enterprises' wireless networks to anyone (friend, foe, hacker, or spy) who cared to enter.

So how do you find these people who would expose your wireless network? Oddly enough, the easiest way is for your company to start using wireless Internetworking—in an organized fashion. That eliminates the need for employees to buy their own access points, and it gives the IT department the tools it needs to detect and eliminate them [8].

10. SUMMARY AND CONCLUSIONS

Attacks on information technology are unsettling and easy to carry out. The means are relatively inexpensive, easy to smuggle, virtually untraceable, and completely deniable. This, coupled with the fact that the civilian wireless networks, which are most attractive to terrorists, are also the most vulnerable, makes infowar the perfect weapon in the terrorist arsenal of the future.

Currently, the security solutions lag far behind the potential threat. This situation is likely to continue until the threat becomes reality, forcing a reassessment of preventive measures. The basic concepts and principles that must be understood and can help realistically guide the process of moving forward in dealing with the surveillance tools for the information warfare of the future are as follows.

- Fortunately, the U.S. military senior leadership is becoming involved in IW, and, in many cases, taking the lead on this perplexing issue. With this emphasis, they must carefully assess the vulnerabilities of the systems they employ.

- Systems proposals must be thoroughly evaluated and prioritized by highest value payoff. This needs to be accomplished through a more balanced investment strategy by the U.S. military that conquers our institutional prejudices that favor killer systems weapons.
- Offensive systems will be at risk if the U.S. military does not apply sufficient defensive considerations in this process.
- The electromagnetic spectrum will be their `Achilles heel' if the U.S. military does not pay sufficient attention to protecting their use of the spectrum and, at the same time, recognize that they must take away the enemy's ability to see themselves and to control his or her forces.
- Interdict opportunities exist for adversaries to intrude on U.S. military systems.
- Other nations have realized the value of offensive applications of the information warfare arsenal of the future; therefore, the U.S. military must attack. The issue from two directions, offensively and defensively, with almost equal accentuation.
- The information warfare arsenal of the future adds a fourth dimension of warfare to those of air, land, and sea. When the Soviets developed a nuclear program after World War II, the United States was caught by surprise. In this new dimension, the U.S. military must stay ahead.
- As with so many other design issues, taking the user's experience into account suggests how to proceed with implementing cookies.
- If the data contains examples of only a single class, extra work may be involved as some popular types of data-mining methods may not be appropriate.
- Although automated understanding of natural language is not available, an increasing number of techniques can be used for exploiting text data.
- An important feature of bandwidth/packet management technology is its stealthy wireless internet security features that render complete invisibility and protection for end users from wireless network hackers and other wireless users sharing the same access points [8].

11. REFERENCES

[1] John R. Vacca, Satellite Encryption, Academic Press, 1999.

[2] John R. Vacca, Net Privacy: A Guide to Developing & Implementing an Ironclad ebusiness Privacy Plan, McGraw-Hill, 2001.

[3] John R. Vacca, Wireless Broadband Networks Handbook, McGraw-Hill, 2001.

[4] John R. Vacca, i-mode Crash Course, McGraw-Hill, 2002.

[5] John R. Vacca, The Essential Guide to Storage Area Networks, Prentice Hall, 2002.

[6] John R. Vacca, The Cabling Handbook (2nd Edition), Prentice Hall, 2001.

[7] John R. Vacca, Wireless Data Demystified, McGraw-Hill, 2003.

[8] John R. Vacca, *Computer Forensics: Computer Crime Scene Investigation, 2nd Edition*, Charles River Media, 2005.

Chapter 28

CIVILIAN CASUALTIES: THE VICTIMS AND REFUGEES OF INFORMATION WARFARE WIRELESS NETWORK SECURITY

1. INTRODUCTION

National information infrastructures are becoming an important vehicle for the generation of national wealth throughout the developed world. National information systems are, among other things, already used to conduct commerce and regulate and control national production. Nations are becoming increasingly dependent on their information infrastructure as the information age evolves. Accordingly, this infrastructure may now be considered to be representing an extension of national sovereignty and any attacks on national information systems may be perceived as attacks on the nation itself. This has resulted in civilian causalities (the victims and refugees of information warfare).

A further argument may be that the national security implications of information attacks make the defense against such attacks a military task. If this is the case, the vast majority of the world's military forces require a significant review of their current doctrine and capabilities. Most are presently not capable of operating in a hostile information environment. An alternative argument may be that attacks against national information systems are criminal in nature and are, therefore, the responsibility of national police forces. Again, most police forces are incapable of defending against such attacks. Indeed, there is probably no organization in the world that can adequately defend national information infrastructures as of yet. Regardless of the capabilities of the various organizations, a clear observation is that the jurisdiction boundaries that separate civilian and military security responsibilities are blurring as the information age evolves.

Separating military and civilian information operations, particularly as they pertain to defending national information systems, is also complicated by the military dependence on the civilian information infrastructure. Significant elements of many of the information systems used by the world's modern military forces are designed, developed, and managed by civilians, primarily for civilian purposes, and make extensive use of the civilian information infrastructure. This is particularly the case with communication systems. The use of unique systems by the military forces for all of their information tasks is not economically viable. Therefore, an attack that targets a military capability via a multiuser information system may inadvertently disrupt civilian users. Likewise, an attack that is directed at a civilian user of an information system may inadvertently affect military users. Is a military or civilian response more appropriate in each of these cases? To many this may appear to be a trivial issue, but distinguishing between civilian and military information operations is important if an appropriate (and legal) national response is to be determined.

The identification of the source of an information attack can be difficult, at times impossible, and can contribute to the problem of determining an appropriate response. Following a skilled information attack, identifying whether the act was calculated and hostile, or simply an accident or a system error may well be impossible. Determining whether a nation, or an individual committed the IW attack or non-nation-state organization may also be impossible, as may be ascertaining the extent of any damage caused to civilians or refugees. Given the embryonic state of international law pertaining to the information domain, pursuing a response through the courts may also be impossible and/or pointless. Therefore, although distinguishing between a military information operation (MIO) and a civilian information operation (CIO) is highly desirable, and, from a legal viewpoint, it may be essential, such distinction is often impossible.

Attempting to resolve tomorrow's information security challenges with today's security infrastructure and culture is unlikely to prove successful. Securing a national information infrastructure presents unique challenges to national security agencies and demands unique and innovative solutions.

The need for macro-level information security in the information domain is becoming more obvious. There is a strong argument for the development of a national information authority that has the responsibility for assuring the integrity of all national information systems, advising on the development of new information systems, sponsoring research and development into information-assurance technologies, and ultimately prosecuting information operations in support of diplomatic, counter-criminal, and conflict-resolution objectives.

A national information authority would offer many strategic opportunities and benefits (including significant efficiencies) and could comfortably address information issues across portfolios, including national security and defense considerations. Such an organization would not deny the individual elements of a nation's armed forces the right to develop their own information strategies; indeed, all arms have both a single-service and joint responsibility to develop robust information strategies now. The further a nation travels down the information age path, however, the more necessary the development of a professional, specialist national information body appears. It is an option that should be considered by any government with a genuine commitment to national security and the protection of its civilians.

2. WHAT THE CYBER MASSES HAVE TO LOSE

As previously explained in the preceding chapters, information warfare (IW) is the latest development in a long list of revolutions in military affairs based on new technology (other examples include the introduction of airplanes, the atom bomb, and long-range missiles). IW is defined as an attack on information systems for military advantage using tactics of destruction, denial, exploitation, and/or deception. The information cycle is vulnerable to these tactics at each step from information gathering to data entry to data transmission to information processing to information dissemination. Current research is searching for robust solutions at each step in the information cycle but the problem is systemic in that for every new solution, a new threat is developed in response to what the cyber masses have to lose.

The rise of IW is linked to widespread diffusion of information technology. The most important enabling feature of the diffusion of information technology is declining cost. Since the 1950s, costs have declined 94% every five years and most experts expect this trend to continue.

The IW threat will continue to grow at the expense of the cyber masses because entry costs are low and decreasing, leading a large number of foreign governments to already organize strategic IW organizations within their military. A second feature of IW that affects IW is that as the technology becomes less expensive, it becomes more efficient to decentralize away from a hierarchical command structure such as is traditional to military tradition.

Information systems are so critical to military operations that it is often more effective to attack an opponent's information systems than to concentrate on destroying its military forces directly. There is a perception within military circles that control of information may become more important than air superiority in previous wars. This has lead to a reevaluation of military doctrine referred to as a revolution in military affairs (RMA).

A revolution in military affairs is a major change in the nature of warfare brought about by the innovative new application of new technologies which, when combined with dramatic changes in military doctrine and operational and organizational concepts, fundamentally alters the character and conduct of military operations.

The United States is potentially vulnerable to IW attack because it is more dependent on information systems than any other country in the world. For example, in the United States, 99% of all military communications is carried over civilian infrastructure, thus intermingling military/civilian targets. In a civilian context, the quality of life of our most basic needs is dependent on automated information- management systems.

The U.S. Department of Defense (DoD) has budgeted billions of dollars to IW and all the military services have formed distinct IW organizations, which are drafting IW military strategies. In January 1997, the Defense Science Board within the Pentagon released a task force report warning of U.S. vulnerability to a "Electronic Pearl Harbor," which puts the cyber masses at great risk with a lot to lose.

The Defense Advanced Research Projects Agency (DARPA) is funding millions of dollars in research to develop an "electronic immune system" that will provide some level of protection to the cyber masses against IW attacks. The Pentagon already spends $8 billion a year to protect its information military systems.

A Presidential Commission was formed by Executive Order 13010 to issue recommendations on how to best protect the cyber masses from IW. Eight critical national infrastructures were considered so vital that their incapacity or destruction would have a debilitating effect on the defense and economic security of the United States. These eight critical national infrastructures as listed by the executive order are:

1. Electric power system (see sidebar, "Electric Power System Vulnerabilities")
2. Gas and oil storage and transportation
3. Telecommunications
4. Banking and finance
5. Transportation

6. Water-supply systems
7. Emergency services (including medical, police, fire, and rescue)
8. Continuity of government services (including federal, state, and local government services) [8]

Electric Power System Vulnerabilities

Nationwide rolling blackouts could have a devastating impact on the economy, but experts also fear that the stress being placed on the nation's power grid could make it more susceptible to disruptions from hackers. In California's Silicon Valley, large Internet data centers have been blamed for stressing the region's power grid beyond what its Korean War-era design can handle. Now, other states, including Oregon, Utah, and Washington, are preparing for possible rolling blackouts.

From a cybersecurity perspective, the electric power grids in the West are now more fragile, and margins for error are significantly less. With diminishing margins and power reserves, the probability for cascading catastrophic effects is higher.

The recent power shortages come as the Critical Infrastructure Assurance Office (CIAO) of the U.S. Department of Commerce delivered to Congress the first status report on private-sector efforts to bolster cyberdefenses for systems that run critical sectors of the economy. Although progress has been made in improving information sharing, officials acknowledged that they still know very little about how failures in one sector could affect other sectors.

In the context of broader infrastructure assurance, the scale and complexities of the energy infrastructure and their impact on infrastructure security and reliability are not fully understood. The energy industry continues to be the target of Internet-based probes and hacker attacks that seek to exploit known vulnerabilities in off-the-shelf software and systems that are increasingly being used to control and manage the power grid.

Likewise, the sector continues to fall victim to poor personnel security practices, ports, and services that are open to the Internet; outdated software without current security patches;

and improperly configured systems. With the system itself teetering on the brink of collapse, it becomes easier for a smaller incident to have a wider impact. For instance, if someone were to find a way to force the shutdown of a single power plant or a section of the power grid, the results would be much more devastating, because there is not enough reserve capacity to take up the slack.

In addition to the technical risks, the publicity generated by the recent crisis in California, and the possibility that hackers may try to exploit known vulnerabilities, there exists the possibility of making a bad situation worse. One risk with a situation like this is that it exposes the flaws of the system to public scrutiny. It shows everyone how vulnerable the cyber masses economy is to a power disruption. Like it or not, there are people in the world who pay attention to such revelations.

Anytime the visibility of a system is raised, it acts as an attack magnet. It is recommended that companies, particularly utility companies, treat the power crisis as a signal to begin stepping up wireless network monitoring and security operations. The link between the stress level on the power grid and its vulnerabilities act "like blood in water to a shark." Hackers smell weakness and a chance for their 15 minutes of fame.

But electric companies have made significant progress in stepping up their security preparedness and have also set up information sharing and analysis centers to enable system administrators to share information with the FBI's National Infrastructure Protection Center. When a transmission system is stressed, the system operators and security coordinators are operating at a heightened level of alert so they can quickly address and return the transmission system to normal from any situation that may occur. The electric system can withstand sudden disturbances such as electric short circuits or unanticipated loss of system elements. This was the case decades ago, and it is still true today [8].

The U.S. government should face the ethical consequences of the new global battleground now before a crisis arises by having a declarative policy concerning IW attacks. During the Cold War, the United States used a policy of strategic nuclear deterrence, warning that any attack on the United States could expect total destruction in return. It is commonly believed that this policy of deterrence was successful but is impossible to prove. By analogy, analysts have wondered if a similar strategy might deter IW attacks on the U.S. National Information Infrastructure (NII). For a strategy of deterrence to work the following must hold:

- The incident must be well defined.
- The identity of the perpetrator must be unambiguous.
- The will and ability to carry out a deterrence strike must be believed.
- The perpetrator must have something of value at stake.
- The deterrence strike must be controllable [8].

This strategy of deterrence must be measured in the context of the inherent vulnerability of large technologically based systems. In what has been called the "complex-system issue," there are axioms:

- Complex systems fail in unpredictable ways from causes that seem to be minor and, often, obvious flaws in retrospect.
- The failure of a complex system may be exceptionally difficult to discover and repair.
- Complex systems fail at inopportune moments—usually during demanding system use when the consequences of failure are highest [8].

It must be possible to determine if an "event" involving one of the United States's' vital infrastructures is the result of an accident, criminal attack, isolated terrorist incident, or an act of war. The damage to the cyber masses from an event may be the same regardless of the cause, but the cause of the event will determine the jurisdiction and nature of the response from the U.S. Government. Possible jurisdictions include private industry, the FBI/Department of Justice, CIA/NSA, or DoD; possible responses range from doing nothing to a nuclear retaliatory strike [8].

2.1 Ethical Challenges of IW to Prevent Cyber Masses Losses

This part of the chapter analyzes the most significant ethical questions of IW as a new form of warfare. Many of the questions have been raised before in previous contexts, but the unique characteristics of IW bring urgency to the search for new relevant answers.

It should be noted that this analysis is also pertinent to other military situations generally referred to as operations other than war (OOTW), such as peacekeeping missions, preludes to conflict, alternatives to conflict, sanctions, and blockades. For example, in an IW analogy to the U.S. blockade of Cuba during the Cuban missile crisis, there are IW techniques (jamming and denial of service attacks) that could be used to block and, thus, isolate rogue nations from international communications without circumventing physical sovereignty—much in the same way the British decided to sever all transatlantic telegraph cables that linked Germany to international communications at the outset of World War I [8].

2.2 What Constitutes an Act of War in the Information Age?

The nation-state combines the intangible idea of a people (nation) with the tangible construct of a political and economic entity (state). A state under international law possesses sovereignty, which means that the state is the final arbiter of order within its physical geographical borders. Implicit to this construct is that a state is able to define and defend its physical geography. Internally, a state uses dominant force to compel obedience to laws, and externally, a state interacts with other states, interaction in either friendly cooperation, competition, or to deter and defeat threats.

At the core view of any nation-state's view of war should be a national information policy that clearly delineates national security thresholds over which another nation-state must not cross. This national information policy must also include options that consider individuals or other non-state actors who might try to provoke international conflicts.

Increasingly, the traditional attributes of the nation-state are blurring as a result of information technology. With IW, the state does not have a monopoly on dominant force, nor can even the most powerful state reliably deter and defeat IW attacks. Non-state actors are attacking across geographic boundaries, eroding the concept of sovereignty based on physical geography. With the advent of the information age, the United States has lost the sanctuary that it has enjoyed for over 200 years. In the past, U.S. citizens and businesses could be protected by government control of our air, land, and sea geographical borders, but now, an IW attack may be launched directly through (or around) these traditional geographical physical defenses.

War contemplates armed conflict between nation-states. Historically, war has been a legal status that could be specified by declaration and/or occur by way of an attack accompanied by an intention to make war. The modern view of war provides a new look at just war tradition, "jus ad bellum," (when it is right to resort to armed force) and "jus in bello," (what is right to do when using force). The six requirements of "jus ad bellum" were developed by Thomas Aquinas in the 13th century:

1. The resort to force must have a just cause.
2. It must be authorized by a competent authority.
3. It is expected to produce a preponderance of good over evil.

4. It must have a reasonable chance of success.
5. It must be a last resort.
6. The expected outcome must be peace [8].

There are two requirements for "jus in bello": The use of force must be discriminate (it must distinguish the guilty from the innocent), and the use of force must be proportional (it must distinguish necessary force from gratuitous force). The application of just war reasoning to future IW conflicts is problematic, but there is a growing voice that there is a place for the use of force under national authority in response to broader national security threats to the values and structures that define the international order.

Looking at one aspect of the application of just war reasoning to IW is the problem of proportionality. It is impossible to respond to every IW action, because there are too many. At what threshold in the lives of the cyber masses and their money, should the United States consider an IW attack an act of war? How many cyber masses live for a certain IW attack or what is the threshold in monetary terms or physical destruction.

Article 51 in the United Nations Charter encourages settlement of international disputes by peaceful means. However, nothing in the Charter impairs the inherent right of individual or collective self-defense if an armed attack occurs.

> **Note:** Infringement of sovereign geographical boundaries by itself is not considered an "armed attack." Experts do not equate "use of force" with an "armed attack.". Thus, certain kinds of data manipulation as a result of IW that are consistent with "use of force" would not constitute an "armed attack" under Article 51.

On the other hand, Article 41 of the United Nations specifically states measures that are not considered to be an "armed attack": complete or partial interruption of economic relations and of rail, sea, air, postal, telegraphic, radio, and other means of communications. IW might still be considered an Act of War, however, if fatalities are involved.

If data manipulation is such that the primary effects are indistinguishable from conventional kinetic weapons, then IW may be considered an "armed attack." The paradigm shift is that weapons are devices designed to kill, injure, or disable people or to damage and destroy property, and have not traditionally included electronic warfare devices.

So, what are the ethical implications of the blurring distinction between acts of war from acts of espionage from acts of terrorism? Let's take a look [8].

2.2.1 Ethical Implications

It is important to be precise in what the cyber masses identify as a crime and what they identify as an act of war. An "armed attack" as stated in Article 51 contemplates a traditional military attack using conventional weapons and does not include propaganda, information gathering, or economic sanctions. Espionage is a violation of domestic and not international law.

The threat analysis section of the 1997 Defense Science Board Report indicates that a significant threat includes activities engaged on behalf of competitor states. This introduces the new concept of low-intensity conflict in the form of economic espionage between corporations. In the age of multinational corporations that view geographical boundaries and political nation-states as historical inconveniences, should economic warfare between multinational corporations involve the military?

The new IW technologies make it difficult to distinguish between espionage and war. If espionage is conducted by computer to probe a nation's databanks and military control systems, when is it an act of war versus an act of espionage? Does it depend on whether the intelligence was passively read versus information actively destroyed in battle and/or manipulated? Does it depend on whether the intelligence was used for military advantage or for political or criminal advantage? Does the answer depend on whether a state of war exists?

A different scenario is modifying internal computer software (via viruses, Trojan horse, or logic bomb) or hardware (chipping) before shipment to cause an enemy's computer to behave in a manner other than they would expect. If during peacetime, gaining entry to a computer's internal operating system could be considered a criminal offense or act of espionage, despite the fact that the action in question took place before the enemy had acquired ownership of the computer. Is this prudent preparation for IW or is this a hostile action that could precipitate a war? If the computer hardware "chip" is commercially manufactured and altered, what are the legal and ethical implications for a company inserting internal hardware hooks as specified in cooperation with national security at the "request" of the government—especially if the company has international sales? Finally, is IW a potential step that might lead to an escalated conventional military conflict that could have been avoided by other means [8]?

2.2.2 Can IW Be Considered Nonlethal?

Nonlethal weapons are defined as weapons whose intent is to nonlethally overwhelm an enemy's lethal force by destroying the aggressive capability of his or her weapons and temporarily neutralizing their soldiers. Nonlethal most often refers to immediate casualty counts, not on downstream collateral effects.

In response to the power of cyber masses opinion and instant global media coverage, the U.S. military has begun to develop a new kind of weaponry designed to minimize bloodshed by accomplishing objectives with the minimum use of lethality. This weaponry includes sticky foam cannons, sonic cannons, and electromagnetic weapons— which effectively temporarily paralyze the enemy without killing them.

Is it more ethical to use a sophisticated smart bomb precisely targeted to kill 30–40 soldiers immediately or is it more ethical to choose a nonlethal weapon that has the same tactical effect with no immediate casualty count, but an indirect collateral effect of 600–700 cyber mass deaths? Ethically, the function of the target against which the weapon is used and the existence or lack of a state of war determines one ethical framework for analysis. For instance, disabling the electronics of a fighter plane or air defense radar during wartime is the goal of a large investment in electronic warfare equipment by the United States, and is considered fair and ethical. However, disabling the electronics of a civilian airliner or air traffic control, during either peacetime or wartime, violates the principles of discrimination of combatants and proportionality of response, and is considered unethical and a act against the cyber masses [8].

2.2.3 Is It Ethical to Set Expectations for a "Bloodless War" Based on IW?

As nonlethal weaponry of all types (especially IW weapons) advance from novelty to norm, however, many potential pitfalls will need to be faced. The most important of these is the expectation that such weapons will ultimately allow wars to be fought without casualties.

Nonlethal military capabilities are not new, although IW weapons are the newest weapons in the nonlethal arsenal. Military forces have used riot-control chemical agents, defoliants, rubber bullets, and electric stun weapons for decades. As U.S. military forces are involved in missions that require extended direct contact with civilians (Somalia, Bosnia), force can no longer be viewed as either on or off, but rather as a continuum with nonlethal weapons on one end and nuclear devices on the other end. In more traditional conventional warfare, IW attacks to disrupt, deny, and destroy

C4I capabilities (command, control, communication, and computer intelligence) are a core part of military tactics and strategy. If IW weapons can be used to remotely blind an opponent to incoming aircraft, disrupt logistics support, and destroy or exploit an adversary's communications, then many of the problems associated with the use of ground forces for these missions can be avoided.

It is important to point out that although nonlethal weapons are not meant to be fatal, they can still kill if used improperly or against people particularly susceptible to their effects. Because these technologies are potentially lethal in these circumstances, the term "nonlethal" has not been universally accepted within the U.S. military. For example, the U.S. Marines Corps uses the term "less lethal" to imply that there is no guarantee of nonlethality.

Asserting that IW will ultimately allow future wars to be fought without casualties is a widespread misconception likely to prove counterproductive and even potentially dangerous to the cyber masses. First, all nonlethal weapons are not equally applicable to all military missions. Second, overselling of nonlethal capabilities without providing a context can lead to operational failures, deaths, and policy failure. Third, unrealistic expectations about nonlethal weapon capabilities inhibit their adoption by military forces who need to build confidence in these weapons.

There is a large asymmetry in global military power when comparing the United States to other nation-states. In 1994, the U.S. DoD budget exceeded that of Russia, China, Japan, France, and Great Britain combined. This asymmetry makes it unlikely another nation-state would challenge the United States in a direct high-technology conventional war, except in circumstances that cyber masses should not depend on (incredible miscalculations and/or ignorant dictators, which were both present in the Gulf War.

Despite the luxury of a bumbling opponent, the success of the Gulf War has lead the U.S. citizenry to expectations of low casualties in all future conflicts. These expectations go against two cardinal rules of military strategy: (1) you do not plan to refight the last war and (2) the future battlefields cannot not be dictated by the United States. The next battlefield for which the U.S. DoD is preparing is a global battlefield with weapons of information warfare targeting the civilian infrastructure. Even in this scenario, military and civilian casualties will be likely from either primary or secondary effects from IW attacks [8].

2.2.4 Is It Ethically Correct to Respond to IW Tactics with IW Tactics?

If the United States is attacked by IW weapons, how should the U.S. government respond? By changing perspectives from defense to offense, what is in the U.S. arsenal to wage IW against an adversary:

- Offensive software (viruses, worms, trojan horses)
- Sniffing or "wiretapping" software (enabling the capture of an adversary's communications)
- Chipping (malicious software embedded in systems by manufacturer)
- Directed energy weapons (designed to destroy electronics & not humans/buildings)
- Psychological operations (sophisticated and covert propaganda techniques) [8]

A strategy that uses these weapons in various combinations has the potential to replace conventional military forces. The questions remains: Is it ethically correct for the United States to defend its security interests by resorting to the same IW tactics that are being used against it? Should information attacks be punished by information counterattacks? The options include maintaining the United States's superpower status at all costs; covertly listening to their adversaries, but not actively disrupting operations; or contracting mercenaries, who are not officially affiliated with the U.S. government, to do their dirty work.

Cracking computers to deter and punish computer cracking erodes any moral basis the United States has for declaring the evils of IW warfare. It is also harder to predict secondary effects due to the globalization of systems. Retaliation may produce effects ranging from nothing to being counterproductive through destruction of U.S. interests. A nation-state or nonstate actor that sponsors an attack on the United States. NII might lack an NII of their own for the United States to attack in punishment, and, thus, not be intimidated by a U.S. IW deterrence strategy.

Short of an official declaration of war, nation-states may seek UN Security Council action authorizing "all necessary means" even in the absence of an "armed attack" in cases of any threat to peace, breach of peace, or act of aggression. Every breach of international law creates a duty to pay for loss or damages; nation-states may seek recompense under "state responsibility doctrine." In additional to recompense, retribution in the form of proportional countermeasures are authorized when an IW attack that does not involve the use of force violates international law. IW may violate multiple international laws depending on the scenario including the following:

- UN Convention on Law of the Sea (prohibits unauthorized broadcasts from the high seas)
- International Telecommunications Convention of 1982 (requires nations to avoid "harmful interference")
- INTELSAT Convention (satellite communications for nonmilitary purposes) [1]
- INMARSAT (maritime satellite communications for "only peaceful purposes")
- Chicago Convention (refrain from endangering safety of flight) [8]

According to DoD policy Directive 5100.77, U.S. military forces are bound by law to follow the rules of engagement of the specific conflict as follows: "The Armed Forces of the United States shall comply with the law of war in the conduct of military operations and related activities, however such conflicts are characterized." The problem is that there are no characterized rules of engagement for IW conflicts, which can take the form of isolated operations, acts of retribution, or undeclared wars.

The most serious problem for using IW retaliation to counter IW attacks is that adversaries could counter and/or copy IW capabilities. Every breakthrough in offensive technology eventually inspires a matching advance in defensive technology, thus escalating an IW weapons race.

A last issue related to retaliation is the ethical dilemma faced by the intermingling of the military and civilian sides of society. Given the uncertainty of deterrence and identifying the enemy, the strategy that is the most ethical for retaliation is a strategy that attempts to separate the military from civilians and, in so doing, having a diminished impact, which potentially prolongs the duration of the conflict; or a strategy that attempts to minimize lethality and duration, but deliberately targets civilian systems [8]?

2.2.5 Can Protection from IW Take Place in the United States Given Our Democratic Rights?

How much government control of the U.S. NII is permissible in a free society? Most of the IW technology is software—which is easy to replicate, hard to restrict, and dual-use by nature (uses for both civilian or military). In the 1997 Defense Science Board report, it states that the DoD is "confused" about when a court order is required to monitor domestic communications. This raises basic questions about the constitutional and ethical balance between privacy [2] and national security in a new IW context.

A "Big Brother" approach that places all of a nation's telecommunications under a single government jurisdiction is improbable given the diffusion and complexity of technology and the shrinking size of government. Most systems were built to serve commercial users who will vehemently object to unfunded mandates (taxes) and new requirements not driven by business demand (CLIPPER chip encryption and key escrow accounts). Regardless, it is critical to the future security of the United States that the cyber masses find a way to protect their infrastructure from IW attack and have contingency plans for potential IW crises. If the IW attack is detected and the enemy identified, but the United States is unable to react promptly due to bureaucratic inefficiency or indifference from private industry, it may be too late to react at all.

Current political discussion has floated tax incentives and direct subsidies to promote industry cooperation. In a related matter that may provide a precedent, the government has pledged to provide telephone companies with at least $1.2 billion to ensure that FBI officials can access telephone conversations over digital circuits (as opposed to accessing telephone conversations over analog circuits, which is technically much easier) [8].

Now, let's take a look at how much damage and/or destruction cyberattacks actually cause.

3. THE DESTRUCTION OF PERSONAL ASSETS IN INFORMATION WARS

The Mounties always get their man—or, when it comes to hackers, their boy. In 2000, Canadian cops announced the arrest of a Montreal-area 15-year-old for disabling CNN's Web site. His father was also nabbed, on unrelated charges of plotting to assault a business associate. The teen suspect, who was identified only by the hacker handle "Mafiaboy," allegedly bragged about his exploits in on-line chats. He was not what one would call a genius. Mafiaboy was charged with two counts of "mischief to data" and faces two years' detention plus a $786 fine. While awaiting trial, he could not enter any public space that hosts wireless networked computers.

The CNN.com incident was part of a rash of "denial of service" attacks that crippled Yahoo!, eBay, and other Internet titans, leading to a manhunt that stretched throughout the United States, Canada, and Germany. The international dragnet was spurred by damage and/or destruction of personal and corporate assets estimates ranging up to $2.3 billion.

Critics charge that companies and prosecutors regularly inflate such numbers and that the Mafiaboy case is no exception. If you're a law enforcement organization, it makes the crime look more serious. If you're a company, it allows you to get more money from insurance (see sidebar, "Hacker Insurance"). And if you're the press, it makes the story more sensational."

Hacker Insurance

In the increasingly competitive hacker insurance market, American International Group is making an offer it hopes prospective clients won't refuse—a free, comprehensive security assessment. AIG, the largest commercial insurance underwriter in the United States, hopes the free on-site security check—which ordinarily can cost tens of thousands of dollars—will encourage more companies to buy insurance coverage from it. AIG is one of the biggest players in a swarm of underwriters and brokers who are rushing into the hacker insurance market, a sector that the Insurance Information Institute estimates could generate $6.9 billion in annual premiums by 2009.

The insurers' sales efforts are being aided by highly publicized events such as the "Anna Kournikova" worm that tied up mail servers around the world. Insurance industry officials indicate their business is doubling every 7 to 13 months, as worries about hacking increase and more information technology professionals realize their companies' standard insurance policies don't cover risks incurred by their Internet-based businesses.

Cyber masses aren't used to spending money on this. The cost of the insurance application in the past included (for almost everyone) an on-site security assessment that would cost upward of $60,000, regardless of whether you bought the insurance.

To help convince qualified prospects (applicants must be seeking $6 million or more in coverage) to buy insurance, AIG will pay independent security firms Global Integrity and Unisys to do the on-site assessments. The firms will do external probes and "ethical hacking" of a prospect's Web site, as well as perform a three-day, on-site analysis to determine what types of security problems the company faces.

At the end of the assessment, if a prospect decides not to buy AIG's coverage, the company can keep the security report and assessments as AIG's gift. Although AIG's assessment is free, some competitors expressed skepticism. AIG's offer may create a false sense of security among insurance buyers. Security is not a product; it's a process.

What's Covered

Companies interested in hacker insurance can buy coverage either as a package or a la carte. Some policies only pay for risks associated with loss or misuse of intellectual property. Others cover liability for misuse of a company's site by a third party, or damage caused by an outside hacker.

Premiums are generally based on a company's revenue, as well as the type and amount of coverage being sought. Rates vary. A package policy that covers a range of risks, including liability, loss of revenue, errors and omissions, and virus protection, can cost from $10,000 to $54,000 per year (or more) for each million dollars of coverage in the policy.

Given the range of costs and coverage, industry officials warn potential buyers to be wary. Some policies cover only the amount of net income lost due to hacking. A better choice for some companies may be coverage for lost revenue.

Numerous variables can affect premiums. Just as a buyer of auto insurance can choose a high dollar deductible to lower the premium, hacker insurance buyers can choose different waiting periods before coverage begins. For instance, a policy that begins paying for business losses just four hours after a hacker shuts down a site may cost more than a policy that begins paying after 24 hours of downtime. These waiting periods, called "time element deductibles," are variable and depend on the kind of business being covered and the amount of risk a business may face.

Companies can also get substantial discounts on their policies if they have managed service contracts with an insurer-certified security firm. Security assessments are critically important for both insurers and insurance buyers. Hacker insurance is such a new product that there are no reliable actuarial tables to determine rates. Therefore, insurance companies rely heavily on the assessments to help them determine the amount of risk they are taking on with a given company. For the companies seeking insurance, assessments should help them find (and immediately fix) holes in their defense systems.

Stiff Competition

Underwriters competing with AIG (the Chubb Group, Fidelity and Deposit Companies, St. Paul Companies, Lloyd's of London and Wurzler) are rolling out a fleet of new products and alliances to help them gain market share. Chubb recently announced new coverages designed for on-line banks, brokerages, and insurance companies. Wurzler has joined with Hewlett-Packard to market its products to a select group of HP's clients.

Insurance brokers and security firms are teaming up to sell branded products and services. Marsh & McClennan Companies, the world's largest insurance brokerage, is selling insurance provided by AIG, Chubb and Lloyd's. The brokerage relies on internet security systems to do its security assessments. Counterpane Internet Security has allied with brokers Safeonline and Frank Crystal & Co. to provide its clients with special policies underwritten by Lloyd's.

It's a wildly growing market, and its primary underwriters are AIG, Fidelity and Deposit, and Wurzler. Hacker insurance has been a small market because people were waiting for e-commerce to hit. Well, now e-commerce has hit.

Insurers are finding a ready market for their products, because companies with Internet operations are increasingly under attack. A survey done in 2004 by the Federal Bureau of Investigation and the Computer Security Institute, an association of computer security personnel from the private and public sectors, found that from March 2002 to March 2003, 32% of the 1,085 governmental agencies and businesses that responded indicated that they experienced denial-of-service attacks. Viruses are also wreaking havoc. Losses from 2000's "Love Bug" virus were estimated to be as high as $20 billion.

AIG's move to lower the cost of obtaining hacker insurance shows the market is beginning to mature, according to industry experts. And security analysts hope it will encourage more Net companies to get insurance coverage.

Companies need to understand that getting hacked is not only an inconvenience. Anything Internet-facing is a point of vulnerability. Companies can be attacked directly or they can be used to attack someone else. There's real exposure and liability. They need to reduce their risk, and the only way to do that is through proper insurance [8].

Pricing cyberintrusions is pretty much a guessing game. If a burglar steals your television set, you know what its replacement value might be. But what's the value of the time of all the people who had to drop their work and deal with this hacker nonsense? More tangential costs are also often tabulated; the $5.6 billion figure associated with recent attacks, calculated by the Yankee Group, includes the expense of security upgrades, consulting fees, and losses in market capitalization from tumbling stock prices.

The implication is that companies wouldn't have had to spend the money if they never had a problem. That's like saying you don't need to get a lock for your front door unless somebody breaks in [8].

3.1 Fair Punishment

Inflated estimates can skew jail terms. A formula should be devised to calculate the severity of hacks. The question is, how serious a societal harm has been done in the hands of these companies, without any real check or balance? If the FBI is going to punish somebody by sending them to prison, they probably don't want to send them to prison for something that's just a nuisance.

In one famous case, an editor at the computer-security webzine *Phrack* was charged with publishing a document stolen from BellSouth's network. Prosecutors valued the 13-page paper at $80,550, which included the $42,000 cost of the computer it was typed on, $7,000 for the printer, and $7,300 for a "project manager." It was revealed at trial, however, that BellSouth sold a nearly identical document to the public for just $14 per copy [8].

4. SHORT- AND LONG-TERM PERSONAL ECONOMIC IMPACT ON CYBER CITIZENS

Cyberattacks cost U.S. organizations and their cyber citizens $600 million in 2003—more than double the average annual losses for the previous three years. The study, released by the San Francisco-based Computer Security Institute (CSI) and the San Francisco FBI Computer Intrusion Squad, found that 95% of survey respondents detected some form of security breach in 2003.

Based on information from 617 of CSI's member organizations, 75% reported serious security attacks, including theft of proprietary information, financial fraud, system penetration from outsiders, denial-of-service attacks, and sabotage of data or wireless networks. This figure, up from 67% in 2001, didn't include data from common security problems caused by computer viruses, laptop theft, and abuse of Internet access by employees.

According to the report, 79% of respondents confirmed that they sustained financial losses due to security attacks, but only 47% said they were willing and able to quantify these costs. The figures are based on responses from 1,087 computer security practitioners in 617 U.S. corporations, government agencies, financial institutions, medical institutions, and universities.

The $600 million in verifiable losses claimed by respondents was more than twice the average annual total of $473 million reported from 2001 to 2004. Eighty respondents reported $100.1 million in losses from theft of proprietary information and 97 organizations listed $90 million in losses from financial fraud.

CSI indicates a continuing trend in the study—that computer security threats to large corporations and government agencies come from both inside and outside the organization. Whereas media reports often focus on outside computer crackers, 85% of respondents were worried about disgruntled employees. Sixty-five respondents indicated that they suffered $51 million in damages from sabotage of data or wireless networks, compared to a combined total of $65 million for previous years.

For the sixth consecutive year, 63% of respondents identified their Internet connection as a frequent point of attack, compared with 36% who cited internal systems as the target. The short- and long-term personal economic impact on cyber citizens continues to be staggering.

Unauthorized access and security attacks are widespread. The private sector and government organizations must increase their focus on sound security practices, deployment of sophisticated defensive technology, and adequate training and staffing of security managers.

Next, let's take look at why corporations are mobilizing against the threat of a federal Internet privacy-protection law, which violates the privacy of their employees during the onset of information warfare [8].

5. THE VIOLATION OF PRIVACY DURING INFORMATION WARS

Privacy—who could possibly be against it? Not IBM, which has vowed to yank all its ads from Web sites that fail to post a clear privacy policy. Not America Online, which promises never to disclose information about members to "outside companies." And certainly not Microsoft, which in 1999 threw its weight behind a plan that could one day let people skip automatically past sites that don't meet their privacy standards. The biggest collectors of information, it seems, are suddenly in the forefront of the campaign for our right to be let alone.

Privacy protection is good for business. But it may not be not quite that simple. True, millions of Americans are wary of the Internet, and surveys suggest that many are hanging back because of confidentiality concerns.

However, the recent frenzy of corporate initiatives is only partly about building public trust. It's also about fending off legislation. Corporate America is mobilizing against the threat of a broad federal privacy-protection law. In particular, businesses are disturbed by one likely element of such a law: a subject access provision that would allow citizens to find out what companies know about them and how the information is being used.

To comply with such a measure, corporate information systems would have to be retrofitted to serve a purpose for which they weren't designed—a vastly expensive undertaking that worried executives liken to the year 2000 problem. The technological costs, however, could be exceeded by the psychological costs [8].

5.1 Junkbusters

If subject access becomes law, Americans will be stunned to discover how much data large corporations have on them. People are going to be horrified.

So far, the United States has addressed the subject on a case-by-case basis. The confidentiality of video rentals is protected, for example, because a reporter got hold of Robert Bork's rental records during the fight over his failed nomination to the Supreme Court. Otherwise, corporate lobbyists have sold Republican and Democratic leaders alike on their view of the Internet economy as a tender, if vital, young thing needing protection from the regulatory mechanisms of the past.

The market can do the job. In addition, companies are banding together to develop privacy guidelines, hoping to show that they can regulate themselves. That premise, however, is under mounting attack on two fronts, domestic and foreign.

The immediate pressure is coming from Europe. A European Union privacy directive that took effect in October not only includes subject access but also requires that, when soliciting information from people, companies clearly spell out what they intend to do with it. This concept is anathema to many large U.S. companies. Accustomed to collecting data for hazy purposes (a "personalized experience"), businesses reserve the right to discover more specific uses or sell the information later on.

But the most annoying element of the EU directive, as far as U.S. corporations are concerned, is a ban on transborder shipment of data to countries that don't offer "adequate" privacy guarantees. The Sabre Group, a Texas-based airline-reservation network, is fighting in Swedish court for the right to maintain in its global data bank such facts as a passenger's wheelchair use or preference for kosher meals. Prodded by Sabre and other large information-oriented companies, the U.S. government is trying to convince European officials that the argument isn't really over the degree of privacy protection, but over two different "cultural perspectives." The Europeans have gone to ridiculous extremes, creating privacy commissions and "privacy czars" to deal with such trivialities as L. L. Bean's decision to send out a catalog of their home products as opposed to their clothing products." Literally interpreted, the EU directive would bar a traveling American business executive from flying home with the names and phone numbers of European clients in his laptop [8].

5.2 Double Standard

Such fears are overwrought. But the European officials point to deep historical reasons (including Nazism) for their view of privacy as a basic human right. The White House is not in any position to cut deals on that, any more than the British are in a position to cut deals on the U.S.'s First Amendment. But if Washington has to make concessions, U.S. multinationals could find themselves in the ticklish position of explaining why they have granted rights to Europeans that they are trying to withhold from Americans. The self-regulation concept has already suffered a series of embarrassments at home. In 1999, Microsoft was discovered to be collecting data on users who had expressly requested anonymity.

Privacy advocates agree that there are informal and technological fixes for many of the problems. On-line privacy protection has the potential to become a significant industry in itself. But it will grow much faster with legal incentives. In the absence of sanctions, the privacy commissioner of Hong Kong claims that self-regulation amounts to putting Count Dracula in charge of the blood bank.

Oddly enough, the concept of subject access originated in the United States, with the Fair Credit Reporting Act of 1971. Credit companies have been living quite profitably with the rule for over 33 years.

Many of the same companies that have been battling against a federal privacy law, have pressed Congress to enact more stringent copyright and patent laws. They're only against regulation when it's something they don't like [8].

Exposed on-line? On the Web, your personal life is merely marketable data. Learn how to protect your personal information on-line next.

6. THE INDIVIDUAL EXPOSED

On the Internet, goes the saying, nobody knows you're a dog. If that ode to on-line anonymity was once true, the notion seems laughable today. The Internet is now more like an unlocked diary, with millions of consumers divulging marketable details of their personal lives, from where they live to what they eat for dinner. Operators of sites on the World Wide Web collect and sell the information, or use it to lure advertising. Software tracks the sites you visit and the pages that catch your eye. If you were a dog, on-line snoops would soon learn that you're a collie who plays a mean game of Frisbee catch and likes your kibble moist.

No one is immune. On-line databases bulge with facts on millions of Americans. "Spammers" cram your e-mailbox with ads. And continuous loopholes in *Netscape Navigator* and Microsoft *Explorer* are giving Web administrators direct access to the hard disk of any browser user visiting their site.

Businesses recognize the Web's potential as a "shopping mall,"—but because of concern over consumer privacy, many stores in that "shopping mall" were forced out of business. It doesn't help matters that many Web sites don't reveal how they intend to use the information collected or whether it might be shared. In a recent survey by the Boston Consulting Group, more than 77% of on-line users worried more about offering up private facts on-line than they did via phone or mail—so they often refused or gave false information. Mounting user fears prompted the Federal Trade Commission to hold a four-day public workshop recently to determine whether the government should step in. Congress is examining the issue, too; several measures to govern the use and sale of personal data, such as Social Security numbers, are pending.

Facing possible regulation, on-line companies vowed during the workshop to help individuals preserve their anonymity and decide whether to reveal personal details. After all, a pro-consumer stance is good for business. It is estimated that on-line consumer commerce will grow from 2001's $68 million to as much as $101 billion by the year 2005—if privacy is addressed, that is.

Of course, you can avoid keying in anything you consider private. But that would bar you from using quite a few sites, and abstinence is not always foolproof. Using a technology called "cookies," some sites, unbeknownst to you, can pick up the address of the site you most recently visited and the Internet service provider (ISP) or on-line service used, and can log your movements within the site. Even companies that advertise there can also drop cookies on your hard drive without your knowledge; some expire only after the year 2005.

Web users do have a few ways to deal with the cookie problem. Surfing through the *Anonymizer* hides your identity but slows you down to some degree. You can also program most popular Web browsers to accept or reject cookies before they are downloaded to your hard disk. Many shareware programs, which can be tried out before being purchased, can help you manage cookies (you might want to permit cookies from a personalized news product, for example) or cut them out entirely [8].

6.1 You're Everywhere! You're Everywhere!

None of these tools, however, will wipe out details about you that are stored in on-line databases ranging from telephone books such as *Switchboard* and *WorldPages* to commercial reference services such *as Lexis-Nexis, CDB Infotek,* and *Information America.* Résumé banks, professional directories, alumni registries, and news archives can all be harvested, as well.

Resourceful thieves can exploit these on-line caches. The Delaware State Police nabbed a couple recently who had obtained birth certificates and drivers' licenses in others' names (thus enabling them to open bank accounts and get credit cards) using information gleaned from sources that included the Internet. Going on-line made it much easier to get at some of the more personal information.

Eight major reference services announced an agreement at the FTC workshop to prevent the misuse of nonpublic data, such as the name, address, and Social Security number found at the top of a credit report. A law enforcement agency, for example, might see all of the data, but a commercial enterprise might not see the Social Security number.

> **Note:** The Fair Credit Reporting Act restricts dissemination of data in the body of a credit report (such as credit card accounts, car loans, or mortgages), but does not cover the material at the top.

Much of the material in on-line databases is culled from public files such as property tax records and drivers' license rolls. That raises questions about the quality of the data. It's a known fact that databases are notoriously inaccurate. Yet major institutions use such services to judge, say, fitness for jobs and insurance. Privacy advocates say consumers should be told if any personal facts are being sold, and should have the right to dispute errors in the databases. In their proposed privacy guidelines, however, reference services agreed merely to tell people the "nature" of the data held on them.

Privacy advocates also argue that consumers should be able to opt out of junk e-mail, or spam. America Online, the largest on-line service, says that up to 34% of the 19 million e-mail messages its members receive each day are junk, and that spamming is members' No. 1 complaint. Although all major on-line services and ISPs prohibit spamming and use filtering programs to weed it out (several have won injunctions barring spammers from their wireless networks), the filters don't always work. Sleazy marketers often also use fake return addresses that are nearly impossible to track down. The FTC recently vowed to prosecute perpetrators of fraud and deception, soliciting the assistance of the Internet E-Mail Marketing Council.

The FTC recently gave on-line businesses and organizations six months to a year to make good on their promises to protect the privacy of on-line consumers. If that does not happen, the FTC will consider taking stronger steps to enable people to browse and buy in confidently as if they were shopping at the local mall [8].

7. UNCOVERING SECRET IDENTITIES

Stolen identity [7]? It can ruin your credit. And that's just the beginning.

When Dee Helus (named changed to protect her privacy) and her husband went to the bank in December to refinance their home, they thought it would be routine. After all, the couple, who live in Kansas City, Kansas, were refinancing with their existing mortgage lender, and they prided themselves on their credit history. So it was quite a shock when the bank officer turned them down, pointing to their credit report, which listed numerous accounts in arrears.

It turns out that a woman in Illinois had applied for credit 55 times using Helus's name and Social Security number. In all, she had made purchases totaling $120,000, leaving a trail of unpaid debt that Banks is desperately trying to prove is not hers: a $61,000 loan for a mobile home, three car loans, credit card bills, and charges for a cellular phone and other services. The perpetrator torched her credit to the point where even the perpetrator herself was denied.

Helus is a victim of a crime of the 21st century—identity theft. It happens when one individual uses another's personal identification (name, address, Social Security number; date of birth; mother's maiden name) to take over or open new credit cards and bank accounts; apply for car and house loans; lease cars and apartments; and, even take out insurance. The perpetrators don't make the payments, and the victim is left to deal with the damage—calls from collection agencies and creditors; the endless paperwork that results from trying to expunge fraudulent accounts from a credit record; and, the agony of waiting to see if more phony accounts pop up. Meanwhile, the proliferation of black marks on a credit report can be devastating. Victims of identity theft are often unable to get loans; some run into trouble applying for a job. A few have even been arrested after the thief committed a crime in the victim's name.

Many identity thieves use stolen personal information to obtain driver's licenses, birth certificates, and professional licenses, making it easier to get credit. Most victims don't even know how the criminals pulled it off. Data have been stolen from desk drawers in the workplace, mailboxes, job application forms, and the Internet. False identification cloaks a thief in anonymity, and the impostor can often use the alias for a prolonged period of time. Thieves typically have the bills sent to an address that is not the victim's, concealing the scheme for months, even years. Most victims aren't aware that their credit has taken a nose-dive until, like Banks, they apply for credit themselves or receive a call from a bill collector [8].

7.1 Proof Positive

In the 1980s, criminals who wanted free plastic simply made up counterfeit credit cards with the correct number of digits. To thwart them, the industry instituted sophisticated security measures involving holograms and algorithms. Now criminals are taking advantage of what some see as the weakest link in the credit system: personal identity. There is nothing in the system that demands proof that you are the person you say you are. Personal identifiers are now, more than ever, a valuable commodity to criminals.

Although no single agency tracks identity fraud, statistics collected by the GAO point to a growing problem. Trans Union, for example, one of the three major credit bureaus, indicates three-fourths of all consumer inquiries relate to identity fraud. Those inquiries numbered 855,255 in 2000; in 2004 there were 1,299,699. The costs of identity fraud can be very high: The Secret Service indicates losses to victims and institutions in its identity-fraud investigations were $5.2 billion in 2004, up from $1.8 billion in 2000. It's a problem that Congress has to address.

Identity fraud is a relatively new phenomenon, and it's not a crime, except in a handful of states, such as Arizona and California, has recently made it so. Legislation is pending to make it a federal crime.

Most victims call the police. But in states with no statute, some police departments refuse to take a report because the law sees the victim in a case of identity fraud as the party that granted the credit (the bank or the merchant, for example), and not the person impersonated. That's frustrating to the victims because they often need a report to prove they are not the bad guys.

Victims need proof because the attitude they often encounter when dealing with creditors is guilty, guilty, guilty. Every person Doyle Comfort (name changed to protect privacy) talks to has been skeptical, condescending, and hostile, and he is still trying to clean his credit report of 26 bounced checks, written to stores in Arizona in 1997 on a bank account opened fraudulently in his name. That really aggravates Smith who has already been turned down for a mortgage. Victims often have to play detective, coming up with clues, leads, and even the basic evidence that a fraud has been committed. You have to do all the footwork yourself [8].

7.2 Even the Kitchen Sink

Typically, creditors ask an identity-theft victim to fill out an affidavit certifying he or she did not incur the debt. Some require much more: One collection agency told Banks it needed a copy of her driver's license, her Social Security card, her birth certificate, and any lease or mortgage contract for the past five years—all for a $87 cable bill. In the end, Banks neither paid the bill nor sent the copies to prove her innocence, opting to explain the $87 item on her credit report to future creditors. As with many victims of identity theft, sensitive documents were the last thing she wanted to send to a stranger.

The belligerence that victims encounter from some creditors is particularly irksome to those who suspect a creditor's negligence in the first place. Many creditors do not take the proper steps to verify the identity of the credit applicant. Dorothy Haskins (name changed to protect privacy) of

Reston, Virginia, points to a credit-card application that started an impostor on a crime spree in her name: The application was preprinted with the impostor's name and address, but the impostor had crossed off her own name (leaving her address) and written in Haskins name, Social Security number, and occupation. The bank gave the impostor a credit card with a $50,000 credit line, leading Haskins to ask: "Wouldn't a reasonable person say something is fishy here?" Bankers, meanwhile, insist they are on the ball. All the banks have systems that detect fraud. They have to modify them for every new scheme that comes up [8].

7.3 Dialing for Dollars

To make matters worse, two weeks after Haskins notified the bank that the account was fraudulent, the bank sold it to a collection agency, and she and her children started receiving threatening phone calls and letters. It didn't stop there. The card triggered an avalanche of preapproved credit offers to the phony Haskins mailbox: One different address on a credit account was all it had taken for one of the credit bureaus to switch Haskins credit-file address to the impostor's. They came to Haskins like candy. Some credit bureaus won't change a file address until three creditors report a new address, but a criminal on a spree can quickly cross that threshold.

It's not easy getting a credit report back on track. The credit bureau says contact the creditor, the creditor says contact the credit bureau, and the consumer just gets ping-ponged back and forth. The bureaucracy can be maddening: Banks recently received a letter from the credit bureau Experian saying it was reinserting a disputed item. The letter did not say which of the 25 accounts it referred to.

One of the few weapons victims have to protect them is a "fraud alert," which credit bureaus will put in consumer credit files. This notifies anyone who pulls the report that the subject is a victim of fraud and that he or she should be called to verify any credit application. The alert isn't foolproof.

Credit bureaus might want to step up their efforts at finding a solution before more aggrieved consumers turn to the courts. Recently, in Clarksdale, Mississippi, a man won a lawsuit against Trans Union for failing to clean up his credit report. The award: $7.8 million.

Meanwhile, the credit-reporting industry has formed a taskforce to tackle identity theft. Among solutions being considered are taking files of theft victims off-line and sharing fraud alerts among credit bureaus more quickly. Individual creditors are also taking steps to stem their losses and prevent future ones. In San Francisco, Cellular One routinely flags suspect applications and compares details with credit reports. That's how Irene Cole (named changed to protect privacy) in San Francisco found out her identity had been compromised: Recently, a woman applied for service using Irene Cole's identity, but Cellular One's fraud department thought the application looked suspicious and phoned her to check. Moreover, it alerted competitors in the area that they might be the next targets.

Identity theft is a crime that comes back to haunt its victims, and many are taking determined measures to prevent its recurrence. Banks and her husband have taken an unusual vow: When they have children, they will not get them Social Security numbers—even though that means no tax deduction. To her way of thinking, safeguarding her children's identity is far more valuable.

Service providers have largely solved the cloning problem, but the monitoring of private affairs in cyber space is still an issue, and e-commerce has barely been addressed. Cellular security is better, but foes still lurk [8].

8. THE MONITORING OF PRIVATE AFFAIRS IN CYBER SPACE

Not being Prince Charles or Newt Gingrich, most of us give little thought to cell-phone eavesdropping. After all, who cares if someone overhears you telling your husband or wife you're stuck in traffic. Of course, if the conversation is of a sensitive nature, then one of your concerns is (or should be) the security of your phone.

Cellular service providers have a different security problem. Their great concern is service theft, through which criminals succeed in using a cell-phone without paying for it.

In the early days of cellular telephony, service theft mostly meant cloning. People with radio scanners would simply "sniff" the cellular frequency bands, pick up cell-phone identification numbers, and program them into other phones. That problem has been reduced by almost two orders of magnitude through the application of some thoughtful technology. But it has been replaced by other problems: subscription fraud (the same problem that bedevils issuers of credit cards) and the misapplication of service provider subsidies on handsets.

Subscription fraud has several forms: pretending to be another, real person; pretending to be a nonexistent person; and even just being yourself and pretending you intend to pay your bill. Subsidy fraud involves taking a phone whose cost has been heavily subsidized by a cellular carrier and activating it on a different carrier's wireless network.

Solutions to these problems exist. However, the newest and best of them cannot be implemented on old handsets, so the technical situation is not without interest. Some of the solutions, particularly those used to fight subscription fraud, tend by their very nature to inhibit sales (after all, the idea is to eliminate deadbeats), which presents the executives of cellular companies with a dilemma. On the one hand, many of them need the revenue stream from a large number of subscribers to help them pay off the huge investments they made when they bid wildly for spectrum space back in 1995. On the other, they have no desire to be cheated.

As the practice of conducting serious business over the Internet continues to grow, other security issues will arise. In particular, someone conducting business on a cell- phone needs to be confident of the identity of the other instrument's user. The technical solutions to be discussed here, such as RF fingerprinting and authentication, do a good job of guaranteeing that the handset is what it claims to be, but they guarantee nothing about the person using it.

Several approaches are being pursued to user identification. The problem, in fact, is not finding solutions, but getting everyone to agree on which to use. To do banking over a cell-phone, your bank, your cellular service provider, and your phone must agree on the same end-to-end solution. And the phone companies, as an industry, must standardize that solution to drive mass-market end-user accessibility [8].

8.1 Analog Yes, Digital No

When it comes to eavesdropping, the situation is pretty simple. Analog phones are easy to bug; digital are hard. Although it is illegal to sell scanners in the United States today that are capable of receiving the frequency bands used for cellular telephony (824-849 MHz, 869-894 MHz, 1.85-1.91 GHz, and 1.93-1.99 GHz), older units that can receive them are readily available. Moreover, it is hardly rocket science to modify a new, compliant receiver to add the extra bands.

> **Note:** The scanners are inherently capable of receiving at least the lower bands; they have just been rigged to block them.

Lest anyone think that analog cellular telephony is an old, dead technology, as of June 2002, over 40% of the subscribers in the United States still used analog handsets, according to Boston's Yankee Group. And many who have dual-mode phones (capable of analog and digital operation) turn to the analog mode when roaming, especially in rural areas.

The latest figures from the Cellular Telecommunications Industry Association (CTIA), Washington, D.C., indicate merely that digital penetration today exceeds 80%. But the CTIA counts dual-mode handsets as digital, so its number may not be so different from the Yankee Group's. Whatever the precise numbers, the message is clear: Eavesdropping is not of only historical interest.

Digital phones, be they of the time- or code-division multiple-access (TDMA or CDMA) variety, are, unlike analog units, quite foolproof against eavesdropping by ordinary mortals. Would-be listeners-in, for one thing, has to know what system they are trying to tap into, because TDMA and CDMA are utterly different. For TDMA, what can be snatched out of the ether is a digital data stream representing one side of each of three multiplexed conversations. Eavesdroppers need to lock onto the correct time slot to get the conversation they want.

In the case of CDMA, what they wind up with is an even thornier problem—a mishmash of half a dozen conversations, each modulated by a different pseudorandom code, all occupying the same band. So, the signal has to be decoded with the same code, which has been obtained in some mysterious fashion.

Plus, in digital systems, voice is vocoded. The sound is not only digitized, but also compressed. As before, someone interested in decompressing it needs to know the compression algorithm used.

In short, eavesdroppers need to build what amounts to the receiving part of a cellular phone base station to have a chance of "overhearing" a call. Small wonder that none of the system operators or phone manufacturers regards eavesdropping on digital cell-phones as a problem [8].

8.2 Ethereal Signatures

The fight against cloning analog handsets has gone a lot better than efforts to combat eavesdropping. Conceived in innocence, early analog phones were almost comically vulnerable to security attacks. For one thing, the signaling between handset and base station takes place in the clear, so anyone with a suitable RF scanner can simply listen-in and learn the phone numbers (called "mobile identity numbers," or MINs) of handsets in the vicinity and the electronic serial numbers (ESNs) that go with them. To program those numbers into another handset is the work of a minute, and behold, another cloned phone is ready for use.

Once the problem manifested itself, service providers began taking steps to protect themselves. Working with the U.S. Secret Service, they persuaded Congress in 1998 to amend the law pertaining to "fraud and related activity in connection with access devices" (Title 18, Section 1029, of the U.S. Code), so as to make it a federal crime to own a scanning receiver or a cell-phone programmer with intent to defraud. That same law also makes it a crime knowingly, and with intent to defraud, to use a counterfeit phone, to traffic in such phones, or to possess 15 or more of them. The law is serious, specifying maximum prison terms of 10 or 15 years (for first-time offenders), depending on the exact nature of the crime.

The service providers also instituted the use of personal identification numbers (PINs) that a user had to key in before a call could be completed. PINs certainly made it tougher for thieves to use stolen phones. But because the PINs were transmitted in the clear, they were not very effective against cloning.

What did help was a technology pioneered by the military for keeping track of enemy troop movements, namely, RF fingerprinting. The technology involves measuring several (unspecified) parameters associated with RF signals and characterizing them (again, in a proprietary manner) to produce a signature unique to the transmitter being studied. Even nominally identical transmitters, manufactured on the same assembly line to the same specifications, have slight differences, which are sufficient (as Corsair named its product) to tell them apart [8].

8.3 Authentication Secrets

With the advent of digital and more advanced analog phones, an even more effective fraud-fighting technology came into use—authentication. A sort of handshaking process, authentication makes use of secret numbers that are stored in the phone and known to the wireless network, but never passed over the air. Every time a call is made, the wireless network sends the handset a random number, which the handset then combines with its secret number using an algorithm designed for the task. The result is another random number that the handset sends back to the wireless network, which has meanwhile performed the same calculations. If the numbers match, the call is completed; if not, it is not.

The algorithm is designed to avalanche very quickly. If the input numbers are off by even a single bit, the resulting number will not even be close to the right answer. Because a different random number is used for each challenge, an eavesdropper would have a hard time figuring out a phone's secret number. This is not to suggest that sophisticated code crackers could not do it (the experts at the National Security Agency would probably consider it a warm-up exercise), but even high-level criminals rarely have access to the required expertise or equipment.

Criminals, by the way, generally clone cell-phones not for economic reasons, but rather in the pursuit of anonymity. Eighty-three percent of narcotics dealers arrested in 2001 were found to be in possession of cloned phones, according to testimony from the Drug Enforcement Administration, Arlington, Virginia.

Call counting is another technique that can be used instead of (more often, in addition to) authentication. Like authentication, it requires a phone capable of performing its part of the process. With call counting, both the handset and the wireless network track the number of calls made by the handset. Those numbers are compared whenever a call is made. If they do not match or if they disagree by more than a specified amount (generally one call), then the call is not allowed. Obviously, if someone has cloned a phone, then both he or she and the legitimate users will be making calls, so the network will have their combined number, whereas each handset will have only its own.

RF fingerprinting and authentication between them have proven extremely effective. Cloning fraud has dropped about 99% over the past four to five years. It has been replaced, however, by another kind of fraud called "identity theft" (as previously discussed), also known as subscription fraud [8].

8.4 Who Are You?

Criminals, like electrons, tend to take the path of least resistance. Make it really hard to steal what they want one way, and they find a different way to get it. In the case of cell-phones (or, more accurately, cell-phone service), the defenses in place against cloning have motivated criminals to adopt the various techniques used by credit- card thieves, which are all lumped together under the rubric of subscriber fraud.

As with cloning, the industry's first defensive move was to persuade Congress to strengthen the relevant statute (in this case Title 18, Section 1028 of the U.S. Code, "fraud and related activity in connection with identification documents and information"). As the law now stands, it is a federal crime merely to steal someone's identity information with intent to defraud. Previously, the government had to wait until fraud was committed before it could act.

The industry became particularly susceptible to subscriber fraud when it started pursuing new customers through such nontraditional channels as telemarketing and the Internet. Previously, cell-phone service was mostly purchased in face-to-face transactions in company-owned stores, and clerks could do things like check photo IDs to verify a customer's identity. Now companies are finding they will have to get back to the basics if they are to keep subscriber fraud losses at a tolerable level. They are going to have to verify addresses against credit-card databases, for example. But, there are legitimate reasons for discrepancies, because people may have just moved or they may maintain multiple residences. Therefore, methods must be developed for screening out bad risks without turning off legitimate customers.

Technology as such is of limited value in this area. One thing computers are being used to do is keep track of subscriber calling patterns—the numbers they tend to call or receive calls from. If a subscriber is terminated for nonpayment of bills, and if a "new" subscriber shows up with pretty much the same calling pattern, then an alarm can be raised calling attention to the possibility that this may be the same person, and the company can look more closely at him [8].

8.5 Subsidy Loss

A major problem, especially in Latin America, are cell-phones moving sideways through the distribution channels. Cellular handsets are often heavily subsidized by service providers, who supply them to subscribers on condition that the subscribers remain with the company for a specific period, typically a year. But what sometimes happens is that the phones wind up being activated on some other carrier's wireless network.

A distributor, for example, who has purchased a batch of subsidized handsets at a low price from one carrier may find that he or she can sell them at a handsome profit to a dealer who is not affiliated with that carrier. In Latin America, that dealer may not even be in the same country as the distributor. As a result, the carrier loses the money it invested in subsidizing the phone.

As with subscriber fraud, the remedy is mostly a matter of running a tighter ship. But some sort of technological fix will also be developed, which can be described as an authentication kind of approach for the activation process. This technology is foreseen as showing up in some second-generation phones, and it will be part of any third-generation deployment [8].

9. THE NEW ORDER AND STATE MEDICAL ID CARDS

The recent hacking of 9,000 administrative patient files from one of the country's top hospitals underscores the lack of firm, clear, universal standards to ensure the security of on-line medical records. But although officials are crafting regulations governing electronic patient records for the health care industry, some analysts and industry players are skeptical about how effective these specifications will be.

In an attempt to remedy the situation, the U.S. government is finalizing and releasing the security and privacy portions of the Health Insurance Portability and Accountability Act (HIPAA), which will define interface and security standards and policies. Unless it is derailed by the new administration, both the regulatory commissions that accredit hospitals and the federal agencies that receive complaints will enforce the HIPAA privacy regulations [8].

9.1 Bumpy Road Ahead

But the industry has a long way to go. The privacy provisions are a quagmire. A lot of it is onerous and expensive, and a lot of it hard to interpret (see sidebar, "New Medical Privacy Rules").

New Medical Privacy Rules

Before President Clinton left office, he announced a sweeping set of federal rules aimed at protecting the privacy of medical records and other personal health information, establishing the potential for penalties to be imposed on executives at health care businesses that breach the new standards.

The regulations, which were prepared by the U.S. Department of Health and Human Services (HHS), are the final version of proposed rules that were issued in 1999 after Congress failed to pass comprehensive medical privacy legislation as required by the 1996 Health Insurance Portability and Accountability Act (HIPAA).

Oral, paper-based, and electronic communications are all covered by the measures. That casts a wider net than the original proposal, which applied to electronic records and to paper ones that at some point had existed in electronic form.

Under the regulations, health care providers are prohibited from releasing most information about individual patients without getting their consent in advance. But in another change from the proposed rules, HHS indicates doctors and hospitals will be given full discretion in determining what personal health information to include when sending patients' medical records to other providers for treatment purposes.

However, the final rules also tighten the consent requirement, mandating that approval be secured from patients for even routine use and disclosure of health records for purposes such as bill payments. Patients also must be given detailed written information about their privacy rights and any planned use of their personal information.

In addition, HHS is calling on hospitals, health insurers, and health care clearinghouses to establish procedures for protecting the privacy of patients, including the appointment of executives to oversee their internal privacy procedures. And companies are prohibited from accessing health records for employment purposes.

Under the HIPAA, civil fines of $100 per violation can be imposed, up to a total of $25,000 per year. Criminal penalties of up to $250,000 and 10 years in prison could also be targeted at individuals who try to profit from the sale of health information. Most health care companies will be given two years to comply with the regulations.

Nothing is more private than someone's medical or psychiatric records. And, therefore, if the government is to make freedom fully meaningful in the Information Age, when most of the stuff is on some computer somewhere, then the government has to protect the privacy of individual health records. The regulations were made necessary by the great tides of technological and economic change that have swept through the medical profession over the last few years.

HHS estimates that complying with the HIPAA rules will cost the health care industry $40.9 billion. But in the long run, government officials claim, the regulations will help achieve savings of almost $60 billion over the next 10 years, as a result of related rules that eliminate paperwork by issuing standards for electronic communication of health insurance claims.

The government is expected to receive a lot of backlash regarding the inclusion of paper and oral communications in the new rules. Originally, the HIPAA was intended to apply solely to electronic communications. It could be virtually impossible to monitor written and oral messages [8].

One of the problems is that the HIPAA is supposed to offer specifications to cover all privacy implementations, from one-doctor offices to giant health care organizations. It's too strict in many respects and too loose in others to offer adequate regulations across the board [8].

9.2 Lessons to Learn

However, there is a whole range of institutions that must be educated on any guidelines to be implemented, including third-party companies that offer electronic patient-record hosting or storage [4]. But, with start-ups, patients face the risk that companies that store their records on-line will go out of business. A bankrupt company could sell its data to a company with a different privacy policy.

Despite the obstacles, on-line medical records will eventually gain more general acceptance [5]. The biggest resistance is fear. Once fear is behind the patients and the companies that store their records, on-line medical records can really take off.

For privacy experts, the beginning of the 21st century is looking more like 1984. Big Brother is watching and listening, and he won't go away anytime soon [8].

10. BIG BROTHER IS HERE AND IS STAYING

Workplace surveillance was the leading privacy concern in 2004, according to an analysis recently released by the Privacy Foundation, a Denver-based nonprofit group that performs research and educates the public on privacy issues. In 2004, millions of Americans were watched at work, as employers became increasingly concerned about employee productivity and their use of the Internet. Three-fourths of major U.S. companies now perform some type of in-house electronic surveillance, according to the American Management Association, and 32% of all companies surveyed now monitor e-mail.

The Big Brother tactic has led to some people losing their jobs. Dow Chemical fired 68 employees and disciplined 679 others in 2004 for allegedly storing and sending sexual or violent images on the company's computers. Xerox, The New York Times Co., and the CIA were others that fired or disciplined employees because of alleged bad behavior.

Employers may be rightly concerned about security and productivity issues, or legal liability arising from e-mailed sexual banter. But pervasive or spot-check surveillance conducted through keystroke monitoring software, reviewing voice-mail messages, and using mini-video cameras will undoubtedly affect morale and labor law, as well as employee recruitment and retention practices.

In the future, the Privacy Foundation predicts that employers, especially so-called "New Economy" companies, may offer "spy-free" workplaces as a fringe benefit. But only as fringe benefit. Big Brother is here and staying—and, it's only going to get worse. "You ain't seen nothing yet [8]!"

10.1 Big Brother Is Watching and Listening

"How the United States Spies on You," read the afternoon headline in Le Monde, enough, certainly, to jolt Parisians on their commute home. Across Europe recently, politicians and the press were in full cry over a vast Anglo-American electronic surveillance system named "Echelon." The system scans billions of private e-mails, faxes, and telephone conversations each hour, according to a report debated by the European Parliament. Echelon, said Parliament President Nicole Fontaine, is "a violation of the fundamental rights" of European Union citizens.

The most incendiary charge is that Echelon represents economic espionage nonpareil, helping the United States and its English-speaking allies steal trade (and jobs) from non-Anglos. But charges cited are mostly old, well-known cases: In 1994, U.S. intelligence discovered that French companies were offering bribes to Saudi Arabia and Brazil for multibillion-dollar contracts. Washington complained, and U.S. firms got the deals.

U.S. officials insisted last week that American intelligence does not steal trade secrets for U.S. firms. Even if it tried, the National Security Agency, which oversees Echelon, is drowning in data thanks to the global communications revolution. In some ways, people's communications have never been safer from becoming intelligence. And France is certainly not a slouch in the industrial espionage arena.

Although it has the added spice of Internet-age privacy concerns, the Echelon flap revealed anew how Europe and the United States are increasingly at odds over matters from defense cooperation to genetically engineered "frankenfoods." Just a few days earlier, a French intelligence report suggested the NSA helped create Microsoft to eavesdrop around the world. The loser this time may be the United Kingdom, whose special intelligence-sharing accord with Washington looks to some like disloyalty to the EU. George Orwell, after all, was English.

Speaking of Orwell, he would be either shocked or pleased on how close his book 1984 is to reality. Meet John Norseen: He's going to read your mind and inject you with smart thoughts [8].

10.2 BioFusion

Buck Rogers, meet John Norseen. Like the comic-strip hero, a 20th-century man stuck in the 25th century, Norseen feels he's not quite in the right time: His brain-research ideas are simply too futuristic. And he admits his current obsession seems to have been lifted from a Rogers saga. The Lockheed Martin neuroengineer hopes to turn the "electrohypnomentalophone," a mind-reading machine invented by one of Buck's buddies, from science fiction into science fact.

Norseen's interest in the brain stems from a Soviet book he read in the mid-1980s, claiming that research on the mind would revolutionize the military and society at large. The former Navy pilot coined the term "BioFusion" to cover his plans to map and manipulate gray matter, leading (he hopes) to advances in medicine, national security, and entertainment. He does not do the research but sees himself as the integrator of discoveries that will make BioFusion a reality and the ultimate IW weapon for Big Brother.

BioFusion would be able to convert thoughts into computer commands, predicts Norseen, by deciphering the brain's electrical activity. Electromagnetic pulses would trigger the release of the brain's own neurotransmitters to fight off disease, enhance learning, or alter the mind's visual images, creating what Norseen has dubbed "synthetic reality."

The key is finding "brain prints." Think of your hand touching a mirror. It leaves a fingerprint. BioFusion would reveal the fingerprints of the brain by using mathematical models. Just like you can find one person in a million through fingerprints, you can find one thought in a million.

It sounds crazy, but Uncle Sam is listening and watching. The National Aeronautics and Space Administration, the Defense Advanced Research Projects Agency, and the Army's National Ground Intelligence Center have all awarded small basic research contracts to Norseen, who works for Lockheed Martin's Intelligent Systems Division. Norseen is waiting to hear if the second stage of these contracts (portions of them classified) comes through.

Norseen's theories are grounded in current science. Mapping human brain functions is now routine. By viewing a brain scan recorded by a magnetic resonance imaging (MRI) machine, scientists can tell what the person was doing at the time of the recording–say, reading or writing.

Emotions from love to hate can be recognized from the brain's electrical activity [8].

10.2.1 Thought Police

So could the murderous thoughts of a terrorist be compromised: applying neuroscience research to antiterrorism? Norseen has submitted a research-and-development plan to the Pentagon, at its request, to identify a terrorist's mental profile. A miniaturized brain-mapping device inside an airport metal detector would screen passengers' brain patterns against a dictionary of brain prints. Norseen predicts profiling by brain print will be in place by 2009.

A pilot could fly a plane by merely thinking, indicates Norseen. Scientists have already linked mind and machine by implanting electrodes into a paralyzed man's brain; he can control a computer's cursor with his mind. Norseen would like to draw on Russian brain-mimicking software and American brain-mapping breakthroughs to allow that communication to take place in a less invasive way. A modified helmet could record a pilot's brain waves. When you say right 090 degrees, the computer would see that electrical pattern in the brain and turn the plane 090 degrees. If the pilot misheard instructions to turn 090 degrees and was thinking "080 degrees," the helmet would detect the error, then inject the right number via electromagnetic waves.

Finally, if this research pans out, you can begin to manipulate what someone is thinking even before they know it. But Norseen feels he is "agnostic" on the moral ramifications, that he's not a mad scientist–just a dedicated one. The ethics don't concern home, but they should concern someone else [8].

11. SUMMARY AND CONCLUSIONS

This chapter has considered the application of civilian information operations (CIO's) to the conventional warfare environment. Although the array of CIO tools and techniques has been presented as discrete elements in a schematic diagram, the CIO environment is complex, multidimensional, interactive, and still developing. Accordingly, the introduction of a CIO capability into an existing military force requires careful consideration and adherence to a series of principles espoused within this chapter. These principles are defined within a framework of concepts including Information Assurance, Information Superiority and Information Dominance. This framework can be applied to both the introduction of a CIO capability and the application of CIO's in information warfare.

CIO's will change the nature of future wars and will eventually evolve into a separate paradigm of warfare—IW. However, CIOs can be applied to today's conventional environment and it is within this context that more urgent attention from military planners is required. CIO's offer both a support capability to existing arms of the military and also an additional dimension to conventional warfare. They may be used to strike enemy systems, control the overall information environment, deter enemy aggression, or support either themselves or other military strategies. Regardless of which tasks they are employed for, CIO's offer a significant addition to the conventional inventory and should be developed as a matter of priority as an essential Joint Force operational capability in dealing with the civilian casualties of information warfare, as follows.

- Information warfare (IW) is the latest development in a long list of revolutions in military affairs based on new technology (other examples include the introduction of airplanes, the atom bomb, and long-range missiles).
- IW is defined as an attack on information systems for military advantage using tactics of destruction, denial, exploitation, and/or deception.
- Information systems are so critical to military operations that it is often more effective to attack an opponent's information systems than to concentrate on destroying its military forces directly.
- In the civilian context, the U.S. economic, social, and political structures are increasingly dependent on complex and extensive systems for financial transactions, telecommunications, electric power, energy distribution, and transportation. Because these systems rely on each other, a serious disruption in any one system will cascade quickly through the other systems potentially causing a national security crisis. Under these circumstances, the ability of the government to respond will be interrupted and severely constrained.
- In addition to outages caused by natural disasters and accidents, these systems present a tempting target for IW attack to those contemplating an action against U.S. interests.
- IW provides a new context for the application of ethical theories.
- The ethical questions about IW are not meant to be a complete set of ethical questions, but rather a subjective assessment of what are the ethical questions derived from the most important issues exposed by IW.
- New unforeseen ethical questions will inevitably arise from IW in the near future.
- It is hoped that this research will begin a dialog on the issues and lay a framework for more substantive work by ethicists.

- IW and its complement (protection from IW) require important new research efforts from both the technology and ethics research communities.
- The threat of IW raises the following ethical challenges: (1) What constitutes an act of war in the information age? (2) What are the ethical implications of the blurring distinction between acts of war from acts of espionage from acts of terrorism? (3) Can IW be considered nonlethal? (4) Is it ethical to set expectations for a "bloodless war" based on IW? (5) Is it ethically correct to respond to IW tactics with IW tactics? (6) Can protection from IW take place in United States, given our democratic freedoms [8]?

12. REFERENCES

[1] John R. Vacca, *Satellite Encryption*, Academic Press, 1999.
[2] John R. Vacca, *Net Privacy: A Guide to Developing & Implementing an Ironclad ebusiness Privacy Plan*, McGraw-Hill, 2001.
[3] John R. Vacca, *Electronic Commerce, Third Edition*, Charles River Media, 2001
[4] John R. Vacca, *The Essential Guide to Storage Area Networks*, Prentice Hall, 2002.
[5] John R. Vacca, *Public Key Infrastructure: Building Trusted Applications and Web Services,* Auerbach Publications, 2004.
[6] John R. Vacca, *Firewalls : Jumpstart for Network and Systems Administrators,* Prentice Hall, 2005.
[7] John R. Vacca, *Identity Theft*, Prentice Hall, 2002.
[8] John R. Vacca, *Computer Forensics: Computer Crime Scene Investigation, 2nd Edition*, Charles River Media, 2005.

PART VII

WIRELESS NETWORK SECURITY SOLUTIONS AND FUTURE DIRECTIONS

Chapter 29

PROVIDING WIRELESS NETWORK SECURITY SOLUTIONS FOR ISP INTRANET, INTERNET AND E-COMMERCE

1. INTRODUCTION

In recent years, wireless ISP Intranets, Extranets, Internet and E-Commerce, have become more widely deployed in many enterprises. Many of these enterprises are now extending their Intranets to reach key customers and business partners via Extranets. Undoubtedly, Intranets and Extranets offer clear cost savings and ease of installation compared with the expensive leased line networks or WANs based on proprietary technology. In addition, they enable highly productive and cost effective new ways for enterprises to communicate with their customers, and collaborate with their enterprise partners. However, Intranets and Extranets build on the public Internet which uses the TCP/IP protocol as their communication language. The success of the Internet is its open design and it is flexible, powerful and well-suited to communication between a variety of platforms with divergent capabilities. However, its success also causes a security problem for information transmitted over it. TCP/IP was designed to be open and had no security consideration in mind at the beginning. There are a number of serious security flaws inherent in the TCP/IP protocol suite, which can be used to launch the malicious attacks. The new security challenges have emerged and become more and more critical for its enterprise employment.

So, with the preceding in mind, this chapter outlines the new security concerns for an enterprise to deploy Intranets and extranets. The chapter takes an in-depth look at some security flaws of the TCP/IP protocol and current security threats. In order to implement the wireless network security in a enterprise's Intranet and Extranets, this chapter also introduces new security technologies and services. Based on these technologies, wireless security solutions are designed for a wireless security sensitive e-commerce enterprise to implement Intranet and Extranet. Finally, the chapter discusses some tests and simulations to verify the integrity of the wireless network security system.

2. NEW WIRELESS NETWORK SECURITY CONCERNS

Information is the key asset in most enterprises. Enterprises gain a competitive advantage by knowing how to use their information. As Intranets and Extranets have become widely deployed by more and more enterprises, the growth in wireless network complexity has increased the potential risks to enterprise confidential information. The threat comes from others who would like to acquire the information or limit business opportunities by interfering with normal enterprise processes. Wireless information security has become a new challenge and a critical problem to these security-sensitive enterprises.

Intranet and Extranet security breaches can take a variety of forms. For example:

- The enterprise information can be interrupted, intercepted, modified and fabricated when it is transmitted from its internal wireless network to the customers, branch offices, partners and remote employees over the public network like the Internet.
- An unauthorized person might gain access to a enterprise's computer system.
- Users may share documents between geographically separated offices over the Internet or Extranet
- Telecommuters accessing the enterprise Intranet from their home computer can expose sensitive wireless data
- Inside users (employees) authorized to use the system for one purpose might use it for another, etc [1].

Therefore, in this chapter, it is assumed that the insiders in an enterprise are fully trustful. This chapter will only consider the treats from outside and address mainly wireless network security [1].

2.1 Security Goal

Enterprise valuable and sensitive wireless information should be secure while making it readily available. Wireless information security can be defined as a combination of confidentiality, integrity, availability and authentication. Therefore, the goal of any wireless security policy is to ensure that each of the four fundamental components are adequately addressed [1]:

2.1.1 Confidentiality

Confidentiality, sometimes referred to as privacy, is the protection of wireless information from unauthorized disclosure. Most enterprises must provide wireless information to some individuals while blocking access for someone else. Only users who need to view sensitive wireless information are authorized to do so. This can be achieved either by restricting access to the information or by encryption to ensure the intercepted information unreadable [1].

2.1.2 Integrity

Integrity refers to the ability to protect information, data or transmissions from unauthorized, uncontrolled or accidental alterations as it travels between senders and recipients. Data integrity is achieved by preventing unauthorized or improper changes to data, ensuring internal and external consistency. Digital signatures validate a user's identity, so that message recipient can be sure that message senders are who they claim they are. Digital signatures also provide strong evidence that the message has not been altered since it was signed [1].

2.1.3 Availability

Availability refers to the resources on the enterprise wireless network that must always be accessible to people. This applies to everything from servers and services to the information stored in a particular database. An unauthorized user can compromise the system and fill mailboxes with unsolicited commercial e-mail or disable large portions of the Intranet. Limiting the number of entry points to the wireless network and restricting the number of people with access to these resources are two key methods for ensuring that your resources remain available [1].

2.1.4 Authentication

Authentication refers to the process of verifying the claimed identity of an individual, station or originator. It also refers to the determination of whether to grant a user permission to connect to the system.

So, why do you need to protect your wireless information? Where do these wireless security threats come from? Where else? The Internet!

In a local LAN, where information is only limited to use in the internal network, the enterprise information is fully secure if not considering the inside attacks. The birth of the Internet brought endless business opportunities to the e-Commerce enterprises. As these enterprises are migrating to the Internet, the wireless security threats also emerge since IP-based wireless network data is wide open to tampering and eavesdropping. Let's first take a deep look at the vulnerabilities and flaws in the TCP/IP protocol suite -- the language of the Internet [1].

2.2 TCP/IP Vulnerabilities

TCP/IP is the most widely used wireless network communication standard that runs over the Internet. It's a connectionless, best-effort delivery protocol. The transmission data is broken into packets, each packet with a destination address for the routing through the Internet. The sequence number in each packet is used to identify the order of the packets, and the TCP port number provides a mechanism to direct data to a specific application. The packets travel over the Internet through the different (best possible) ways and assemble when they reach the destination. Thus, the packets are easily to be incepted, modified and fabricated when they are transmitted over the public network. There are a number of security flaws in TCP/IP protocol suite. These security flaws incur some attacks like packet sniffing, IP spoofing, sequence number spoofing, routing attacks, source address spoofing, authentication attacks, and Denial of Service, etc. [1]

2.2.1 Packet Sniffing

Packet Sinffing is the action of intercepting and reading wireless network traffic that is being transmitted across a shared wireless network communication channel. This is usually done by using utility software checking for interfaces working in promiscuous mode, or physical checking of all the connections. A common target for packet sniffers is user account and password. Since in a normal wireless networking environment, the user accounts, passwords and data information for most protocols (such as HTTP,

FTP, Telnet, SNMP, POP3, SMTP, IMAP) are sent across the wireless network in plain-text. Once equipped to eavesdrop, it is not difficult for an attacker to capture packets that contain readable user account and password [1].

2.2.1.1 How Does It Work?

Ethernet is the most popular wireless computer network. The Ethernet protocol works by sending packets to all wireless networked computers on the same network segment. The packet header contains the proper address of the destination host. The wireless network interface card (NIC) hardware in a networked computer receives every piece of packets that is transmitted across the wireless network. Usually, the wireless network device driver software will only process incoming packets which contain the address of its host computer, or broadcast packets. However, the wireless network adapter hardware can be configured to operate in an altered state that is to be in promiscuous mode.

The promiscuous mode is the biggest security flaw of the Ethernet. In promiscuous mode, the wireless network device can process all traffic transmitted across the network and forward the packets to its operating system, no matter whether these packets are addressed to itself or not. This option is used as a debugging tool for wireless network administration. However, it could be misused by an attacker to eavesdrop the traffic transmitted over the wireless network. Sniffing software just works by placing a system's wireless network interface into promiscuous mode. After the wireless network traffic is processed by the Network Interface Card, software mechanisms are used to filter the captured packet, extract and reconstruct the data portion of the packets, and to display it in a readable format [1].

2.2.1.2 Defenses

There are several ways to prevent this kind of attacks. They are as follows:

- Wireless Network Segmentation
- Encryption
- Secure Socket Layer (SSL) [1]

2.2.1.2.1 Wireless Network Segmentation

A wireless network segmentation can separate the network part with high-level security from the low-level security network part, and minimizes the amount of information that can be collected with a network sniffer or an analyzer which operate in promiscuous mode. Repeaters and passive hubs used for wireless network segmentation do not provide security because the flow of data arrives to any of its interfaces. Instead, routers, switches and bridges can limit the flow of traffic, and allow the traffic only go to its destination interface [1].

2.2.1.2.2 Encryption

Encryption schemes can be used to prevent the contents from being read even though an attacker capture the packets. E.g. Public key encryption programs PGP uses various forms of encryption and combines messages with a simple packet format to provide a simple and efficient security mechanism for the transmission of electronic mail (E-mail) [1].

2.2.1.2.3 Secure Socket Layer (SSL)

SSL can be built into popular web browser and web servers. It allows encrypted web surfing, and is almost always used in e-commerce when users enter their credit card information [1].

2.2.2 IP Spoofing

The spoofing attack exploits the fact that IP does not perform a robust mechanism for authentication. It believes that a packet comes from where it claims. Since many systems (such as router access control lists) define which packets may and which packets may not pass based on the sender's IP address, this is a useful technique to an attacker, and is also a strong weakness of the TCP/IP protocol. IP addresses can be configured in software, and it is usually easy to configure one address for a machine, as for another. A packet simply claims to originate from a given address, and there is no way to know if it is or not.

To engage in IP spoofing, a hacker first uses a variety of techniques to find an IP address of a trusted port, which is permitted access through the packet-filtering router or firewall [3], then modifies IP address in the packet header on the external wireless network to gain access to the internal network. It is possible to route packets through the packet-filtering router or firewall if they is not configured to filter incoming packets whose source address is in the local domain. It is possible to spoof even if no reply packets can reach the attacker [1].

2.2.2.1 Detection

If monitor packets using wireless network-monitoring software such as netlog is used, find a packet on your external interface that has both its source and destination IP addressed in the internal network. This usually means the internal network is currently under attack.

Another way to detect IP spoofing is to compare the process accounting logs between systems on the internal network. If the IP spoofing attack has succeeded on one of the systems, the log entry on the victim machine shows a remote access; on the apparent source machine, there is no corresponding entry for initiating that remote access. So, with the preceding in mind, the following defenses must be used:

1. Avoid reliance on address-based authentication and trust mechanisms
2. Disabling source routing
3. Using a screening router or firewall which can intelligently filter network packets based on configurable rules. First of all, this can thwart inbound attacks that originate from external wireless networks implemented by configuring the router to discard the incoming datagram with a source address belonging to the internal network. Second, it can also thwart outbound attacks that originate inside of your own wireless networks, implemented by discarding the outgoing datagram with a source address from an external wireless network [1].

2.2.3 TCP Sequence Number Attack

The TCP Sequence Number Attack uses TCP sequence number prediction to construct a TCP packet sequence without ever receiving any responses from the server. This allows an attacker to spoof a trusted host on a wireless local network.

The connection-oriented transport control protocol (TCP) uses a 3-way handshake to establish the connection before transmitting data between the server and client. Telnet is just such an application. When a Telnet session is started, the application layer will request TCP as its transport service in order to insure reliability of the connection.

The client sends a connection request to the server by setting the SYN bit and selecting an initial sequence number ISNc; the server acknowledges the request by setting the ACK bit and its own sequence number ISNs; and, the client acknowledges the acknowledges by setting the ACK bit. After the 3-way handshake, data transmission can take place.

Before data transmission, the client must first receive the sequence number ISNy from the Server. The 3-way handshake protocol can be easily exploited by the attacker. By monitoring a wireless network connection, a hacker can record the exchange of sequence numbers and predict the next sequence number.

For example, the sequence number attack works this way: In this case, the attacker could send the following sequence to impersonate trusted Client (C). Now, even though the message 2 from the Server (S) to the Client(C) didn't go to the Attacker(X), the attacker was able to know the contents, and could send data at the right time.

In some systems, for example, the initial sequence number variable is incremented by a constant amount once per second, and by half that amount each time a connection is initiated. Thus, if the attacker initiates a legitimate connection and observes the ISNs used, he or she can calculate the ISNs used on the next connection attempt. The defenses used in this case are as follows: Randomize the increment of the sequence number attacks, and make it difficult to guess or calculate; and, use a cryptographic algorithm (or device) for ISNs generation, such as DES [1].

2.2.4 Denial-of-Service

Denial of service occurs when a hostile entity uses a critical service of the computer system in such a way that no service or severely degraded service is available to others. The DoS attack is simple: send more requests to the machine than it can handle. DoS causes the datagram to be discarded before final delivery, thus effectively blocking the communication path. Some DoS attacks are: SYN Flooding attack, The Ping of Death (ICMP attack), E-mail Bombing, Spoofing Attack DNS, Windows Nuke (newk)OBB Attack on Port 139, etc. [1]

2.2.4.1 SYN Flooding Attack

SYN attacks take advantage of a flaw in the TCP "three-way handshake." In a normal connection, the client sends a SYN message requesting the connection to the server; the server acknowledges the request with a SYN/ACK message; and then, the client sends the ACK message acknowledging the approval. The connection between the client and the server is then open, and the service-specific data can be exchanged between the client and the server. The problem arises at the point where the server system has sent an acknowledgment (SYN-ACK) back to the client, but has not yet received the ACK message (it is called "half-open connection"). When the server receives the SYN request from the client, it must keep track of the partially opened connection in a "listen queue" for a period of time (typically 75 seconds). The server has built in its system memory a data structure describing all pending connections. This data structure is of finite size, and it can be made to overflow by intentionally creating too many half-open connections.

In a DoS attack, an attacker can send a bundle of SYN requests whose source address are set to a routable, but unreachable (false return address), which can easily be done by IP spoofing. The final ACK messages will never be sent to the victim server. The server's queue is filled up with pending connections. When the queue limit is reached, TCP drops all new incoming requests until time-out and closes the connections. However, the attacker continues sending a new patch of IP-spoofed packets requesting new connections faster than the server can expire the pending connections, and the process begins again. The service is disabled indefinitely. So, with te preceding in mind, the following defenses can be initiated:

- Deploying system operating patches: several vendors have released operating system patches to compensate and react to SYN attacks
- Not running the visible-to-the-world servers at a level too close to its capacity.
- Using packet filtering to guard against attacks by checking the pattern of information or request, prevent obviously forged packets from entering into the wireless network address space.
- Using Proxy Server to protect the main server. A proxy server stands between both the client and the server during the connection. A proxy server acts as the "man in the middle," so that there is no direct contact between a client and the server [1].

Denial of service is very easy to launch, but difficult (sometimes impossible) to track. It is becoming a greater problem for the Internet in recent years. It grows at a rate of about 50% per year greater than the rate of growth of Internet hosts, although the total number of incidents is small [1].

2.2.4.2 Smurf Attack

Not to be confused by the recent Belgium children's video showing smurfs being blown up in a terrorist attack, the smurf attack is another kind of DoS attack. It's named after the source code employed to launch the attack (smurf.c). The smurf attack employs forged ICMP echo request packets and the direction of those packets to IP network broadcast address.

For example, the attacker (128.100.153.9) issues the ICMP ECHO_REQUEST to the broadcast address of the intermediary wireless network (the network mask is 102.100.36.255). The attacker spoofs the source address using the IP address (207.125.64.39) of the system it wishes to target. When the intermediate wireless network receives the packet with the falsified source address, they respond. The targeted victim system then receives flooding echo replies from all systems on the intermediary wireless network. This flood can overwhelm the targeted victim's wireless network. Both the intermediate and victim's wireless networks will see degraded performance, and eventually results as being unavailable for the service [1].

2.2.4.3 The Ping of Death (ICMP Attack)

The Ping of Death attack is also known as a "Ping Flood" attack. It employs a flaw in many vendors' implementations of ICMP. ICMP is part of the IP protocol using the IP datagram to deliver messages. PING is a TCP/IP command that simply sends out an IP packet to a specified IP address or host name to see if there is a response from the address or host. Normally a PING (echo request) packet's size is about 32 to 64 bytes. However, the attacker can send a constant stream of over-sized forged PING packets to the target system. In many cases, this flood of traffic can cause an overflow in system's internals, and result in system crashes. So, with the preceding in mind, the defense here includes blocking ICMP packets on the wireless network firewalls to prevent the traffic from effecting the internal system [1].

2.2.5 IP Session Hijacking

IP Session Hijacking is an action when the attacker takes over the control of the client's session (Telnet). The hijacking attack is usually launched after the user's authentication is complete. An attacker first needs to attack the connection by either closing it or messing up the user's SEQ/ACK. Then the attacker takes over the session without being noticed by the user.

Session hijacking is a higher level attack. This occurs when a TCP link is established between the client and the server after the client authenticated itself. You cannot be sure that it will be the same person for the rest of the session. There are techniques to take over the connection. These tools send a message to the client to cut the connection so that the attacker can communicate via the same TCP link to the server. Next, are some techniques that an attacker usually uses to launch an IP Session Hijacking attack:

- Using reset (RST)
- Using FIN to close a connection
- SEQ/ACK mess up
- Defenses [1]

2.2.5.1 Using Reset (RST)

TCP segments have flags which indicate the status of the packet. RST is one flag that tells the receiving TCP module to abort the connection because of some abnormal condition. To be accepted, only the sequence number has to be correct since there is no ACK in a RST packet. When A and B are in the connection, the attacker H watches the traffic between A and B, and calculates the sequence number for A's next sent packet from B's ACK's packet. When B is waiting for A's response, H launchs a forged RST packet to B as if it comes from A [1].

2.2.5.2 Using FIN To Close A Connection

Another flag in TCP segments is FIN, which tells the receiver that the sender does not have any more data to send. This flag is used when closing a connection in a normal legal way. This works almost the same as the former one. Instead of sending an RST packet, it sends the FIN packet. The attacker H can pretend to be either A or B, and sends a FIN packet to the other host, thus closing the connection between A and B [1].

2.2.5.3 SEQ/ACK Mess Up

Since TCP separates good and forged packets by their SEQ/ACK numbers, B trusts the packets from A because of its correct SEQ/ACK numbers. So, if there is a way to mess up A's SEQ/ACK, B would stop believing A's real packet. The attacker H could then impersonate to be A, but using correct SEQ/ACK numbers to connect with B. To mess up A's SEQ/ACK numbers, the attacker simply insert a data packet with correct SEQ/ACK number for B at the right time. Host B would accept the packet, and update ACK numbers. When A continues to send packets to B, the real packets would be dropped [1].

2.2.5.4 Defenses

The preceding can be solved by using an encryption scheme. In this case, the attacker can still take over the session, but he can't see anything because the session is encrypted. The attacker dosen't have the needed cryptographic key to decrypt the data stream, therefore, be unable to do anything with the hijacked session. In addition, a scheme that authenticates the data's source throughout the transmission is also needed [1].

2.2.6 Source Routing Attack

The biggest security hole in TCP/IP protocol is IP source routing. Briefly, IP source routing is an option that can be used to specify a direct route to a destination and return path back to the origination. The route can involve the use of other routers or hosts that normally would not be used to forward packets to the destination. This means if the originator of the connection wishes to specify a particular path for some reason, replies may not reach the originator if a different path is followed. The attacker can then exploit this flaw and use any IP source address desired, including that of a trusted machine on the target's local network. Any facilities available to such machines become available to the attacker. The following example shows how this can be used, such that an attacker's system could masquerade as the trusted client of a particular server is as follows:

1. The attacker would change his or her host's IP address to match that of the trusted client.
2. The attacker would then construct a source route to the server that specifies the direct path the IP packets should take to the server and should take from the server back to the attacker's host, using the trusted client as the last hop in the route to the server.
3. The attacker sends a client request to the server using the source route.
4. The server accepts the client request as if it came directly from the trusted client and returns a reply to the trusted client.
5. The trusted client, using the source route, forwards the packet on to the attacker's host [1].

2.2.6.1 Defenses

It is rather hard to defend against this sort of attack. The best idea would be for the gateways into the wireless local network to refuse the source routing protocol; and, to also be configured to reject external packets that claim to be from the internal local network [1].

2.2.6.2 Man-in-the-Middle Attack

The "man in the middle" is a rogue program that intercepts all communication between the client and a server with which the client is attempting to communicate via SSL. The rogue program intercepts the legitimate keys that are passed back and forth during the SSL handshake, substitutes its own, and makes it appear to the client that it is the server, and to the server that it is the client.

The encrypted information exchanged at the beginning of the SSL handshake is actually encrypted with the rogue program's public key or private key, rather than the client's or server's real keys. The rogue program ends up establishing one set of session keys for use with the real server, and a different set of session keys for use with the client. This allows the rogue program not only to read all the data that flows between the client and the real server, but also to change the data without being detected [1].

3. NEW TECHNOLOGIES

Though there are a variety of wireless network attacks, there also exist a variety of security technologies available to address these security holes and provide security services over IP-based wireless networks. This part of the chapter will first briefly give an introduction of these security technologies that will be covered in the next part for deriving the solution to the enterprise's Intranet and Extranet. These technologies include IPSec, VPN, SSL (now TLS), Kerberos, Firewall, Certificate, etc. [1]

3.1 Internet Protocol Security (IPSec)

IPSec is a security framework of open standards designed by the Internet Engineering Task Force (IETF) to secure private communications over the IP wireless network. IPSec applies at the IP layer and thus offers protection for IP and all upper layer protocols. Security services provided by IPSec are data authentication, confidentiality, integrity and replay prevention. The greatest advantage of IPSec is that it is completely transparent for the applications [1].

3.1.1 IPSec Overview

IPSec relies on two mechanisms (or protocols), Authentication Header (AH) and Encapsulating Security Payload (ESP). The parameters necessary for the use of these protocols are managed by security associations (SA), an association containing the parameters used to protect a given part of the traffic. SAs are stored in the Security Association Database (SAD) and are managed using the IKE (Internet Key Exchange) protocol. The protection offered by IPSec is based on choices defined in the Security Policy Database (SPD). This database allows you to decide, for each packet, if it will be afforded some security services, will be authorized to pass by or will be rejected.

IPSec has two modes: transport mode, which protects only the transported data, and tunnel mode, which also protects the IP header. IPSec can be used either on a terminal host or on a security gateway, which allows for both link-by-link and end-to-end security. IPSec can thus be used, in particular, for the creation of virtual private networks (VPNs) or for the protection of remote accesses [1].

3.1.1.1 How It Works With Outbound Traffic

When the IPSec "layer" receives data to be sent, it starts by consulting the policy database (SPD) to determine what processing is required for the packet. If the packet must be afforded security services, the IPSec engine recovers the characteristics of the corresponding SA(s) and consults the SA database (SAD). If the necessary SA already exists, it is used to process the traffic in question. If not, IPSec calls IKE to establish a new SA with the necessary characteristics [1].

3.1.1.2 How It Works With Inbound Traffic

When the IPSec "layer" receives a packet from the wireless network, it examines the header to determine if this packet was afforded IPSec protection and if so what are the SA references. It then consults the SAD to determine the parameters to use for checking and/or deciphering the packet. Once the packet is checked and/or deciphered, the SPD is consulted to determine if the required IPSec processing was applied. If the received packet is a "normal" IP packet, the SPD makes it possible to know if it can bypass IPSec [1].

3.1.2 Enterprise Deployments

By placing IPSec-enabled hardware at different points in the wireless network (routers, firewalls, hosts, as BITW "crypto boxes"), different security deployments can be realized. End-to-end security can be achieved by deploying IPSec-enabled stacks on hosts. A VPN can be constructed by IPSec-enabled routers protecting traffic between protected subnets. Three basic configurations are possible:

The first situation occurs when two distant private networks are to be connected using an unreliable wireless network such as the Internet. In such a case, a virtual private network (VPN) is established between the security gateways.

The second situation corresponds to the case where mobile users are to securely access the Intranet. The unreliable wireless network can be the Internet, the telephone network, etc. Lastly, the third situation occurs when two parties communicate in a secure way, but do not have any confidence in the wireless network that separates them.

So, the disadvantage of end-to-end security is that various applications such as QoS solutions, traffic stopping, firewalling and traffic monitoring (which require the ability to "inspect" or modify a transient packet), will be unable to make the decisions that they are supposed to make. In addition, Network Address Translation (NAT) will also fail to modify a packet that has been secured. There are also more complex configurations, where several security associations can possibly afford different security services:

In the preceding example, the first association can ensure that the security services required by the external security policy (authentication and confidentiality for example), and the second SA, can ensure the services required by the internal security policy (authentication of the terminal host for example) [1].

3.2 Virtual Private Network (VPN)

A Virtual Private Network (VPN) is a collection of technologies that creates secure connections between two locations over the Internet or any insecure wireless network that uses the TCP/IP protocol suite for communication. It usually achieves this by employing some combination of tunneling, encryption, authentication, access control, and auditing. It provides a virtual "tunnel" through the Internet or other public networks in a manner that provides the same security and features formerly available only in private networks.

VPN technology also allows the branch offices or enterprise partners to connect to the enterprise over a public network, while still maintaining secure communications. The VPN connection across the Internet logically operates as a Wide Area Network (WAN) link between the sites. It also allows the employees working at home or on the road to connect in a secure fashion to a remote enterprise server using the routing infrastructure provided by the Internet. From the user's perspective, the VPN is a point-to-point connection between the user's computer and a enterprise server [1].

3.2.1 Tunneling Concept

Tunneling is a method of using an Internetwork infrastructure to transfer data for one wireless network over another wireless network. The data to be transferred (or payload) can be the frames (or packets) of another protocol. Instead of sending a frame as it is produced by the originating node, the tunneling protocol encapsulates the frame in an additional header. The additional header provides routing information so that the encapsulated payload can traverse the intermediate Internetwork.

The encapsulated packets are then routed between tunnel endpoints over the Internetwork. The logical path through which the encapsulated packets travel through the Internetwork is called a tunnel. Once the encapsulated frames reach their destination on the Internetwork, the frame is unencapsulated and forwarded to its final destination. Tunneling includes this entire process (encapsulation, transmission, and unencapsulation of packets).

New tunneling technologies have been introduced in recent years. These newer technologies include:

- Point-to-Point Tunneling Protocol (PPTP)
- Layer 2 Forwarding (L2F)
- Layer 2 Tunneling Protocol (L2TP)
- IP Security Protocol (IPSec) [1]

3.2.1.1 Point-to-Point Tunneling Protocol (PPTP)

PPTP is a Layer 2 protocol that uses a TCP connection for tunnel maintenance and Generic Routing Encapsulation (GRE) encapsulated PPP frames for tunneled data. The payloads of the encapsulated PPP frames can be encrypted and/or compressed [1].

3.2.1.2 Layer 2 Forwarding (L2F)

L2F is a tunnel technology proposed by Cisco. It is a transmission protocol that allows dial-up access servers to frame dial-up traffic in PPP and transmit it over WAN links to an L2F server (a router). The L2F server then unwraps the packets and injects them into the wireless network. Unlike PPTP and L2TP, L2F has no defined client. L2F functions in compulsory tunnels only [1].

3.2.1.3 Layer 2 Tunneling Protocol (L2TP)

L2TP is a combination of PPTP and L2F. L2TP encapsulates PPP frames to be sent over IP, X.25, Frame Relay, or Asynchronous Transfer Mode (ATM) networks. When configured to use IP as its datagram transport, L2TP can be used as a tunneling protocol over the Internet. L2TP can also be used directly over various WAN media (such as Frame Relay) without an IP transport layer [1].

3.2.1.4 IP Security Protocol (IPSec)

IPSec is a Layer 3 protocol standard that supports the secured transfer of information across an IP wireless network. In addition to its definition of encryption mechanisms for IP traffic, IPSec defines the packet format for an IP over IP tunnel mode, generally referred to as an IPSec Tunnel Mode. An IPSec tunnel consists of a tunnel client and a tunnel server, which are both configured to use IPSec tunneling and a negotiated encryption mechanism.

IPSec Tunnel Mode uses the negotiated security method to encapsulate and encrypt entire IP packets for secure transfer across a private or public IP wireless network. The encrypted payload is then encapsulated again with a plain-text IP header and sent on the wireless network for delivery to the tunnel server. Upon receipt of this datagram, the tunnel server processes and discards the plain-text IP header, and then decrypts its contents to retrieve the original payload IP packet. The payload IP packet is then processed normally and routed to its destination on the target wireless network [1].

3.2.2 How Tunneling Works

For Layer 2 tunneling technologies, a tunnel is similar to a session. Both of the tunnel endpoints must agree to the tunnel and must negotiate configuration variables, such as address assignment or encryption or compression parameters. In most cases, data transferred across the tunnel is sent using a datagram-based protocol. A tunnel maintenance protocol is used as the mechanism to manage the tunnel.

Layer 3 tunneling technologies generally assume that all of the configuration issues have been handled out of band, often by manual processes. For these protocols, there may be no tunnel maintenance phase. For Layer 2 protocols (PPTP and L2TP), however, a tunnel must be created, maintained, and then terminated.

Once the tunnel is established, tunneled data can be sent. The tunnel client or server uses a tunnel data transfer protocol to prepare the data for transfer. For example, when the tunnel client sends a payload to the tunnel server, the tunnel client first appends a tunnel data transfer protocol header to the payload. The client then sends the resulting encapsulated payload across the Internetwork, which routes it to the tunnel server. The tunnel server accepts the packets, removes the tunnel data transfer protocol header, and forwards the payload to the target wireless network [1].

3.3 Secure Socket Layer (SSL)

SSL is a security scheme proposed by Netscape Communication Enterprise for providing a secure channel between two application hosts. SSL can provide data encryption, server authentication, message integrity, and optional client authentication for a TCP/IP connection.

SSL is the technology used to encrypt and decrypt messages sent between the browser and server. By encrypting the data, you protect messages from being read while they are transferred across the Internet. SSL encrypts a message from the browser, then sends it to the server. When the message is received by the server, SSL decrypts it and verifies that it came from the correct sender (a process known as authentication).

SSL consists of software installed on both the browser and server. Several enterprises, including Verisign, SSL.com, and Equifax, offer SSL encryption and authentication tools. Verisign's digital certificates, are already installed in most recent versions of the major browsers.

Digital certificates are used by the SSL security protocol to encrypt, decrypt, and authenticate data. The certificate contains the owner's enterprise name and other specific information that allows recipients of the certificate to identify the certificate's owner. The certificate also contains a public key used to encrypt the message being transported across the Internet. SSL uses two kinds of certificates: root certificates and server certificates. Root certificates are installed on the browser, and server certificates exist on the Web server. A root certificate tells the browser that you will accept certificates signed by the owner of the root certificate. A server certificate is installed on the Web server. It works much like the root certificate and is in charge of encrypting the messages sent to browsers and decrypting messages received from browsers [1].

3.3.1 How Does SSL Work?

SSL uses the RSA public key cryptography, which is widely used for authentication and encryption in the computer industry. The public key encryption is a technique that uses two asymmetric keys for encryption and decryption. Each pair of keys consists of a public key and a private key. The public key is made public by distributing it widely. The private key is never distributed; it is always kept secret. Data that is encrypted with the public key can be decrypted only with the private key. Conversely, data encrypted with the private key can be decrypted only with the public key.

SSL handles the scrambling of messages so that only the intended recipient can read it. The encryption/decryption process goes something like this:

1. The user browses to the secure Web server's site.
2. The user's SSL secured session is started and a unique public key is created for the browser (using the certificate authority's root certificate).
3. A message is encrypted and then sent from the browser using the server's public key. The message is scrambled during the transmission so that nobody who intercepts the message can read it.
4. The message is received by the Web server and is decrypted using the server's private key [1].

The process of SSL encryption relies upon two keys: the server's public key and private key. The private key only exists on the Web server itself and is used by the Web server to encrypt and decrypt secure messages. The public key exists on any client computer that has installed a root certificate for that Web server. Once the public key is installed, the user can send encrypted messages and decrypt messages received from the Web server.

SSL doesn't prevent the message from being intercepted. However, it does make the message useless to the rogue interceptor. In other words, someone could capture the message on its way to the secure Web server, but could not decrypt it because they do not have the server's private key.

The encryption process can be either symmetric or asymmetric. Symmetric encryption uses a single key by both parties to encrypt and decrypt secure messages. The problem is that the key itself has to be passed along as part of the conversation. With asymmetric encryption, the keys are never transported over the public wireless network, there is never a risk of them being stolen by a attacker [1].

3.3.2 Secure HTTP (SHTTP)

SHTTP is the scheme designed by Enterprise Integration Technologies (EIT). It is a higher level protocol that only works with the HTTP protocol, but is potentially more extensible than SSL. S-HTTP is backwards compatible with HTTP. It is designed to incorporate different cryptographic message formats into WWW browsers and servers. This will include PEM, PGP, and PKCS-7. Non S-HTTP browsers/servers should be able to communicate with S-HTTP without a discernible difference, unless they request protected documents

SHTTP provides a wide variety of mechanisms to provide for confidentiality, authentication, and integrity. SHTTP is not tied to any particular cryptographic system, key infrastructure, or cryptographic format [1].

3.4 Kerbros

Kerberos was developed by MIT as a solution to the wireless network security problems. It is designed to provide strong authentication for client/server applications by using secret-key cryptography.

Password based authentication is not secure for use over the public wireless network, since passwords sent across the network can be intercepted and used by eavesdroppers to impersonate the user. In addition to the security concern, password based authentication is inconvenient; users need to enter a password each time they access a wireless network service.

The Kerberos protocol uses strong cryptography, so that a client can prove its identity to a server (and vice versa) across an insecure wireless network connection. After a client and server has used Kerberos to prove their identity, they can also encrypt all of their communications to assure privacy and data integrity. Thus, Kerberos basically provides entity authentication in a centralized environment (authentication between both principles comes from their respective trusts in the central server) and provides a shared secret key agreement mechanism [1].

3.4.1 Overall Functionality

At a basic level, the Kerberos protocol brings four entities on the scene: the client, the server with which the client wishes to communicate, the Autehtication server (AS) and the Ticket Granting Server(TGS). These last two, which usuallty reside on the same host, make up the Key Distribution Center (KDC). The whole system is based on the use of tickets, which are issued by the KDC and enable the clients to authenticate when connecting to servers.

The AS contains a database of all the principles) clients and servers) for which it is responsible in its realm. For each such principal, the AS stores information such as the principal's name and secret key. Hence the AS shares a different secret key with each principal located in its realm. Authentication between the client and the server takes place in three steps. First the client obtains a special Ticket Granting Tag (TGT) from the AS. Next the client obtains a ticket from the TGS for use with the server (this exchange between the client and the TGS is authenticated thanks to the TGT). Finally, the client authenticates with the server using the ticket obtained from the TGS. The reason for the TGS is that it avoids using the client's secret key each time a ticket is required for a server. The client uses the session secret key generated by the AS to authenticate when tickets will be obtained [1].

3.5 Symmetric Encryption Versus Asymmetric Encryption (Private Key Verseus Public Key)

Symmetric, or private-key, encryption (also known as conventional encryption) is based on a secret key that is shared by both communicating parties. The sending party uses the secret key as part of the mathematical operation to encrypt (or encipher) plain text to cipher text. The receiving party uses the same secret key to decrypt (or decipher) the cipher text to plain text. Examples of symmetric encryption schemes are the RSA RC4.

Asymmetric, or public-key, encryption uses two different keys for each user: one is a private key known only to this one user; the other is a corresponding public key, which is accessible to anyone. The private and public keys are mathematically related by the encryption algorithm. One key is used for encryption and the other for decryption, depending on the nature of the communication service being implemented.

Public key encryption technologies allow digital signatures to be placed on messages. A digital signature uses the sender's private key to encrypt some portion of the message. When the message is received, the receiver uses the sender's public key to decipher the digital signature to verify the sender's identity [1].

3.6 Certificates

With symmetric encryption, both sender and receiver have a shared secret key. The distribution of the secret key must occur (with adequate protection) prior to any encrypted communication. However, with asymmetric encryption, the sender uses a private key to encrypt or digitally sign messages, while the receiver uses a public key to decipher these messages. The public key can be freely distributed to anyone who needs to receive the encrypted or digitally signed messages. The sender needs to carefully protect the private key only.

To secure the integrity of the public key, the public key is published with a certificate. A certificate (or public key certificate) is a data structure that is digitally signed by a certificate authority (CA)—an authority that users of the certificate can trust. The certificate contains a series of values, such as the certificate name and usage, information identifying the owner of the public key, the public key itself, an expiration date, and the name of the certificate authority. The CA uses its private key to sign the certificate. If the receiver knows the public key of the certificate authority, the receiver can verify that the certificate is indeed from the trusted CA and, therefore, contains reliable information and a valid public key. Certificates can be distributed electronically (through Web access or email), on smart cards, or on floppy disks.

Public key certificates provide a convenient, reliable method for verifying the identity of a sender. IPSec can optionally use this method for end-to-end authentication. Remote access servers can use public key certificates for user authentication [1].

3.7 Extensible Authentication Protocol (EAP)

Most implementations of PPP provide very limited authentication methods. EAP is an IETF-proposed extension to PPP that allows for arbitrary authentication mechanisms for the validation of a PPP connection. EAP was designed to allow the dynamic addition of authentication plug-in modules at both the client and server ends of a connection. This allows vendors to supply a new authentication scheme at any time. EAP provides the highest flexibility in authentication uniqueness and variation [1].

3.8 Digital Signature

As more and more e-Commerce enterprises try to increase the security of transactions on the Internet and intranets, more and more use is being made of digital certificates, making them a key component of Internet security [1].

3.8.1 How It Works

A digital certificate can be thought of as the digital equivalent of an employee badge or a driver's license. The certificate identifies its owner to someone who needs proof of the bearer's identity. The private cryptographic key corresponding to a certificate also can be used to digitally sign documents before they are distributed; correspondents or enterprise partners then use a copy of the digital certificate to confirm the sender's digital identity.

Digital certificates are often used to establish a user's identity for electronic transactions, including browsing Web sites, engaging in electronic commerce, signing E-mail and remotely accessing wireless network resources.

For example, a Web browser and a Web server using the SSL protocol authenticate each other with digital certificates to ensure that each party is who he or she claims to be. The public key contained in a Web server's digital certificate also is used by browsers to encrypt data sent back to the server.

Similarly, the Secure Electronic Transaction (SET) protocol for electronic commerce requires that a digital certificate be issued for each credit card, and that both merchants and customers have digital certificates to prove their identity.

> **Note:** In the case of customers, these are the digital certificates issued for their credit cards.

Digital certificates also play a major role in guaranteeing digital signatures for such purposes as verifying the authenticity of E-mail. A sender can generate a digital signature for a message using a private key, but recipients of the signed message need the sender's corresponding public key to verify the digital signature. Obtaining a copy of the sender's digital certificate is one way of doing this.

Enterprises also can issue digital certificates to their employees, making it possible to control access to wireless network resources based on those certificates. This eliminates the need to remember log-on names and passwords for each workgroup server, printer and other resource. And when employees are on the road, they can use their digital certificates to identify themselves to the enterprise firewall when trying to access the enterprise's wireless network [1].

4. ENTERPRISE MODEL AND WIRELESS NETWORK SECURITY IMPLEMENTATION

This part of the chapter will take an E-commerce enterprise (it will be called the Enterprise in the following) as part of an implementation scenario. A wireless network security solution will be derived, by employing the technologies previously mentioned [1].

4.1 3.1 E-commerce Enterprise Requirements

Most enterprises use Internet-based services to provide enhanced communications and a cost-saving means of automating enterprise processes. Typically an E-commerce (the Enterprise) enterprise has wireless network components. On one hand, the Enterprise has some servers (SQL server, 2003 server, etc.) containing sensitive enterprise information, such as financial data or proprietary technology. All these data should be as secure as possible, to make sure that no unauthorized person can get to it. On the other hand, the Enterprise would like to employ inexpensive Internet to communication with its branch offices, distributed enterprise partners and the remote employees. It provides some enterprise services for its customers to access over the Internet to its web server, the ftp server and the e-mail server. It also provides the wireless messaging services over the Internet for the mobile users [1].

4.2 Wireless Security Infrastructure: Wireless Network Topology Security

When developing a secure Intranet and Extranet, the first concern should be wireless network topology security. The internal wireless network is a typical Ethernet with a hub connecting all the machines. The problem here is that any device can monitor communications between two machines on the same wireless network segment. As previously mentioned, the Enterprise has some confidential information contained in the servers that are needed to keep it fully secure. This can be achieved by limiting the access of unauthorized users. One good way is segmenting the Internal wireless network from both a security and performance point of view.

Thus, the Enterprise's internal wireless network is segmented into two subnet parts: One subnet is the trusted wireless network which contains the confidential enterprise resources like financial data, proprietary technology or acquisition and merger information, etc. Another is the publicly opened and less trusted DMZ (Demilitarized Zone) part, which allows web browsing, file transfer and mobile access.

In this scenario, a firewall is used here between the Enterprise and the Internet to separate the untrusted wireless network (defined as the External wireless network) from the Enterprise's trusted network and DMZ part (defined as the Internal wireless network). For this deployment, all of the internal system is protected by the firewall from Internet-based attacks. The router is used between the DMZ and the trusted internal wireless network to provide the second-line of protection of the trusted wireless network because the router has the ability to filter out certain traffic. Segmenting wireless networks effectively prevents packets from traversing the entire network, thus reducing the amount of information that can be collected with a network monitor or an analyzer through physical accessing to a segment.

The DMZ part lies between the trusted wireless network and the untrusted public wireless network, and serves as an additional protection for the trusted internal wireless network behind the firewall. The problem of this solution is that all the traffic to the trusted network must travel through the DMZ part, which causes overloaded traffic in the DMZ part. The performance of the wireless network is greatly reduced, especially when you apply the security solutions like packet filtering, access control and encryption. In addition, it increases the difficulty for the firewall to control the traffic.

In the next solution, the trusted network and the DMZ part arecompletely separated. This solution solves problems and has the following advantages:

- Balance the traffic between two subnets, thus unloading the traffic of the DMZ subnet. The traffic going to the trusted wireless network can be routed directly, no need to pass through the DMZ;
- Since each subnet has its different security policy, it is better for the firewall to separate them. It is easy for the firewall to control the traffic and also easy to be configured.
- Greatly increase the wireless network performance when a variety of security technologies are implemented [1].

So, based on the preceding scenario, you need to design your security policies for the Enterprise with the following solutions:

- Internal users can access the entire wireless network, including the trusted network, DMZ and the Internet
- External users can access only the DMZ network (a network that provides Mail, FTP and HTTP services)

- With any traffic coming into the enterprise, the trusted network must be securely authenticated and encrypted
- Allow the traffic from the trusted wireless network to the DMZ part, but block the traffic initiated from DMZ to the trusted wireless network
- Remote employee and branch office or enterprise partner must use VPN or IPSec to access to the Enterprise trusted wireless network
- Authentication.
- The router that is used between DMZ and the trusted subnet can be configured to allow users from the trusted subnet access to DMZ, but doesn't permit some traffic from DMZ go through the trusted subnet.
- The firewall should follow the principle "That which is not Expressly Permitted is prohibited. Deny any service unless it is implicitly permitted" (the firewall denies all services by default, unless they are explicitly permitted).
- The basic rules of the firewall:
  - Allow all outgoing packets from the internal wireless network except the packets with a source address from an external wireless network (prevent IP Spoofing).
  - Allow incoming packets with port 80 (HTTP), port 25 (SMTP), port 110 (POP3), port 20-21 (FTP), port 53 (DNS), VPN, IPSec and SSL service ports go to the DMZ, but block the packets with a source address belong to the internal network (prevent IP Spoofing)
  - Allow ICMP packets, and block ICMP nukes
  - Allow broadcast packets
  - Allow ARP and RARP [1]

4.2.1 Network Address Translation (NAT)

NAT is applied in this scenario to enable the Enterprise to maintain unregistered IP addressing schemes and provide Internet access to all internal hosts utilizing a single Enterprise IP address. This would thus conceal the internal IP address from the untrusted Internet [1].

4.3 Remote Access VPN Solution

To provide remote users with the ability to connect to the internal wireless network, regardless of their locations, the Enterprise must deploy a secure remote access solution. The solution must allow roaming or remote clients to connect to LAN resources, and the solution must allow remote offices to connect to each other to share resources and information (LAN-to-LAN connections). In addition, the solution must ensure the privacy and integrity of data as it traverses the Internet [1].

4.3.1 Remote User Access

Building up a remote access VPN connection provides remote access to corporate resources over the public Internet, while still maintaining the privacy of the information. Rather than making a long distance (or 1-800) call to a enterprise or outsourced Network Access Server (NAS), the user calls a local ISP. Using the connection to the local ISP, the VPN software creates a virtual private network between the dial-up user and the enterprise VPN server across the Internet [1].

4.3.2 Connecting Two Remote Sites

In order to implement the secure communication between the Enterprise and its branch offices or enterprise partners, VPNs are used to connect the local area networks at two remote sites. The Enterprise gateway acts as a VPN server connect to a local ISP with a dedicated line, while branch offices or enterprise partners can connect to the local ISP by either using a dedicated link or a dial-up link [1].

4.4 IPSec VPN Employment

A dial VPN consists of an IPsec gateway and a PC equipped with an IPsec client. The IPsec gateway is located between the enterprise office and the Internet. The IPsec gateway is often part of the firewall. It permits users that are connected to the Internet to reach resources on the enterprise network in a very secure way. The PC can use the Internet at the same time that it is connected to the enterprise network. Only packets destined for the enterprise network are encrypted via Ipsec [1].

4.4.1.1 Router With IPsec Software

Instead of each PC being equipped with IPsec client software, the encryption can also be done in the router that is connected to the public wireless network. The advantage is that you don't need to change the configuration of the PC's, the only thing that is necessary is to configure the IP address of the IPsec gateway and the destination IP addresses that are behind the gateway [1].

4.4.1.2 LAN-to-LAN VPNs

When using IPsec in tunnel mode, it allows for two private wireless networks to communicate over the public Internet in a very secure way. Both private wireless networks can use private IP addresses; these IP addresses are not visible on the Internet [1].

4.5 Firewall

For a firewall to function as the enterprise desires, the wireless network service access policy should exist prior to the implementation of the firewall. The policy must be realistic and sound. A realistic policy provides a balance between protecting the wireless network from known risks on the one hand and providing users reasonable access to the wireless network resources on the other. If a firewall system denies or restricts services, only a strong wireless network service access policy will prevent the firewall's access controls from being modified or circumvented on an ad hoc basis. A sound, management-backed policy can provide this defense against user resistance.

A router is a network traffic-managing device that sits in between sub-networks and routes traffic intended for, or emanating from, the segments to which it's attached. Naturally, this makes them sensible places to implement packet filtering rules, based on your security polices that you've already developed for the routing of wireless network traffic.

A firewall insulates a private wireless network from a public wireless network using carefully established controls on the types of request they will route through to the private network for processing and fulfillment. For example, an HTTP request for a public Web page will be honored, whereas an FTP request to a host behind the firewall may be dishonored. Firewalls typically run monitoring software to detect and thwart external attacks on the site, and are needed to protect internal enterprise networks. Firewalls appear primarily in two flavors; application level gateways and proxy servers. Other uses of firewalls include technologies such as Virtual Private Networks that use the Internet to tunnel private traffic without the fear of exposure [1].

4.6 Defining Firewalls

A firewall is defined as a collection of components or a system placed between two wireless networks. It possesses the following properties:

- All traffic from inside to outside, and vice-versa, must pass through it
- Only authorized traffic, as defined by the local security policy, is allowed to pass through it
- The system itself is highly resistant to penetration [1]

Put simply, a firewall is a mechanism used to protect a trusted wireless network from an untrusted wireless network, usually while still allowing traffic between the two. Typically, the two wireless networks in question are an enterprise's internal (trusted) wireless network and the (untrusted)

Internet. However, nothing in the definition of a firewall ties the concept to the Internet. Traditionally, the Internet is defined as the worldwide network of networks that uses TCP/IP for communications. The Internet is further defined as any connected set of networks. Although many firewalls are currently deployed between the Internet and internal wireless networks, there are good reasons for using firewalls in any Internet, or intranet, such as an enterprise's WAN. There will be more about this use of firewalls later in this chapter.

Another approach to firewalls, views them as both policy and the implementation of that policy in terms of wireless network configuration. Physically, a firewall comprises one or more host systems and routers, plus other security measures such as advanced authentication in place of static passwords. A firewall may consist of several different components, including filters, or screens, that block transmission of certain classes of traffic, and a gateway, which is a machine or set of machines relaying services between the internal and external wireless networks by means of proxy applications. The intermediate area occupied by the gateway is often referred to as the demilitarized zone (DMZ). These terms will all be explained in more detail, starting with traffic [1].

4.7 Firewalls As Filters

When TCP/IP sends data packets on their merry way, the packets seldom go straight from the host system that generated them to the client that requested them. Along the way, they normally pass through one or more routers. In this, TCP/IP transmissions differ from LAN communications, which broadcast over a shared wire.

To look at how TCP/IP routes packets, and how this allows sites to filter for security, let's first examine old-fashioned LAN communications. Suppose five PCs reside on a LAN. If PC #2 wants to send some data to PC #4, it shouts out over the wireless network and hopes that PC #4 hears it. The other three systems on the same wireless network will also hear the same data. This is true of both Ethernet and Token Ring, the two most widely used LAN protocols. This method of communication, in which a number of computers share the same wiring, increases efficiency, limits distance and scope. It also limits the number of computers that can talk on the same wireless network.

Early efforts to enable computers to communicate with each other over long distances, used telephone lines and switches to connect calls from one specific computer to another in a remote location (the X.25 protocol was developed for this). A connection between two computers might pass through several switches until it reached its final destination. When LANs emerged, it made sense for all the computers on one LAN to have access to the machine that had access to the remote connection, thus creating a WAN. LAN protocols, however, were incompatible with X.25, and the machine hosting the connection to the WAN tended to get overworked.

Next came a special type of switch called a router, which could take over the work of making external connections, and could also convert LAN protocols, specifically IP, into WAN protocols. Routers have since evolved into specialized computers. The typical router is about the same size as a VCR, although smaller models and rackmounted units for major interconnections have entered the market.

Basically, routers look at the address information in TCP/IP packets and direct them accordingly. Data packets transmitted over the Internet from the Web browser on a PC in Florida to a Web server in Pennsylvania, will pass through numerous routers along the way, each of which makes decisions about where to direct the traffic.

Suppose the Web browser is on a PC, which is on a LAN with a PPP connection to an Internet Service Provider (ISP). A router, or a computer acting as a router, will likely direct the packets out from the LAN to the ISP. Routers at the ISP will send the data to a backbone provider, which will route it, often in several hops, to the ISP that serves the machine that hosts the Web site.

Routers make their routing decisions based on tables of data and rules. It is possible to manipulate these rules by means of filters, so that, for example, only data from certain addresses may pass through the router. In effect, this turns a router that can filter packets into an access-control device, or firewall. If the router can generate activity logs, this further enhances its value as a security device [1].

4.8 Firewalls As Gateways

Internet firewalls are often referred to as secure Internet gateways. Like the gates in a medieval walled city, they control access to and from the wireless network.

In firewall parlance, a gateway is a computer that provides relay services between two wireless networks. A firewall may consist of little more than a filtering router as the controlled gateway. Traffic goes to the gateway instead of directly entering the connected wireless network. The gateway machine then passes the data, in accordance with an access-control policy, through a filter, to the other wireless network or to another gateway machine connected to the other network.

In some configurations, called dual-homed gateways, one computer containing two wireless network connectors acts as the gateway. Alternatively, a pair of machines can create a miniature wireless network referred to as the DMZ. Typically, the two gateways will have more open communication through the inside filter than the outside gateway has to other internal hosts. The outside filter can be used to protect the gateway from attack, while the inside gateway is used to guard against the consequences of a compromised gateway [1].

4.9 Firewalls As Control Points

By concentrating access control, firewalls become a focal point for the enforcement of a security policy. Some firewalls take advantage of this to provide additional security services, including traffic encryption and decryption. In order to communicate in encryption mode, the sending and receiving firewalls must use compatible encrypting systems. Current standards efforts in encryption and key management have begun to allow different manufacturers' firewalls to communicate securely, but these efforts have a ways to go before the customer can assume compatibility. Firewall-to-firewall encryption is thus used for secure communication over the public Internet between known entities with prior arrangement, rather than for any-to-any connections. Nevertheless, it is a powerful feature, enabling the creation of a virtual private networks (VPN) as a lower-cost alternative to a leased line or a value-added network (VAN).

Verifying the authenticity of system users is another important part of wireless network security, and firewalls can perform sophisticated authentication, using smart cards, tokens and other methods. Firewalls can also protect other external wireless network connections, such as remote dial-in. An enterprise can apply the same traffic-restricting protections, enhanced by authentication [1].

4.10 Internal Firewalls

While the phenomenal growth of Internet connections has understandably focused attention on Internet firewalls, modern enterprise practices continue to underscore the importance of internal firewalls. Mergers, acquisitions, reorganizations, joint ventures and strategic partnerships all place additional strains on security as the scope of the wireless network's reach expands. Someone outside the enterprise may suddenly need access to some, but not all, internal information. Multiple wireless networks designed by different people, according to different rules, must somehow trust each other. In these circumstances, firewalls play an important role in enforcing access-control policies between wireless networks and protecting trusted wireless networks from those that are untrusted.

Consider a manufacturing enterprise that has, over time, developed separate wireless networks within the sales, marketing, payroll, accounting and production departments. Although users in one department may wish to access certain other wireless networks, it is probably unnecessary and undesirable for all users to have access to all networks. Consequently, when connecting the wireless networks, the enterprise may choose to limit the connection, either with packet-filtering routers or with a more complex firewall.

In a WAN that must offer any-to-any connectivity, other forms of application-level security can protect sensitive data. However, segregating the wireless networks by means of firewalls, greatly reduces many of the risks involved; in particular, firewalls can reduce the threat of internal hacking--that is, unauthorized access by authorized users, a problem that consistently outranks external hacking in information-security surveys. By adding encryption to the services performed by the firewall, a site can create very secure firewall-to-firewall connections. This even enables wide-area networking between remote locations over the Internet. By using authentication mechanisms on the firewall, it is possible to gain a higher level of confidence for persons outside the firewall who request data from inside the firewall (for example, salespersons on the road needing access to an inventory database), and are indeed who they claim to be [1].

4.11 Firewalls And Policy

Diagrams of the various configurations of filters and gateways help when planning a firewall defense, but the system administrator must not lose sight of the broader definition of a firewall as an implementation of security policy. A firewall is an approach to security; it helps implement a larger security policy that defines the services and access to be permitted. In other words, a firewall is both policy and the implementation of that policy in terms of wireless network configuration, host systems and routers; as well as, other security measures such as advanced authentication in place of static passwords.

Firewall design policy is a lower-level policy that describes how the firewall will actually go about restricting the access and filtering the services as defined in the wireless network service access policy. An examination of both levels of policy follows [1].

4.12 Firewall Design Policy

The firewall design policy is specific to the firewall and defines the rules used to implement the wireless network service access policy. The enterprise must design the policy in relation to, and with full awareness of, issues such as the firewall's capabilities and limitations, and the threats and vulnerabilities associated with TCP/IP. As mentioned earlier, firewalls generally implement one of two basic design policies: Permit any service unless it is expressly denied; or deny any service unless it is expressly permitted.

Firewalls that implement the first policy (the permissive approach) allow all services to pass into the site by default, with the exception of those services that the service-access policy has identified and disallowed. Firewalls that implement the second policy (the restrictive approach) deny all services by default, but then pass those services that have been identified as allowed. This restrictive second policy follows the classic access model used in all areas of information security.

The permissive first policy is less desirable, since it offers more avenues for circumventing the firewall. With this approach, users could access new services not currently addressed by the policy. For example, they could run denied services at non-standard TCP/UDP ports that are not specifically mentioned by the policy.

This is where firewall design comes in. Certain firewalls can implement either a permissive or a restrictive design policy. An enterprise can also choose to locate those systems requiring services that should not be passed through the firewall on screened subnets, separated from other site systems. Some use this approach for Web servers, which are partially shielded by packet filtering but are not sheltered behind the firewall [1].

> **Note:** If the Web server calls information from, or feeds data to, internal database systems, then that connection between the Web server and the internal machines should be well protected.

4.12.1 Security Policy

The drawback of the Firewall is that it presents a single point of failure. A firewall generally acts as a choke point, a single point through which all incoming and outgoing wireless network traffic is passed. If the firewall is configured wrong or is down, then the whole internal wireless network is compromised. This concern can be addressed by building some redundancy into the choke point. Since a router has the function of filtering traffic and access control, it can be a reasonable consideration.

In order to connect from the LAN to the outside world, a proxy is often installed on the Firewall machine. A proxy is a small program that can see both sides of the firewall. Requests for information from the Web server are intercepted by the proxy, forwarded to the server machine, and the response is then forwarded back to the requester. A proxy server mediates traffic between a protected wireless network and the Internet. Many proxies contain extra logging or support for user authentication. Since proxies must "understand" the application protocol being used, they can also implement protocol specific security (an FTP proxy might be configurable to permit incoming FTP and block outgoing FTP). Another way of contacting the outside world from behind a firewall is allowing the firewall to pass requests for port 80 that are bound to or returning from the WWW server machine. This has the effect of poking a small hole in the dike through which the rest of the world can send and receive requests to the WWW server machine.

Use network switches instead of network hubs. Using hubs, when a packet arrives at one port of the hub, it is copied to the other ports, so that all segments of the LAN can see all packets. With switches, the packets are only forwarded to the destination port, not to all ports. This also improves network throughput [1].

4.12.2 Router

Routers are often used to connect two or more wireless networks. They control the flow of data packets on a wireless network and determine the best way to reach the appropriate destination. In the enterprise wireless network, you should use a router to separate the network segments. You can then configure it to route the traffic based on predetermined rules to deny or block unauthorized traffic. For example, you can use filtering commands to limit certain protocols (snmp) or employ access lists to control the IP addresses that are allowed through. The router provides the second-line of defense for the trusted wireless network [1].

4.12.3 Password

A password is the most widely used scheme to authenticate and identify users to the system. Most system or wireless networks use passwords as the only means of authentication and identification. Normally, passwords are transmitted in plain-text over the public wireless network, for example, when you access a system over a network using Telnet, FTP or rlogin. Passwords can be intercepted and displayed in plain-text by using packet sniffer.

Even if a password is encrypted before transmission, it still can be captured and retransmitted at a later time, and is referred to as a "reply attack." It can also be cracked by some password cracking software. The defenses used here include the following:

- Making a password longer with a mix of letters, numbers and special characters. The longer the password, the more secure.
- Using one-time password scheme, like smart card or token card.
- Employing Kerberos scheme [1]

It is as important to have a secure Web server, as it is to maintain a secure mail and database system. So, if you are concerned that your data is being intercepted and read, consider getting a Secure Socket Layer (SSL) server. Examples of SSL servers include the Netscape Secure Server and the Oracle Secure Server. SSL servers encrypt data before sending it, which makes it almost impossible for anyone to intercept and read the communication going back and forth between your server and the client on the other end.

Any browser that can accept SSL will decrypt your information as it is sent back. Your server has a public key and a private key, and your browser has a public key and a private key, and when the browser makes a connection with a server, the two exchange keys. In this sense, your browser and your server mesh by talking to one another cryptically.

You can do something similar with mail. Pretty Good Privacy (PGP) was developed for Internet mail by Phil Zimmerman a few years ago and is another option for maintaining security on an intranet. It can be used for any text transmission over the Internet.

PGP functions in much the same way as browsers do in that PGP exchanges keys with an individual or a site on the Internet. The information is encrypted before it is sent and is then decrypted on the receiving end. Should someone intercept it in transmission, the information wouldn't be useful.

The Trusted Network is the wireless network that internal employees use when at the office or via a secure controlled dial-in mechanism. On the other hand, a DMZ (Demilitarized Zone) is an isolated wireless network placed as a buffer area between an enterprise's Trusted Network and the Nontrusted Network. It can be used to hide the design and configuration of the Trusted Network. The DMZ prevents the following attacks by outside users form gaining direct access to the Trusted Network:

- Filter and manage DoS attacks
- Scan e-mail messages for virus, content and size
- Passive eavesdropping /packet sniffing
- Application-layer attack
- Port scans
- Limit access to the Trusted Network via a single protocol

- IP address spoofing [1]

One of the most common rules is that a single protocol cannot transverse the DMZ. This means if the client is entering into the DMZ via http on port 80, he or she can't continue into the trusted network on the same port and protocol. The DMZ is also used to control outbound traffic via proxy servers and filter servers by:

- Controlling e-mail messages based on destination, size and content
- Scanning for virus going out of the DMZ
- Monitoring and limiting access to unauthorized access sites or web sites [1]

4.13 Wireless Network Security Policy

The Wireless Network Security Policy identifies the threats against which protection is required, and defines the required level of protection. The Wireless Network Security Policy will itself contain several different policies, for example a Wireless Network Service Access Policy and System Specific Policies.

The Wireless Network Security Policy will be based on a security strategy such as Least Privilege, Defense In Depth, Choke Point, Weakest Link ,Fail Safe Stance etc. The role of the security strategy can be illustrated with a small example: Strategy 1--Everything is forbidden unless explicitly permitted; and, Strategy 2--everything is permitted unless explicitly forbidden [1].

Implementations of both of these strategies can be found in enterprises. They adopt philosophically opposing views of how to implement security. So, some understanding of the services available on the Internet, and the risks these present, is required before an effective wireless network security policy can be developed [1].

4.14 Port Scanning

A port scanner is a program that listens to well-known port numbers to detect service running on a system that can be exploited to break into the system. If the port is listening, then the scan will succeed, otherwise the port isn't reachable. The hacker could access those open ports with some type of application. They could possibly even flood those ports with requests from several sources. Some port scanning attack can be detected by monitoring the system log files using intrusion detection software. However, some scanning programs use a SYN scan, and are difficult to detect. These programs employs a SYN packet to create a half-open connection that the victim system can't be logged into. They include the following:

- Identification: Is simply the process of identifying one's self to another entity or determining the identity of the individual or entity with whom you are communicating.
- Access Control(Authorization): Refers to the ability to control the level of access that individuals or entities have to a wireless network or system and how much information they can receive.
- Non-repudiation: The ability to prevent individuals or entities from denying (repudiating) that information, data or files that were sent or received, or that information or files were accessed or altered, when in fact they were. This capability is crucial to e-commerce [1].

4.15 Certificates

The oldest form of security, is to ask for a password. A password is a classic "what you know" type of security. Of course, the problem is that anyone can access your information if he or she knows the password. A certificate (or public-key certificate, or Digital ID) is a "What you know and why you have type of security." In order to access information, you need to have a specific file in your disc, that will authenticate you. Those files are encrypted to provide a high level of security.

There are many standards of certificates. The most popular one is X.509 (by ITU). A X.509 certificate holds the following information:

- Name and identifying information (enterprise, for example) of the certificate holder (CH)
- The public key of the CH
- Issuer's name: The name of the enterprise that issued the certificate (VerSign, Netscape, etc)
- Issuer's digital signature
- Expiration date
- Serial number [1]

The issuer is an entity that attests to the identity of the holder of the certificate. The issuer is usually an external enterprise (like VeriSign) that all it does is to verify the identity.

Certificates are very useful when extra security is needed. It's your Digital ID, and it can be used to identify you in cyberspace (your electronic network). The certificate proves several important services:

- Real-Time encryption over SSL
- Single user login: You can log in once, through your browser (if it supports certificates, that is), and the browser will use that login every time a certificate is needed
- Secure E-mail (Over S/MIME)

- Strong authentication [1]

Certificates work in both ways. If you connect to some server, you can view it's certificate, so you'll be sure to whom you're talking to. The new browsers also have client-side certificates, so the server can know who you are [1].

4.16 Secure E-Mail (S/MIME)

S/MIME is a new standard for secure E-Mail. It is an open standard (the specifications are open for all, which means many enterprises can issue a S/MIME compatible E-Mail client), which is used for encrypted and signed mail. S/MIME has these basic features:

- Encryption
- Authentication (Digital signatures)
- Cross-Platform messaging
- Tamper detection (it uses a secure hashing function to detect message tampering) [1]

The main advantage of S/MIME is it's interoperability. In other words, the fact that it's an open standard, and it has a good chance of becoming the De-Facto standard for secure E-Mail [1].

5. DEVELOPING WIRELESS NETWORK SECURITY POLICES AND CONTROLS

An enterprise's wireless network security plan consists of security policies. Wireless security policies give specific guidelines for areas of responsibility, and consist of plans that provide steps to take and rules to follow to implement the policies.

Policies should define what you consider valuable, and should specify what steps should be taken to safeguard those assets. Policies can be drafted in many ways. One example is a general policy of only a few pages that covers most possibilities. Another example is a draft policy for different sets of assets, including e-mail policies, password policies, Internet access policies, and remote access policies. Two common problems with enterprise policies are: The policy is a platitude rather than a decision or direction; and, the policy is not really used by the enterprise. Instead, it is a piece of paper to show to auditors, lawyers, other enterprise components, or customers, but it does not affect behavior.

A good risk assessment will determine whether good wireless network security policies and controls are implemented. Vulnerabilities and weaknesses exist in wireless network security policies because of poor security policies and the human factor. Wireless security policies that are too stringent are often bypassed because people get tired of adhering to them (the human factor), which creates vulnerabilities for security breaches and attacks.

For example, specifying a restrictive account lockout policy increases the potential for denial of service attacks. Another example is implementing a security keypad on the server room door. Administrators may get tired of entering the security PIN number and stop the door from closing by using a book or broom, thereby bypassing the security control. Specifying restrictive password policy can actually reduce the security of the wireless network. For example, if you require passwords longer than seven characters, most users have difficulty remembering them. They might write their passwords down and leave them where an intruder can find them.

To be effective, a policy requires visibility. Visibility aids the implementation of policy by helping to ensure that the policy is fully communicated throughout the enterprise. This is achieved through a plan for each policy, that is a written set of steps and rules. The plan defines when, how, and by whom the steps and rules are implemented. Management presentations, videos, panel discussions, guest speakers, question/answer forums, and newsletters increase visibility. If the enterprise has computer security training and awareness, it is possible to effectively notify users of new policies. It also can be used to familiarize new employees with the enterprise's policies.

Wireless security policies should be introduced in a manner that ensures that management's unqualified support is clear, especially in environments where employees feel inundated with policies, directives, guidelines, and procedures. The enterprise's policy is the vehicle for emphasizing management's commitment to wireless security and making clear their expectations for employee performance, behavior, and accountability.

People like confidentiality and privacy, however attackers can eavesdrop or steal information that is sensitive to a person or enterprise. If an enterprise comes up with a new innovative product and would like to store the ideas on a computer system, it is going to want protection for that the data on the system and the transferring of data from one system to another. Wireless networks and data communication channels are often insecure, subjecting messages transmitted over the channels to passive and active threats. With a passive threat, an intruder intercepts messages to view the data. This intrusion is also known as eavesdropping. With an active threat, the intruder modifies the intercepted messages. An effective tool for protecting messages against the active and passive threats inherent in data communications is cryptography.

Cryptography is the science of mapping readable text, called plaintext, into an unreadable format, called ciphertext, and vice versa. The mapping process is a sequence of mathematical computations. The computations affect the appearance of the data, without changing its meaning.

To protect a message, an originator transforms a plaintext message into ciphertext. This process is called encryption. The ciphertext is transmitted over a wireless network or data communications channel. If the message is intercepted, the intruder only has access to the unreadable ciphertext. Upon receipt, the message recipient transforms the ciphertext into its original plaintext format. This process is called decryption.

The mathematical operations used to map between plaintext and ciphertext are cryptographic algorithms. Cryptographic algorithms require the text to be mapped, and at a minimum, require some value that controls the mapping process. This value is called a key. Given the same text and the same algorithm, different keys produce different mappings.

Cryptography is used to provide the following services: authentication, integrity, non-repudiation, and secrecy. In an e-mail message, for example, cryptography provides:

- Authentication: Allows the recipient of a message to validate its origin. It prevents an imposter from masquerading as the sender of the message.
- Integrity: Assures the recipient that the message was not modified en route.

> **Note:** The integrity service allows the recipient to detect message modification, but not to prevent it.

- Non-repudiation: There are two types of non-repudiation service. Non-repudiation with proof of origin provides the recipient assurance of the identity of the sender. Non-repudiation with proof of delivery provides the sender assurance of message delivery.
- Secrecy: Also known as confidentiality, prevents disclosure of the message to unauthorized users [1].

5.1 Public Key Infrastructures

Public key cryptography can play an important role in providing needed security services including confidentiality, authentication, digital signatures, and integrity. Public key cryptography uses two electronic keys: a public key and a private key. These keys are mathematically related, but the private key cannot be determined from the public key. The public key can be known by anyone while the owner keeps the private key secret.

A Public Key Infrastructure (PKI) provides the means to bind public keys to their owners and helps in the distribution of reliable public keys in large heterogeneous wireless networks [2]. Public keys are bound to their owners by public key certificates. These certificates contain information such as the owner's name and the associated public key, and are issued by a reliable certification authority (CA). Digital certificates, also called Digital IDs, are the electronic counterparts to driver licenses, passports, or membership cards. A digital certificate can be presented electronically to prove your identity or your right to access information or services online. Digital certificates are used not only to identify people, but also to identify Web sites (crucial to secure e-business) and software that is being sent over the Web. Digital certificates bring trust and security when you are communicating or doing business on the Internet.

A PKI is often composed of many CAs linked by trust paths. The CAs may be linked in several ways. They may be arranged hierarchically under a "root CA" that issues certificates to subordinate CAs. The CAs can also be arranged independently in a wireless network. This makes up the PKI architecture [1].

5.2 Digital Signatures

Electronic transactions are becoming increasingly important. Many enterprises offering online services and e-commerce would like to have mechanisms in place to increase confidence in electronic transactions. When a buyer buying a product from a seller hands a bank check (bill of exchange) to the seller, he or she has to sign the check verifying his or her identity and making the transaction legal.

The widespread use of PKI technology to support digital signatures can help increase confidence in electronic transactions. For example, the use of a digital signature allows a seller to prove that goods or services were requested by a buyer and therefore demand payment. The use of a PKI allows parties without prior knowledge of each other to engage in verifiable transactions.

For example, a buyer interested in purchasing goods electronically would need to obtain a public key certificate from a CA. The process of obtaining a certificate from a CA is to generate a public-private key pair. The buyer sends the public key with valid information about the enterprise to a registration authority (RA), and asks for a certificate. The RA verifies the buyer's identity based on the information provided and vouches for the identity of the buyer to a CA, who would then issue the certificate.

The newly certified buyer can now sign electronic purchase orders for the goods. The goods vendor receiving the purchase order can obtain the buyer's certificate and the certificate revocation list (CRL) for the CA that issued the buyer's certificate, check that the certificate has not been revoked, and verify the buyer's signature. By verifying the validity of the certificate, the vendor ensures receipt of a valid public key for the buyer; by verifying the signature on the purchase order, the vendor ensures the order was not altered after the buyer issued it.

Once the validity of the certificate and the signature are established, the vendor can ship the requested goods to the buyer with the knowledge that the buyer ordered the goods. This transaction can occur without any prior business relationships between the buyer and the seller [1].

5.3 Secure Sockets Layer

Secure Sockets Layer (SSL) is a protocol that protects data sent between Web browsers and Web servers. SSL also ensures that the data came from the Web site it is supposed to have originated from and that no one tampered with the data while it was being sent. Any Web site address that starts with "https" has been SSL-enabled.

SSL provides a level of security and privacy for those wishing to conduct secure transactions over the Internet. SSL protocol protects HTTP transmissions over the Internet by adding a layer of encryption. This ensures that your transactions are not subject to "sniffing" by a third party.

SSL provides visitors to your Web site with the confidence to communicate securely via an encrypted session. For enterprises wishing to conduct serious e-commerce, such as receiving credit card numbers or other sensitive information, SSL is a must. Web users can tell when they've reached an SSL-protected site by the "https" designation at the start of the Web page's address. The "s" added to the familiar HTTP (the Hypertext Transfer Protocol) stands for secure.

Enterprises that want to conduct business via the Internet, use the capabilities of SSL to contact a certificate authority, such as VeriSign Inc., which is a third-party enterprise that confirms an enterprise is indeed what it claims to be. Once that is complete, the enterprise can set up its Web servers for SSL connections. Users don't have to do anything to trigger an SSL connection. The client portion of SSL is built into the Web browser [1].

5.4 Secure E-mail

Standard Internet e-mail is usually sent as plain text over wireless networks. Intruders can monitor mail servers and wireless network traffic to obtain sensitive information.

There are currently two actively proposed methods for providing secure e-mail security services: Pretty Good Privacy (PGP) and

Secure/Multipurpose Internet Mail Extensions (S/MIME). These services typically include authentication of the originator and privacy for the data. They can also provide a signed receipt from the recipient. At the core of these capabilities is the use of public key technology and the large-scale use of public keys requires a method of certifying that a given key belongs to a given user.

PGP is a military grade encryption scheme available to all computer users. It works using paired sets of keys. The public key can be used to encode a message that can only be decoded with the matching private key. Likewise, an e-mail "signed" with a private key can be verified as authentic with its matching public key.

S/MIME is the same cryptographic method used for secure e-mail, adopted by every major e-mail vendor in the industry. S/MIME uses public key cryptography to digitally sign and encrypt each message sent between trading partners. This ensures that not only can the message not be read, but also that the message came only from the sender and was not altered in transport [1].

5.5 Authentication

Modern computer systems provide a service to multiple users and require the ability to accurately identify the user making a request. In traditional systems, the user's identity is verified by checking a password typed during login; the system records the identity and uses it to determine what operations may be performed. The process of verifying the user's identity is called authentication. Password-based authentication is not suitable for use on wireless networks. Passwords sent across the wireless network can be intercepted and subsequently used by eavesdroppers to impersonate the user.

Verifying the identity of someone or something is important. Administrators do not want unauthorized users or imposters to impersonate users. Administrators want to be able to verify that whoever is logging on to a system is who they say they are. Microsoft Windows XP and 2003 supports two types of authentication protocols: Kerberos authentication protocol and NTLM authentication protocol. The Kerberos authentication protocol is the default authentication protocol for computers running Windows XP and 2003. NTLM authentication protocol is provided for backward compatibility with other Microsoft operating systems. This next part of the chapter outlines the various features of each protocol and the application of each protocol [1].

5.6 Kerberos Authentication

Kerberos is designed to provide strong authentication for client/server applications by using secret-key cryptography. The Kerberos protocol uses strong cryptography so that a client can prove its identity to a server (and vice versa) across an insecure wireless network connection. Kerberos is a trusted third-party authentication system, whose main purpose is to allow people and processes (known to Kerberos as principals) to prove their identity in a reliable manner over an insecure wireless network. Instead of transmitting secret passwords in the clear, where they may be intercepted and read by unauthorized parties, principals obtain special Kerberos vouchers (known as session tickets) from Kerberos, which they can use to authenticate themselves to each other. The session ticket lasts only for the session a user is logged on to.

Kerberos authentication requires the existence of a trusted wireless network entity that acts as an authentication server for clients and servers requesting authentication information. This authentication server is known the key distribution center (KDC). It has access to a database consisting of a list of users and client services, their default authentication parameters, their secret encryption keys, and other data. Authentication is typically a one-way process. This is the process by which a service authenticates the client. An advantage of Kerberos over NTLM is that it allows for mutual authentication, where the client authenticates the service.

Kerberos authentication occurs when special authentication model messages and session tickets are passed among client applications, server applications, and one or more KDCs. Client processes acting on behalf of users authenticate themselves to servers by means of the session ticket. The KDC generates tickets, which are sent to the requesting client processes. Kerberos maintains a set of secret keys, one for every entity to be authenticated within a particular realm (a realm is the Protocols equivalent of a Windows XP and 2003 domain) or domain. A client presents a ticket to the server as evidence that the principal is who it claims to be. The ticket presented to the server "proves" that a KDC authenticated the client.

Kerberos streamlines the process of logging on and accessing resources as opposed to NTLM. In Kerberos authentication, the computer first contacts the KDC for authentication to the wireless network. Then, when the user is ready to access a resource for the first time, the computer contacts the KDC for a session ticket to access the resource. On each subsequent attempt, the computer can simply contact the resource directly, using the same ticket, without having to go to a domain controller first. In this way, unnecessary communication with the domain controller is eliminated. This new process allows users to log on faster and gain access to wireless network resources more quickly [1].

5.7 NTLM Authentication

In NTLM authentication, to avoid revealing passwords directly over an untrusted wireless network, a challenge-response system is used. At its simplest, the server sends the user some sort of challenge, which would typically be some sort of random string. The user would then compute a response, usually some function based on both the challenge and the password. This way, even if an intruder captures a valid challenge-response pair, it will not help the intruder gain access to the system since future challenges are likely to be different and thus require different responses.

In Microsoft Windows NT, the client contacts a primary domain controller (PDC) or a backup domain controller (BDC) to log on to the domain. Then, when the client is ready to establish a session with a particular resource, such as a printer share, it will contact the server that maintains the resource. The server, in turn, will contact the domain controller that maintains the resource in order to give it the client's required credentials or access token. NTLM is used in Windows XP and 2003 for backward compatibility with other Windows products such as Windows NT. NTLM is also used with the Telnet service in Windows XP and 2003 so users do not transmit their passwords in clear text to the Telnet service. The Telnet service is only implemented on Windows XP and 2003 when Services for Unix is installed [1].

5.8 Smart Cards

Smart Cards are typically credit card type cards that contain a small amount of memory and sometimes a processor. Since smart cards contain more memory than a typical magnetic stripe and can process information, they are being used in security situations where these features are a necessity. They can be used to hold system logon information such as the user's private key along with other personal information on the user including passwords. In a typical smart card logon environment, the user is required to insert his or her smart card into a reader device connected to the computer. Then, the software uses the information stored on the smart card for authentication. When paired with a password and/or a biometric identifier, the level of security is increased. For example, requiring the user to simply enter a password for logon is less secure than having them insert a smart card and enter a password. File encryption utilities which use the smart card as the key to the electronic lock is another security use of smart cards [1].

5.9 Secure Code

Electronic software distribution over any wireless network involves potential security problems. Software can contain programs such as viruses and Trojan horses. To help address some of these problems, you can associate digital signatures with the files. A digital certificate is a means of establishing identity via public key cryptography; code signed with a digital certificate verifies the identity of the publisher and ensures that the code has not been tampered with after it was signed. Certificates and object signing establish identity and let the user make decisions about the validity of a person's identity. When the user executes the code for the first time, a dialog box comes up. The dialog box provides information on the certificate and a link to the certificate authority.

Microsoft developed the Microsoft Authenticode technology, which enables developers and programmers to digitally sign software. Before software is released to the public or internal to the enterprise, developers can digitally sign the code. If the software is modified after digitally signing the software, the digital signature becomes invalid. In Internet Explorer, you can specify security settings that prevent users form downloading and running unsigned software from any security zone. Internet Explorer can be configured to automatically trust certain software vendors and authorities so that software and other information is automatically accepted [1].

5.10 Technologies To Secure Wireless Network Connectivity

Enterprises and other organizations use the Internet because it provides useful services. An enterprise could choose to support or not support Internet-based services based on a business plan or an information technology strategic plan. In other words, enterprises should analyze their business needs, identify potential methods of meeting the needs, and consider the security ramifications of the methods along with cost and other factors.

Most enterprises use Internet-based services to provide enhanced communications between enterprise units, or between the enterprise and its customers, or provide a cost-savings means of automating enterprise processes. Security is a key consideration—a single security incident can wipe out any cost savings or revenue provided by Internet connectivity.

Some of the ways to protect the enterprise from outside intrusions include firewalls and virtual private networks (VPNs) [1].

5.11 Firewalls

Many enterprises have connected or want to connect their private LANs to the Internet so that their users can have convenient access to Internet services. Since the Internet as a whole is not trustworthy, their private systems are vulnerable to misuse and attack. A firewall is a safeguard that one can use to control access between a trusted wireless network and a less trusted one. A firewall is not a single component; it is a strategy for protecting an enterprise's Internet-reachable resources. A firewall serves as the gatekeeper between the untrustworthy Internet and the more trustworthy internal wireless networks.

The main function of a firewall is to centralize access control. If outsiders or remote users can access the internal wireless networks without going through the firewall, its effectiveness is diluted. For example, if a traveling manager has a modem connected to his or her office computer that he or she can dial into while traveling, and that computer is also on the protected internal wireless network, an attacker who can dial into that computer has circumvented the firewall. If a user has a dial-up Internet account with a commercial ISP, and sometimes connects to the Internet from his or her office computer via modem, he or she is opening an unsecured connection to the Internet that circumvents the firewall. Firewalls provide several types of protection:

- They can block unwanted traffic.
- They can direct incoming traffic to more trustworthy internal systems.
- They hide vulnerable systems that cannot easily be secured from the Internet.
- They can log traffic to and from the private wireless network.
- They can hide information such as system names, wireless network topology, wireless network device types, and internal user IDs from the Internet.
- They can provide more robust authentication than standard applications might be able to do [1].

As with any safeguard, there are trade-offs between convenience and security. Transparency is the visibility of the firewall to both inside users and outsiders going through a firewall. A firewall is transparent to users if they do not notice or stop at the firewall in order to access a wireless network. Firewalls are typically configured to be transparent to internal wireless network users (while going outside the firewall); on the other hand, firewalls are configured to be non-transparent for the outside wireless network coming through the firewall. This generally provides the highest level of security without placing an undue burden on internal users. Types of firewalls include packet filtering gateways, application gateways, and hybrid or complex gateways [1].

5.11.1 Packet Filtering Gateways

Packet filtering firewalls use routers with packet filtering rules to grant or deny access based on source address, destination address, and port. They offer minimum security, but at a very low cost, and can be an appropriate choice for a low-risk environment. They are fast, flexible, and transparent. Filtering rules are not often easily maintained on a router, but there are tools available to simplify the tasks of creating and maintaining the rules. Filtering gateways do have inherent risks, including:

- The source and destination addresses and ports contained in the IP packet header are the only information that is available to the router in making a decision whether or not to permit traffic access to an internal wireless network.
- They do not protect against IP or DNS address spoofing.
- An attacker will have a direct access to any host on the internal wireless network once access has been granted by the firewall.
- Strong user authentication isn't supported with some packet filtering gateways.
- They provide little or no useful logging [1].

5.11.2 Application Gateways

An application gateway uses server programs (called proxies) that run on the firewall. These proxies take external requests, examine them, and forward legitimate requests to the internal host that provides the appropriate service. Application gateways can support functions such as user authentication and logging. Because an application gateway is considered as the most secure type of firewall, this configuration provides a number of advantages to the medium-high risk site:

- The firewall can be configured as the only host address that is visible to the outside wireless network, requiring all connections to and from the internal wireless network to go through the firewall.
- The use of proxies for different services prevents direct access to services on the internal wireless network, protecting the enterprise against insecure or badly configured internal hosts.
- Strong user authentication can be enforced with application gateways.

- Proxies can provide detailed logging at the application level [1].

5.11.3 Hybrid Or Complex Gateways

Hybrid gateways combine two or more of the above firewall types and implement them in series rather than in parallel. If they are connected in series, then the overall security is enhanced; on the other hand, if they are connected in parallel, then the wireless network security perimeter will be only as secure as the least secure of all methods used. In medium to high-risk environments, a hybrid gateway may be the ideal firewall implementation [1].

5.12 Virtual Private Networks And Wireless Wide Area Networks

Many enterprises have wireless local area networks and information servers spread across multiple locations. When enterprise-wide access to information or other WLAN-based resources is required, leased lines are often used to connect the WLANs into a Wireless Wide Area Network. Leased lines are relatively expensive to set up and maintain, making the Internet an attractive alternative for connecting physically separate LANs.

The major shortcoming to using the Internet for this purpose is the lack of confidentiality of the data flowing over the Internet between the LANs, as well as the vulnerability to spoofing and other attacks. Virtual private networks use encryption to provide the required security services. Typically encryption is performed between firewalls, and secure connectivity is limited to a small number of sites.

One important consideration when creating virtual private networks is that the security policies in use at each site must be equivalent. A VPN essentially creates one large wireless network out of what were previously multiple independent wireless networks. The security of the VPN will essentially fall to that of the lowest common denominator—if one LAN allows unprotected dial-up access, all resources on the VPN are potentially at risk [1].

5.13 Remote Access

Increasingly, enterprises require remote access to their information systems. This may be driven by the need for traveling employees to access e-mail, sales people to remotely enter orders, or as an enterprise decision to promote telecommuting. By its very nature, remote access to computer systems adds vulnerabilities by increasing the number of access points [1].

5.14 Dial-in

Typically the remote computer uses an analog modem to dial an auto answer modem at the enterprise location. Security methods for protecting this connection include:

- Controlling knowledge of the dial-in access numbers
- Username/password pairs
- Advanced authentication [1]

5.14.1.1 Controlling Knowledge Of The Dial-In Access Numbers

Controlling then knowledge of the dial-in access numbers is vulnerable to automated attacks by "war dialers." These are simple pieces of software that use auto-dial modems to scan blocks of telephone numbers to locate and log into modems [1].

5.14.1.2 Username/Password Pairs

Since an attacker would need to be tapping the telephone line, dial-in connections are less vulnerable to password sniffer attacks that have made reusable passwords almost useless over public wireless networks. However, the use of wireless network sniffers on internal wireless networks, the lack of password discipline, and social engineering, make obtaining or guessing passwords easy [1].

5.14.1.3 Advanced Authentication

There are many methods that can be used to supplement or replace traditional passwords. A few examples are:

- Dial-back modems: These devices require the user to enter a username/password upon initial connection. The enterprise modem then disconnects and looks up the authorized remote telephone number for the connecting user. The enterprise modem then dials the remote modem and establishes a connection.
- Public key certificates: The use of public key certificates described earlier when logging on.
- Microsoft Challenge Handshake Authentication Protocol (MS-CHAP): This is a variant of CHAP that does not require a plaintext version of the password on the authenticating server.
- Microsoft Challenge Handshake Authentication Protocol version 2 (MS-CHAP): This provides mutual authentication, stronger initial data encryption keys, and different encryption keys for sending and receiving.

- Extensible Authentication Protocol (EAP): This is an extension to the Point-to-Point protocol (PPP) that works with dial-up clients [1].

The enterprise's ability to monitor the use of remote access capabilities can also become an issue. The most effective approach is to centralize the modems into remote access servers or modem pools. There should be control in allowing users to connect their own modems to their work computers. In most cases, this should not be allowed due to the fact that it becomes difficult to monitor modems that are not accessed through the firewall and are distributed throughout the enterprise. They are potential security risks.

Information regarding access to enterprise computer and communication systems, such as dial-up modem phone numbers, should be considered confidential. This information should not be posted on electronic bulletin boards, listed in telephone directories, placed on business cards, or made available. The Network Services Manager should periodically scan direct dial-in lines to monitor compliance with policies and should periodically change the telephone numbers to make it more difficult for unauthorized parties to locate the enterprise [1].

5.14.1.3.1 Choose The Right Level Of Encryption: Performance

Finally, the stronger the encryption algorithm, the larger the delay that will be introduced form the encryption and decryption processes. Private/Public encryption schemes, such as RSA, can be 10 to 100 times slower than secret key encryption, such as DES. While the original DES standard has been broken in brute force attacks of only three days, Triple DES standard increases the number of keys and is more secure [1].

6. SUMMARY AND CONCLUSIONS

This chapter began with some fundamental concepts of wireless network security. The chapter outlined the main security risks for an enterprise to deploy intranets and extranets. The chapter further described some security vulnerabilities in the TCP/IP protocol suite that gives attackers many opportunities for malicious attacks on Internet connected computers. Finally, the chapter also presented the methods and techniques that attackers use to circumvent wireless network security.

7. REFERENCES

[1] "General Network Security Concepts," GeoCities, Copyright © 2005 Yahoo! Inc. All rights reserved. Yahoo! Inc., 701 First Avenue, Sunnyvale, CA 94089, 2004.

[2] John R. Vacca, *Public Key Infrastructure: Building Trusted Applications and Web Servicess*, Auerbach Publications, 2004.

[3] John R. Vacca, *Firewalls : Jumpstart for Network and Systems Administrators*, Digital Press, 2004.

Chapter 30

ENHANCING WIRELESS WEB SERVER SECURITY

1. INTRODUCTION

Regardless of the software you're using, wireless Web servers are vulnerable because they're generally "out there" on the public Internet. This opens up the computer that hosts the wireless Web server software to access from those outside the enterprise. For this reason, it's especially important to protect your public wireless Web servers and isolate them from other systems on your internal wireless network. For example, this would involve placing them in a DMZ or perimeter wireless network (also sometimes called a sc[illegible]d subnet) with both front and back end firewalls [3] to protect the [illegible] from the Internet and to protect the internal LAN from the DMZ.

1.1 What Are The Security Issues?

It is important that only authorized users be able to add, change or delete wireless Web server content. Some components, such as Java applets and scripts, can present security risks.

Another security risk occurs when wireless Web server users connect to a database to access information. This chapter will discuss what you can do to protect your wireless Web server from these risks.

2. THE VULNERABILITIES OF WIRELESS WEB SERVERS

Because wireless web servers are one of the few system components on a target wireless network that typically communicates with untrusted third parties, they are frequently the targets of malicious attacks by intruders. Intruders can easily launch automated attacks against thousands of systems simultaneously to identify the relatively few vulnerable systems. New attacks can be set up and launched quickly from remote locations, foiling attempts by enterprises to develop effective countermeasures. Once wireless web servers have been compromised, the enterprise's other wireless network resources are at greater risk. Intrusions can be very costly to the enterprise in terms of money, time, and damage to reputation. The confidentiality and/or integrity of the stored data can be jeopardized. Availability may also be affected, making the information on the enterprise's web site effectively unobtainable. In addition, a compromised wireless web server could be used to distribute illegally copied software, attack tools, and pornography or as a base from which to attack other wireless networks, possibly exposing the enterprise to legal liability.

With good planning and rigorous implementation of secure configurations and operational procedures, enterprises can operate successful web sites while protecting their wireless networks and information resources.

2.1 What Can Be Done To Protect Wireless Web Servers?

Enterprises need a security plan and a policy for implementing the plan, monitoring its effectiveness, and updating it. All those [illegible]ved with or affected by the information processing systems have a role [illegible]otecting the security and the privacy of information assets. Security pla[illegible]hould include an overview of the security requirements of the system, the controls needed to meet those requirements, and the responsibilities of all individuals who access the system. With this basic planning as the foundation for secure systems, enterprises should apply the following specific recommendations to improve the security of their web servers:

- Plan carefully and address the security aspects of deployment of wireless web servers.
- Implement appropriate security management practices and controls to maintain and operate a secure website.

- Deploy, configure, and manage wireless web server operating systems to meet the security requirements of the enterprise.
- Wireless web server applications should be deployed, configured, and managed to meet the security requirements of the enterprise.
- Ensure that only appropriate content is published on the website and that the content is adequately protected from unauthorized alteration.
- Take appropriate steps to protect web content from unauthorized access or modification.
- Active content should be used only after careful consideration of the benefits to be gained and the associated risks.
- Authentication and cryptographic technologies should be used appropriately to protect certain types of sensitive data.
- Use the wireless network infrastructure to help protect public wireless web servers.
- An ongoing process must be used to maintain the continued security of public wireless web servers [1].

2.1.1 Plan Carefully And Address The Security Aspects Of Deployment Of Wireless Web Servers

Careful planning is essential before the installation, configuration, and deployment of wireless web servers. It is more difficult to address security issues once deployment and implementation have been completed. A detailed and well-designed deployment plan facilitates the enterprise's decisions about tradeoffs between usability, performance, and risks. A deployment plan makes it possible to maintain secure configurations and to identify security vulnerabilities. The deployment plan should address: First, the purpose of the wireless web server, the information to be stored on or processed through the server, and the security requirements of the information and of related systems, networks, and services. Second, should be the human resource requirements for the deployment and operational phases of wireless web servers and their supporting infrastructures, including the types of personnel, their skills and training, and levels of effort required.

2.1.2 Implement Appropriate Security Management Practices And Controls To Maintain And Operate A Secure Website

Appropriate management practices are critical to operating and maintaining secure wireless web servers. Enterprises should identify their information system assets and determine the policies, standards, procedures, and guidelines that are needed to support the confidentiality, integrity, and availability of information system resources. All management controls that are required to protect information system assets should be developed, documented, and implemented. It is recommended here that enterprises apply the following practices to ensure the security of wireless web servers and their supporting wireless network infrastructure:

- An enterprise-wide information system security policy
- Configuration/change control and management
- Risk assessment and management
- Standardized software configurations that satisfy the information system security policy
- Security awareness and training
- Contingency planning, continuity of operations, and disaster recovery
- Certification and accreditation
- Incident response policy and procedures [1]

2.1.3 Deploy, Configure, And Manage Wireless Web Server Operating Systems To Meet The Security Requirements Of The Enterprise

The first step in securing a wireless web server is securing the underlying operating system. Most commonly available wireless web servers operate on a general-purpose operating system. Many security issues can be avoided if the operating systems supporting the wireless web servers are configured appropriately. The default hardware and software configurations of wireless web servers may be set by vendors to emphasize features, functions, and ease of use, rather than the security of the system. Since each enterprise's security requirements are very different, web administrators should configure new servers to reflect their enterprise's security requirements. When these requirements change, the web servers should be reconfigured. The steps needed to secure the operating system include:

- Patch and upgrade the operating system.
- Remove or disable unnecessary services and applications.
- Configure operating system user authentication.
- Configure resource controls.
- Test the security of the operating system [1].

2.1.4 Wireless Web Server Applications Should Be Deployed, Configured, And Managed To Meet The Security Requirements Of The Enterprise

In many respects, the requirements for secure installation and configuration of web server applications are the same as for the operating systems. First and foremost, only the minimal and necessary portion of wireless web server services should be installed. If vulnerabilities are identified, they should be eliminated through patches or upgrades. Unnecessary applications, services, and scripts should be removed immediately after the installation process has been completed. The steps that should be taken to secure the web server application include:

- Patch and upgrade the wireless web server application.
- Remove or disable unnecessary services, applications, and sample content.
- Configure wireless web server user authentication.
- Configure wireless web server resource controls.
- Test the security of the wireless web server application and web content [1].

2.1.5 Ensure That Only Appropriate Content Is Published On The Website And That The Content Is Adequately Protected From Unauthorized Alteration

Enterprises should develop a web publishing process or a policy that determines what information may be published openly, what information may be published with restricted access, and what information should not be published in any publicly accessible repository. Websites are vulnerable to individuals who mine an enterprise's website in search of valuable information. In general, the following kinds of information should be carefully examined and reviewed before publication on a public website:

- Classified information
- Proprietary information
- Information on the composition or preparation of hazardous materials or toxins
- Sensitive information relating to homeland security
- Detailed physical and information security safeguards
- Details about wireless network and information system infrastructure (address ranges, naming conventions, access numbers)
- Information that specifies or implies physical security vulnerabilities
- Detailed plans, maps, diagrams, aerial photographs, and architectural drawings of enterprise buildings, properties, or installations [1]

2.1.6 Take Appropriate Steps To Protect Web Content From Unauthorized Access Or Modification

Enterprises should control the information that is made available on public websites through their publishing processes or policies. Websites should be protected to assure that the information is not modified without authorization (see sidebar, "URL Authorization").

URL Authorization

Wireless web servers, by their very nature, are usually exposed to outsiders and thus are vulnerable to compromise and attack. Web sites are set up for many different purposes. A popular use of the web (especially on intranets) is to allow a group of people to access information they need to do their jobs, quickly and easily via a standard interface (the web browser) from any location (office, home, on the road). However, this information is often proprietary or sensitive, and you don't want just anyone to be able to access it just because it's on the wireless web server.

That's where URL authorization comes in. In these cases, you need a mechanism for controlling who can or can't gain access to a particular web page or site. There are several ways to do this – for instance, by setting password protection on the page. The problem with this solution is that if an unauthorized person discovers (or guesses) the password, he or she will be able to access the page. A more secure way to control access is to tie it to the user's account in the domain and/or role in the enterprise. For example, Windows Server 2003 gives you a way to do this, by implementing URL authorization and configuring role-based authentication via Authorization Manager. Web applications written within the ASP.NET framework can use URL authorization to control access [2].

Users rely on the integrity of the information made available to them. Because the information on public websites is easily accessible, it is more vulnerable to tampering and change than the information that is made available by the enterprise in other ways. Public web content must be protected through the appropriate configuration of web server resource controls. Some of the resource control practices that should be applied include:

- Install or enable only necessary services.
- Install web content on a dedicated hard drive or logical partition.
- Limit uploads to directories that are not readable by the wireless web server.

- Define a single directory for all external scripts or programs executed as part of web content.
- Disable the use of hard or symbolic links.
- Define a complete web content access matrix to identify the folders and files within the wireless web server document directory that are restricted and those that are accessible. People who have access to both the restricted and accessible folders and files should be identified.
- Disable directory listings.
- Use user authentication, digital signatures, and other cryptographic mechanisms as appropriate.
- Use host-based intrusion detection systems and/or file integrity checkers to detect intrusions and verify web content [1].

2.1.7 Active Content Should Be Used Only After Careful Consideration Of The Benefits To Be Gained And The Associated Risks

Interactive elements, supported by technologies such as ActiveX, Java, VBScript, and JavaScript, enable users to interact with websites in new ways. No longer confined just to accessing text-based documents, users can carry out a wide range of applications. These interactive elements introduce new web-related vulnerabilities since they involve moving code from a wireless web server to a client application for execution. Users are at risk because active content can take actions on the user's computer without the permission or knowledge of the user. Content generation technologies on the wireless web server pose a similar risk because, when accepting input from users, they may be induced to take actions that could harm the server. One such risk is accepting large amounts of information that can overflow buffers and be used to execute commands or gain unauthorized access to the wireless web server. All content must be protected, and close attention should be given to proper programming of browsers and servers. The different active content technologies have different vulnerabilities associated with them, and all must be carefully considered to balance benefits and risks.

2.1.8 Authentication And Cryptographic Technologies Should Be Used Appropriately To Protect Certain Types Of Sensitive Data

Enterprises should examine all of the information available on their public wireless web servers and determine their requirements to protect the integrity and confidentiality of that information. Wireless web servers can support a range of authentication and encryption technologies, which can be used to identify and authenticate users with different privileges for accessing information. Using appropriate user authentication techniques, enterprises can selectively restrict access to specific information. Otherwise, all information on a public wireless web server could be accessed by anyone with access to the server. Certain user authentication processes protect the user as well by enabling the user to verify the server being accessed is the "authentic" wireless web server and not a counterfeit version operated by a malicious entity.

Technologies based on cryptographic functions can provide an encrypted channel between a web browser client and a wireless web server that supports encryption. Wireless web servers may be configured to use different cryptographic algorithms, providing varying levels of security and performance.

2.1.9 Use The Wireless Network Infrastructure To Help Protect Public Wireless Web Servers.

The wireless network infrastructure that supports the wireless web server plays a significant role in the security of the web server. With careful configuration and deployment, the wireless network infrastructure can be used to protect the public wireless web server. Wireless network design is influenced by factors such as cost, performance, and reliability, as well as by security. But wireless network design alone cannot protect a web server. The frequency, sophistication, and variety of web attacks carried out today reinforce the need for layered and diverse defense mechanisms. Some of these defense-in-depth mechanisms include selection of a relatively safe wireless network on which the public wireless web server will be located and configuration of the network to support and protect the web server.

2.1.10 An Ongoing Process Must Be Used To Maintain The Continued Security Of Public Wireless Web Servers

Finally, maintaining a secure wireless web server requires constant effort, resources, and vigilance. After a wireless web server has been deployed, web administrators must monitor it on a daily basis to assure the continuing level of security. The following steps are essential to maintaining the security of a wireless web server:

- Configuring, protecting, and analyzing log files
- Backing up critical information frequently
- Maintaining a protected authoritative copy of the enterprise's web content

- Establishing and following procedures for recovering from compromise
- Testing and applying patches in a timely manner
- Testing security periodically [1].

3. SUMMARY AND CONCLUSIONS

Enterprises and users benefit when access to public wireless web servers is safe and convenient and when the enterprise's electronic information resources are secure, reliable, and available. As is the case with all other aspects of remote access to enterprise resources, the use of public wireless web servers entails risks as well as benefits. Finally, these risks and benefits must be managed through careful planning, and through the implementation of guidelines, for the secure operation of public wireless web servers.

4. REFERENCES

[1] Shirley Radack, "Security Of Public Web Servers," Computer Security Division, Information Technology Laboratory, National Institute of Standards and Technology, U.S. Department of Commerce, Gaithersburg, MD 20899-8930, 2003.

[2] Deb Shinder, "How URL Authorization Increases Web Server Security," WindowSecurity.com, Copyright © 2005 TechGenix Ltd. All rights reserved. TechGenix Ltd., Las Vegas/Malta, April 6, 2005.

[3] John R. Vacca, *Firewalls : Jumpstart for Network and Systems Administrators*, Digital Press, 2004.

Chapter 31

WIRELESS NETWORK SECURITY SOLUTIONS FOR CONSIDERATION

1. INTRODUCTION

All wireless networks, which by definition consist of autonomous computing nodes, are potentially subject to security problems. In the absence of central control of the sort offered by traditional mainframe computers, users with malicious intentions may be able to: undermine the confidentiality or integrity of your wireless network-accessible data; use the capabilities of your networked components for illegal or unsavory purposes; or, interfere with the legitimate activities of your wireless network users. As the Internet and its protocols have come to dominate wide area connectivity, enterprises wishing to secure their wireless networks have established perimeter defenses (firewalls, VPNs, antivirus systems, intrusion prevention and intrusion detection systems) to protect their wireless networks, data, and applications from remote intruders and other threats.

The availability of inexpensive, easily installed WLAN equipment opens up new pathways for attacks and other security breaches. Unlike wired networks, where eavesdropping on network traffic will be apparent to watchful network administrators (unless the attacker has elaborate, military-grade detection equipment on site), WLAN data streams can be passively observed using ordinary WLAN cards without being detected by administrators. Furthermore, the eavesdropper's device may even be in the parking lot or on the sidewalk. Unlike eavesdroppers on wired LANs, WLAN eavesdroppers need not be on site making an electrical connection to the network.

In fact, the spread of WLANs made possible a new sport known as "war driving." War drivers have WLAN-equipped laptops, and sometimes, high gain antennas for the radio frequencies WLANs employ. In city after city, war drivers find they can connect to dozens or hundreds of WLANs; observe all the traffic on the network (which may include an entire enterprise network); use the enterprise network's Internet connection for any purpose (which may include anti-social activities such as broadcasting SPAM or releasing viruses); and, damage or deface data and software on the exposed network. Wi-Fi WLANs have limited, though not entirely useless, security facilities. But, war drivers consistently find that half or two-thirds of the wireless networks they discover have all their security features disabled. One large retailer, which implemented its point-of-sale system on a WLAN without enabling security features, found that conniving customers were changing the prices of items they purchased.

In addition, WLANs installed for the benefit of an enterprise's internal users may not be protected by the wireless network's perimeter defense. Perimeter defenses focus on wide area connections, and assume that internal devices aren't pathways for attacks and eavesdropping. Unprotected internal WLANs serve as a soft underbelly that is extremely vulnerable to attacks despite the careful deployment of security solutions at the perimeter. Since WLANs allow access from outside the physical security barriers of your office, it is really a form of remote access to your wireless network. Industry best practices for remote access include implementing a strong 2-factor authentication.

This chapter describes the various security challenges and solutions for consideration of Wi-Fi wireless LANs; attempts the industry has made to address those challenges; shortcomings of those initial attempts; and, the best possible practices, for enterprises and residential users who want to take advantage of the real benefits of WLANs. The solutions for consideration in this chapter integrate strong 2-factor authentication with standards-based WLAN security. This chapter also discusses solutions for consideration with regards to whether you choose to implement a VPN (Layer 3) or an 802.1X system (Layer 2) in order to provide the degree of authentication and access control your wireless network demands.

2. UNDERSTANDING AUTHENTICATION AND ACCESS CONTROL

Authentication is the foundation technology for protecting wireless networks, servers, client systems, data, and applications from improper disclosure, tampering, destruction, and other forms of interference. The essence of an authentication system is discovering and confirming the identity of a person, an enterprise, a device, or more generally, of any software process on the wireless network. In the non-digital world, you readily authenticate people you know personally by their appearance or the sound of their voice on the phone, and you authenticate people you don't know personally by examining their documents, such as photo IDs. In the digital world, software processes exchange data at a sort of least common denominator level without these physical clues, and authenticating the identification of a person bound to a software process is a tricky problem.

Users can be authenticated by something they know, something they have, or something they are. The most common example of "something you know" is the traditional user ID and password combination. A common example of "something you have" is an access card that is swiped through a card reader. "Something you are" can be established with fingerprint readers, retinal scanners, facial recognition systems, and hand geometry analyzers.

Because all of these authentication techniques can be circumvented to one degree or another, many enterprises that want higher levels of security combine two or more of these methods to create "two-factor" or "three-factor" authentication. For example, getting money from your ATM machine involves two-factor authentication that involves presenting your card—something you have—and entering a PIN—something you know.

Managing and deploying authentication will require you to commit resources, including management attention and employee training along with the monetary costs of the system. Important considerations that must be traded off include the protective strength of the solution, the up-front and ongoing costs of the components and system support, simplicity and transparency for users, and coverage for various operating systems and other infrastructural elements. A universally deployed public key infrastructure (PKI) [2] can play a critical role in authentication for enterprises that are capable of making the necessary commitments of time, money, and effort. Specially chosen pairs of keys, where one of the keys is kept secret and the other one is exposed publicly, can help provide strong authentication. However, the public key must be bound to an identity, usually either a person or an enterprise, or there is a potential for misuse. The most common form of this binding is via the issuance of digital certificates, which perform a function similar to that of a notary public. The administration of hierarchies of certificates can be time consuming.

Authentication provides a greater or lesser degree of assurance that users are who they say they are, but in itself it doesn't control access to wireless network resources. Access control is the job of authorization systems. Authorization can be thought of as a grid with each wireless network resource along the x-axis and each user (or other entity) along the y-axis. At each intersecting cell, a list of privileges is created.

Thus user Bert may be allowed to run and back up all the systems administration servers and applications, but blocked from reading personnel files and running the payroll application. Enterprise directories are the primary repositories of access control information, though it's not uncommon for enterprises to have numerous standalone databases, Web applications, and security systems in addition to the directory or directories it has deployed [1].

3. DANGERS OF UNCONTROLLED AND UNAUTHENTICATED ACCESS

Pretty much every potential security disaster can begin with failures of authentication and authorization. A wireless LAN installed by a naive user or even by an inexperienced wireless network administrator may, for the sake of simplicity, refrain from forcing users to log in. Unauthenticated users may include: serious, money-motivated attackers or extortionists; vandals with time on their hands who treat unauthorized access as a sport; passers-by who would like to take advantage of free Internet access; spammers who are always seeking bogus source addresses for their material; fraud perpetrators; pornographers; and, even, without stretching the point too far, terrorists [1].

> **Note:** Permitting unauthenticated and unauthorized users on the wireless network opens up all these potential threats even if network traffic is encrypted and files and applications require authentication and authorization to be accessed. Of course, opening the entire wireless network up to unauthorized users gives them opportunities to undermine the hitherto protected parts of the network. They can attack the encryption methods, run dictionary and brute-force attacks on passwords, and cast about for public or unprotected servers and clients where they can install Trojan Horse programs or so-called zombie software, even though the usual protections of wired networks remain intact, at least for a while.

4. STANDARDIZED ATTEMPTS TO MANAGE AUTHENTICATION- AND ACCESS-CONTROL-BASED DANGERS

In order for wireless clients and access points to interchange data, they must be *associated*. The wireless nodes can only become associated once they are mutually authenticated. However, the standard default authentication method for 802.11 wireless networks, called open system authentication, performs authentication in name only—any node that requests authentication in an open system authentication environment will be authenticated. Obviously such a system can't provide a base for useful access control or authorization. As a security mechanism, open system authentication is, totally insecure. Open system authentication is, however, quite easy to use.

The 802.11 standard also provides for an alternative authentication method, known as shared key authentication. The standard 802.11 encryption method, known as Wireless Equivalent Privacy or WEP, uses 40-bit or 104-bit secret keys on access points and stations. Shared key authentication makes use of these shared secrets to implement a challenge and response exchange between potential authenticating stations and an access point.

Though it's not part of the actual 802.11 standard, most access point vendors support another kind of access control by refusing to exchange data with stations whose MAC address is not on a list of welcome nodes. As with the other authentication and access control features of 802.11 networks, by default this feature is turned off, and only administrators who know about it in advance and want to employ it will be able to take advantage of it [1].

5. SHORTCOMINGS OF STANDARD AUTHENTICATION AND ACCESS-CONTROL TECHNIQUES

This part of the chapter will put off the discussion of the shortcomings of WEP to data privacy for later. But, shared key authentication can't be any stronger than the encryption algorithms underlying it. Furthermore, it's considered poor security practice to employ the same keys for data encryption and authorization. One reason for this principle is that authentication can expose an algorithm's shortcomings in a different fashion than ordinary data encryption does. For example, although shared authentication never passes keys in the clear, management frames aren't encrypted, so information that can be captured from the authentication process may be useful in defeating encryption. In fact, there are specific attacks on WEP-based shared authentication. An attacker who manages to capture the management frames of a successful authentication session will have the raw material to authenticate repeatedly, even though he or she doesn't actually capture the WEP key (see sidebar, "In Detail: How WEP Authentication Operates").

In Detail: How WEP Authentication Operates

When a station requests authentication, the access point sends it a random 128-byte message. The station encrypts the message with its WEP key and sends the ciphertext result to the access point. The access point decrypts the ciphertext with its own copy of the WEP key. If the resultant plaintext is the same as that originally sent by the station, then the access point authenticates the station. Thereupon the station is associated with the access point [1].

As mentioned earlier in this chapter, there are numerous effective authentication systems with varying degrees of security assurance, including smart cards, cryptographic tokens, biometric readers, Kerberos, and others. However, none of these mechanisms is supported for 802.11 WLANs as they were initially deployed.

MAC address access control is very weak. Because WEP encryption protects only the data payload of each frame, MAC addresses are always in the clear. Anyone who can capture traffic (that is, anyone with a WLAN card-equipped PC and protocol analysis software) can simply read the addresses that ostensibly serve as secret keys. In addition, practically all WLAN cards can have their MAC addresses changed via software. So attackers who have captured the MAC addresses can reset their wireless network interfaces to a value that will give them full access. No decryption or brute force effort is required—this spoofing operation is practically foolproof.

A problem related to WEP keys is the difficulty of key management with 802.11 networks. There is no built-in mechanism for distributing keys to stations—they must be entered manually as a string of hexadecimal numbers or they can be derived from a password or other seed at the expense of some of the protective capabilities of the keys. In addition, sharing a single key among all users means that if a laptop or PDA that connects to the wireless network is lost or stolen, then all the users of that network must change their keys, the pain of which is multiplied by the need to do the job manually [1].

6. REVISED STANDARDIZED APPROACHES TO AUTHENTICATION

There has been a substantial amount of effort by standards organizations since 1999 towards improving the authentication picture for WLANs. One important development was the IEEE's approval of the 802.1X standard in June 2001. This standard extends authentication and authorization systems that already exist within the enterprise to the WLAN. It does this function by, in effect, establishing a temporary wireless network that carries only 802.1X traffic, which consists primarily of authentication requests and responses. The access point, or a wireless network access server connected directly to the access point, relays requests and responses between the client station and an authentication, authorization, and accounting server, typically implemented using Remote Dial-in User Service (RADIUS), originally designed to authenticate dial-up sessions over modems. Once the user has been authenticated over the dedicated, limited 802.1X link, then traffic can flow over the regular wireless network. The beauty of the system is that the RADIUS server can be located anywhere that the wired network goes, but the connection to the authentication server is secure because any traffic other than 802.1X packets from the authenticating station (known as a "supplicant"), will be blocked. Furthermore, the access point, the access server, and other intervening devices (known as "authenticators") need no knowledge of the details of the authentication operation—they simply relay requests and responses between the supplicant and the authentication server.

An 802.1X solution can be created with various types of key derivation systems used for authentication and encryption. The Extensible Authentication Protocol (EAP) is the foundation for 802.1X transactions (see Table 31-1) [1]. Some of these systems include Transport Layer Security (TLS, the successor to SSL or Secure Sockets Layer, the widely implemented protocol for authenticating Web transactions); MD5, a basic level challenge and response method that only authenticates in one direction; Tunneled TLS or TTLS, which performs TLS functions via an encrypted tunnel; and, Protected EAP (PEAP), which operates much like TTLS. A proprietary Cisco version of EAP is Lightweight EAP (LEAP). LEAP has served as something of a stopgap for Cisco customers without PKI in place. It's worth noting that a dictionary attack tool known as anwrap.pl is available from numerous Web sites, so if you must maintain a LEAP environment, use every effort to enforce the use of random-appearing passwords [1].

Table 31-1. Comparison: 802.1X Authentication Protocols

Characteristic	**MD5**	**Cisco LEAP**	**EAP-TLS**	**PEAP and TTLS**
Key length	None	128	128	128
Mutual authentication	No. Authenticates user, but not authentication device	Yes	Yes	Yes
Overall security	Weak	Stronger than MD5; weaker than other EAP solutions. Vulnerable to dictionary attack from "anwrap.pl". Strongest		Strong
Client software support	Native support in Windows XP. Other operating systems require client software.	Requires proprietary features in the NIC and AP. Wide range of operating-system support with Cisco 802.11 wireless card.	Native support available in Windows XP and Windows 2000. Other operating systems may require additional client software.	PEAP support available natively in Windows XP and Windows 2000. TTLS support available via third party software.

7. TODAY'S REAL SOLUTIONS: STRONG AUTHENTICATION

Some enterprises want to offer free, unimpeded wireless Internet access to visiting customers and vendors. Thus, they use open system authentication and disable WEP on a wireless network accessible to these users. It's critically important that such a wireless network be connected to the enterprise network outside the firewall [3] or in the DMZ, between a firewall that blocks all unrequested traffic and one that permits inbound traffic for Web servers and perhaps a few other sorts of outward facing server applications. Open wireless LANs outside the firewall are still subject to some of the problems listed earlier: your wireless network will be implicated in the bad behavior of any unauthenticated user. But all things being equal, the internal wireless network will be as secure from attacks over the WLAN as it is from Internet-based attacks.

The most direct way to extend the services of an outside-the-firewall WLAN to internal users is to employ a VPN. A Virtual Private Network (VPN) is an encrypted link into the internal wireless network that runs over a public wireless network, such as the open WLAN that's been described here.

Accessing the VPN requires authentication, which can be gauged to the necessary degree of security. For non-critical facilities, a simple ID/password logon may be sufficient. Best practices for remote access security suggest deploying smart cards or tokens to raise the degree of protection. These techniques may be combined with biometric readers or other more esoteric systems for the most critical networks. Practically speaking, VPNs require software on the client and on a VPN server inside the network.

VPNs operate at Layer 3 (the wireless network layer of the ISO networking model), so a client needs an IP address and IP connectivity in order to log in to the wireless network. An 802.1X system can kick off the authentication process without previously established IP connectivity. This capacity can also simplify deployment.

One of the advantages of an 802.1X authentication system compared to a VPN is that the wireless network need not be located outside the firewall. Because the access points won't forward any data aside from the authentication process itself, there is next-to-no opportunity for wireless attackers to access the wired network, even if they can receive the wireless signals in the parking lot. However, this fact is not an argument for deploying wireless networks without solid encryption. Those parking lot intruders could still observe unencrypted wireless traffic and capture e-mail messages, passwords, and any other sensitive data that traverses the wireless network.

An 802.1X authentication system will sometimes be easier to deploy than a VPN is, if users sometimes need access at different sites (see Table 31-2) [1]. With VPNs, users may need different passwords or smart cards away from their home sites, while 802.1X and RADIUS can simplify multi-site logins.

Table 31-2. Comparison of link layer (802.1x) and network layer (VPN) protection

Characteristic	**Link Layer (802.1X) Layer 2**	**Network Layer (VPN) Layer 3**
Authentication Services	Authenticates interface to the network. Normally based on user of the system.	Authenticates an IP address to the network. Normally based on user of the system.
Authentication Vulnerabilities	Dictionary, Man-in-the-middle, Replay	Dictionary
Data protection	Protects all data frames into and out of the nic.	Protects all IP datagrams based on the source or destination address.
Unprotected data	Management frames	Other IP addresses directed to NIC. Non-IP datagrams (ARP)

Scope of data protection	Link only	From system to gateway or endpoint
Interaction with other security layers	None	Potential problem if same layer (IPsec within IPsec)
Mobility Support	Re-authentication typically needed for each new link	Authentication stability across links and link state changes
Wireless System vulnerabilities	To other authenticated systems	To any other wireless system, authenticated or not
Provider Service theft	None practical	Authenticated system providing proxy services
Availability Now:	WPA	IPsec, L2TP, PPTP

For medium size and large enterprises deploying large scale WLANs, recent developments in hardware may be worth considering. So-called WLAN switches take many of the smarts out of access points, leaving them to handle only the basic physical layer functions. Authentication, encryption, and key management may be performed at the switch or passed along to the AAA server. WLAN switches are also useful for consolidating multiple access points, which can be tricky with standalone access points [1].

8. THREATS TO DATA PRIVACY AND INTEGRITY

While failures of authentication and authorization can lead to dramatic and embarrassing outside attacks, failures of data privacy and integrity can be insidious. If the attacker is careful, you may not notice that he or she is intercepting messages, capturing credit card or medical account data, or collecting information that could make it easy to access otherwise protected wireless network components. Of course, data privacy and integrity failures can also be dramatic and embarrassing—for instance, when an attacker attempts to extort hundreds of thousands of dollars in exchange for not releasing your customers' credit card information, or when a competitor gets a copy of an e-mailed proposal before it's sent to a customer.

Because radio signals are useful for networking precisely because they radiate out from an antenna into the surrounding space, there is always a potential for data on the WLAN to be visible to a sufficiently sensitive receiver in an unexpected place outside the physical control of the network operators. Any data transiting the network (file transfers, mail, application transactions, Web browsing, instant messaging, music downloading, videoconferences, VoIP phone calls, and much more) is subject to being observed, copied, or conceivably, modified and re-injected onto the wireless network [1].

9. STANDARDIZED ATTEMPTS TO ENSURE PRIVACY

The original privacy and integrity mechanism developed and deployed by the WLAN industry was WEP, for Wired Equivalent Privacy. The idea was that within the physically secure premises of an ordinary Ethernet network, most enterprises didn't worry much about violations of data privacy and integrity. Thus the bar the industry set for itself wasn't some kind of ideal hyper-security, but just a plain old wired network. You can't walk through a building that has a wired Ethernet network with a laptop and an Ethernet card and capture all the data going by without physically connecting to the wireless network. Physically connecting to the wireless network would be visible to network administrators if they were paying attention, or if they had implemented a system of alarms when rogue nodes attached themselves to the network—not an especially difficult or uncommon thing to do. Except for the fact that an eavesdropper attempting to intercept Ethernet communications necessarily exposes his or her MAC address to the wireless network and WLAN eavesdroppers don't (as long as they don't transmit), WLAN eavesdroppers have much the same access to network data as Ethernet eavesdroppers have.

However, the equivalence between WEP and wired networks breaks down quickly. WEP is primarily an encryption algorithm (see sidebar, "In Detail: How WEP Encryption Operates"). When it is turned on, it encrypts the data payload of each frame that traverses the wireless network. WEP also supports a form of data integrity checking by calculating a checksum for the encrypted data, and appending the checksum to the data payload prior to the encryption operation. After the frame is received and decrypted, the checksum is recalculated and compared to the received checksum. If there is a match, the frame is accepted; if the checksums don't match, the frame is rejected or otherwise flagged.

The first version of WEP used 40-bit shared secrets, partly because of the then-current export limits on strong encryption and partly to minimize the processing burden. The primary difference between WEP 1 and WEP 2 was the step up to 104-bit secret keys [1].

In Detail: How WEP Encryption Operates

Here's how WEP encryption works: the plaintext frame data payload is XOR-ed bitwise with an RC4 keystream. The WEP RC4 algorithm operates by providing a pseudo random number generator (PNRG) with a 24-bit value known as an initialization vector or IV and the 40-bit or 104-bit secret key value. The PNRG then puts out a keystream the same length as the data payload for the XOR operation.

Note: Without the participation of the IV in the creation of the keystream, the keystream would always be the same and therefore useless for encryption.

The IV is appended to the data payload and sent in the clear. At the receiving end, the local PNRG is initialized with the IV and the secret key it already knows, producing a keystream that is XOR-ed with the ciphertext data payload to decrypt it. RC4 is a reputable security algorithm, employed successfully with Secure Sockets Layer (SSL), the protocol that ensures your confidentiality when you buy online from a https:// site [1].

As processing power has gotten cheaper, breaking 40-bit keys by brute force has become nearly trivial. Pure 104-bit keys are still beyond the capabilities of brute force attacks [1].

> **Note:** The cryptographic strength of a key increases exponentially with the width of the key.

10. REVISED STANDARDIZED APPROACHES FOR PROTECTING CONFIDENTIALITY AND INTEGRITY

Wi-Fi Protected Access (WPA) is the Wi-Fi Alliance response to the shortcomings of earlier encryption, integrity, and authentication mechanisms. It is a subset of the 802.11i wireless security standard. In fact, it makes up the parts of 802.11i that existing products are able to upgrade to without replacing any hardware. The part of 802.11i that will only be possible with a hardware upgrade is the Advanced Encryption Standard (AES) encryption algorithm.

On the shared key front, WPA embraces a dynamic key distribution protocol, the Temporal Key Integrity Protocol (TKIP). After 802.1X authentication is successfully concluded, the authentication server provides the user with a pair of unique session keys. These keys are employed to create a range of other keys that are used to protect various associations between authorization servers, access points, and stations. One of these types of derived keys, the temporal key, is combined with a 48-bit IV and used with the RC4 cipher to encrypt a frame.

TKIP also includes a new Message Integrity Code (MIC), which operates alongside the ICV but has a much higher degree of cryptographic strength. With its own 64-bit keys and a nonlinear algorithm, the MIC provides much stronger assurances that data has not been modified. In fact, the MIC assures the integrity of frame headers as well as that of the payload [1].

11. TODAY'S REAL SOLUTIONS: ENCRYPTION

WPA-certified products are on the market now. In many cases, there are software and firmware upgrades for older network interface cards and access points. The issues related to authentication and authorization can be addressed to the required level of security you need by deploying 802.1X and one or more of the flavors of EAP. These authentication solutions support greatly enhanced systems for secure encryption and reliable integrity assurance, even though these systems wrap around the core RC4 cipher engine from the WEP days. It's worth noting that WEP and WPA systems can coexist in "mixed mode," which is actually the least-common-denominator mode—namely, WEP. Clearly it would not make sense to invest in an 802.1X infrastructure and upgraded access points and clients only to be held hostage to a non-upgraded part of the WLAN. The Wi-Fi Alliance claims that WPA addresses the encryption weaknesses of WEP-based products (see sidebar, " In Detail: Weaknesses of WEP Encryption"), and so far, there is little evidence to dispute these claims [1].

In Detail: Weaknesses of WEP Encryption

RC4 security relies on using a different key for every encryption. The IV method of changing 24 bits of the key while leaving the other 40 or 104 bits constant means that keys will repeat after every 224 frames, that is, every 16,777,216 frames. It may take your home network a long time to see this much traffic, but a busy office or campus might wind up repeating IVs, and therefore repeating complete keys, every few hours. Furthermore, because of the "birthday paradox" (which demonstrates that the odds are better than even that two or more members of a random group of people will have the same birthday), there are better than even odds of an IV collision (that is, an instance of key re-use) after as few as 5,000 frames. By XOR-ing the two ciphertexts with the same key, you produce the XOR of the two corresponding plaintexts. If you can use pattern detection or otherwise guess one of the plaintexts, you can begin to create a decryption dictionary for every IV. Furthermore, the Integrity Check Value (ICV) will tell you whether your guesses are correct. Widely available tools, such as AirSnort automate this encryption breaking process. Once AirSnort has captured five or ten million frames, it can find the secret key in less than one second. WEPcrack is another publicly available key discovery tool.

Note: Can you dumb it down at all? Too techie.

The main problem with WEP encryption is not that RC4 has shortcomings, but that WEP does a bad job of distributing keys to stations and access points. Because key distribution is out-of-band (that is, not performed over the WLAN), it's difficult to change the keys, especially in a sizable enterprise with intermittently available users. In addition, features some vendors have added, such as deriving WEP secret keys from passwords, make it easier for attackers to recover the keys.

The ICV, the checksum intended to ensure the integrity of WEP-encrypted data, is even easier to defeat than WEP encryption. There are well known ways to manipulate checksums using the CRC32 method so that modified data can be compensated for, thus producing a bogus guarantee of the integrity of the ICV [1].

12. ROGUE ACCESS POINTS

There are two kinds of rogue access points worth noting. One is the employee-initiated internal kind. Employees who have WLANs at home decide they'd like to surf the Web from the lunch room, so they buy their own access point and plug it into an active Ethernet jack. Needless to say, this user won't be likely to turn on any sort of authentication or encryption.

In spite of the probable innocence of these WLAN deployers, the danger is exactly the same as that of a purposely installed unencrypted, null-authenticated WLAN located inside the firewall.

The other sort of rogue access point is the external sort. It could be an inadvertent installation by another tenant in your building or a someone in a neighboring building. One possibility in this situation is that the neighboring AP will associate with some of your users. In theory, someone on the neighboring wireless network could capture such things as passwords and Web transactions from your users, but that prospect seems unlikely if the overlap of WLANs is accidental. More likely, your captured users will complain to the wireless network manager because they lack expected access to the internal wireless network's servers and applications, and the problem can be solved.

The other external possibility is a true attacker who wants to capture your WLAN traffic in order to get some kind of confidential data. As with an inadvertent external rogue access point, users will soon observe odd behavior and ask the IT group for help. This sort of attacker would have to be extremely patient and stealthy to carry out such an attack for long, though they can do a lot of harm by simply collecting logon information from users as they attach to the wireless network.

The primary danger of rogue access points appears when unencrypted and unauthenticated wireless networks are installed inside the perimeter defenses. Such wireless networks constitute about as severe a security breach as can be imagined, especially if their reception areas extend beyond the secure premises. In fact, open WLANs inside the firewall are subject to abuse by disgruntled, dishonest, or bored employees as well as by outsiders [1].

13. SHUTTING DOWN ROGUE ACCESS POINTS

Even WEP goes a long way toward preventing damage from external attackers installing rogue access points. Full WPA implementation on enterprise-configured client stations and access points will make it practically impossible for external attackers using unauthorized access points to capture data. The potential harm an open internal unauthorized access point could cause, can't be mitigated much by WEP or WPA, because the worst vulnerability is to the wired part of the network, which hardly anyone ever encrypts.

In order to eliminate the gross risks of unauthorized internal access points, you need an auditing tool that can detect the devices and track them down. The most basic form these tools take is that of a wireless-capable protocol analyzer available as software or as handheld devices with useful features for finding rogues. There are many options available on the market. A free downloadable program for finding rogue access points is Network Stumbler. Some of these tools integrate with a GPS to pinpoint the physical location of identified access points. The problem is that there's no substitute for walking around with your laptop or handheld auditing tool for anyone with more than a few hundred square feet of space in their facility. And you can't just walk around once—you have to do it regularly.

Some vendors have come up with products designed to be installed permanently and watch for rogue access points and prevent authentication outside an administrator-defined boundary. These products may be able to limit the amount of walking around required to be sure there are no holes created by rogue WLANs [1].

14. DENIAL OF SERVICE

Denial of service (DoS) attacks are possible on any kind of wireless network, not just WLANs. In their least sophisticated form, they consist of blasting traffic at a target wireless network or component until the link is so full of traffic that no more can be accepted or until a server or other resource becomes overloaded and refuses additional data. Some attackers manage to take over hundreds or thousands of devices on unprotected wireless networks and use these "zombie" nodes to focus spurious traffic on a target. WLANs could be attacked this way just as any other wireless network could.

However, WLANs have some DoS vulnerabilities that they don't share with other wireless networks. The most brute-force DoS attack is an attack on WLAN radios. The 802.11 networks operate on unlicensed Industrial, Scientific, and Medical (ISM) bands at 2.4GHz and 5.8GHz. The unlicensed bit is crucial to the cheapness and easy setup of WLAN equipment. The 2.4GHz band in particular must support a large number of other radio applications, including cordless phones, microwave ovens, Bluetooth devices, burglar alarms, RFIDs, and others. All of these devices, including WLAN components, are required to limit their power output in the interest of peaceful coexistence. However, it's still possible for these devices to interfere with one another. WLANs are also vulnerable to more targeted and stealthy DoS attacks such as:

- Simulated out-of-sequence frames may result in terminating all current connections and limiting re-attachment;
- Spoofed EAP logoff frames can be employed to terminate connections;

- Nodes configured to have more than their fair share of access time can prevent other nodes from accessing the wireless network;
- Spoofed control frames can interfere with power-saving sleep mode to interrupt connections [1].

15. TODAY'S SOLUTIONS: DENIAL OF SERVICE

Finally, as with wired networks, tracking down WLAN DoS attacks can be difficult, slow, and inefficient. Radio jamming can be tracked down fairly easily with the handheld or laptop tools used for detecting rogue access points, and some of the permanently placed detection systems will also be useful to track these problems down. Most of the other attacks can only be unmasked by careful monitoring and analysis with a protocol analyzer. Some of these vulnerabilities could be mitigated with changes to the standards, but none of them are on a standards track for now [1].

> **Note:** Enterprises don't care much whether their employees have secure WLANs at home, unless the employees have access to the enterprise wireless network. Then the issue of security at home becomes an extension of the enterprise issue. Even employees connecting through VPNs can't be considered secure if there is a WLAN connecting the employee's laptop to a cable modem or DSL line. A competitor or attacker parked on the street or located on another floor of the building may be able to capture the VPN login or other enterprise data. Preventing employees from creating home networks is probably unrealistic. Convincing them that the data they can access on the enterprise wireless network must be protected is a good first step. Educating them on the types of attack they may encounter can help encourage them to at least turn on the security features of their WLANs.

16. SUMMARY AND CONCLUSIONS

Wireless local area networks (WLANs) based on the Wi-Fi (wireless fidelity) standards are one of today's fastest growing technologies in enterprises, schools, and homes, for good reasons. They provide mobile access to the Internet and to enterprise wireless networks so users can remain connected away from their desks. These wireless networks can be up and running quickly when there is no available wired Ethernet infrastructure. They can be made to work with a minimum of effort without relying on specialized enterprise installers. Some of the enterprise advantages of WLANs include:

- Mobile workers can be continuously connected to their crucial applications and data.
- New applications based on continuous mobile connectivity can be deployed.
- Intermittently mobile workers can be more productive if they have continuous access to email, instant messaging, and other applications.
- Impromptu interconnections among arbitrary numbers of participants become possible [1].

But having provided these attractive benefits, most existing WLANs have not effectively addressed security-related issues. For instance, WLANs should only be deployed with full awareness of the potential security breaches they can introduce. In fact, an enterprise's security policy should define: who is allowed to use WLANs; who is permitted to install and configure WLANs; what standards of authentication, access control, encryption, and integrity assurance must be met in varying types of facilities; what current and future features must be available in wireless products that will be deployed; how unauthorized access points will be discovered and corrected; and, numerous other wireless-oriented issues. If you don't have a written, up-to-date security policy, putting one together ought to be your highest priority.

Finally, if there are unprotected WLANs connected to an enterprise network, it's crucial that these WLANs be located outside the firewall and other perimeter defenses. Wherever WLANs are attached to the enterprise network, it's crucial to install and maintain a secure authentication system that is commensurate with the security risks the enterprise faces. In addition, it's crucial to find and secure any unauthorized access points. In most cases, enterprises will want to update their existing access point firmware and software, client driver software, and authentication servers to the WPA standards, and only purchase WPA-compliant products going forward. These steps will go a long way towards ensuring that you don't sacrifice essential security in order to reap the benefits of WLANs.

17. REFERENCES

[1] "Security Problems and Solutions for Wireless LANs," Copyright © 2004 ActivCard, Inc., All rights reserved., ActivCard, Inc., 6623 Dumbarton Circle, Fremont, California 94555 USA. 2004.

[2] John R. Vacca, Public Key Infrastructure: Building Trusted Applications and Web Servicess, Auerbach Publications, 2004.

[3] John R. Vacca, Firewalls : Jumpstart for Network and Systems Administrators, Digital Press, 2004.

Chapter 32

SUMMARY, CONCLUSIONS AND RECOMMENDATIONS

1. INTRODUCTION

It is now possible to implement secure wireless networks at a reasonable cost. Much has been written about the security issues surrounding the use of wireless networks in this book and many enterprises have postponed implementation of wireless networks due to security concerns. Unfortunately, this reaction causes enterprises that postpone deployment to miss out on the many benefits of wireless networks provide new challenges to security and network administrators that are outside of the wired network.

Due to the inherent nature of wireless transmission and the availability of published attack tools downloaded from the Internet, security threats must be taken seriously. Best practices dictate a well thought out layered approach to wireless network security. Access point configuration, firewalls [3], and VPNs should be considered. Wireless security policies should be defined for acceptable wireless network thresholds and performance. Wireless network intrusion detection systems complement a layered approach and provide vulnerability assessment, network security management, and ensure that what you think you are securing is actually secured.

With the preceding in mind, great strides have been made in the area of wireless network security. Effective, standards-based wireless network security solutions are now available. For enterprises that have postponed wireless network deployments due to security concerns, there is no longer a need to wait. Technologies such as 802.1X/EAP, TKIP, and WLAN gateways provide robust security while enabling enterprises to realize the benefits of wireless networks.

Enterprises that are already using wireless networks need to reexamine their security measures in light of these new, standards-based security technologies. These enterprises should look to implement these technologies to address security issues. Enterprises that are using proprietary wireless network security solutions should examine the benefits of moving to a standards-based approach.

All enterprises need to be concerned about rogue access points. Wireless networks analyzers or monitoring tools can be used to proactively address the problem of rogue access points. Even though an enterprise may have chosen not to implement wireless networks, the threat of rogue access still exists and should be monitored. A number of vendors and industry groups have been working on the challenge of wireless network security and have developed effective WLAN safeguards that reduce the risk to predictable, acceptable and manageable levels.

2. SUMMARY

In summary, the wireless network as a computing paradigm has brought unprecedented access, flexibility and usability to the IT environment in a relatively short time. When one considers that the "conventional" computing environment of an enclosed mainframe system accessed exclusively by hard wired terminals, evolved over a period of 32+ years. Such slow growth of an IT system allowed for a maturing of the implementation plan and the time to develop adequate security measures.

By contrast, the wireless network paradigm has exploded just within the last seven years. From an infancy in which wireless IT access was the stuff of theory, experimentation and science fiction, enterprises have come to a point where wireless network access is common to the general public on the same level of widespread use as the television or the automobile. While such success is amazing and a testimony to the ingenuity of its developers, it also brings with it unprecedented security risks. Because of the rapid nature of industry growth, such core necessities as protocol standardization and development and administration of security agreements has had to occur at a rapid pace. Enterprises must review where the industry is in the development of wireless security, what problems still daunt the industry and the direction of addressing these problems that is currently showing the most promise.

With the preceding in mind, this part of the chapter addresses at a summary level, the most significant security risks in the wireless computing environment. The purpose of this part of the chapter is to introduce in a centralized fashion, the scope of the problem and the most significant talking points on the issue of wireless security and to summarize where the industry is in addressing these problems and where it is going [1].

2.1 Most Significant Sources of Security Risk in the Wireless Environment

The heart of the security problem in the wireless security setting is that networks are at heart an amalgam of independent processing units and the architecture itself is resistant to controls. The very ease of access, flexibility of expansion and evolution, that makes the wireless world so successful, makes it highly vulnerable from a security standpoint.

Ease of access itself represents the greatest security challenge. The very navigation protocols that make it possible for users of a wireless network to find their destination nodes, leave vulnerabilities for those who would use the network in an unethical, harmful or illegal manner. To date, the strongest response to this form of threat is access controls, which will be discussed shortly.

A second level of threat to the wireless network are rogue access points both from within the network or department, or from outside of the network firewall. Rogue access points will also be discussed in greater depth momentarily as well.

Unauthorized use of wireless network services represents a threat to the security of the system as well. Such wireless network utilization not only creates unproductive network traffic, but unauthorized users are the ones who will attempt to use network resources in a harmful way or attempt to break into data resources either for entertainment or illegal reasons. “Hacking” of this nature is a very serious problem, particularly for wireless network clients whose software and data repositories are of a sensitive and/or financial nature.

Another devious mythology used by wireless network attackers is spoofing and session hijacking. Using the open nature of the wireless network and sophisticated detection tools, the network hacker can simulate legitimate network directional information--“spoofing” those commands and in that way redirect traffic within the network in a way that disrupts performance and causes unauthorized and unexpected results for network users. Such activity represents forgery and is a serious attack on the wireless network.

In addition to these threats, attackers will often "spy" on the wireless network infrastructure, tracking traffic flow and eavesdropping on the network in operation. Such information can be valuable to anyone planning an attack or unauthorized use of wireless network functionality [1].

2.2 Rogue Access Points

A rogue access point is an unauthorized point of entry into the wireless network that is used to access the infrastructure either from within the enterprise environment or from outside the firewall. The ease of access to the wireless network combined with broad availability of standardized equipment, makes it very easy for an intruder to purchase the necessary equipment and "plug in" to the network with virtually no resistance.

From within an enterprise, an employee can bring unauthorized equipment and plug into the network from a break room, an unused cubicle or any open access point. Such ports to the wireless network are routinely built into the structure of the office complex when the offices are constructed to allow the enterprise maximum flexibility over time. In the same way, this gives employees maximum flexibility for unauthorized access.

The solution for internal introduction of rogue access points is rigorous wireless network topology audits to assure that all internal nodes to the network are by design. While external intrusion using rogue access to the wireless network is most likely to represent criminal intent, internal access is the greater threat because it can occur inside the firewall. It is likely that internal access is done for benign reasons such as desire to use the Internet for chat or other recreational activates. However controls are necessary because an internal break in can occur by someone outside the enterprise as long as they can access the port access points anywhere in the WAN or by disgruntled employees or employees with criminal intent. The good news is that rigorous implementation and compliance with WPA standards, dramatically reduces the external rogue access intrusion. The WPA standards will be discussed later in this chapter [1].

2.3 Access Control Methods

In light of what is known know about rogue access to the wireless network, access control becomes the primary defensive mechanism for reducing or eliminating unauthorized access to the network. And, of the access control tools and weapons, authentication holds the highest promise as a security approach that can provide effective protection. Authentication is so effective because it can be a response to both internal and external rogue intrusion to the wireless network. Other access control methods are limited to internal attack, but they still deserve to be included in the wireless network security plan.

Physical access control however cannot be overlooked in importance. Internal rogue access of the wireless network occurs within the firewall and utilizes access points supported by the infrastructure of the enterprise. So, two primary controls should be reviewed for feasibility and security at those points; and, be improved to cut down or eliminate the potential of a security breach. First of all, securing those access points by physical restraints or other technological resources that the IT department can introduce will cut down on the temptation to internal users to access the wireless network from an unused entry point.

Secondly, rigorous monitoring of network use is in order. Oftentimes software wireless network "sniffers" or other system level intelligence can be integrated into the wireless infrastructure and appropriate alerts or automated responses programmed into the day to day operation of the network. If such precautions are utilized, security assurance in the wireless network will improve noticeably [1].

2.4 Authentication

Authentication is an access control method that is well understood by the wireless network users. By utilizing the login and password system, authentication not only permits wireless network security to know who is utilizing the system and control access easily, it also affords the ability to control with precision how each user can use the system, the level of security they can be granted and the level of impact each user is permitted to have on the data resources and network performance.

Wireless security has long been regimented to make available security levels that range from low risk access to very highly secured networks, such as those used by financial or military enterprises. Such disciplines worked out under the mainframe paradigm, do provide a structure for the enterprise to implement access control systems in the wireless landscape.

Authentication similarly can be introduced at a relatively loose level or by increasingly more strict controls imposed, depending on the potential risk and the criticality of the resources and services offered by the wireless network. User name and password controls are common, but for more tightly controlled wireless networks, an access card (such as used in the ATM setting layers) is placed into the security plan for greater control. Furthermore, with the rate of revision of passwords, the level of complexity of passwords are required to maintain and raise the standard of security significantly, but also to introduce "ease of use" difficulties that must be considered. Over all however, how authentication is used (a program for routine review and improvement of authentication), is critical to the creation and ongoing operation of a wireless security scheme [1].

2.5 Encryption

Encryption addresses a different level of security concerns than access control and authentication are designed to defend against. Encryption stated simply is the encoding of sensitive data within the wireless network to be decoded upon arrival at the destination point. In this way, if the information is accessed by an unauthorized agent, decryption would be difficult or impossible.

Encryption has its value in a wireless security plan because it defends against spoofing, session hijacking or external unauthorized monitoring of the network. Encryption of the user name and password does enter into the authentication plan, but only by way of securing those codes against possible identity theft [4]. Within the operation of the wireless network, encryption of packet data traversing the network does provide a higher defense against network compromise coming from a hacker learning of network data flows; as well as, access to sensitive data.

The trade off of encryption is the high cost of keeping abreast of encryption technologies and standards, and the impact on wireless network performance due to the overhead encryption and decryption must have on the movement and access to data packets. Before deciding on the use of or the level of encryption to utilize performance evaluation, throughput, and response time, capacity studies should be completed in order to have a firm grasp of what the potential customer impact of such security will have [1].

3. CONCLUSIONS

In conclusion, the diligent management of security is essential to the operation of wireless networks, regardless of whether they have wireless segments or not. It's important to point out here that absolute security is an abstract, theoretical concept--it does not exist anywhere. All wireless networks are vulnerable to insider curiosity, outsider attack, and eavesdropping. No one wants to risk having the wireless networks data exposed to the casual observer or open to malicious mischief. Regardless of whether the network is wired or wireless, steps can and should always be taken to preserve network security and integrity.

It should be clear from the previous discussion that wireless networks can take advantage of all of the security measures available on wired LANs, and then add additional security features not available in the wired world. What would be the result? Well, that would be the surprising conclusion that wireless networks can be, in fact, more secure than their wired counterparts.

Wireless networks provide new challenges to security and network administrators that are outside of the wired network. Due to the inherent nature of wireless transmission and the availability of published attack tools downloaded from the Internet, security threats must be taken seriously. Best practices dictate a well thought out layered approach to wireless network security. Access point configuration, firewalls, and VPNs should be considered. Security policies should be defined for acceptable wireless network thresholds and performance. Wireless network intrusion detection systems complement a layered approach and provide vulnerability assessment, network security management, and ensure that what you think you are securing is actually secured.

Nevertheless, with the preceding in mind, great strides have been made in the area of wireless network security. Effective, standards-based wireless network security solutions are now available. For enterprises that have postponed wireless network deployments due to security concerns, there is no longer a need to wait. Technologies such as 802.1X/EAP, TKIP, and WLAN gateways provide robust security while enabling enterprises to realize the benefits of wireless networks.

Enterprises that are already using wireless networks need to reexamine their security measures in light of these new, standards-based security technologies. These enterprises should look to implement these technologies to address security issues. Enterprises that are using proprietary wireless network security solutions should examine the benefits of moving to a standards-based approach.

All enterprises need to be concerned about rogue access points. Wireless network analyzers or monitoring tools can be used to proactively address the problem of rogue access points. Even though an enterprise may have chosen not to implement wireless networks, the threat of rogue access still exists and should be monitored.

3.1 The Wi-Fi Protected Security Specification – The WPA

The most significant leap forward toward a universally applicable wireless security specification was the release of the Wi-Fi Protected Access (WPA) Security Specification. As with any important step toward greater control in IT history, it has been the implementation of industry standards that brought the ability to impose controls on an other wise out of control situation.

The WPA addressed in detail and put standardized protocols in place for the highest level security measures that needed to become stabilized and supported across the board. Amongst those were user authentication standards, a data encryption protocol that came to be identified as TKIP and data validation methodologies.

While WPA became an industry standard in early 2003, the first drafts of each of the protocols were not well developed enough to represent a strong response to the security threats that faced the wireless industry. However, throughout 2003 and on into 2005, each discipline within WPA has continued to be strengthened, made more intelligent and bolstered with improved technology and moved forward both in sophistication and in representing a potent security resource for all concerned with wireless security.

The efforts that have gone into implementing and developing the WPA protocols are beginning to provide a realizable security resource for all members of the wireless community. With WPA serving as a standard under which continued innovation and research can thrive, the industry is very close to becoming a secure environment. In that WPA has become the de facto standard, most wireless products in 2007 will be WPA compliant. Smaller enterprises that chose not to comply with WPA directly will fall into line to remain competitive with a market that is not tolerant of products that are out of the standard security agreements. In this way, the natural competitiveness of the marketplace will impose a discipline that will lead all participants to a more secure environment, that will assure that the revolution in computing represented by the wireless architecture will continue to spread and grow both in breath of acceptance and depth of services offered [1].

4. RECOMMENDATIONS

Wireless networking has great potential for improving access to services in many enterprises. For this reason, it has been spreading rapidly around the world. Unfortunately, many implementations are being done without attention to issues of security and authentication. As a result, most wireless networks in many enterprises are set up so that anyone with the proper equipment can access the wireless network, even from outside the enterprise. Anyone with the proper equipment can also spy on traffic. They can see users' passwords as well as other data. As enterprises move more and more services online, the amount of damage that can be done by having unauthorized people who learn passwords of the enterprises' users is increasing.

These dangers are not just theoretical. Tools to tap nearby wireless networks are widely available, even for palmtop devices. A whole subculture has sprung up of people going around, scanning for open wireless nodes, and publicizing them to people who want free wireless access. Thus, the underlying problem with wireless networking is that anyone in the vicinity of your enterprise can watch everything that happens on the network, and use the network for themselves.

> **Note:** There are two primary security issues: The first is access--making sure that only authorized people can use the wireless network. Without proper access control anyone in the vicinity of the building can use the wireless network, and thus get access to RUnet. The second is privacy--making sure that no one can watch your communications. Without this, anyone in the vicinity of the building can watch everything you do on a wireless network. This will let them steal your passwords and look at everything you are doing.

You will need to adopt solutions that cover both of these issues. Some of the approaches address only one, other address both. As a quick summary, there are two approaches commonly used at most enterprises: WPA or possibly WEP for access control and some level of privacy; supplemented by end to end encryption for privacy; and, end to end encryption (typically SSL) for privacy, and special gateway systems for access control.

Generally, small enterprises use the first approach, and larger ones the second approach. This chapter will present WPA, then end to end encryption, and finally the services appropriate for larger enterprises.

A few enterprises use MAC address checks for access control and end to end encryption for privacy. While this probably meets the requirements of the wireless policy, it is not recommended. Managing MAC addresses is difficult. It's generally better to use one of the two approaches previously listed.

It would also be possible to use WPA in "enterprise" mode, or 802.1x, for access control. It has only become reasonable to consider it, because it requires all clients to support WPA or 802.1x, and those technologies are just now widely available [2].

4.1 WPA And WEP

WPA and WEP are technologies that "encrypt" the traffic on your wireless network. That is, they scramble it so that an attacker can't make any sense of it. To unscramble it at the other end, all systems using it must know a "key" or password.

WPA and WEP provide both access control and privacy. Privacy comes from the encryption. Access control comes from the fact that someone must know the password to use your wireless network.

For this reason, for small wireless networks, using WPA is enough to meet the requirements of the wireless policy. However, you will still want to make sure that any services that use a password or other private information use SSL or some other type of end to end encryption.

WPA has two modes, personal and enterprise. For small enterprises, you'll want to use the personal mode. It just requires a password. Enterprise mode is for larger enterprises that have a Radius server that will support WPA.

WEP is significantly less secure than WPA, but can be used until your equipment can be upgraded to support WPA. While WEP is widely regarded as insecure, it is still a lot better than nothing [2].

4.1.1 Choosing A Good Password

It is critical to use a good password. There are attacks against WPA that will break your security if your password uses words or any other well-known sequences. WPA allows passwords as long as 63 characters. It is strongly recommend that you use a long random password, or at the very least a long phrase (at least 20 characters, but preferably longer). The phrase should not bc taken from any web site or published work. Most software saves the password, so you only need to type it once on each system.

Even better than a long phrase, is a truly random password. For example, consider using http://anyenterprise.org/random.php3. This generates a random 32-character hex string. You can combine two of them (and leave off one character) to get a 63-character password [2].

4.1.2 Background

WEP "Wireless Equivalent Privacy" was part of the original wireless specifications, so almost all hardware supports it. Unfortunately it is also fairly insecure. Hackers can easily find out the password and then do anything they want with your wireless network. The software for doing this is widely available.

WPA "Wifi Protected Access" is a modified version of WEP, which changes the effective key quite often. It is much more secure than WEP. While there are theoretical attacks, it is currently believed that they are not practical as long as you use a good password.

As of Sept 31, 2003, all new 802.11b and g hardware that is tested for Wi-Fi certification must implement WPA. Thus, WPA is now fairly widely available.

There is yet another standard that will be available shortly: WPA with Advanced Encryption Standard (AES). This is also known as WPA2. This should help the remaining issues with WPA. If all of your hardware supports WPA with AES or WPA2, use it! Certification testing for products with WPA2 started Sept 1, 2004, so some products do exist, but at the moment they aren't yet common [2].

4.1.3 Other Common Precautions

There are two other security precautions available in most wireless hardware: MAC address control and disabling SSID broadcast. First of all, MAC address control is a feature (available on most access points) that causes the access point only to talk to specific wireless devices. Each device (the wireless card that you put in your laptop) has a "MAC address" assigned to it. This address should be shown on the card. It will be a 12-digit hex number. Sometimes it is shown with punctuation (00:07:50:03:6b:c0). Other times it is simply 12 hex digits (000750036bc0). You register this address with the access point. The access point will only talk to cards that are registered.

It is recommended here that this is not necessary if you are using WPA with reasonably good passwords, at least not if you have just a few users. If you have a large user community, your password is likely to get out. For larger enterprises, it is recommended that a gateway box be used with login controls rather than MAC address control.

Second, "SSID" is the technical term for the wireless network name. You enter this into the access point when you set it up. Access points normally advertise this name by broadcasting it regularly.

> **Caution:** Don't leave the name as the default, or you are likely to conflict with other access points in your area. Normally you use your name or some name associated with your department.)

This broadcast is used by people as a starting point to find wireless networks to attack. For that reason, most access points let you turn off broadcast of the SSID. Often this is called making it a "closed wireless network." In theory, closed wireless networks are more secure, because an attacker won't notice them.

However, in practice, it is probably not a good idea. First, hacker software has other ways of locating closed wireless networks. Anyone who can break WEP or WPA will be using tools that do not depend upon SSID broadcasts.

Second, disabling SSID broadcasts makes it more difficult for other people to set up and manage nearby wireless networks. It's useful to see how many wireless networks are operating on each channel, without having to go to hacker tools [2].

4.2 End to End Encryption: SSL, SSH, Etc

The approaches described previously, all work at the wireless network level. That is, they are based on features of the wireless network cards and access points. This is extremely useful protection, particularly if you can use WPA or WPA2.

However, encrypting at the wireless network level isn't enough. Thus, it is strongly recommended that you to use end to end encryption. In fact, most larger enterprises depend upon end to end encryption, and do not use WPA.

> **Tip:** However, they then need a separate method to control access.

End to end encryption means that the whole conversation is encrypted, from your PC to the service you're talking to. The most commonly used end to end encryption is SSL. As long as you are using SSL to talk to a web server, your conversation is private. It doesn't matter whether someone can watch your wireless network. They won't be able to make sense of what they are seeing, because it is encrypted. It is possible to use SSL with other services such as email.

SSH is another common end to end method of encryption. SSH is often used as a replacement for telnet: there are SSH clients that let you log into systems just as telnet would. The difference is that the SSH connection is encrypted. SSH can also be used with other services, such as copying files (a replacement for FTP in many circumstances).

Thus, one way to deal with the privacy problem is to make sure that users only use services that are secured with SSL, SSH, or some other encryption technology. In particular, you would need to make sure that any service that requires a password uses encryption.

While possible, this solution is going to require great care. You will need to:

1. Examine every wireless network service used by your users
2. Verify that they can all be accessed using a secure protocol such as SSL and SSH
3. Verify that your users have configured their systems to use the secure protocol. In many cases access is insecure by default -- users need to enable encryption [2].

> **Tip:** However, if you are going to depend upon end to end encryption, you will need to implement similar precautions for your services, and any other services used by your users.

4.2.1 Using Valid Certificates

If you are going to depend upon end to end encryption, you will need to caution users to pay attention to security warnings. As noted in the preceding, it is particularly easy to impersonate a well-known server on a wireless network, or to do what is called a "man in the middle attack." The best protection against this is the fact that SSH and SSL use digital certificates to verify the identify of the server.

Unfortunately the user has to pay attention in order to get the benefits of this protection. If you are talking to a fake server, the user will normally see an error message saying that there is something wrong with the certificate presented by the server. If users commonly click through error messages, they may end up talking to a rogue server. That server can then steal their password. If the rogue server passes the transaction through to the real server, the user may never know that their password has been stolen. More ambitious attacks could steal any information that the user has access to, and even change data on servers.

If you're going to ask users to start paying attention to messages about certificates, this means you are going to have to start using valid certificates on all servers. However, experimental or small-scale services may use "self-signed" certificates. These certificates will generate warning messages. So, it is recommended that all enterprises use commercial certificates for any service that is going to be used widely, particularly by wireless users [2].

4.3 Access Control

Access control means limiting who can use your wireless network. This is important for two primary reasons: If others are using your wireless network, it is likely to be slower for your authorized users; and, most enterprises periodically receive reports of abuse from systems on the wireless network (someone may be sending spam or threatening email). Those reports will normally be sent to the department within the enterprise that operates the wireless network from which the problem originated. If you don't have any way to limit access to your wireless network, you have no way to know who might have done it, nor any way to prevent reoccurrence.

WPA is intended primarily to protect privacy. However, it also provides a reasonable degree of access control, as long as you have few enough users that you can control who have access to the password.

For larger enterprises, it is recommended that you use a wireless gateway, or WPA operating in enterprise mode. Both of these require the user to login before they get access.

Wireless gateways are boxes placed between the wireless network and the rest of network. They keep track of active users (by MAC address), and require a login for a new user. Normally they use a web interface. They trap attempts to access any web site, and put up a login screen [2].

4.4 Basic Security

The previous discussion has focused on issues that are specific to wireless technology. You also need to make sure that all systems you use have good basic security. If you are using general-purpose systems (Windows or Linux) in any part of your infrastructure, they need to be hardened. If you are using "appliances" or other sorts of dedicated hardware, look carefully at their security. Do they have any remote management (a web interface)? If so, make sure you change any default passwords, and restrict access as much as possible (just to workstations from which you will actually be doing management, if this is possible).

These days even access points are "intelligent," so you need to do the same kind of security checks that you would with any other component. Make sure you have complete documentation for all ways to manage your access points.

Many devices, including both access points and appliances, permit management using Simple Network Management Protocol (SNMP). This is a standard protocol for managing a variety of devices. There are common tools for dealing with SNMP devices. SNMP management has both good points and dangers. You can sometimes do things using SNMP that you can't using the vendor's GUI interface (with some products you can disable SSID broadcasts using SNMP but not with their normal interface). However SNMP is also a security issue: it is often shipped using a default password. When you are changing passwords, make sure you change the SNMP password (often called the "community"); as well as, the more visible ones. Indeed, it is recommended that you disable SNMP, unless you are actually using it [2].

4.5 VPNs

Unfortunately SSL and SSH are not yet sufficiently widespread, that you can be sure all activity is covered by one or the other of them. Another approach is to use virtual private network (VPN) technology. This creates another layer of networking on top of the wireless network. This layer is encrypted. Because VPNs are implemented in software (at least on the user's end), they are independent of any weaknesses in the wireless network technology, and they can be used with any vendor's wireless network cards. Windows and several other operating systems come with the software needed to use a VPN. However you will need to supply something on the other end.

There are similarities between SSL/SSH and a VPN: both encrypt your communications. The difference is that SSL and SSH are used for individual connections (when you are talking to a web server using SSL, the specific connection between you and the web server is encrypted using SSL). With a VPN, all of your traffic goes through a single encrypted connection. If every service you talk to supports SSL or SSH, the result is the same. A typical VPN setup will look like the one shown in Fig. 32-1 [2].

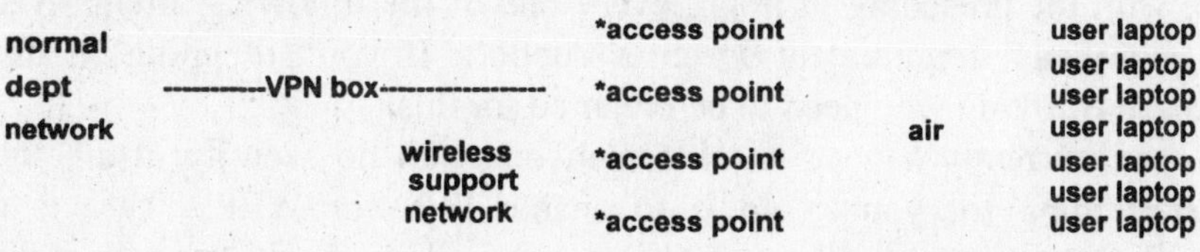

Figure 32-1. A Typical VPN setup.

The two vertical lines represent wired networks. The one on the left is a normal wireless network. The one to its right is the wireless support network, a separate wired network or VLAN is used just to connect access points. Its only connection is through the VPN box. The wireless access points are connected to that wireless support network.

In order to access any system on the Internet, users must first establish a VPN connection to the box. Once the VPN is established, all of their communications are encrypted. In fact all of their communications go over a single connection to the VPN box. That box then separates out the various connections and sends them to appropriate places within the Internet. Because all access goes through the VPN box, intruders can't access the wireless network.

The VPN box should be configured so that users must establish a VPN from their laptop to it. In the process of establishing the VPN, they need a username and a password. Thus, this approach deals with both access and privacy issues. It deals with access issues, because users must specify a username and password in order to establish the VPN. It deals with privacy issues because all traffic is encrypted. The significant drawbacks to VPNs are: First, all users will need to set up VPN software. This may require significant user support. And second, there are situations where this approach is probably not practical. The primary one involves guests.

If you take this approach, you need to think about how you will handle guests (if you host a conference, and attendees expect to use laptops, it may not be practical to get all of them to set up VPN software). You may also have people from other enterprises that need to access their home enterprises using an enterprise firewall. It is normally not practical to use two VPN's at once. So, if someone needs to use VPN software to get into their wireless home network, they probably will not be able to use your wireless VPN. For these reasons, people implementing this approach should plan on having the ability for selected users to bypass the VPN. However, this bypass needs to be controlled so that normal users don't use it. Otherwise, their passwords will be compromised.

So, with the preceding in mind, every one of the following alternatives has a significant requirement for user support. If you are going to use wireless networking, you need to be prepared for this:

1. If you currently have a wireless network with no security at all, the easiest thing for you to do is to enable WPA or WEP. There is a significant support implication to this approach: you will need to make sure that all users have the proper key or password configured on their system. If you also restrict on the basis of NIC address, you will need to maintain a list of NIC addresses of every one of your users.
2. For larger enterprises, it is recommended that you put a wireless gateway between your wireless network and the rest of the network.
3. In either case, it is recommended that you make sure that all services you provide to your users are protected by end to end encryption (SSL, SSH, etc). This is required if you a password for the service [2].

Finally, although there are some things that can be done to make a wireless network more secure, the overall state of wireless security is still seriously lacking. Wireless is still very much an emerging technology, and more research needs to be done in several fields, one of which is the wireless security field. The biggest problem, as always, is people who use the default installation out of the box for their wireless routers. This lack of security should not be taken lightly though, as it can have serious ramifications on the Internet as a whole.

5. REFERENCES

[1] Jerry Malcolm, "Security Risks In The Wireless Computing Environment," Copyright © 1998-2005 HNS, All rights reserved., HNS, [Malcolm Systems Services, 8302 S. Jamestown Ave., Tulsa, Oklahoma 74137], December 22, 2004.

[2] "Wireless Security Recommendations For Rutgers," © 2005 Rutgers, The State University of New Jersey. All rights reserved. Office of Information Technology, Davidson Hall 181, New Brunswick/Piscataway, New Jersey 732/445-2729, May 18, 2005.

[3] John R. Vacca, *Firewalls : Jumpstart for Network and Systems Administrators,* Digital Press, 2004.

[4] John R. Vacca, *Identity Theft,* Prentice Hall, 2002.

PART VIII

APPENDICES

Appendix A

ENSURING BUILT-IN FREQUENCY HOPPING SPREAD SPECTRUM WIRELESS NETWORK SECURITY

1. INTRODUCTION

The Sensors Directorate sponsored new technology developed by Robert Gold Comm Systems, Inc. (RGCS) under a Phase II Fast Track Small Business Innovation Research program. This technology provides powerful security protection for wireless computer networks, cell phones, and other radio communications. Benefits include highly secure communications with the overhead of encryption and selective addressability of receivers, individually or in groups.

2. ACCOMPLISHMENT

Dr. Gold developed a built-in self-synchronizing and selective addressing algorithm based on times-of-arrival (TOA) measurements of a frequency-hopping radio system. These algorithms allow a monitor to synchronize to a frequency-hopping radio in a wireless network by making relatively brief observations of the TOAs on a single frequency. RGCS designed the algorithms for integration into spread-spectrum, frequency-hopping systems widely used for wireless communications such as wireless fidelity computer networks, cellular phones, and two-way radios used by the military, police, firefighters, ambulances, and commercial fleets [1].

3. BACKGROUND

Although very convenient for users, wireless communication is extremely vulnerable to eavesdropping. For example, hackers frequently access wireless computer networks (laptop computers linking to the wireless network).

Encrypting the data increases the security of these wireless networks, but encryption is complex, inconvenient, time consuming for users, and adds a significant amount of overhead information that reduces data throughput. In frequency-hopping (spread-spectrum) wireless networks now in wide use, users protect the data by sending it in brief spurts, with the transmitter and receiver skipping in a synchronized pattern among hundreds of frequencies. An intruder without knowledge of the synchronization pattern would just hear static.

A major vulnerability of many spread-spectrum wireless networks involves compromising the network security by intercepting unprotected information. Originators must send the sync pattern information to authorized receivers, often unprotected.

The Gold algorithms support code-division multiple access, frequency-hopping multiple access, and ultra-wide-band spread-spectrum communication systems. They are designed for incorporation into enhanced versions of existing products, most of which already include circuitry that manufacturers can adapt to implement the technology [1].

4. ADDITIONAL INFORMATION

To receive more information about the preceding or other activities in the Air Force Research Laboratory, contact TECH CONNECT, AFRL/XPTC, (800) 203-6451, and you will be directed to the appropriate laboratory expert. (03-SN-21).

5. REFERENCES

[1] "New Technology Provides Powerful Security Protection For Wireless Communications," Air Force Research Laboratory AFRLAir AFRL, 1864 4th St., Bldg. 15, Room 225, WPAFB, OH 45433-7131, 2004.

Appendix B

CONFIGURING WIRELESS INTERNET SECURITY REMOTE ACCESS POINTS

1. INTRODUCTION

This Appendix describes how to configure and add wireless remote access points (APs) as RADIUS clients of the Microsoft 2003 Internet Authentication Service (IAS) servers.

2. ADDING THE ACCESS POINTS AS RADIUS CLIENTS TO IAS

You must add wireless remote APs as RADIUS clients to IAS before they are allowed to use RADIUS authentication and accounting services. The wireless remote APs at a given location will typically be configured to use an IAS server at the same location for their primary RADIUS server and another IAS server at the same or a different location as the secondary RADIUS server. The terms "primary" and "secondary" here do not refer to any hierarchical relationship, or difference in configuration, between the IAS servers themselves. The terms are relevant only to the wireless remote APs, each of which has a designated primary and secondary (or backup) RADIUS server. Before you configure your wireless remote APs, you must decide which IAS server will be the primary and which will be the secondary RADIUS server for each wireless remote AP.

The following procedures describe adding RADIUS clients to two IAS servers. During the first procedure, a RADIUS secret is generated for the wireless remote AP; this secret, or key, will be used by IAS and the AP to authenticate each other. The details of this client along with its secret are logged to a file. This file is used in the second procedure to import the client into the second IAS [1].

> **Tip:** You must not use this first procedure to add the same client to two IAS servers. If you do this, the client entries on each server will have a different RADIUS secret configured and the wireless remote AP will not be able to authenticate to both servers.

2.1 Adding Access Points To The First IAS Server

This part of the appendix describes the adding of wireless remote APs to the first IAS server. A script is supplied to automate the generation of a strong, random RADIUS secret (password) and add the client to IAS. The script also creates a file (defaults to Clients.txt) that logs the details of each wireless remote AP added. This file records the name, IP address, and RADIUS secret generated for each wireless remote AP. These will be required when configuring the second IAS server and wireless remote APs [1].

> **Tip:** The RADIUS clients are added to IAS as "RADIUS Standard" clients. Although this is appropriate for most wireless remote APs, some APs may require that you configure vendor–specific attributes (VSA) on the IAS server. You can configure VSAs either by selecting a specific vendor device in the properties of the RADIUS clients in the Internet Authentication Service MMC or (if the device is not listed) by specifying the VSAs in the IAS remote access policy.

2.1.1 Scripting The Addition Of Access Points To IAS Server (Alternative Procedure)

If you do not want to add the wireless remote APs to the IAS server interactively using the previous procedure, you can just generate the RADIUS client entries output files for each wireless remote AP without adding them to IAS. You can then import the RADIUS client entries into both the first IAS server and the second IAS server. Because you can script this whole operation, you may prefer to add your RADIUS clients this way if you have to add a large number of wireless remote APs [1].

Tip: This procedure is an alternative method for adding RADIUS clients in a scripted rather than an interactive fashion.

3. CONFIGURING THE WIRELESS ACCESS POINTS

Having added RADIUS clients entries for the wireless remote APs to IAS, you now need to configure the wireless remote APs themselves. You must add the IP addresses of the IAS servers and the RADIUS client secrets that each AP will use to communicate securely with the IAS servers. Every wireless remote AP will be configured with a primary and secondary (or backup) IAS server. You should perform the procedures for the wireless remote APs at every site in your enterprise.

The procedure for configuring wireless remote APs varies depending on the make and model of the device. However, wireless remote AP vendors normally provide detailed instructions for configuring their devices. Depending on the vendor, these instructions may also be available online.

Prior to configuring the security settings for your wireless remote APs, you must configure the basic wireless network settings. These will include but are not limited to:

- IP Address and subnet mask of the wireless remote AP
- Default gateway
- Friendly name of the wireless remote AP
- Wireless Network Name (SSID) [1]

The preceding list will include a number of other parameters that affect the deployment of multiple wireless remote APs: settings that control the correct radio coverage across your site for example, 802.11 Radio Channel, Transmission rate, and Transmission power, and so forth. Discussion of these parameters is outside the scope of this appendix. Use the vendor documentation as a reference when configuring these settings or consult a wireless network services supplier.

The guidance in this appendix assumes that you have set these items correctly and are able to connect to the wireless remote AP from a WLAN client using an unauthenticated connection. You should test this before configuring the authentication and security parameters listed later in this appendix [1].

3.1 Enabling Secure WLAN Authentication On Access Points

You must configure each wireless remote AP with a primary and a secondary RADIUS server. The wireless remote AP will normally use the primary server for all authentication requests, and switch over to the secondary server if the primary server is unavailable. It is important that you plan the allocation of wireless remote APs and carefully decide which server should be made primary and which should be made secondary. To summarize:

- In a site with two (or more) IAS servers, balance your wireless remote APs across the available servers so that approximately half of the wireless remote APs use server 1 as primary and server 2 as secondary, and the remaining use server 2 as primary and server 1 as secondary.
- In sites where you have only one IAS server, this should always be the primary server. You should configure a remote server (in the site with most reliable connectivity to this site) as the secondary server.
- In sites where there is no IAS server, balance the wireless remote APs between remote servers using the server with most resilient and lowest latency connectivity. Ideally, these servers should be at different sites unless you have resilient wide area network (WAN) connectivity [1].

Table B-1 [1] lists the settings that you need to configure on your wireless remote APs. Although the names and descriptions of these settings may vary from one vendor to another, your wireless remote AP documentation helps you determine those that correspond to the items in Table B-1.

Table B-1. Wireless Access Point Configuration

Item	**Setting**
Authentication Parameters	
Authentication Mode	802.1X Authentication
Re-authentication	Enable
Rapid/Dynamic Re-keying	Enable
Key Refresh Time-out	60 minutes
Encryption Parameters (these settings usually relate to static WEP encryption)	(Encryption parameters may be disabled or be overridden when rapid re–keying is enabled)
Enable Encryption	Enable
Deny Unencrypted	Enable
RADIUS Authentication	
Enable RADIUS authentication	Enable
Primary RADIUS authentication server	Primary IAS IP Address
Primary RADIUS server port	1812 (default)

Secondary RADIUS authentication server	Secondary IAS IP Address
Secondary RADIUS server port	1812 (default)
RADIUS authentication shared secret	XXXXXX (replace with generated secret)
Retry Limit	5
Retry timeout	5 seconds
RADIUS Accounting	
Enable RADIUS accounting	Enable
Primary RADIUS accounting server	Primary IAS IP Address
Primary RADIUS server port	1813 (default)
Secondary RADIUS accounting server	Secondary IAS IP Address
Secondary RADIUS server port	1813 (default)
RADIUS accounting shared secret	XXXXXX (replace with generated secret)
Retry Limit	5
Retry timeout	5 seconds

Tip: The Key Refresh Time-out is set to 60 minutes for use with dynamic WEP. The Session Timeout value set in the IAS remote access policy is the same or shorter than this. Whichever of these has the lower setting will take precedence, so you only need to modify the setting in IAS. If you are using WPA, you should increase this setting in the AP to eight hours. Consult your vendor's documentation for more information.

Use the same RADIUS secrets procedure to add wireless remote APs to IAS. Although you may have not yet configured a secondary IAS server as a backup to the primary server, you can still add the server's IP address to the wireless remote AP now (to avoid having to reconfigure it later).

Depending on the wireless remote AP hardware model, you may not have separate configurable entries for Authentication and Accounting RADIUS servers. If you have separate configurable entries, set them both to the same server unless you have a specific reason for doing otherwise. The RADIUS retry limit and timeout values given in Table B-1 are common defaults but these values are not mandatory [1].

Note: If you are currently using wireless remote APs with no security enabled or only static WEP, you need to plan your migration to an 802.1X–based WLAN.

3.2 Additional Settings To Secure Wireless Access Points

In addition to enabling 802.1X parameters, you should also configure the wireless remote APs for highest security. Most wireless network hardware is supplied with insecure management protocols enabled and administrator passwords set to well-known defaults, which poses a security risk. You should configure the settings listed in Table B-2 [1]; however, this is not an exhaustive list. You should consult your vendor's documentation for authoritative guidance on this topic. When choosing passwords and community names for Simple Network Management Protocol (SNMP), use complex values that include upper and lowercase letters, numbers, and punctuation characters. Avoid choosing anything that can be guessed easily from information such as your domain name, company name, and site address.

Table B-2. Wireless Access Point Security Configuration

Item	Recommended Setting	Notes
General		
Administrator password	XXXXXX	Set to complex password.
Other management passwords	XXXXXX	Some devices use multiple management passwords to help protect access using different management protocols; ensure that all are changed from the defaults to secure values.
Management Protocols		
Serial Console	Enable	If no encrypted protocols are available, this is the most secure method of configuring wireless remote APs although this requires physical serial cable connections between the wireless remote APs and terminal and hence cannot be used remotely.
Telnet	Disable	All Telnet transmissions are in plaintext, so passwords and RADIUS client secrets will be visible on the network. If the Telnet traffic can be secured using Internet Protocol security (IPsec) or SSH, you can safely enable and use it.
HTTP	Disable	HTTP management is usually in plaintext and suffers from the same weaknesses as unencrypted telnet. HTTPS, if available, is recommended.
HTTPS (SSL or TLS)	Enable	Follow the vendor's instructions for configuring keys/certificates for this.

SNMP Communities		SNMP is the default protocol for network management. Use SNMP v3 with password protection for highest security. It is often the protocol used by GUI configuration tools and network management systems. However, you can disable it if you do not use it.
Community 1 name	XXXXXX	The default is usually "public." Change this to a complex value.
Community 2 name	Disabled	Any unnecessary community names should be disabled or set to complex values.

You should not disable SSID (WLAN network name) broadcast since this can interfere with the ability of Windows XP to connect to the right network. Although disabling the SSID broadcast is often recommended as a security measure, it gives little practical security benefit if a secure 802.1X authentication method is being used. Even with SSID broadcast from the AP disabled, it is relatively easy for an attacker to determine the SSID by capturing client connection packets. If you are concerned about broadcasting the existence of your WLAN, you can use a generic name for your SSID, which will not be attributable to your enterprise [1].

3.3 Replicating RADIUS Client Configuration To Other IAS Servers

Typically, the wireless remote APs in a given site are serviced by an IAS server at that site. For example, the site A IAS server services wireless remote APs in site A, while the site B server services wireless remote APs in site B and so on. However, other server settings such as the remote access policies will often be common to many IAS servers. For this reason the export and import of RADIUS client information is handled separately by the procedures described in this appendix. Although you will find relatively few scenarios where replicating RADIUS client information is relevant, it is useful in certain circumstances (for example, where you have two IAS servers on the same site acting as primary and secondary RADIUS servers for all wireless remote APs on that site) [1].

4. REFERENCES

[1] "Securing Wireless LANs with PEAP and Passwords, Chapter 5: Building the Wireless LAN Security Infrastructure," © 2005 Microsoft Corporation. All rights reserved. Microsoft Corporation, One Microsoft Way, Redmond, WA 98052-6399, April 2, 2004.

Appendix C

WIRELESS NETWORK SECURITY MANAGEMENT, RESILIENCY AND SECURITY WITH CDMA

1. INTRODUCTION

With the advent of Wireless Broadband, wireless users are enjoying many rich application and services over the Internet virtually anywhere they go. Wireless Broadband services offer a new way of life where information is at the fingertips users whenever it's needed reducing the stress of anticipation. Enterprise transactions can be carried out in real-time, minimizing delay and maximizing productivity. These benefits outweigh some of the potential security concerns of doing business over the Internet.

Wireless Broadband service providers are facing wireless public network risks such as viruses, worms, blended threats, and software vulnerabilities that can impact their infrastructure, service offering and their end user experiences. Security measures must be taken in order to minimize service disruptions, avoid loss of revenue from theft and maintain a high level of customer satisfaction.

A combination of deploying security solutions and putting the appropriate processes in place will help face these challenges and proactively protect Wireless Broadband service offerings. This appendix discusses the layered end-to-end wireless network approach to security that not only ensures wireless network security, but overall wireless network reliability, management and resiliency to allow business continuity. The discussion begins first with CDMA.

2. THE INHERENT WIRELESS SECURITY OF THE CDMA AIR INTERFACE

Code Division Multiple Access (CDMA) technology originated from military applications and cryptography, and to date, do not have any report of highjacking or eavesdropping on a CDMA call in a commercially deployed wireless network. CDMA air interface is inherently secure and is clearly superior to first-generation analog and Time Division Multiple Access (TDMA) systems. The inherent security of CDMA air interface comes from spread spectrum technology and the use of Walsh codes.

CDMA utilizes specific spreading sequences and pseudo-random codes for the forward link (the path from the base station to the mobile) and on the reverse link (the path from the mobile to the base station). These spreading techniques are used to form unique code channels for individual users in both directions of the communication channel. Because the signals of all calls in a coverage area are spread over the entire bandwidth, it creates a noise-like appearance to other mobiles or detectors in the wireless network as a form of disguise, making the signal of any one call difficult to distinguish and decode.

CDMA also has a unique soft handoff capability that allows a mobile to connect to as many as six radios in the wireless network, each with its own Walsh code. This means that someone attempting to eavesdrop on a subscriber's call has to have several devices connected at exactly the same time in an attempt to synchronize with the intended signal. In addition, CDMA employs a fast power control, 800 times per second, to maintain its radio link. It is difficult for a third party to have a stable link for interception of a CDMA voice channel, even with a full knowledge of a Walsh code. Synchronization is critical, as without this synchronization, the listener only hears noise.

For CDMA 1xEV-DO (the high speed data technology), the forward link utilizes rate control instead of power control and time division multiplexing instead of spreading codes. However, it still has inherent security that protects the identity of users and makes interception very difficult. In addition, the Media Access Control

Identification (MACID) assigned to users is encrypted. User packets are assigned variable time slots and the data rate is controlled by the access terminal based on radio conditions. Packets are divided into sub-packets using Hybrid Automatic Repeat Request (HARQ) and early termination mechanisms. These attributes makes it virtually impossible to identify the user or correlate user packets. 1xEV-DO standard specification supports a security protocol layer ready for implementation of future security protocols [1].

> **Note:** Computer viruses grew by 45% in 2005, with 112,438 to a total of 163,035. Losses from insecure wireless networks are expected to reach $800K per day.

3. LAYERED DEFENSE APPROACH TO SECURE MANAGEMENT, USER AND CONTROL PLANES

With wireless air interface secured, focus can now be placed on other areas of the User, Control, and Management planes to ensure they are secured. The User plane consists of all user data/voice traffic from the air interface. The Control plane consists of signaling messages, session management and setup and subscriber/device authentication. The Management plane consists of Operations, Administration, and Maintenance (OA&M) platform, interfaces, access control and remote access. All three planes are composed of multiple platforms and interface connections that require a holistic approach to protect from external and internal threats. Fig. C-1 [1] shows a unique layered defense model to wireless network security, such that solutions can be placed across the wireless network to complement CDMA security standards [1].

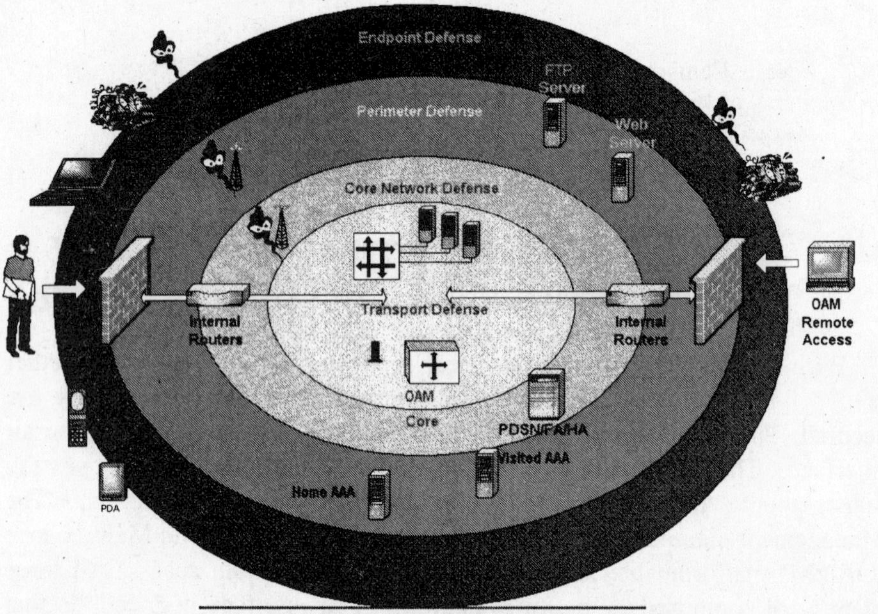

Figure C-1. Layered Defense Model

4. CORE WIRELESS NETWORK SECURITY

Securing the management plane is a critical part of an end-to-end security solution. Wireless network management nodes contain management policies and databases that are critical to the operation of the network. To ensure robustness of the management plane, security must address platform, internal and external threats [1].

4.1 Provisioning

Older cellular technology transmits subscriber identity information over the air interface during registration and call set-up in a format that can be easily detected and read by radio scanner devices, making it susceptible to fraudulent activities such as cloning and tumbling. CDMA avoids these issues by using a 64-bit authentication key (A-key) and the Electronic Serial Number (ESN) of the mobile. The A-key is used to generate sub-keys that provide voice privacy and message encryption. CDMA allows several distribution methods of A-keys to valid users for acquiring subscription-related information to communicate with the wireless network providing service.

For all distribution methods, security data is provided electronically in an encrypted format. The most secured distribution method uses handsets that are pre-programmed with the A-key and ESN by the mobile vendor, and then the wireless provider or dealer assigns ESN with Mobile Identification Number (MIN). This approach ensures that neither the equipment manufacturer nor the dealer has all three pieces of security information [1].

4.2 Subscriber Authentication

Subscriber authentication is a key control mechanism to protect the infrastructure and to prevent unauthorized access to wireless network resources. CDMA 1X access authentication is accomplished by means of an 18-bit authentication signature that is verified by the wireless network's databases of user information, the Home Location Register and Authentication Center. The 1xEV-DO also uses the same 512-bit algorithm in OTASP to exchange keys between the mobile device and the Access Node-Authentication Authorization Accounting (AN-AAA) server. Both technologies utilize strong authentication key exchange protocols to ensure identity.

For CDMA2000 1X data sessions and EV-DO, users are authenticated using the Challenge Handshake Authentication Protocol (CHAP) by the Packet Data Serving Node-Authentication Authorization Accounting (PDSN-AAA) server. CHAP is a proven Internet authentication protocol that is leveraged in the wireless network to verify identity [1].

4.3 Packet Core

In CDMA2000 architecture, the wireless packet core network is leveraged for both 1X and 1xEV-DO. The wireless packet core is the ideal place for applying IP services -- especially security services, common across the CDMA2000 access wireless network.

So, with the preceding in mind, Mobile IP Foreign Agent (FA) to Home Agent (HA) and HA outbound connections must also be protected. These connections can be protected via IP Security (IPSec) encrypted Virtual Private Routed Network (VPRN) capabilities on FAs and Has [1].

5. TRANSPORT WIRELESS SECURITY – PROTECTING TRAFFIC IN TRANSIT

Protecting mobile user or management information is an important role of the layered security defense. Prevailing VPN protocols (IPSec and Secure Socket Layer (SSL)) are complementary. IPSec VPN provides encryption, authentication protection at the IP layer, thus protecting all IP application traffic. SSL VPN provides secure communications for the web application.

IPSec is usually best for remote access that requires full use of wireless network resources from laptops. Support for telecommuters and branch sites where multiple applications will be used, for example, are ideal applications for IPSec.

SSL VPNs are easy to deploy because no client is required on the mobile device. Using the Web browser as a universal client makes it easy to upgrade from legacy application interfaces to a Web interface. SSL VPNs provide security above the TCP/IP layer, thus ensuring compatibility with existing Network Address Translation (NAT) services, firewall configurations and proxy settings. SSL VPNs are good options for securing applications or services for an enterprise partner access to a single or limited set of applications, online customer access to services or mobile employee on-demand access to e-mail from any Web browser while at the airport or on the road [1].

6. PERIMETER SECURITY

Perimeter security has been around for some time and is generally referred to as a firewall. With the continuous increase in Internet viruses and worms, the complexity of updating firewall rules and ensuring detection/prevention of unknown threats can be a burdensome task. Fig. C-2 [1] shows a threat protection system (TPS) that uses a Snort-based™ intrusion detection method that is easy to use with outstanding support for rules maintenance and report generation.

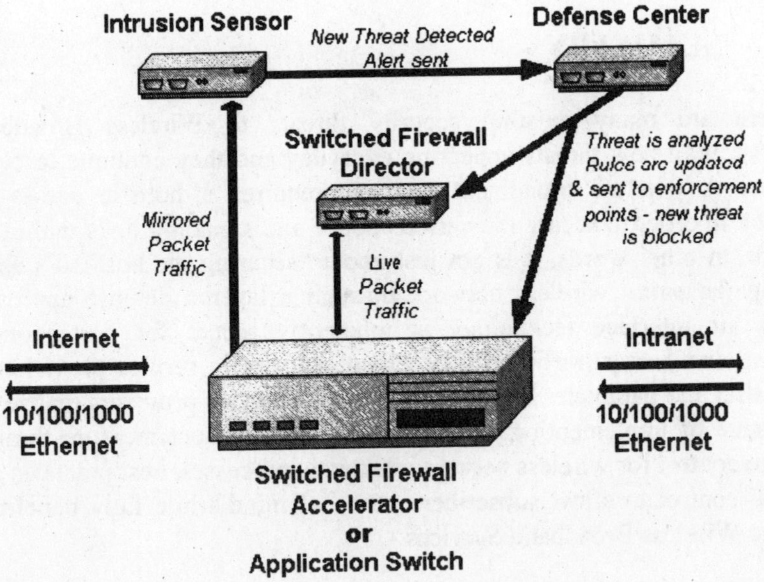

Figure C-2. Threat protection system.

The TPS Intrusion Sensor uses a rule-based, deep-packet inspection method to examine protocol fields on mirrored packet traffic for attacks. The sensors detects anomalies such as port scan, IP stack fingerprinting, Denial of Service attack and Address Resolution Protocol (ARP) spoofing. Threats are analyzed at the Defense Center and automatically update rules that are sent to firewalls in real-time to block the detected threat. Policy management and control along with threat analysis, reports, trap and trace capabilities, and event database for analysis are provided by the Defense Center which supports a hierarchical grouping of sensors for centralized management [1].

7. END POINT COMPLIANCE

Mobile enterprise users using laptop or smart devices are increasingly taking advantage of the speed and services of EV-DO. End points, either mobile devices or management consoles, can be potential sources of threats. These devices may be infected with viruses or worms which if allowed to connect to the wireless network, may become the source of an attack. Unprotected end points can carry a Distributed Denial of Service (DDOS) attack handler, which launches the attack to service provider resources such as Web, email or Domain Name System (DNS) servers [1].

8. SUMMARY

There are many existing security threats to Wireless Broadband networks, with new threats appearing everyday and they continue to be on the rise. A wireless broadband network requires a holistic end-to-end approach in order to secure the various nodes and signaling links within the network. In other words, it is not just about "securing the box"; it's about securing the entire wireless network through a layered defense approach. CDMA air interface technology is inherently secure for protection of signaling and bearer traffic with excellent security in service provisioning for handset and parameter distribution. Today's service providers realize the importance of implementing security policy enforcement measure through software control for wireless network elements, processes, best practices and physical control to allow subscribers peace of mind while fully benefiting from the Wireless Broadband Services.

9. REFERENCES

[1] "CDMA End-to-End Security," Copyright © 2005 Nortel Networks. All rights reserved. Nortel Networks, 35 Davis Drive, Research Triangle Park, NC 27709, USA, 2005.

Appendix D

LIST OF TOP WIRELESS NETWORK SECURITY IMPLEMENTATION AND DEPLOYMENT ENTERPRISES

1. INTRODUCTION

This appendix provides a list (as shown in Table D-1) of the top wireless network security implementation and deployment enterprises. This list will help wireless network administrators and security managers to design, implement, deploy and enforce wireless network security.

Table D-1. List of the top wireless network security implementation and deployment enterprises.

Enterprise Name	Contact Information	URL
1. AirDefense, Inc.	4800 North Point Parkway Suite 100 Alpharetta, GA 30022 USA TEL: 770-663-8115 FAX: 770-453-9601 info@airdefense.net	http://www.airdefense.net/
2. AirMagnet, Inc.	894 Ross Drive Suite 200 Sunnyvale, CA 94089 USA TEL: 408-400-0200 FAX: 408-744-1250 sales@AirMagnet.com	http://www.airmagnet.com
3. iAnywhere	One Sybase Drive Dublin, CA 94568-7902 USA	http://www.ianywhere.com

	TEL: 800-801-2069 FAX: 519-747-4971 contact_us@ianywhere.com	
4. Aventail	808 Howell St. 2nd Floor Seattle, WA 98101 USA TEL: 877-283-6824 FAX: 206-215-1120 aventailcustomerservice@aventail.com	http://www.aventail.com/
5. BlackBerry	185 Columbia Street Waterloo, ON N2L 5Z5 Canada TEL: 877-255-2377 FAX: 519-888-7884 https://www.blackberry.com/contact/	http://www.blackberry.com
6. CREDANT Technologies	15305 Dallas Parkway Suite 1010 Addison, TX 75001 USA TEL: 866-273-3268 FAX: 972-458-5454 info@credant.com	http://www.credant.com/index.php
7. Extended Systems	5777 N. Meeker Avenue Boise, ID 83713 USA TEL: 800-235-7576 FAX: 406-587-9170 info@extendsys.com	http://www.extendsys.com/
8. Fluke Networks	PO Box 777 Everett, WA 98206 USA TEL: 1-800-283-5853 FAX: 425-446-5043 info@flukenetworks.com	http://www.flukenetworks.com
9. Good Technology, Inc.	4250 Burton Drive Santa Clara, CA 95054 USA	http://www.good.com/
10. Hewlett-Packard Company	3000 Hanover Street Palo Alto, CA 94304-1185 USA TEL: (650) 857-1501 FAX: (650) 857-5518 info@hp.com	http://www.hp.com/
11. Intel	2200 Mission College Blvd. Santa Clara, CA 95052-8119 USA TEL: 408-765-8080 FAX: 408-765-9904 www.intel.com	http://www.intel.com

12. Intermec Technologies Corporation	6001 36th Avenue West P.O. Box 4280 Everett, WA 98203-9280 USA TEL: 425-348-2600 FAX: 425-355-9551 info@intermec.com	http://www.intermec.com/
13. RSA Security	174 Middlesex Tpke. Bedford, MA 01730 USA TEL: 1-800-732- 8743 FAX: 781-301-5170	http://www.rsasecurity.com
14. StillSecure	100 Superior Plaza Way Suite 200 Superior, CO 80027 USA TEL: 303-642-4530 FAX: 650-846-1005 sales@stillsecure.com	http://www.stillsecure.com
15. TESSCO Technologies, Inc.	TESSCO Technologies, Inc. 11126 McCormick Road Hunt Valley, MD 21031-1494 USA TEL: 800-508-5444 FAX: 410-527-0005	http://www.tessco.com
16. Utimaco Safeware	10 Lincoln Road Suite 102 Foxborough, MA 2035 USA TEL: 508-543-1008 FAX: 508-543-1009 info@utimaco.us	http://www.utimaco.us/
17. VIACK Corporation	One Metro Center 700 12th Street NW Suite 700 Washington, DC 20005 USA TEL: 202-558-5116 FAX: 202-558-5128	http://www.viack.com
18. Zebra Technologies Corporation	333 Corporate Woods Parkway Vernon Hills, IL 60061-3109 USA TEL: +1-847-634-6700 FAX: +1-847-913-8766 www.zebra.com/contact	http://www.zebra.com/

Appendix E

LIST OF WIRELESS NETWORK SECURITY PRODUCTS

1. INTRODUCTION

This appendix provides a list (as shown in Table E-1) of 23 wireless network security products from 17 vendors.

Table E-1. List of wireless network security products.

Wireless adapters			
Vendor	**Product**	**802.11 flavors supported**	**URL**
3Com	3Com Wireless PC Card with XJACK Antenna	802.11a/b/g/i	http://www.3com.com/index2.html
Actiontec	Wireless PC Card	b/g/i	http://www.actiontec.com/
Apple	Airport Extreme NIC	b/g/i	http://www.apple.com/
Belkin	F5D7011 High Speed Wireless Notebook Card	b/g/i	http://world.belkin.com/
Buffalo	AirStation WLI-CB-G54 High Speed Wireless Adapter	b/g/i	http://www.buffalotech.com/
Cisco	Aironet 802.11a/b/g Wireless CardBus Adapter	a/b/g/i	http://www.cisco.com/
Linksys	WPC55AG Dual Band Adapter	a/b/g/i	http://www.linksys.com/servlet/Satellite?childpagename=US%2FLayout&packedargs=c%3DL_Content_C1%26cid%3D1115417027773&pagename=Linksys%2FCo

			mmon%2FVisitorWrapper
SMC	SMC2536W-AG EliteConnect Universal High Power Wireless Adapter	a/b/g/i	http://www.smc.com/
Wireless access points			
Vendor	**Product**	**802.11 flavors supported**	**URL**
3Com	3Com Wireless LAN Access Point 8750	a/b/g/i	http://www.3com.com/index2.html
Actiontec	GT701WG Wireless DSL Gateway	b/g/i	http://www.actiontec.com/
Belkin	F5D7230 High-Speed Wireless G Router	b/g/i	http://world.belkin.com/
Buffalo	AirStation WBR2-G54 High Speed Mode Wireless Cable/DSL Router	b/g/i	http://www.buffalotech.com/
Cisco	Aironet 1100 Access Point	b/g/i	http://www.cisco.com/
Compex	NetPassage WPE54G-SMA Access Point	b/g/i	http://www.cpx.com/
HP	ProCurve 520wl dual radio Access Point	b/g/i	http://www.hp.com/
Linksys	WAP54G Access Point	b/g/i	http://www.linksys.com/servlet/Satellite?childpagename=US%2FLayout&packedargs=c%3DL_Content_C1%26cid%3D1115417027773&pagename=Linksys%2FCommon%2FVisitorWrapper
Netgear	WG302 Prosafe Access Point	b/g/i	http://www.netgear.com/
Netopia	3347W ADSL 3-D Reach Wi-FI Gateway	b	http://www.netopia.com/
Proxim	Orinocco AP-4000	a/b/g/i	http://www.proxim.com/
SMC	SMC 2555W-AG Wireless Access Point	a/b/g/i	http://www.smc.com/
Wireless switches			
Vendor	**Product**	**802.11 flavors supported**	**URL**
Airespace	Airespace 4000 with AP-1200	a/b/g/i	http://www.airespace.com/
Aruba	Aruba 800 wireless	a/b/g/i	http://www.arubanetworks.com/

	switch		
Trapeze	Mobility Exchange 20 switch and Mobility Point -252 Access Point	a/b/g/i	http://www.trapezenetworks.com/homepage.html

Appendix F

LIST OF WIRELESS NETWORK SECURITY STANDARDS

1. INTRODUCTION

This appendix provides a list (as shown in Table F-1) of wireless network security standards. Table F-1 details the various wireless network security standards that are currently available for installation of wireless networks and their merits.

Table F-1. Comparative table depicting the merits of various wireless security standards introduced by the Institute of Electrical and Electronic Engineers (IEEE).

Feature	**Wireless Standard**			
	802.11b	**802.11a**	**802.11g**	**802.16a**
Status	Widely adapted. Readily available everywhere.	New Technology	New technology with rapid growth expected	Newer technology that will surpass all other standards
Speed	11Mbps	Up to 54Mbps	Up to 54Mbps	Up to 75 Mbps
Frequency	2.4 GHz. Conflicts may occur with other 2.4GHz devices.	5 GHz. Can co-exist with 2.4 GHz bandwidths	2.4 GHz. Conflicts may occur with other 2.4GHz devices.	2-11 GHz. Compatible with 802.11 b or g
Range	100-150 feet indoor	25-75 feet indoor	100-150 feet indoor	Up to 30 miles with a cell radius of 4-6 miles.
Public Access	Accessible via public hotspots	None	Accessible via public hotspots using 802.11b/802.11g	Accessible via public hotspots that use 802.11b/802.11g
Compatibility	Widely adopted	Incompatible with 802.11b or 802.11g	Interoperates with 802.11b networks (at 11Mbps).	Interoperates with 802.11 wireless LANs
			Incompatible with 802.11a.	

Appendix G

LIST OF MISCELLANEOUS WIRELESS NETWORK SECURITY RESOURCES

1. INTRODUCTION

This appendix provides a list (as shown in Table G-1) of miscellaneous wireless network security resources.

Table G-1. List of miscellaneous wireless network security resources.

Resource	Description	URL
1. Wireless Networks and Mobile Computing	The latest wireless and mobile news, analysis and research links from Network World Fusion. Site Resources. Product, Tests, Buyer's Guides, Careers, Newsletters Strong wireless security for the SOHO network.	http://www.networkworld.com/topics/wireless.html
2. Security	Papers, tutorials and other resources on wireless security technologies.	http://www.networkworld.com/links/Research/Wireless_and_mobile/Security/
3. Tech Blogs on ZDNet | blogs.ZDNet.com	Blogs on technology, for business and IT professionals. Updated daily, ZDNet Blogs provides analysis, expertise, and a filter on the news. RSS feeds available for all blogs. ...	http://blogs.zdnet.com/
4. SecurityFocus	Home Bugtraq Vulnerabilities Mailing Lists Security Jobs Tools ... Pen-test. Security Basics. Vuln Dev ... IP Port Security. Creating a Test Network. Re: Wireless Security ...	http://www.securityfocus.com/

5. SANS Institute - Information and Computer Security Resources	SANS provides a Varity of resources to the community to help promote good security practices. The resources include consensus projects, white paper, forums, PDF documents, webcasts, and links. ... Top 20 List. Information and Computer Security Resources ... Malicious Code Managed Services Network Devices Wireless Access Work Monitoring	http://www.sans.org/resources/
6. Library Computer and Network Security: Network Security - Securing Wireless Networks	Library Computer and Network Security: A resource provided by the California State Library's Infopeople Project.	http://infopeople.org/resources/security/network/wireless.html
7. Web Security Resources - Web Security	You'll find a list of resources that cover web security, computer security, network security, free software ... Help Desk Software. Web Security Resources. Wireless Network Security ...	http://www.jamesmaurer.com/web-security-resources.asp?cat_id=43
8. Wireless Network Security	Wireless Network Security. List of Figures	http://csrc.nist.gov/publications/nistpubs/800-48/NIST_SP_800-48.pdf
9. System Administration Resources	Resources for departmental IT staff. Connecting to the Network ... a list of Rutgers IP address space and other technical information. Wireless Policy. Wireless security recommendations ...	http://techdir.rutgers.edu/index.shtml
10. Talisker Security Wizardry Computer Security Portal Home Page	Security Wizardry. Computer Network Defence Ltd. Home.	http://www.networkintrusion.co.uk/
11. ITPRC - Network Security	ITPRC Network Security - Links to information on hacking, exploits, firewalls, intrusion, PKI, and other Network Security topics ... Access Control List Resources (Cisco) Attack, Hacking, and Exploit	http://www.itprc.com/security.htm

	Information. DNS Security ... Government Agencies and Security Organizations ... Mailing Lists. Misc. Security Sites. Virus ...	
12. Wireless Security Issues: Infrastructure and Network Security: Best Web Links: SearchSecurity.com	News and advice for information security managers. ... browsing your wireless network? This tip addresses wireless LAN security basics and includes a list of tools you ... a WLAN and a list of resources for more information.	http://whatis.techtarget.com/bestWebLinks/0,289521,sid14_tax283845_idx61,00.html
13. EDUCAUSE | Security Task Force | Security Architecture Design	A Resource For Higher Education. Partnerships. Security Architecture Design ... security between client and network resources. Furthermore, both only support limited roaming.	http://www.educause.edu/content.asp?page_id=1261&bhcp=1
14. Wireless Attacks Primer	Computing & Technology>Internet / Network Security> Wireless Security> Wireless Attacks Primer by .,. Articles & Resources. Tools ... among other features (a list of supported adaptors)	http://netsecurity.about.com/od/secureyourwifinetwork/a/aa081604.htm
15. Linux & Wireless LANs	Wireless LAN resources for Linux. The Linux Wireless LAN Howto is an Open Source project sponsored by Hewlett Packard	http://www.hpl.hp.com/personal/Jean_Tourrilhes/Linux/
16. Computer Networking - Wireless Networks - Home Networking - WinMX News	Learn about wireless networks, Internet and computer networking. Study tutorials, network practice tests, and other reference material. Computer networking topics - home network, hardware, wireless, P2P. ... Articles & Resources. Networking Basics WiFi Wireless	http://compnetworking.about.com/
17. PracticallyNetworked.com	Get The Newsletter! Find a Network Term. internet.commerce. Find a Hotspot... • Wireless AP. • Wireless NIC. • Network Storage ... • Wi-Fi Network Security SMBs	http://www.practicallynetworked.com/
18. IBM Research | Press Resources | IBM Research Demonstrates Industry's First Auditing Tool For Wireless Network ...	Project list. Paper search ... to assess companies' wireless network security. For ease of use, the ... tool for security experts to maintain wireless network security	http://domino.research.ibm.com/comm/pr.nsf/pages/news.20010712_wireless.html
19. Network Security Software, Hardware, Services and Research Papers - Fast Company	Fast 50. Magazine Resources. subscribe " ... Gramm-Leach-Bliley Act on Wireless Network Security by AirMagnet	http://knowledgestorm.fastcompany.com/fastco/search/browse/121/121.jsp
20. Improving Security in Wireless Networks	Improving Security in Wireless Networks.	http://support.3com.com/infodeli/tools/wireless/11mbpswirelesslan/3crwe5

		1196/improving_your_wireless_security.pdf
21. Wardriving Tools, Wardriving Software, Wardriving Utilities (Wireless LAN Security & Wardriving - 802.11)	Website dedicated to Wireless LAN Security and Wardriving. Includes lots of whitepapers, presentations, tools, and resources.	http://www.wardrive.net/wardriving/tools
22. OSCaM Technical - Your Firewall/Router/Wireless Solution Provider	OSCaM Technical provides comprehensive content filtering, anti-spam, data security and antivirus solutions for network and desktop computing environments.	http://www.oscamtechnical.net/firewallvpn.html
23. Mike's List Of Computer Resources	Home Wireless Security ... Network Abuse Clearinghouse (ISP contact addresses, forwarding of spam complaints) Directory of Spam-Fighting Information and Resources	http://www.noc.utoronto.ca/~mikep/rescomp.html
24. Resources - The Community's Center for Security	Security discussion resources	http://www.linuxsecurity.com/content/view/101892/155
25. Microsoft PowerPoint - Wireless Network Security 2	Wireless Network Security. Technology. Wireless Security Concerns	http://www.cse.msu.edu/~cse491/Sect601/lecturesF03/KimWirelessNetworkSecurity2.pdf
26. ZDNet: Tech News and White Papers for IT Professionals	Where Technology Means Business: ZDNet delivers the best tech news, and resources for IT hardware, software, networking and services. It's the top site for IT managers and tech-savvy business people.	http://www.zdnet.com/
27. SAFE: Wireless LAN Security in Depth - version 2 [SAFE Blueprint] - Cisco Systems	Guest. The 2.4-GHz ISM band (used by 802.11b) makes use of spread-spectrum technology. Spread spectrum dictates that data transmissions are spread across numerous frequencies. ... Security and attack mitigation based on policy. Authentication and authorization of users to wired network resources. Wireless ... needs to have access to this list.	http://www.ciscosystemsnetwork.com/en/US/netsol/ns340/ns394/ns171/ns128/networking_solutions_white_paper09186a008009c8b3.shtml
28. Wireless Security Resources	Wireless Network Security - Best Practices.	http://cyberforge.com/weblog/aniltj/archive/2004/03/17/396.aspx
29. Network Security Glossary: DNS	A definition of the network security term DNS (also known as Domain Name System) ... Technology>Internet / Network Security> Information Resources> Computer Security ... Scanner Software Free Network Monitoring Free Wireless Security Tools ... maintains a list of domain names ...	http://netsecurity.about.com/cs/generalsecurity/g/def_dns.htm

30. IT Architect | Wireless Channel	News, reviews and thinking on the future of wireless and mobility.	http://www.itarchitect.com/channels/wireless.jhtml
31. SANS Institute Free Webcast: Wireless - The Ever Changing Landscape of Network Security	Research Projects. Resources. Press Room. Sample Policies. Top 20 List ... Webcast Overview: Wireless - The Ever Changing Landscape of Network Security	https://www.sans.org/webcasts/show.php?webcastid=90431
32. South Carolina Bar | Member Resources | Practice Management | Wireless	Best Practices for Wireless Network Security ... list of answers to frequently asked questions. The Strongest Links. Cutting the Tether: Great Resources on Wireless ...	http://www.scbar.org/PMAP/wirelesslinks.asp
33. Wireless Networking	Maintain pages on Resources for Internet Security and Wireless, Wireless Community Network List and Community Wireless list	http://bengross.com/wireless.html
34. WebDeveloper.com - Resources for web pages on wireless devices	WebDeveloper.com Forums, Where Web Developers and Designers Learn How to Build Web Sites, Program in HTML, Java and JavaScript and More! ... Solution Center. WHITEPAPER : Network Security Policy: Best Practices ... ASP Sample: Filtered Directory List	http://www.webdeveloper.com/forum/showthread.php?threadid=23688
35. WiFi Security Checklist	Security enhancements: one must generally operate a homogeneous wireless network.	http://www.securitytechnique.com/2003/11/wsc.html

Appendix H

GLOSSARY

1. INTRODUCTION

This appendix or glossary provides a list of wireless network security terms and acronyms.

2. TERMS AND ACRONYMS

1G: First Generation

2G or second generation: A term for analog and digital current networks operating on 800 MHz or 1900 MHz spectrums. These include Advanced Mobile Phone Service (AMPS), Time Division Multiple Access (TDMA), Global System for Mobile Communications (GSM), and Code Division Multiple Access (CDMA).

2.5G: Two-and-a-Half Generation

3DES: Triple Data Encryption Standard

3G or third generation: Also known as Universal Mobile Telecommunications System (UMTS). The 3rd Generation standards boost additional services, more extensive roaming capabilities and transfer rates 26x faster than those allowed by current CDMA networks.

802.11: Refers to a family of specifications developed by the IEEE for wireless LAN (Local Area Network) technology. 802.11 specifies an over-the-air interface between a wireless client and a base station or between two wireless clients. The IEEE accepted the specification in 1997.

802.11a (also referred to as Wi-Fi5): An extension to 802.11 that applies to wireless LANs and provides up to 54 Mbps (Megabits per second) in the 5 GHz (Gigahertz) band.

802.11b (also referred to as Wi-Fi): An extension to 802.11 that applies to wireless LANS and provides 11 Mbps transmission (with a fallback to 5.5, 2 and 1 Mbps) in the 2.4 GHz band. 802.11b uses only DSSS. 802.11b was a 1999 ratification to the original 802.11 standard, allowing wireless functionality comparable to Ethernet.

802.11g: Applies to wireless LANs and provides 20+ Mbps in the 2.4 GHz band.

802.11i: IEEE specification for enhanced WLAN security through the use of stronger encryption protocols such as the TKIP and AES (Advance Encryption Standard). These protocols provide replay protection, cryptographically keyed integrity checks, and key derivation based on the IEEE 802.1X port authentication standard.

802.1x: If your wireless clients happen to run Windows XP, a strong alternative is IEEE 802.1x. 802.1x provides a carrier for secure delivery of session keys used to encrypt traffic between the supplicant and authenticator, addressing a serious omission in the WEP standard. For example, session keys might be created "on the fly" by the access point or supplied by a RADIUS server. If a hacker recovered keys from WEP session traffic, the keys would be of no value for other sessions.

Accreditation: The authorization and approval granted to an ADP system or network to process sensitive data in an operational environment; and, made on the basis of a certification by designated technical personnel of the extent to which design and implementation of the system meet prespecified technical requirements for achieving adequate data security.

ACL: Access Control List

ACO: Authenticated Cipher Offset

Ad-Hoc Mode: Each client communicates directly with the other clients within the network. Ad-hoc mode is designed such that only the clients within transmission range (within the same cell) of each other can communicate. If a client in an ad-hoc network wishes to communicate outside of the cell, a member of the cell MUST operate as a gateway and perform routing.

Advanced Encryption Standard (AES): The Advanced Encryption Standard (AES) is an encryption algorithm for securing sensitive but unclassified material by U.S. Government agencies.

AES: Advanced Encryption Standard

AH: Authentication Header

AMPS: Advanced Mobile Phone System

AP: Access Point

API: Application Programming Interfaces

Application System: An application system is a software package that processes, transmits, or disseminates information according to established internal procedures. An application system is run at an automated information system facility. A word processor usually runs only one application system. A mainframe computer may run thousands of application systems.

ATM: Automatic Teller Machine

Automated Information System (AIS): An AIS is the organized collection, processing, transmission, and dissemination of information in accordance with defined procedures.

Automated Information System (AIS) Facility: An AIS facility is an organizationally defined set of personnel, hardware, software, and physical facilities, a primary function of which is the operation of an automated information system(s) and an application system(s). AIS facilities range from large centralized computer centers to individual stand-alone microprocessors such as personal computers and word processors.

BSS: Basic Service Set

CDMA: CDMA stands for Code Division Multiple Access. It is the underlying technology that enables millions of wireless phones to communicate where wireless carriers have implemented the technology in their wireless networks.

CDPD: Operates as an overlay on analog cellular networks (AMPS) to allow packetized data transfer, which increases the speed and efficiency of old analog networks.

CERT: Computer Emergency Response Team

Certification: A technical evaluation made as part of and in support of the accreditation process, that establishes the extent to which a particular computer system or network design and implementation meet a prespecified set of security requirements.

CIO: Chief Information Officer

Computer Security: Computer Security is the protection of a computer system against internal failures, human errors, attacks, and natural catastrophes that might cause improper disclosure, modification, destruction, or denial of service.

Computer System Security Plan (CSSP): This plan is a document describing the security and privacy requirements of a given system and the agency's plan to meet these requirements.

CRC: Cyclic Redundancy Check

Data Encryption Standard (DES): A National Institute of Standards and Technology (NIST) standard secret key cryptography method that uses a

56-bit key encryption. DES is based on an IBM algorithm, which was further developed by the U.S. National Security Agency. It uses the block cipher method, which breaks the text into 64-bit blocks before encrypting them. There are several DES encryption modes. The most popular mode exclusive-OR-s each plain-text block with the previous encrypted block. DES decryption is very fast and widely used. The secret key may be kept completely secret and reused again, or a key can be randomly generated for each session, in which case, the new key is transmitted to the recipient using a public key cryptography method such as RSA. Triple DES (3DES) is an enhancement of DES that provides considerably more security than standard DES, which uses only one 56-bit key. There are several 3DES methods. EEE3 uses three keys and encrypts three times. EDE3 uses three keys to encrypt, decrypt, and encrypt again. EEE2 and EDE2 are similar to EEE3 and EDE3, except that only two keys are used, and the first and third operations use the same key.

DDoS: Distributed Denial of Service

DES: Data Encryption Standard

DHCP: Stands for Dynamic Host Configuration Protocol--a protocol for assigning dynamic IP addresses to devices on a network. With dynamic addressing, a device can have a different IP address every time it connects to the network. In some systems, the device's IP address can even change while it is still connected. DHCP also supports a mix of static and dynamic IP addresses. DHCP client support is built into Windows XP, 2003 and NT workstations.

digital cellular systems: Standards which utilize digital transmissions such as TDMA, CDMA, IS, and GSM.

DMZ (Demilitarized Zone): A networking term for a specially designed network segment where external users are allowed to access resources without getting any access to internal networks.

DoD: Department of Defense

DoS: Denial of Service

DSSS: Direct Sequence Spread Spectrum

Dynamic Host Configuration Protocol (DHCP): The protocol used to assign Internet Protocol (IP) addresses to all nodes on the network.

EAP (Extensible Authentication Protocol): EAP is an 802.1X standard that allows passing of security authentication data between client, access point or switch and RADIUS server to authenticate users. There are a number of EAP variants including EAP-TLS, EAP-TTLS, LEAP and PEAP.

EAP-TLS (EAP Transport Layer Security): EAP-TLS provides for certificate-based, mutual authentication of the client and the network. EAP-TLS relies on client-side and server-side certificates to perform authentication, using dynamically generated user and session-based WEP keys distributed to secure the connection.

EAP-TTLS (EAP Tunneled Transport Layer Security): EAP-TTLS is an extension of EAP-TLS. Unlike .EAP-TLS, EAP-TTLS requires only server side certificates, eliminating the need to configure certificates for each WLAN client.

ECC: Elliptic Curve Cryptography

EDGE: Enhanced Data GSM Environment

EM: Electromagnetic

ESN: Electronic Serial Number

ESP: Encapsulating Security Protocol

ESS: Extended Service Set

Ethernet: A local-area network (LAN) architecture developed by Xerox Corporation in cooperation with DEC and Intel in 1976. Ethernet uses a bus or star topology and supports data transfer rates of 10 Mbps. The Ethernet specification served as the basis for the IEEE 802.3 standard. It is one of the most widely implemented LAN standards. A newer version of Ethernet, called 100Base-T (or Fast Ethernet), supports data transfer rates of 100 Mbps. The newest version, Gigabit Ethernet supports data rates of 1 gigabit (1,000 megabits) per second.

ETSI: European Telecommunications Standard Institute

FCC: Federal Communications Commission

FDMA: Frequency Division Multiple Access

FEC: Forward Error Correction

FH: Frequency Hopping

FHSS: Frequency Hopping Spread Spectrum

FIPS: Federal Information Processing Standard

GFSK: Gaussian Frequency Shift Keying

GHz: Gigahertz

GPRS: General Packet Radio System

GPS: Global Positioning System

GSM: Global System for Mobile Communications

Hash Function: A computationally efficient algorithm that maps a variable-sized amount of text into a fixed-sized output (hash value). Hash functions are used in creating digital signatures.

HTML: HyperText Markup Language

HTTP: HyperText Transfer Protocol

I&A: Identification and Authentication

IBSS: Interdependent Basic Service Set

ICAT: Internet Categorization of Attack Toolkit

IDC: International Data Corporation

IDS: Intrusion Detection System

IEC: International Electrotechnical Commission

IEEE: The IEEE (Eye-triple-E) is a nonprofit, technical professional association of more than 375,000 individual members in 150 countries. The full name is the Institute of Electrical and Electronics Engineers, Inc., although the organization is most popularly known and referred to by the letters I-E-E-E. Through its members, the IEEE is a leading authority in technical areas ranging from computer engineering, biomedical technology and telecommunications, to electric power, aerospace and consumer electronics, among others.

IETF: Internet Engineering Task Force

IKE: Internet Key Exchange

IMT-2000: International Mobile Telecommunication 2000

Industrial, Scientific, and Medical (ISM) Band: The ISM band refers to the government-allotted bandwidth at 2.450 ± .050 gigahertz (GHz) and 5.8 ± 0.75 GHz.

Information Technology Utility (ITU): An ITU is an organizationally defined set of personnel, hardware, software, and physical facilities, a primary function of which is to coordinate the operation of geographically dispersed automated information systems and automated information system facilities. ITUs range in size from wide area networks covering widely dispersed geographical areas to local area networks covering a single office.

Infrared (IR): An invisible band of radiation at the lower end of the electromagnetic spectrum. It starts at the middle of the microwave spectrum and extends to the beginning of visible light. Infrared transmission requires an unobstructed line of sight between transmitter and receiver. It is used for wireless transmission between computer devices, as well as for most handheld remotes for TVs, video, and stereo equipment.

Infrastructure mode: Each client sends all of its communications to a central station, or access point (AP). The access point acts as an Ethernet bridge and forwards the communications onto the appropriate network--either the wired network, or the wireless network.

Institute of Electrical and Electronics Engineers (IEEE): A worldwide professional association for electrical and electronics engineers that sets standards for telecommunications and computing applications.

International Electrotechnical Commission (IEC): An organization that sets international standards for the electrical and electronics fields.

International Organization for Standardization (ISO): A voluntary organization responsible for creating international standards in many areas, including computers and communications.

IP: Internet Protocol

IPSEC: Short for IP Security> It is a set of protocols to support secure exchange of packets at the IP layer. IPsec has been deployed widely to implement Virtual Private Networks (VPNs).

IPX: Internet Packet Exchange

IR: Infrared

ISM: Industrial, Scientific, and Medical

ISO: International Organization for Standardization

ISS: Internet Security Systems

IV: Initialization Vector

Jini: An approach to instant recognition that would enable manufacturers to make devices that can attach to a network independently of an operating system. Jini can be viewed as the next step after the Java programming language toward making a network look like one large computer. Each pluggable device in a network defines itself immediately to a network device registry. Using the Jini architecture, users are able to plug printers, storage devices, speakers, and any other kind of device directly into a network, and every other computer, device, and user on the network will know that the new device has been added and is available through the network registry. When a user wants to use or access the resource, his/her computer will be able to download the necessary programming from it to communicate with it. In this way, devices on the network may be able to access and use other devices without having the drivers or other previous knowledge of the device.

Kbps: Kilobits per second

KG: Key Generator

KHz: Kilohertz

KSG: Key Stream Generator

L2CAP: Logical Link Control and Adaptation Protocol

L2TP: Layer 2 Tunneling Protocol

LAN: Local Area Network

LDAP: Lightweight Directory Access Protocol

LEAP (Lightweight Extensible Authentication Protocol): LEAP is primarily used in Cisco WLAN access points. LEAP provides security during credential exchange, encrypts data transmission using dynamically generated WEP keys, and supports mutual authentication.

LFSR: Linear Feedback Shift Register

Local Area Network (LAN): A data network, located on a user's premises, within a limited geographic region. Communication within a local area network is not subject to external regulation; however, communication across the network boundary may be subject to some form of regulation.

MAC Address: Short for Media Access Control address. The MAC address is a 12-character code that is unique to every single piece of network interface hardware. MAC codes are applied at the time of production by the manufacturer, therefore, it is possible to limit 802.11 users according to the device's unique MAC address.

Mbps: Megabits per second

Medium Access Control (MAC): On a local area network, the sublayers that control which device has access to the transmission medium at a particular time.

MHz: Megahertz

MW: Milliwatt

NIC: Network Interface Card

NIST: National Institute of Standards and Technology

OFDM: Orthogonal Frequency Division Multiplexing

OMB: Office of Management and Budget

Open Systems Interconnection (OSI): A model developed by ISO to allow computer systems made by different vendors to communicate with each other.

OSI: Open Systems Interconnection

OTP: One-Time Password

P2P: Peer to Peer

PAN: Personal Area Network

PC: Personal Computer

PCMCIA: Personal Computer Memory Card International Association

PCS spectrum: Personal communication services (PCS) operates at the 1900 MHz spectrum and is entirely digital (includes digital cellular standards such as TDMA, CDMA, and GSM but not analog standards).

PDA: Personal Digital Assistant

PEAP (Protected Extensible Authentication Protocol): PEAP is an extension of EAP-TLS. Unlike EAP-TLS, PEAP requires only server side certificates, eliminating the need to configure certificates for each WLAN client.

Personal Digital Assistant (PDA): A handheld computer that serves as an organizer for personal information. It generally includes at least a name-and-address database, a to-do list, and a note taker. PDAs are pen-based and use a stylus to tap selections on menus and to enter printed characters. The unit may also include a small on-screen keyboard that is tapped with the pen. Data is synchronized between a user's PDA and desktop computer by cable or wireless transmission.

Personnel Security: Personnel security refers to a program that determines the sensitivity of positions and screens individuals who participate in the design, operation, or maintenance of automated information systems or who have access to such systems.

PHY: Physical Layer

Physical Security: Physical security refers to the combination of devices that bar, detect, monitor, restrict, or otherwise control access to sensitive areas. Physical security also refers to the measures to protect a facility that houses AIS assets and its contents from damage by accident, malicious intent, fire, loss of utilities, environmental hazards, and unauthorized access.

PIN: Personal Identification Number

PKI: Public Key Infrastructure

PPTP: Point-to-Point Tunneling Protocol

RADIUS (Remote Authentication Dial-In User Service): A client/server protocol and software that enables access servers to communicate with a central server to authenticate users and authorize their access to the requested system or service. RADIUS allows organizations to maintain user profiles in a central database that all remote servers can share.

Request for Comments (RFC): A series of numbered documents (RFC 822, RFC 1123, etc.) developed by the Internet Engineering Task Force (IETF) that set standards and are voluntarily followed by many makers of software in the Internet community.

RF: Radio Frequency

RFC: Request for Comment

ROM: Read Only Memory

RSA: Rivest-Shamir-Adelman

RSN: Robust Security Networks

Sensitive Information: Sensitive information is any information, the loss, misuse, disclosure, or unauthorized access to or modification of which could adversely affect the national interest or the conduct of Federal programs, or the privacy to which individuals are entitled under section 552a of Title 5, United States Code (the Privacy Act), but which has not been specifically authorized under criteria established by an Executive Order or an Act of Congress to be kept secret in the interest of national defense or foreign policy.

SIG: Special Interest Group

Smart Card: A credit card with a built-in microprocessor and memory that is used for identification or financial transactions. When inserted into a reader, the card transfers data to and from a central computer. A smart card is more secure than a magnetic stripe card and can be programmed to self-destruct if the wrong password is entered too many times.

SMS: Short Message Service

SNMP: Simple Network Management Protocol

Spoofing: Otherwise known as IP spoofing, which refers to sending a network packet that appears to come from a source other than its actual source.

SRES: Signed Response

SSH: Developed by SSH Communications Security Ltd., Secure Shell is a program to log into another computer over a network, to execute commands in a remote machine, and to move files from one machine to another. It provides strong authentication and secure communications over insecure channels. SSH protects a network from attacks such as IP spoofing, IP source routing, and DNS spoofing. An attacker who has managed to take over a network can only force SSH to disconnect. He or she cannot play back the traffic or hijack the connection when encryption is enabled. When using SSH's login the entire login session, including transmission of a password, it is encrypted; therefore, it is almost impossible for an outsider to collect passwords.

SSID: Service Set Identifier

SSL: Short for Secure Sockets Layer, a protocol developed by Netscape for transmitting private documents via the Internet. SSL works by using a public key to encrypt data that's transferred over the SSL connection. Both Netscape Navigator and Internet Explorer support SSL and many Web sites use the protocol to obtain confidential user information, such as credit card numbers. By convention, URLs that require an SSL connection start with https: instead of http.

switching center: Mobile system sends radio signals to switching center. Switching center chooses the appropriate network link to propagate a signal (public/private network via telephone or other high speed line). The signal is then passed onto an organization's existing LAN/WAN infrastructure.

TCP: Transmission Control Protocol

TDMA: Time Division Multiple Access

TGI: Task Group I

TKIP (Temporal Key Integrity Protocol, pronounced tee-kip): TKIP is an encryption standard that provides per-packet key mixing, a message integrity check and keying mechanism, that fixes the flaws of WEP.

TTP: Trusted Third Party

UMTS: Universal Mobile Telecommunications Service

USB: Universal Serial Bus

USC: United States Code

UWC: Universal Wireless Communications

Virtual Private Network (VPN): A means by which certain authorized individuals (such as remote employees) can gain secure access to an organization's intranet by means of an extranet (a part of the internal network that is accessible via the Internet).

VLAN: Short for virtual LAN, a network of computers that behave as if they are connected to the same wire even though they may actually be physically located on different segments of a LAN. VLANs are configured through software rather than hardware, which make them extremely flexible. One of the biggest advantages of VLANs is that when a computer is physically moved to another location, it can stay on the same VLAN without any hardware reconfiguration.

WAP: Wireless Application Protocol. A universal specification to facilitate the delivery and presentation of Web-based data and services on mobile phones and devices with small displays and limited input facilities.

War Driving: War driving is when hackers drive around in a car equipped with wireless gear looking for illicit access to unsecured wireless networks.

WECA: Wireless Ethernet Compatibility Alliance. WECA's mission is to certify interoperability of Wi-Fi (IEEE 802.11) products and to promote Wi-Fi as the global wireless LAN standard across all market segments.

WEP: Short for Wired Equivalent Privacy. A security protocol for wireless local area networks (WLANs) defined in the 802.11b (Wi-Fi) standard. WEP is designed to provide the same level of security as that of a wired LAN. LANs are inherently more secure than WLANs because LANs are somewhat protected by the physical nature of their structure, having some or all parts of the network inside a building that can be protected from unauthorized access. WLANs, which are over radio waves, do not have the same physical structure and therefore are more vulnerable to tampering. WEP aims to provide security by encrypting data over radio waves so that it is protected as it is transmitted from one end point to another. However, it has been found that WEP is not as secure as once believed.

WEP2: Wired Equivalent Privacy 2

WG: 1000 Wireless Gateway 1000.

Wide Area Network (WAN): A WAN is an arrangement of data transmission facilities that provides communications capability across a broad geographic area (DIMES).

Wi-Fi (Wireless Fidelity (synonymous with 802.11b): A WLAN industry group tasked with defining and certifying interoperability of WLAN devices.

Wireless Application Protocol (WAP): A standard for providing cellular telephones, pagers, and other handheld devices with secure access to e-mail and text-based Web pages. Introduced in 1997 by Phone.com, Ericsson, Motorola, and Nokia, WAP provides a complete environment for wireless applications that includes a wireless counterpart of TCP/IP and a framework for telephony integration, such as call control and telephone book access. WAP features the Wireless Markup Language (WML) and is a streamlined version of HTML for small-screen displays. It also uses WMLScript, a compact JavaScript-like language that runs in limited memory. WAP also supports handheld input methods, such as keypad and voice recognition. Independent of the air interface, WAP runs over all the major wireless networks in place now and in the future. It is also device-independent, requiring only a minimum functionality in the unit to permit use with a myriad of telephones and handheld devices.

Wired Equivalent Privacy (WEP): Wired Equivalent Privacy (WEP) is a security protocol, specified in the IEEE Wireless Fidelity (Wi-Fi) standard, 802.11, that is designed to provide a wireless local area network (WLAN) with a level of security and privacy comparable to what is usually expected of a wired LAN.

WISP: Wireless Internet Service Provider

WLAN: Wireless Local Area Network

WML: Wireless Markup Language

WPA (Wi-Fi Protected Access): WPA is a specification of standards-based, interoperable security enhancements that increases the level of data protection and access control for existing and future wireless LAN systems.

WPAN: Wireless Personal Area Networks

WTA: Wireless Telephony Application

WTLS: Wireless Transport Layer Security

WTP: Wireless Transaction Protocol

WWAN: Wireless Wide Area Network

INDEX